Handbook of herbs and spices

Related titles from Woodhead's food science, technology and nutrition list:

Handbook of herbs and spices Volume 1 (ISBN 1 85573 562 8)
Herbs and spices are among the most versatile and widely used ingredients in food processing. As well as their traditional role in flavouring and colouring foods, they have been increasingly used as natural preservatives and for their potential health-promoting properties, for example as antioxidants. Edited by a leading authority in the field, and with a distinguished international team of contributors, the *Handbook of herbs and spices* provides an essential reference for manufacturers wishing to make the most of these important ingredients. A first group of chapters looks at general issues including quality indices for conventional and organically produced herbs, spices and their essential oils. The main body of the handbook consists of over twenty chapters covering key spices and herbs from aniseed, bay leaves and black pepper to saffron, tamarind and turmeric. Chapters cover key issues from definition and classification to chemical structure, cultivation and post-harvest processing, uses in food processing, functional properties, regulatory issues, quality indices and methods of analysis.

Antioxidants in food (ISBN 1 85573 463 X)
Antioxidants are an increasingly important ingredient in food processing, as they inhibit the development of oxidative rancidity in fat-based foods, particularly meat, dairy products and fried foods. Recent research suggests that they play a role in limiting cardiovascular disease and cancers. This book provides a review of the functional role of antioxidants and discusses how they can be effectively exploited by the food industry, focusing on naturally occurring antioxidants in response to the increasing consumer scepticism over synthetic ingredients.

Natural antimicrobials for the minimal processing of foods (ISBN 1 85573 669 1)
Consumers demand food products with fewer synthetic additives but increased safety and shelf-life. These demands have increased the importance of natural antimicrobials which prevent the growth of pathogenic and spoilage micro-organisms. Edited by a leading expert in the field, this important collection reviews the range of key antimicrobials together with their applications in food processing. There are chapters on antimicrobials such as nisin and chitosan, applications in such areas as postharvest storage of fruits and vegetables, and ways of combining antimicrobials with other preservation techniques to enhance the safety and quality of foods.

Details of these books and a complete list of Woodhead's food science, technology and nutrition titles can be obtained by:

- visiting our web site at www.woodhead-publishing.com
- contacting Customer Services (e-mail: sales@woodhead-publishing.com; fax: +44 (0) 1223 893694; tel.: +44 (0) 1223 891358 ext.30; address: Woodhead Publishing Ltd, Abington Hall, Abington, Cambridge CB1 6AH, England)

Selected food science and technology titles are also available in electronic form. Visit our web site (www.woodhead-publishing.com) to find out more.

If you would like to receive information on forthcoming titles in this area, please send your address details to: Francis Dodds (address, tel. and fax as above; e-mail: francisd@woodhead-publishing.com). Please confirm which subject areas you are interested in.

Handbook of herbs and spices

Volume 2

Edited by
K. V. Peter

CRC Press
Boca Raton Boston New York Washington, DC

WOODHEAD PUBLISHING LIMITED
Cambridge England

Published by Woodhead Publishing Limited, Abington Hall, Abington
Cambridge CB1 6AH, England
www.woodhead-publishing.com

Published in North America by CRC Press LLC, 2000 Corporate Blvd, NW
Boca Raton FL 33431, USA

First published 2004, Woodhead Publishing Ltd and CRC Press LLC
© 2004, Woodhead Publishing Ltd
The authors have asserted their moral rights.

This book contains information obtained from authentic and highly regarded sources. Reprinted material is quoted with permission, and sources are indicated. Reasonable efforts have been made to publish reliable data and information, but the authors and the publishers cannot assume responsibility for the validity of all materials. Neither the authors nor the publishers, nor anyone else associated with the publication, shall be liable for any loss, damage or liability directly or indirectly caused or alleged to be caused by this book.

Neither this book nor any part may be reproduced or transmitted in any form or by any means, electronic or mechanical, including photocopying, microfilming and recording, or by any information storage or retrieval system, without permission in writing from the publishers.

The consent of Woodhead Publishing and CRC Press does not extend to copying for general distribution, for promotion, for creating new works, or for resale. Specific permission must be obtained in writing from Woodhead Publishing or CRC Press for such copying.

Trademark notice: Product or corporate names may be trademarks or registered trademarks, and are used only for identification and explanation without intent to infringe.

British Library Cataloguing in Publication Data
A catalogue record for this book is available from the British Library.

Library of Congress Cataloging in Publication Data
A catalog record for this book is available from the Library of Congress.

Woodhead Publishing ISBN 1 85573 721 3 (book) 1 85573 835 X (e-book)
CRC Press ISBN 0-8493-2535-8
CRC Press order number: WP2535

The publisher's policy is to use permanent paper from mills that operate a sustainable forestry policy, and which have been manufactured from pulp which is processed using acid-free and elementary chlorine-free practices. Furthermore, the publisher ensures that the text paper and cover board used have met acceptable environmental accreditation standards.

Typeset by Ann Buchan (Typesetters), Middx, England
Printed by TJ International, Padstow, Cornwall, England

Contents

List of contributors x

1 Introduction 1
K. V. Peter, *Kerala Agricultural University, India* and K. Nirmal Babu, *Indian Institute of Spices Research, India*
1.1 Introduction to herbs and spices 1
1.2 Uses of herbs and spices 3
1.3 Active plant constituents 4
1.4 The structure of this book 5
1.5 References 8

Part I General issues 9

2 The functional role of herbal spices 11
M. R. Shylaja and K. V. Peter, *Kerala Agricultural University, India*
2.1 Introduction 11
2.2 Classification 12
2.3 Production, consumption and processing 13
2.4 Functional properties 15
2.5 Sources of further information 21

3 Herbs and spices and antimicrobials 22
C. C. Tassou, *National Agricultural Research Foundation, Greece*, and G.-J. E. Nychas and P. N. Skandamis, *Agricultural University of Athens, Greece*
3.1 Introduction 22
3.2 Barriers to the use of herb and spice essential oils as antimicrobials in foods 22
3.3 Measuring antimicrobial activity 23
3.4 Studies *in vitro* 26
3.5 Applications in food systems 27
3.6 Mode of action and development of resistance 32
3.7 Legislation 34

3.8 Future prospects and multifactorial preservation 34
3.9 References 34

4 Screening for health effects of herbs 41
R. Rodenburg, TNO Pharma, The Netherlands
4.1 Introduction 41
4.2 Types of assays 42
4.3 Throughput vs content assays 44
4.4 Assay quality 45
4.5 Screening bio-active compounds 45
4.6 Screening experiments for anti-inflammatory properties 46
4.7 Future trends 49
4.8 Sources of further information 51
4.9 References 51

5 Under-utilized herbs and spices 53
P. N. Ravindran and Geetha S. Pillai, Centre for Medicinal Plants Research, India and K. Nirmal Babu, Indian Institute of Spices Research, India
5.1 Introduction 53
5.2 Sweet flag 53
5.3 Greater galangal 60
5.4 Angelica 64
5.5 Horseradish 69
5.6 Black caraway 72
5.7 Capers 74
5.8 Asafoetida 77
5.9 Hyssop 81
5.10 Galangal 83
5.11 Betel vine 85
5.12 Pomegranate 89
5.13 Summer savory 91
5.14 Winter savory 94
5.15 Other 95
5.16 References 98

Part II Particular herbs and spices 105

6 Ajowan 107
S. K. Malhotra and O. P. Vijay, National Research Centre on Seed Spices, India
6.1 Introduction and description 107
6.2 Production 107
6.3 Cultivation 108
6.4 Chemical structure 109
6.5 Main uses in food processing 111
6.6 Functional properties and toxicity 112
6.7 Quality issues 113
6.8 References 115

7 Allspice ... 117
B. Krishnamoorthy and J. Rema, Indian Institute of Spices Research, India
- 7.1 Introduction and description ... 117
- 7.2 Production and trade ... 119
- 7.3 Chemical composition ... 120
- 7.4 Cultivation ... 125
- 7.5 Uses ... 131
- 7.6 Functional properties ... 132
- 7.7 Quality issues and adulteration ... 134
- 7.8 References ... 138

8 Chervil ... 140
A. A. Farooqi and K. N. Srinivasappa, University of Agricultural Sciences, India
- 8.1 Introduction and description ... 140
- 8.2 Cultivation and production technology ... 141
- 8.3 Uses ... 143
- 8.4 Sources of further information ... 144

9 Coriander ... 145
M. M. Sharma and R.K. Sharma, Rajasthan Agricultural University, India
- 9.1 Introduction and description ... 145
- 9.2 Origin and distribution ... 146
- 9.3 Chemical composition ... 146
- 9.4 Cultivation and post-harvest practices ... 147
- 9.5 Uses ... 149
- 9.6 Diseases, pests and the use of pesticides ... 149
- 9.7 Quality issues ... 151
- 9.8 Value addition ... 157
- 9.9 Future research trends ... 159
- 9.10 References ... 159
- Appendix I ... 160
- Appendix II ... 161

10 Geranium ... 162
M. T. Lis-Balchin, South Bank University, UK
- 10.1 Introduction ... 162
- 10.2 Chemical composition ... 162
- 10.3 Production and cultivation ... 163
- 10.4 Main uses in food processing and perfumery ... 166
- 10.5 Functional properties ... 167
- 10.6 Quality issues and adulteration ... 171
- 10.7 References ... 173

11 Lavender ... 179
M. T. Lis-Balchin, South Bank University, UK
- 11.1 Introduction ... 179
- 11.2 Chemical composition ... 179
- 11.3 Production ... 180

	11.4	Uses in food processing, perfumery and paramedical spheres	182

11.4 Uses in food processing, perfumery and paramedical spheres 182
11.5 Functional properties and toxicity 183
11.6 Quality issues and adulteration 188
11.7 References 190

12 Mustard 196
J. Thomas, K. M. Kuruvilla and T. K. Hrideek, ICRI Spices Board, India
12.1 Introduction and description 196
12.2 Chemical composition 198
12.3 Production and cultivation 200
12.4 Uses 201
12.5 Properties 202
12.6 Quality specifications 204
12.7 References 204

13 Nigella 206
S. K. Malhotra, National Research Centre on Seed Spices, India
13.1 Introduction and description 206
13.2 Chemical structure 207
13.3 Cultivation 208
13.4 Main uses in food processing 209
13.5 Functional properties and toxicity 210
13.6 Quality specifications and adulteration 212
13.7 References 214

14 Oregano 215
S. E. Kintzios, Agricultural University of Athens, Greece
14.1 Introduction and description 215
14.2 Chemical structure 216
14.3 Production and cultivation 219
14.4 Main uses in food processing and medicine 222
14.5 Functional properties 223
14.6 Quality specifications and commercial issues 225
14.7 References 226

15 Parsley 230
D. J. Charles, Frontier Natural Products, USA
15.1 Introduction and description 230
15.2 Chemical composition 232
15.3 Production and cultivation 233
15.4 Organic farming 235
15.5 General uses 238
15.6 Essential oils and their physicochemical properties 239
15.7 References 241

16 Rosemary 243
B. Sasikumar, Indian Institute of Spices Research, India
16.1 Introduction and description 243
16.2 Chemical composition 244
16.3 Production and cultivation 245

16.4	Post-harvest technology	246
16.5	Uses	248
16.6	Toxicology and disease	251
16.7	Conclusion	252
16.8	References	252

17 Sesame 256
D. M. Hegde, Directorate of Oilseeds Research, India

17.1	Introduction	256
17.2	Chemical composition	259
17.3	Production	268
17.4	Processing	275
17.5	Uses	279
17.6	Future research needs	283
17.7	References	284

18 Star anise 290
C. K. George, Peermade Development Society, India

18.1	Introduction, morphology and related species	290
18.2	Histology	292
18.3	Production and cultivation	293
18.4	Main uses	294
18.5	References	295

19 Thyme 297
E. Stahl-Biskup, University of Hamburg, Germany and R. P. Venskutonis, Kaunas University of Technology, Lithuania

19.1	Introduction	297
19.2	Chemical structure	298
19.3	Production	303
19.4	Main uses in food processing	306
19.5	Functional properties and toxicity	310
19.6	Quality specifications and issues	313
19.7	References	318

20 Vanilla 322
C. C. de Guzman, University of the Philippines Los Baños, Philippines

20.1	Introduction and description	322
20.2	Production and trade	325
20.3	Cultivation	326
20.4	Harvesting, yield and post-production activities	330
20.5	Uses	338
20.6	Vanilla products	338
20.7	Functional properties	340
20.8	Quality issues and adulteration	340
20.9	Improving production of natural vanillin	346
20.10	Future outlook	348
20.11	References	349

Index 355

Contributors

(* = main point of contact)

Chapter 1

Professor K. V. Peter*
Kerala Agricultural University
KAU – PO, Vellanikkara
Thrissur, Kerala State
India – 680656

Tel: 0487 2370034
Fax: 0487 2370019
E-mail: vckau@sancharnet.in
       kvptr@yahoo.com

Dr K. Nirmal Babu
Indian Institute of Spices Research
Calicut – 673 012
India

Tel: 0495 2731410
Fax: 0495 2730294
E-mail: nirmalbabu30@hotmail.com

Chapter 2

M. R. Shylaja and Professor K. V. Peter*
Kerala Agricultural University
P O KAU 680656, Vellanikkara
Thrissur, Kerala State
India – 680656

Tel: 0487 2370034
Fax: 0487 2370019
E-mail: vckau@sancharnet.in
       kvptr@yahoo.com
       mrshyla@rediffmail.com

Chapter 3

Dr C. C. Tassou
National Agricultural Research Foundation
Institute of Technology of Agricultural
  Products
S Venizelou 1
Lycovrisi 14123
Greece

Tel: +30 210 2845940
Fax: +30 210 2840740
E-mail: microlab.itap@nagref.gr

Professor G.-J. E. Nychas* and
  Dr P. N. Skandamis
Agricultural University of Athens
Department of Food Science and
  Technology
Iera Odos 75
Athens 11855
Greece

Tel/Fax: +30 10 529 4693
E-mail: gjn@aua.gr

Chapter 4

Dr R. Rodenburg
TNO Pharma
Utrechtseweg 48
3704HE Zeist
The Netherlands

Tel: +31 30 6944844
Fax: +31 30 6944845
E-mail: pharma-office@pharma.tno.nl

Chapter 5

P. N. Ravindran* and G. S. Pillai
Centre for Medicinal Plants Research
Arya Vaidya Sala
Kottakkal – 676 503
Kerala
India

Tel: 0483 2743430
Fax: 0483 2742572/2742210
E-mail: avscmpr@sify.com
avscmpr@yahoo.co.in

Dr K. Nirmal Babu
Indian Institute of Spices Research
Calicut – 676 012
India

Tel: 0495 2731410
Fax: 0495 2730294
E-mail: nirmalbabu30@hotmail.com

Chapter 6

Dr S. K. Malhotra* and Dr O. P. Vijay
National Research Centre on Seed Spices
Ajmer – 305 206
Rajasthan
India

Tel: +91 145 2680955
Fax: +91 145 2443238
E-mail: malhotraskraj@yahoo.com

Chapter 7

Mr B. Krishnamoorthy* and Dr J. Rema
Indian Institute of Spices Research
Calicut 673 012
Kerala
India

E-mail: bkrishnamoorthy@rediffmail.com
remachaithram@yahoo.co.in

Chapter 8

Dr A. A. Farooqi* and K. N. Srinivasappa
Division of Horticulture
University of Agricultural Sciences
GKVK
Bangalore
India

E-mail: azharfarooqi@sify.com

Chapter 9

Dr M. M. Sharma* and Dr R. K. Sharma
Rajasthan Agricultural University
Bikaner
India

E-mail: mmohanrau@yahoo.com

Chapter 10

Dr M. T. Lis-Balchin
School of Applied Science
South Bank University
103 Borough Road
London SE1 0AA

E-mail: lisbalmt@lsbu.ac.uk

Chapter 11

Dr M. T. Lis-Balchin
School of Applied Science
South Bank University
103 Borough Road
London SE1 0AA

E-mail: lisbalmt@lsbu.ac.uk

Chapter 12

Dr J. Thomas*, K. M. Kuruvilla and
  T. K. Hrideek
ICRI Spices Board
Kailasanadu PO
Kerala, India – 685 553

E-mail: jtkotmala@hotmail.com

Chapter 13

Dr S. K. Malhotra
National Research Centre on Seed Spices
Ajmer – 305 206
Rajasthan
India

Tel: +91 145 2680955
Fax: +91 145 2443238
E-mail: malhotraskraj@yahoo.com

Chapter 14

Professor S. Kintzios
Laboratory of Plant Physiology
Agricultural University of Athens
Iera Odos 75
11855 Athens
Greece

Tel: +3210 5294292
Fax: +3210 5294286
E-mail: skin@aua.gr

Chapter 15

Dr D. J. Charles
Frontier Natural Products Co-op
3021 78th Street
Norway, IA
52318
USA

E-mail: denys.charles@frontiercoop.com

Chapter 16

Dr B. Sasikumar
Indian Institute of Spices Research
Marikunnu (PO)
Calicut – 673 012
Kerala
India

Tel: 91 495 2731410
Fax: 91 495 2730294
Email: bhaskaransasikumar@yahoo.com

Chapter 17

Dr D. M. Hegde
Directorate of Oilseeds Research
Rajendranagar
Hyderabad – 500 030
Andhra Pradesh
India
Tel: +91 040 24015222
Fax: +91 040 24017969

E-mail: dmhegde@rediffmail.com

Chapter 18

C.K. George
Peermade Development Society
Post Box 11
Peermade – 685531
Idukki Dist.
Kerala
India

E-mail: ckgeorge@vsnl.com

Chapter 19

Professor E. Stahl-Biskup*
University of Hamburg
Institute of Pharmacy
Department of Pharmaceutical Biology and Microbiology
Bundesstrasse 45
D-20146 Hamburg
Germany

Tel: +49 (0)40 42838 3896
Fax: +49 (0)40 42838 3895
E-mail: elisabeth.stahl-biskup
@uni-hamburg.de

Professor R. P. Venskutonis
Head of Department of Food Technology
Radvilenu pl. 19
Kaunas
LT – 3028
Lithuania

Tel: +370 37 456426
Fax: +370 37 456647
E-mail: rimas.venskutonis@ktu.lt

Chapter 20

Dr C. C. de Guzman
Department of Horticulture
College of Agriculture
University of the Philippines Los Baños
Los Baños
Laguna 4031
Philippines

Tel: (63-49) 536 2448
Fax: (63-49) 536 2478
E-mail: tanchodg@lb.msc.net.ph

1

Introduction

K. V. Peter, Kerala Agricultural University, India and K. Nirmal Babu, Indian Institute of Spices Research, India

1.1 Introduction to herbs and spices

The history of herbs and spices is as long as the history of mankind. People have used these plants since earliest times. No other commodity has played a more pivotal role in the development of modern civilization as spices. The lives of people and plants are more entwined than is often realized. Some herbs have the power to change our physiological functioning, they have revolutionized medicine, created fortunes for those who grow, process and treat them, and in many cases have assumed social and religious significance. Herbs have changed the course of history and in economic terms have greater importance as ingredients in food and medicine, perfumery, cosmetics and garden plants. The knowledge of herbs has been handed down from generation to generation for thousands of years (Brown, 1995). Wars have been fought and lands conquered for the sake of these plants. Even today we continue to depend on herbs and spices for many of our newest medicines, chemicals and flavours and they are used in culinary preparations, perfumery and cosmetics. Many medicinal herbs are also food, oil and fibre plants and have always been grown for a range of purposes (Parry, 1969; Rosengarten, 1973; Andi *et al.*, 1997).

The term 'herb' has more than one definition. In the most generally accepted sense, herbs are plants valued for their medicinal and aromatic properties and are often grown and harvested for these unique properties. Some of the earliest of herb gardens were planted about 4000 years ago in Egypt. Herb growing was often associated with temples, which required herbs and sacred flowers for daily worship and rituals. Both horticulture and botany began with the study of herbs. The earliest gardens were herb gardens. The present-day concept of a herb garden has developed largely from ancient Egyptian, Christian and Islamic traditions. In most parts of the world, herbs are grown mainly as field crops or on a small scale as a catch-crop among vegetables and ornamentals as they were thousands of years ago. The cultivation requirements of some of the most important herbs are given in Table 1.1.

Table 1.1 Cultivating requirements and uses

Plant	Propagation	Common uses
Anise	Annual. Seeds are sown in a dry, light soil in early summer. Seedlings should be thinned to inches apart. Anise needs 120 frost-free days to produce fully ripened seed heads.	The aromatic seeds are used in cooking, in pot-pourris and in some simple home remedies.
Basil	Perennial. Grows easily from seed. It is frost sensitive. Basil needs medium-rich, well-drained soil and full sun. Pinch off tips and flower buds to promote bushiness.	The leaves are a classic complement to tomatoes; they are also used to flavour salads, sauces and vegetables.
Chervil (*Anthriscus cerefolium*)	Annual and resembles parsley. Seeds are sown in spring. Thin to 15 cm (6 inches) apart. Likes moist, well-drained soil and partial shade. Will self-sow.	The leaves, with their delicate anise-like flavour, are often used in soups and salads.
Lavender	Perennial, with many varieties. English lavender is the hardiest. Mulch it over the winter. Propagation is easiest by root division. Likes full sun and alkaline, gravelly soil.	Grown for its fragrance in the garden and to be used in pot-pourris and sachets.
Oregano	Perennial. Prefers well-drained, slightly alkaline soil and full sun. Propagate by seed, root division or cuttings.	The leaves are a favorite seasoning for pizza and other Italian dishes.
Parsley (*Petroselinum crispum*)	Biennial, usually grown as an annual. Both types like a rich, well-drained soil and full sun or partial shade. Parsley seeds seeds germinate slowly. Be patient; keep the soil moist. Thin to (20 cm) 8 inches apart.	Curly leaved parsley is popular as garnish, but flat leaved (Italian) parsley is more flavourful and is used as addition to salads and sauces. Parsley tea makes a healthful tonic.
Rosemary	Perennial, grown indoors in cold climates. Rosemary needs full sun, and a sandy well-limed soil. Cut it back after flowering to prevent it from becoming leggy.	Propagate by layering or cuttings. This is an aromatic flavouring for meat and poultry dishes. Also used for making wreaths.
Savory	Winter savory, a perennial, has a peppery, pungent flavour. Summer savory, an annual, is similar but more delicate. Plant seeds of summer savory in a rich, light, moist soil; thin to 20 cm (8 inches) apart. Winter savory thrives in poorer soil and with less water. It can be propagated by seed, division or cuttings.	Savory is used to flavour sausages and other meats and is sometimes included in a bouquet garni.
Thyme	Perennial. There are many species and varieties including lemon, English, golden and garden. The garden variety is the most popular for cooking. Thyme grows well in dry sloping sides; pruning after flowering will keep it from getting woody. Propagated by cuttings.	The leaves add pungent taste to meats and vegetables; thyme sprigs are a main ingredient in bouquet garnishing for soups and stews.

Source: Reader's Digest (1990).

1.2 Uses of herbs and spices

Herbs and spices have tremendous importance in the way we live, as ingredients in food, alcoholic beverages, medicine, perfumery, cosmetics, colouring and also as garden plants. Spices and herbs are used in foods to impart flavour, pungency and colour. They also have antioxidant, antimicrobial, pharmaceutical and nutritional properties. In addition to the known direct effects, the use of these plants can also lead to complex secondary effects such as salt and sugar reduction, improvement of texture and prevention of food spoilage. The basic effects of spices when used in cooking and confectionery can be for flavouring, deodorizing/masking, pungency and colouring (Table 1.2). They are also used to make food and confectionery more appetizing and palatable. Some spices, such as turmeric and paprika, are used more for imparting an attractive colour than for enhancing taste. The major colour components of spices are given in Table 1.3. Because of their antioxidant and

Table 1.2 Basic uses of herbs and spices

Basic function	Major function	Subfunction
Flavouring	Parsley, cinnamon, allspice, dill, mint, tarragon, cumin, marjoram, star anise, basil, anise, mace, nutmeg, fennel, sesame, vanilla, fenugreek, cardamom, celery	Garlic, onion, bay leaves, clove, thyme, rosemary, caraway, sage, savory, coriander, pepper, oregano, horseradish, Japanese pepper, saffron, ginger, leek, mustard
Deodorizing/ masking	Garlic, savory, bay leaves, clove, leek, thyme, rosemary, caraway, sage, oregano, onion, coriander	
Pungency	Garlic, savory, bay leaves, clove, leek, thyme, rosemary, caraway, sage, oregano, onion, coriander, Japanese pepper, mustard, ginger, horseradish, red pepper, pepper	Parsley, pepper, allspice, mint, tarragon, cumin, star anise, mace, fennel, sesame, cardamom, mustard, cinnamon, vanilla, horseradish, Japanese pepper, nutmeg, ginger
Colouring	Paprika, turmeric, saffron	

Source: Ravindran *et al.* (2002).

Table 1.3 Colour components in spices

Colour component	Tint	Spice
Carotenoid		
β-carotene	Reddish orange	Red pepper, mustard, paprika, saffron
Cryptoxanthin	Red	Paprika, red pepper
Lutin	Dark red	Paprika, parsley
Zeaxanthin	Yellow	Paprika
Capsanthin	Dark Red	Paprika, red pepper
Capsorbin	Purple red	Paprika, red pepper
Crocetin	Dark red	Saffron
Neoxanthin	Orange yellow	Parsley
Violaxanthin	Orange	Parsley, sweet pepper
Crocin	Yellowish orange	Saffron
Flavonoids	Yellow	Ginger
Curcumin	Orange yellow	Turmeric
Chlorophylls	Green	Herbs

Source: Ravindran *et al.* (2002).

Table 1.4 Spices and herbs used in alcoholic beverages

Alcoholic beverages	Spices and herbs used
Vermouth	Marjoram, sage, coriander, ginger, cardamom, clove, mace, peppermint, thyme, anise, juniper berry
Gin	Coriander, juniper berry
Aquavit	Anise, fennel, dill, caraway
Curaçao	Cinnamon, clove, nutmeg, coriander
Kummel	Caraway, fennel, coriander
Anisette	Anise, fennel, nutmeg
Ganica	Cinnamon, cardamom, coriander, mint, fennel, clove, pepper
Geme de cumin	Cumin
Geme de cacao	Clove, mace, vanilla
Geme De menthe	Peppermint
Peppermint schnapps	Peppermint

Source: Ravindran *et al*. (2002).

antimicrobial properties, spices have dual function – in addition to imparting flavour and taste, they play a major role in food preservation by delaying the spoilage of food. Many herbs and spices have been used in cosmetics, perfumery and beauty and body care since ancient times. The toiletries and allied industries use spices and herbs and their fragrant oils for the manufacture of soaps, toothpastes, face packs, lotions, freshness sachets, toilet waters and hair oils. They are essential ingredients in beauty care as cleansing agents, infusions, skin toners, moisturizers, eye lotions, bathing oils, shampoos and hair conditioners, cosmetic creams, antiseptic and antitanning lotions and creams, improvement of complexion and purifying blood (Pamela, 1987; Ravindran *et al.*, 2002). Spices form an important component in quite a few alcoholic beverages and beers (Table 1.4).

1.2.1 Medicinal uses

Herbs and spices have been an essential factor in health care through the ages in all cultures. They are prepared in number of ways to extract their active ingredients for internal and external use. There are a number of different systems of herbal medicine, the most important of which are Chinese and Indian (*Ayurvedic*) systems of medicine. All spices are medicinal and are used extensively in indigenous systems of medicine. Some of the important uses of major medicinal spices in *Ayurveda*, according to Mahindru (1982), are given in Table 1.5. Extracts from herbs and spices are used as infusions, decoctions, macerations, tinctures, fluid extracts, teas, juices, syrups, poultices, compresses, oils, ointments and powders.

Many medicinal herbs used in *Ayurveda* have multiple bioactive principles. It is not always easy to isolate compounds and demonstrate that the efficacy can be attributed to any one of the active principles. However, the active principles and their molecular mechanism of action of some of the medicinal plants are being studied (Tables 1.6 and 1.7).

1.3 Active plant constituents

Herbs and spices are rich in volatile oils, which give pleasurable aromas. In addition, herbs may contain alkaloids and glycosides, which are of greater interest to pharmacologists. Some of the main active constituents in herbs are as follows (Brown, 1995; De Guzman and Sienonsma, 1999):

- Acids – these are sour, often antiseptic and cleansing.
- Alkaloids – these are bitter, often based on alkaline nitrogenous compounds. They affect the central nervous system and many are very toxic and addictive.
- Anthraquinones – these are bitter, irritant and laxative, acting also as dyes.
- Bitters – various compounds, mainly iridoides and sesquiterpenes with a bitter taste that increases and improves digestion.
- Coumarines – are antibacterial, anticoagulant, with a smell of new-mown hay.
- Flavones – these are bitter or sweet, often diuretic, antiseptic, antispasmodic and anti-inflammatory. Typically yellow, and present in most plants.
- Glycosides – there are four main kinds of glycosides.
  cardiac: affecting heart contractions;
  synogenic: bitter, antispasmodic sedative, affecting heart rate and respiration;
  mustard oil: acrid, extremely irritant;
  sulphur: acrid, stimulant, antibiotic.
- Gums and mucilages – these are bland, sticky or slimy, soothing and softening.
- Resins – often found as oleo-resins or oleo-gum resins – they are acrid, astringent, antiseptic, healing.
- Saponins – are sweet, stimulant hormonal, often anti-inflammatory, or diuretic, soapy in water.
- Tannins – are astringent, often antiseptic, checking bleeding and discharges.
- Volatile oils – are aromatic, antiseptic, fungicidal, irritant and stimulant.

1.3.1 Genetic erosion in herbs and spices

People all over the world have picked and uprooted herbs from the wild since ancient times. Medicinal herbs in particular have always been mainly collected from the wild and the knowledge of where they grow and the best time to gather them has formed an important oral tradition among healers of many different countries in many different cultures. These ancient traditions successfully balance supply and demand, allowing plant stock to regenerate seasonally. Owing to the strong commercial pressures of food and pharmaceutical industries of today, the balance now has been disrupted by unregulated gathering, leading to severe genetic erosion. Some of the most commonly used culinary herbs such as chilli peppers (*Capsicum annuum* var. *annuum*) and basil (*Ocimum basilicum*) have such a long history of use and cultivation that truly wild plants have never been recorded. They presumably became extinct because of over-collection.

1.4 The structure of this book

This book is the second volume for the series on Herbs and Spices and has two parts. The first part deals with health benefits of herbs and spices and the use of herbs and spices as antimicrobials and antioxidants. The second part deals with detailed information on individual spices. This covers a brief description, classification, production, cultivation, post-harvest handling, uses in food processing, chemical structure and functional properties of important compounds extracted and quality specifications. The crops covered are tree spices such as allspice and star anise, and important herbs such as chervil, coriander, oregano, parsley, rosemary and thyme. A few other spices such as vanilla and sesame are also included.

Though individual chapters vary in structure and emphasis, depending on the importance

Table 1.5 Use of major medicinal spices in *Ayurveda*

Standard medicine	Turmeric	Ginger	Pepper	Cardamom	Cinnamon/cassia	Nutmeg	Others
1. *Murchchha-paka* of ghee, sesame, mustard, castor oil	✓	—	—	—	—	—	Coriander
2. *Gandha-paka*	—	✓	✓	✓	Tej patra (Cinna-tamla)	✓	Clove, saffron
3. *Anupan*	✓	✓	✓	—	✓	—	*Ocimum sanctum*
4. *Chaturbhadraka*	—	✓	—	—	—	—	—
5. *Panchkala*	✓	—	—	—	—	—	—
6. *Valli Panchamula*	—	✓	✓	—	—	—	—
7. *Trikatu*	✓	✓	✓	—	—	—	Long pepper
8. *Lekniya Varga*	—	—	—	—	—	—	—
9. *Dipaniya Varga*	—	✓	✓	—	—	—	—
10. *Triptaighna Varga*	—	—	—	—	—	—	—
11. *Kushthaghma Varga*	✓	✓	—	—	—	—	—
12. *Vishaghan Varga*	✓	—	✓	—	—	—	—
13. *Stunyasodhanna Varga*	—	✓	—	—	—	—	—
14. *Srouirechanopaga*	—	✓	✓	—	—	—	—
15. *Trishna nigraha Varga*	—	✓	✓	—	—	—	—
16. *Sitaprasemana Varga*	✓	—	—	—	—	—	—
17. *Sulaprasemena Varga*	✓	✓	—	—	—	—	—
18. *Haridradigana*	✓	—	—	—	—	—	—
19. *Mustadigana*	✓	—	—	—	—	—	—
20. *Lakshadigana*	—	—	—	✓	—	—	—
21. *Rasnadi* group	—	✓	—	—	—	—	—
22. *Pippalyadi* group	—	✓	✓	—	—	—	—
23. *Guruchayadi* group	—	✓	—	—	—	—	Coriander
24. *Sunthayadi* group	—	✓	—	—	—	—	—
25. *Duralabhadi* group	—	✓	—	—	—	—	—
26. *Vishwadi* group	—	✓	—	—	—	—	—
27. *Kanadi* group	—	✓	—	—	—	—	—
28. *Granthyadi* group	—	✓	—	—	—	—	—
29. *Kakolyadi* group	—	✓	—	—	—	—	—
30. *Sriphaladi* group	—	✓	—	—	—	—	—
31. *Bhunimvadi* group	—	✓	✓	—	—	—	—
32. *Marichadi* group	✓	—	—	—	—	—	—
33. *Katurikadya* group	✓	✓	—	—	—	—	—
34. *Nimbadi* group	—	✓	—	—	—	—	—
35. *Katurikadya* group	—	✓	—	—	—	—	—
36. *Trikodi* group	—	✓	✓	—	—	—	—
37. *Nidigdhikadi* group	—	✓	—	—	—	—	—
38. *Katphaladi* group	—	✓	✓	—	—	—	—
39. *Navanga* group	—	✓	—	—	—	—	—

No.	Name						Coriander
40.	Pancha bhadra group	–	♪	–	–	–	–
41.	Kiratatiktadi group	–	–	–	–	–	–
42.	Kiratadi group	–	♪	–	–	–	–
43.	Aragbadhadi group	–	–	–	–	–	–
44.	Mustadi group	–	♪	♪	–	–	–
45.	Pathasaptaka group	–	♪	–	–	–	–
46.	Amritashtaka group	–	♪	♪	–	–	–
47.	Kantakaryadi group	–	♪	♪	–	–	–
48.	Swachchlanda Bhairirava	–	♪	♪	–	–	–
49.	Agnikumara Rasa	–	♪	♪	–	–	–
50.	Sri-Mrityunjaya Rasa	–	♪	–	–	–	–
51.	Sarvajwarankusa Vatika	–	♪	♪	–	–	–
52.	Chanderswara	–	♪	♪	–	–	–
53.	Chadrasekhara Rasa	–	♪	♪	–	–	–
54.	Nanajwarchha-Sinha	–	♪	♪	–	–	–
55.	Mritunjaya Rasa	–	♪	♪	–	–	–
56.	Prachamdeswara Rasa	–	♪	–	–	–	–
57.	Tripurabhahairava Rasa	–	♪	♪	–	–	–
58.	Kaphaketu	–	♪	–	–	–	–
59.	Jwara kesari	–	♪	–	–	–	–
60.	Jwara murari	–	♪	–	–	–	–
61.	Situ bhanjdrosa	–	♪	♪	–	–	–
62.	Nawa-Jwarari Rasa	♪	♪	–	–	–	Coriander
63.	Sarwanga Sundara	–	♪	–	–	–	–
64.	Jayabati	–	♪	–	–	–	–
65.	Srirama rasa	–	♪	–	–	–	–
66.	Udakamanjiri	–	♪	–	–	–	–
67.	Kshudradi	–	♪	–	–	–	–
68.	Nagaradi group	–	♪	–	–	–	–
69.	Chaturdasanga	–	♪	–	–	–	–
70.	Ashtadasanga	–	♪	–	–	–	–
71.	Bhargyadi group	–	♪	–	–	–	–
72.	Sathyadi group	–	♪	–	–	–	–
73.	Mustadya group	–	♪	♪	–	–	–
74.	Vyashadi group	♪ ♪	♪ ♪	♪	–	–	Coriander
75.	Watringa Sanga group	–	♪	–	–	–	–
76.	Kankakaryadi group	–	♪	–	–	–	–
77.	Vrihatkatphatedi group	♪	♪	♪	–	–	–
78.	Unmatha Rasa	–	–	–	–	–	–
79.	Vnihat Kasturi Bhairava	–	4	4	–	–	–
80.	Sleshma-kalawala	–	–	–	–	–	–

Source: Mahindru (1982).

Table 1.6 Ayurvedic modes of administration

Modality	Mode/vehicle	Effects	Potential
Cinnamon oil	Volatiles	Antimicrobial	Infections
Curcuma longa	'Band-aid'	Wound healing	Global scope
Asafoetida	Umbilicus	Antiflatulent	Post-operative
Asparagus racemosus	With milk/boiled	Phagocytosis	Rasayana
Centella asiatica	Brahmighrita	Nootropic	Alzheimer's disease

Source: Vaidya (2002).

Table 1.7 Molecular phytopharmacology of a few herbs and spices

Plant	Active principle	Molecular action	Uses
Piper longum	Piperine	RNA synthesis	Antiviral
Curcuma longa	Curcumin	Protein synthesis	Against Alzheimer's
Mangifera indica	Mangiferin	Macrophage activation	Immunostimulant
Coleus forskohlii	Forshlin	cAMP increase	Against glaucoma

Source: Vaidya (2002).

of the spice and the body of research surrounding it, the matter is organized in the same format as in the first volume. It is hoped that this book will form a good reference book for all those who are involved in the study, cultivation, trade and use of spices and herbs.

1.5 References

ANDI C., KATHERINE R., SALLIE M. and LESLEY M. (1997), *The Encyclopedia of Herbs and Spices*. Hermes House, London.
BROWN D. (1995), *The Royal Horticultural Society – Encyclopedia of Herbs and Their Uses*. Dorling Kindersley Limited, London.
DE GUZMAN C.C. and SIENONSMA J.S. (1999), *Plant Resources of South East Asia. No. 13. Spices*. Backhuys Publishers, Leiden, The Netherlands.
MAHINDRU S.N. (1982), *Spices in Indian Life*. Sultanchand and Sons, New Delhi.
PAMELA W. (1987), *The Encyclopedia of Herbs and Spices*. Marshall Cavendish Books Ltd, London.
PARRY J.W. (1969), *Spices Volumes I & II*. Chemical Publishing Co., New York.
RAVINDRAN P.N, JOHNY A.K and NIRMAL BABU K. (2002), Spices in our daily life. *Satabdi Smaranika 2002* Vol. 2. Arya Vaidya Sala, Kottakkal.
READER'S DIGEST (1990), *Magic and Medicine of Plants*. Readers Digest Association, Inc., USA.
ROSENGARTEN F. (1973), *The Book of Spices*, Revised Edition Pyramid, New York.
VAIDYA A.D.B. (2002), Recent trends in research on Ayurveda. *Satabdi Smaranika 2002* Vol. 1. Arya Vaidya Sala, Kottakkal.

Part I

General issues

2
The functional role of herbal spices

M. R. Shylaja and K. V. Peter, Kerala Agricultural University, India

2.1 Introduction

Herbal spices or leafy spices are annual/biennial/perennial plants, the leaves of which (fresh or dry) are primarily used for flavouring foods and beverages. Apart from being used as flavouring agents, herbal spices are also known to possess nutritional, antioxidant, antimicrobial and medicinal properties. Because of the attractive foliage, a few herbs are also used as garnishing spices in many food preparations. The essential oils extracted from tender stems, leaves and flowering tops are used in cosmetics, perfumeries and toiletries and for flavouring liquors, soft drinks, beverages and pharmaceutical preparations. ISO document 676 lists 38 leafy spices (Table 2.1).

Table 2.1 Leafy spices in ISO document 676

Sl No.	Botanical name	Family	Common name	Plant part used as spice
1.	*Allium tuberosum*	Liliaceae	Indian leek, Chinese chive	Bulb, leaf
2.	*Allium fistulosum*	Liliaceae	Stony leek, Welsh onion, Japanese bunching onion	Leaf and bulb
3.	*Allium porrum*	Liliaceae	Leek, winter leek	Leaf and bulb
4.	*Allium schoenoprasum*	Liliaceae	Chive	Leaf
5.	*Anethum graveolens*	Apiaceae	Dill	Fruit, leaf, top
6.	*Anthriscus cereifolium*	Apiaceae	Chevril	Leaf
7.	*Apium graveolens*	Apiaceae	Celery, garden celery	Fruit, root, leaf
8.	*Apium graveolens* var. *rapaceum*	Apiaceae	Celeriac	Fruit, root, leaf
9.	*Artemisia dracunculus*	Asteraceae	Tarragon, estragon	Leaf
10.	*Cinnamomum aromaticum*	Lauraceae	Cassia, Chinese Cassia	Bark, leaf
11.	*Cinnamomum tamala*	Lauraceae	Tejpat, Indian Cassia	Leaf, bark
12.	*Cinnamomum zeylanicum*	Lauraceae	Srilankan cinnamon, Indian cinnamon	Bark, leaf
13.	*Coriandrum sativum*	Apiaceae	Coriander	Leaf, fruit
14.	*Foeniculum vulgare*	Apiaceae	Bitter fennel	Leaf, twig, fruit
15.	*Foeniculum vulgare*	Apiaceae	Sweet fennel	Leaf, twig, fruit
16.	*Hyssopus officinalis*	Lamiaceae	Hyssop	Leaf

17.	*Laurus nobilis*	Lauraceae	Laurel, true laurel, bay leaf, sweet flag	Leaf
18.	*Levisticum officinale*	Apiaceae	Garden lovage, lovage	Fruit, leaf
19.	*Lippia graveolens* and *Lippia berlandieri*	Verbenaceae	Mexican oregano	Leaf terminal shoot
20.	*Melissa officinalis*	Lamiaceae	Balm, lemon balm, melissa	Leaf, terminal shoot
21.	*Mentha arvensis*	Lamiaceae	Japanese mint, field mint, corn mint	Leaf, terminal shoot
22.	*Mentha citrata*	Lamiaceae	Bergamot	Leaf, terminal shoot
23.	*Mentha piperita*	Lamiaceae	Peppermint	Leaf, terminal shoot
24.	*Mentha spicata*	Lamiaceae	Spearmint, garden mint	Leaf, terminal shoot
25.	*Murraya koenigii*	Rutaceae	Curry leaf	Leaf
26.	*Ocimum basilicum*	Lamiaceae	Sweet basil	Leaf, terminal shoot
27.	*Origanum majorana*	Lamiaceae	Sweet marjoram	Leaf, floral bud
28.	*Origanum vulgare*	Lamiaceae	Oregano, origan	Leaf, flower
29.	*Petroselinum crispum*	Apiaceae	Parsley	Leaf, root
30.	*Pimenta dioica*	Myrtaceae	Pimento	Fruit, leaf
31.	*Pimenta racemosa*	Myrtaceae	West Indian bay	Fruit, leaf
32.	*Rosmarinus officinalis*	Lamiaceae	Rosemary	Terminal shoot, leaf
33.	*Salvia officinalis*	Lamiaceae	Garden sage	Terminal shoot, leaf
34.	*Satureja hortensis*	Lamiaceae	Summer savory	Terminal shoot, leaf
35.	*Satureja montana*	Lamiaceae	Winter savory	Terminal shoot, leaf
36.	*Thymus serpyllum*	Lamiaceae	Wild thyme, creeping thyme	Terminal shoot, leaf
37.	*Thymus vulgaris*	Lamiaceae	Thyme, common thyme	Terminal shoot, leaf
38.	*Trigonella foenumgraecum*	Fabaceae	Creeping thyme, fenugreek	Seed, leaf

2.2 Classification

Herbal spices can be classified based on botanical families, crop duration and growth habit.

2.2.1 Classification based on botanical families

Family	Crop
Apiaceae	Dill, celery, fennel, lovage, parsley, etc.
Lamiaceae	Hyssop, mint, basil, marjoram, oregano, rosemary, sage, thyme, etc.
Liliaceae	Leek, chive

2.2.2 Classification based on duration of crop

Annual	Basil, coriander, dill, etc.
Biennial	Caraway, leek, parsley, etc.
Perennial	Sage, laurel, pimenta, curry leaf, chive, mint, oregano, tarragon, thyme, etc.

2.2.3 Classification based on growth habit

Herbs	Caraway, coriander, mint, oregano, marjoram
Shrubs	Rosemary, sage, thyme
Trees	Pimenta, curry leaf, laurel

2.3 Production, consumption and processing

Most of the herbal spices originated in Mediterranean countries and have been used since ancient Egyptian and Roman times mainly for the purpose of embalming. Even today, the Mediterranean zone is the major source of herbal spices, and Germany, France and the USA are the major producers of high-quality cultivated herbs.

Curly parsley, chives and dill are widely grown in Germany, while flat parsley and tarragon are widely grown in France. The USA has cultivation of high-quality herbs such as parsley, tarragon, oregano and basil. The Mediterranean countries of Egypt and Morocco cultivate parsley, chives and dill. East European countries such as Poland, Hungary, Greece and the former Yugoslavia grow herbs on a limited scale. The countries of origin of herbal spices and major areas of cultivation are given in Table 2.2.

The European and American markets are the major consumers of herbal spices. Oregano is the most consumed herb in Europe and USA, followed by basil, bay leaf, parsley, thyme and chives. Herbs such as mint, rosemary, savory, sage and marjoram are consumed only to a limited extent in major markets. Consumption of different herbs vary according to the local food habits. Marjoram is the most sold herb in Gemany, while sage is popular in the USA but less so in Europe.

Egypt, Turkey, Spain and Albania are major exporters of herbal spices. The mild sunny climate and rocky landscape favour production and processing of herbal spices in these countries. Turkey is the biggest oregano and bay leaf exporter, Egypt is the biggest basil, marjoram and mint exporter and Spain is the biggest thyme and rosemary exporter.

Herbal spices can be used either fresh or dried or in the form of extractives such as oils and oleoresins. Herbs have traditionally been traded as dried products. With the advent of modern methods of preservation, frozen herbs and fresh herbs have become available but the industry remains dominated by the trade in dried products.

Different methods are used to dry herbs and spices. Sun drying and shade drying are still widely used. Since natural sun/shade drying leads to quality deterioration by way of contamination, artificial methods such as using circulation of hot air in a specially constructed drying room or drying with the help of hot air or microwave oven have been widely adopted. Freeze drying by applying a vacuum is a method that has proved to be the best method for preserving the delicate flavour and aroma of chives and leek. As sun drying destroys chlorophyll, artificially dried leaves have a better appearance and high market preference.

Organic spices are gaining in market share. The major consumers of organic spices in the world are the USA, Europe and Japan, which are also the major consumers of herbal spices. There is great potential for the cultivation of organic herbal spices to enjoy the premium price in the international market and to improve the quality and appearance of the produce without any pesticide or chemical residues. The spice extracts such as essential oils and oleoresins from leaves and flowering tops of various herbal spices can be recovered using steam distillation, water cum steam distillation, supercritical carbon dioxide extraction and solvent extraction using low-boiling organic solvents. Of the different methods, extraction using compressed carbon dioxide gas or supercritical fluid is the most effective and is currently used on a commercial scale. In steam distillation the plant material is exposed to high temperatures from steam vapour, leading to the degradation of important components of essential oil, while extraction with organic solvents leaves residues of the solvent in spice extracts. In supercritical carbon dioxide extraction the energy cost associated with the process is lower, the extracts are free of solvent and there is no degradation of important components. The important compounds responsible for flavour in various herbal spices are listed in Table 2.3.

Table 2.2 Origin and major areas of cultivation of herbal spices

Sl No.	Spice	Origin	Major areas
1.	Allspice	Central America, Mexico and West Indies	Jamaica, Honduras, Guatemala, Leeward Islands
2.	Basil, sweet	India, Iran, Africa	Belgium, France, Bulgaria, Hungary, India, Italy, Poland, Spain and USA
3.	Bay leaves (laurel)	Countries bordering the Mediterranean	Cyprus, France, Greece, Italy, Israel, Morocco, Portugal, Spain, Turkey and Yugoslavia
4.	Caraway	Europe	Netherlands, Bulgaria, Canada, Germany India, Morocco, Poland, Romania, Russia, Syria, UK and USA
5.	Celery	Europe, Africa	France, Hungary, India, Japan, Netherlands, UK and USA
6.	Chervil	Russia and Western Asia	France, Italy, Russia, Spain, UK and USA
7.	Chive	Northern Europe	Austria, Canada, France, Germany, Italy, Netherlands, Switzerland, UK and USA
8.	Coriander	Africa, Europe	Argentina, Bulgaria, China, France, India, Italy, Morocco, Mexico, Netherlands, Romania, Russia, Spain, Turkey, UK, USA and Yugoslavia
9.	Dill	France, Spain and Russia	Canada, Denmark, Egypt, Germany, Hungary, India, Netherlands, Mexico, Pakistan, Romania, UK and USA
10.	Fennel	Europe and Asia Minor	Bulgaria, China, Denmark, Egypt, France, Germany, India, Italy, Japan, Morocco, Netherlands, Romania, Russia, Syria, UK and USA
11.	Fenugreek	Europe and West Asia	Algeria, Argentina, Cyprus, Egypt, France, Germany, Greece, India, Italy, Lebanon, Morocco, Portugal, Spain, USA and Yugoslavia
12.	Leek	Mediterranean region	Europe, Africa, Near East and USA
13.	Marjoram	Saudi Arabia and Western Asia	France, Germany, Grenada, Hungary, Italy, Morocco, Portugal, Spain, South America, Tunisia, UK and USA
14.	Mint (peppermint)		Argentina, Australia, Brazil, France, Germany, India, Italy, Japan, Taiwan, Yugoslavia, UK and USA
15.	Mint (spearmint)	England and UK	Germany, Japan, Netherlands, Russia
16.	Oregano	Greece, Italy and Spain	Albania, France, Greece, Italy, Mexico, Spain, Turkey and Yugoslavia
17.	Parsley	Sardinia	Algeria, California, Louisiana, Belgium, Canada, France, Germany, Greece, Italy, Japan, Lebanon, Netherlands, Portugal, Spain, Turkey and UK
18.	Rosemary	Europe	Algeria, France, Germany, Italy, Morocco, Portugal, Romania, Russia, Spain, Tunisia, Turkey, Yugoslavia and USA
19.	Sage	Albania and Greece	Albania, Cyprus, Dalmatian Islands, Canada, Southern France, Italy, Portugal, Spain, Turkey, Yugoslavia, UK and USA
20.	Tarragon	Russia	Russia, France and USA
21.	Thyme	China and East Indies	Bulgaria, Canada, France, Germany, Greece, Italy, Morocco, Portugal, Russia, Spain, Tunisia, Turkey, UK and USA

Table 2.3 Compounds responsible for flavour in herbal spices

Spice	Major component	Others
Allspice	Eugenol	Cineol, phellandrene, caryophyllene
Basil, sweet	D-Linalool	Methyl chavicol, eugeneol and cineole
Bay (laurel) leaves	Cineole	L-Linalool, eugenol, methyl eugenol, geraniol, geranyl and eugenyl esters, L-α-terpineol, α-pinene and β-phellandrene
Caraway	Carvone	D-Limonene, carveol, D-dihydrocarveol, L-neodihydro carveol
Celery	D-Limonene	Selinene, sesquiterpene alcohol, sedanolide
Coriander	D-Linalool	D-α-pinene, β-pinene, α and γ-terpinene, gerciniol, borneol, *p*–cymene
Dill	Carvone	Dihydrocarvone, D-Limonene, α-phellandrene, α-pinene and dipentene
Fennel	Anethole	Fenchine, α-pinene, camphene, D-α-phellandrene, dipentene, methyl chavicol and *p*-hydroxyphenyl acetone
Marjoram	Carvacrol	D-Linalool, eugenol, chavicol, methyl chavicol, D-terpineol and carpophyllene limonene, cineol
Mint (peppermint)	Menthol	Menthone, menthyl acetate, β-pinene, α-pinene, sabinene acetate
Mint (spearmint)	L-Carvone	Terpene, carveol, dihydrocarveol acetate
Oregano	Thymol	Carvacrol, α-pinene, cineole, linalyl acetate, linalool, dipentene, *p*-cymene and β-caryophyllene
Parsley	Apiole	Myristicin, α-pinene
Rosemary	Cineole	Borneol, linalool, eucalyptol, camphor, bornyl acetate, α-pinene, camphene, sabinene, phellandrene, α-terpinene
Sage	Thujone	Borneol, cineole, bornylesters, α-pinene, salvene, D-camphor phellandrene, ocimene
Tarragon	Methyl chavicol	L-Pinitol, α-benzopyrene and eugenol
Thyme	Thymol	Carvacol, linalool, L-borneol, geraniol, amyl alcohol, β-pinene, camphene, *p*-cymene, caryophyllene, 1,8-cineole

2.4 Functional properties

In addition to adding flavour to foods and beverages, herbal spices are valued for their nutritional, antioxidant, antimicrobial, insect repellent and medicinal properties.

2.4.1 Nutritional properties

Most of the herbal spices are rich sources of protein, vitamins, especially vitamins A, C and B, and minerals such as calcium, phosphorus, sodium, potassium and iron.

Parsley is the richest source of vitamin A, while coriander is one of the richest sources of vitamins C and A. Parsley and chervil are also rich sources of vitamin K. The nutritive values of various herbal spices are presented in Table 2.4.

2.4.2 Antioxidant properties

Antioxidants are added to foods to preserve the lipid components from quality deterioration. Synthetic antioxidants such as butylated hydroxy anisole (BHA), butylated hydroxy toluene (BHT), propyl gallate (PG) and *tert*-butyl hydroquinone (TBHQ) are the commonly used synthetic antioxidants. Owing to their suspected action as promoters of carcinogenesis, there is growing interest in natural antioxidants.

Table 2.4 Nutritive value of herbal spices (approximate composition/100 g of edible portion)

Spice	Energy (k cal.)	Protein (g)	Fat (g)	Total CHO (g)	Fibre (g)	Ash (g)	Calcium (mg)	Fe (mg)	Mg (mg)	P (mg)	K (mg)	Na (mg)	Zn (mg)	Ascorbic acid (mg)	Thiamin (mg)	Riboflavin (mg)	Niacin (mg)	Vitamin A (IU)
Sweet basil	251	14.4	4.0	61.0	17.8	14.3	2113	42	422	490	3433	34	6	61.2	0.1	0.3	6.9	9375
Bay	313	7.6	8.4	75.0	26.3	3.6	834	43	120	113	529	23	4	–	–	–	2	6185
Chervil	237	23.2	3.9	49.1	11.3	16.6	1346	32	130	450	4740	83	9	NA	NA	NA	NA	NA
Marjoram	271	12.7	7.0	60.6	18.1	12.1	1990	83	346	306	1522	77	4	51	–	–	4	8068
Oregano	306	11.0	10.3	64.4	15.0	7.2	1576	44	270	200	1669	15	4	–	–	–	6	6903
Parsley	276	22.4	4.4	51.7	10.3	12.5	1468	98	249	351	3805	452	5	122	–	1	8	23340
Rosemary	331	4.9	15.2	64.1	17.7	6.5	1280	29	220	70	955	50	3	61	–	–	1	3128
Sage	315	10.6	12.7	60.7	18.1	8.0	1652	28	428	91	1070	11	5	32	–	–	6	5900
Tarragon	295	22.8	7.2	50.2	7.4	12.0	1139	32	347	313	3020	62	4	–	–	1	9	4200
Thyme	276	9.1	7.4	63.9	18.6	11.7	1890	124	220	201	814	55	6	–	–	–	5	3800

Source: Farrel (1990).

Many herbal spices are known as excellent sources of natural antioxidants, and consumption of fresh herbs in the diet may therefore contribute to the daily antioxidant intake. Phenolic compounds are the primary antioxidants present in spices and there is a linear relationship between the total phenolic content and the antioxidant properties of spices. Essential oils, oleoresins and even aqueous extracts of spices possess antioxidative properties.

The plants of the Lamiaceae family are universally considered as an important source of natural antioxidants. Rosemary is widely used as an antioxidant in Europe and the USA. Oregano, thyme, marjoram, sage, basil, fenugreek, fennel, coriander and pimento also possess antioxidant properties, better than that of the synthetic antioxidant butylated hydroxy toluene. Phyto constituents such as carvacrol, thymol, rosmarinic acid and carnosic acid are responsible for the antioxidative property. Important natural antioxidants and components responsible for the property are presented in Table 2.5. Information on the relative antioxidative effectiveness (RAE) of various herbal spices is given in Tables 2.6 and 2.7.

Table 2.5 Antioxidants isolated from herbal spices

Spice	Antioxidants
Rosemary	Carnosic acid, carnosol, rosemarinic acid, rosmanol
Sage	Carnosol, carnosic acid, rosmanol, rosmarinic acid
Oregano	Derivatives of phenolic acid, flavonoids, tocopherols
Thyme	Carvacrol thymol, p-cymene, caryophyllene, carvone, borneol
Summer savory	Rosmarinic acid, carnosol, carvacrol, thymol
Marjoram	Flavonoids
Allspice	Pimentol

Table 2.6 Relative antioxidative effectiveness (RAE) of herbal spices evaluated as whole plant material in different substrates

Spice/herb	Substrate	RAE
Marjoram, rosemary, sage, coriander	Lard	Rosemary>sage>marjoram
32 different plant materials	Lard	Rosemary>sage>oregano>thyme
19 different plant materials	Oil-in-water emulsion	Sage>oregano
32 different plant materials	Oil-in-water emulsion	Allspice>rosemary
Allspice, savory, marjoram, coriander	Sausage, water	Allspice>savory>marjoram
15 different plant materials	Sausage, water	Sage>rosemary>marjoram>aniseed
12 different plant materials	Ground chicken meat	Marjoram>caraway>peppermint

Table 2.7 Relative antioxidative effectiveness (RAE) of herbal spice extracts

Substrate, conditions	RAE
Lecithin emulsion, daylight, room temperature, 26 days	Rosemary>sage
Lard, 50°C	Rosemary>sage>marjoram
Chicken fat, 90°C	Sage>rosemary
Methyl linoleate, 100°C	Sage>deodorized rosemary> untreated rosemary
Lard, 75°C	Oregano>thyme>marjoram> spearmint>lavender>basil
TGSO, 100°C	Summer savory>peppermint> common balm>spearmint> oregano>common basil
Low-erucic rapeseed oil, 60°C, 23 days	Sage>thyme>oregano
Methanol	Oregano>cinnamon= marjoram>caraway
Minced chicken meat, 4°C and –18°C	Caraway>wild marjoram
Raw pork meats, pretreated with NaCl, 4°C and –18°C	Sage>basil>thyme
Microwave cooked pork patties treated with NaCl, –18°C	Basil=thyme

2.4.3 Antimicrobial properties

Herbal spices are important sources of antimicrobials, and the use of spices, their essential oils or active ingredients for controlling microbial growth in food materials constitutes an alternative approach to chemical additives.

Some of the spice essential oils (individual or combinations) are highly inhibitory to selected pathogenic and spoilage micro-organisms. The fractionation of essential oils and further application help to improve the level of activity in some cases. The optical isomers of carvone from *Mentha spicata* and *Anethum sowa* (Indian dill) were more active against a wide spectrum of human pathogenic fungi and bacteria than the essential oils as such. Mixing compounds such as carvacrol and thymol at different proportions may exert total inhibition of *Pseudomonas aeruginosa* and *Staphylococcus aureus*. The inhibition is due to damage in membrane integrity, which further affects pH homeostasis and equilibrium of inorganic ions. Such knowledge on the mode of action helps spice extracts/ingredients to be applied successfully in foods. Also, application of active ingredients instead of essential oil will not change the food's flavour very much.

Plant extracts or seed diffusates could be used for the control of seed-borne pathogens and can be a substitute for costly chemicals for seed treatment. Plant extracts of pimento can be used for controlling fungal growth during storage of wheat grains. Likewise, the seed diffusates of *Anthem graveolens* and *Coriandrum sativum* gave a high level of growth inhibition against seed-borne fungi such as *Alternaria alternata* and *Fusarium solani*.

Of the various herbal spices, oregano and thyme show the highest antimicrobial activity. Carvacrol, present in the essential oils of oregano and thyme, has been proved to be the most important fungitoxic compound. The activity of herbal spices against fungi and bacteria and the mode of application are given in Table 2.8.

Table 2.8 Antimicrobial activity of herbal spices

Spice	Mode of application	Activity against bacteria	Activity against fungus
Basil	Essential oil		Ascophaera apis
Basil	Methyl chavicol	Aeromonas hydrophylla, Pseudomonas fluorescens	Ascophaera apis
Coriander	Essential oil		
Fenugreek	Seed saponins		Fusarium oxysporum f. sp. lycopersici
Fenugreek	Essential oil	Bordetella bronchiseptica, Bacillus cereus, Bacillus pumilus, Bacillus subtilis, Micrococcus flavus, Staphylococcus aureus, Sarcina lutea, Escherichia coli, Proteus vulgaris	
Cumin	Essential oil		Penicillium notatum, Aspergillus niger, Aspergillus fumigatus, Microsporum canis
Fennel	Essential oil	Staphylococcus aureus, Bacillus subtilis	
Ajowan	Seed extracts		Pythium aphanidematum, Macrophomina phaseolina, Rhizactonia solani
Allspice	Plant extract		Fusarium spp., Alternania spp. and Cladosporium spp.
Oregano, coriander and basil	Essential oil	Listeria monocytogenes, Staphylococcus aureus, Escherichia coli, Yersinia enterocolitica, Pseudomonas aeruginosa, Lactobacillus plantarum	Apsergillus niger
Anethum graveolens, coriander	Seed diffusates		Alternaria alternata, Fusarium solani, Macrophomina phaseolina
Pepper mint, thyme, caraway	Essential oil	Agrobacterium tumefaciens, Ralstonia solanacearum, Erwinia carotovora	
Spearmint, basil, parsley	Essential oil	Staphylococcus aureus, Escherichia coli	Candida albicans, Aspergillus niger
Oregano and mint	Essential oil		Aspergillus ochraceus
Oregano	Essential oil or carvacrol		Candida albicans
Oregano, thyme	Essential oil or carvacrol	Streptococcus pneumoniae R36 A, Bacillus cereus	

Table 2.9 Insect repellent properties of herbal spices

Spice	Mode of application	Insects
Fenugreek	Seed extract	*Tribolium castaneum, Acanthoscelides obtectus*
Fennel	Direct contact and fumigation	*Callosobruchus chinensis, Lasioderma serricorne*
Indian dill	Essential oil	*Callosobruchus maculatus*
Dill	Essential oil	*Lucilia sericata*
Peppermint and basil	Powdered aerial parts	*Sitophilus granaricus*
Basil	Fumigation of essential oil	*Callosobruchus maculatus*
Mint	Essential oil	*Drosophila melanogaster*
Peppermint	Leaf powder	*Callosobruchus analis*
Cumin and anise	Vapour of essential oil	*Tetranychus cinnabarinus, Aphis gossypii, Tribolium confusum, Ephestia kuehniella*
Oregano	Essential oil	*Acanthoscelides obtectus, Tetranychus cinnabarinus, Aphis gossypii*

2.4.4 Insect repellent properties

The herbal spices have good insect repellent properties. Powdered plant parts or extracts of seed or essential oils or active ingredients separated from essential oils and oleoresins of spices are used as insect repellents.

The repellent action is noticed against many storage pests of grains and pulses. Herbal spices can also be used as mosquito repellents. The essential oil of basil and piperidine alkaloid separated from long pepper repels mosquitoes. The details of insect repellent properties of herbal spices are presented in Table 2.9.

2.4.5 Medicinal properties

Herbs and spices are known for their medicinal properties and have been used in traditional medicines from time immemorial. Powdered spices are either externally applied or taken internally for various ailments.

The essential oils of many herbs and spices are used in pharmaceutical preparations. The essential oil of coriander is reported to be analgesic, dill and anise oils as antipyretic, coriander, celery, parsley and cumin oils as anti-inflammatory. Recently, anticarcinogenic property has been reported for essential oils of cumin and basil and these can be used as protective agents against carcinogenesis. Also, methanol extracts of allspice, marjoram, tarragon and thyme strongly inhibited platelet aggregation induced by collagen in humans. The important medicinal properties of herbal spices are given in Table 2.10.

Table 2.10 Medicinal properties of herbal spices

Spice	Medicinal properties
Allspice	Stimulant, digestive and carminative
Basil, sweet	Stomachic, anthelmintic, diaphoretic, expectorant, antipyretic carminative, stimulant, diuretic, demulcent
Bay leaves (laurel)	Stimulant, narcotic
Caraway	Stomachic, carminative, anthelmintic, lactagogue
Celery	Stimulant, tonic, diuretic, carminative, emmenagogue, anti-inflammatory
Chive	Stimulant, diuretic, expectorant, aphrodisiac, emmenegogue, anti-inflammatory
Coriander	Carminative, diuretic, tonic, stimulant, stomachic, refrigerent, aphrodisiac, analgesic, anti-inflammatory
Dill	Carminative, stomachic, antipyretic
Fennel	Stimulant, carminative, stomachic, emmenagogue
Fenugreek	Carminative, tonic, aphrodisiac
Leek	Stimulant, expectorant
Marjoram	Carminative, expectorant, tonic, astringent
Mint (peppermint)	Stimulant, stomachic, carminative, antiseptic
Mint (spearmint)	Stimulant, carminative and antispasmodic
Oregano	Stimulant, carminative, stomachic, diuretic, diaphoretic and emmenagogue
Parsley	Stimulant, diuretic, carminative, emmenagogue, antipyretic, anti-inflammatory
Rosemary	Mild irritant, carminative, stimulant, diaphoretic
Sage	Mild tonic, astringent, carminative
Tarrgon	Aperient, stomachic, stimulant, febrifuge
Thyme	Antispasmodic, carminative, emmenagogue, anthelmintic, spasmodic, laxative, stomachic, tonic, vermifuge

2.5 Sources of further information

ANON. (1998), *New Horizons: Challenges Ahead*, Proceedings of World Spices Congress 1998, Spices Board India and All India Spices Exporters Forum.

CSIR (1998), *The Wealth of India – a Dictionary of Indian Raw Materials and Industrial Products*, National Institute of Science Communication, CSIR, New Delhi, India.

FARRELL, K. T. (1990), *Spices, Condiments and Seasonings*, 2nd edition. AVI book, Van Nostrand Reinhold, New York.

GUENTHER, E. (1975), *The Essential Oils*, Robert E. Krieger Publishing Company, Huntington, New York.

PETER, K. V. (ed.) (2001), *Handbook of Herbs and Spices*, Woodhead Publishing Limited, Abington.

POKORNY, J., YANISHLIEVA, N. and GORDON, M. (2001), *Antioxidants in food: practical applications*. Woodhead Publishing Limited, Abington.

PRUTHI, J. S. (2001), *Minor Spices and Condiments – Crop Management and Post Harvest Technology*, ICAR, New Delhi, India.

3
Herbs and spices and antimicrobials

C. C. Tassou, National Agricultural Research Foundation, Greece, and G.-J. E. Nychas and P. N. Skandamis, Agricultural University of Athens, Greece

3.1 Introduction

Herbs and spices are used widely in the food industry as flavours and fragrances. However, they also exhibit useful antimicrobial and antioxidant properties. Many plant-derived antimicrobial compounds have a wide spectrum of activity against bacteria, fungi and mycobacteria and this has led to suggestions that they could be used as natural preservatives in foods (Farag *et al.*, 1989; Ramadan *et al.*, 1972; Conner and Beuchat, 1984a,b; Galli *et al.*, 1985). Although more than 1300 plants have been reported as potential sources of antimicrobial agents (Wilkins and Board, 1989), such alternative compounds have not been sufficiently exploited in foods to date.

In this chapter, the antimicrobial compounds from herbs and spices are reviewed and the barriers to the adoption of these substances as food preservatives are discussed. The mode of action of essential oils and the potential for development of resistance are also discussed. The focus is primarily on bacteria and fungi in prepared foods.

3.2 Barriers to the use of herb and spice essential oils as antimicrobials in foods

Since ancient times, spices and herbs have not been consciously added to foods as preservatives but mainly as seasoning additives due to their aromatic properties. Although the majority of essential oils from herbs and spices are classified as Generally Recognized As Safe (GRAS) (Kabara, 1991), their use in foods as preservatives is limited because of flavour considerations, since effective antimicrobial doses may exceed organoleptically acceptable levels. This problem could possibly be overcome if answers could be given to the following questions:

- Can the inhibitory effect of an essential oil (a mixture of many compounds) be attributed to one or several key constituents?
- Does the essential oil provide a synergy of activity, which simple mixtures of components cannot deliver?

- What is the minimum inhibitory concentration (MIC) of the active compound(s) of the essential oil?
- How is the behaviour of the antimicrobial substance(s) affected by the homogeneous (liquid, semisolid) or heterogeneous (emulsions, mixtures of solids and semisolids) structure of foodstuffs?
- Could efficacy be enhanced by combinations with traditional (salting, heating, acidification) and modern (vacuum packing, VP, modified atmosphere packing, MAP) methods of food preservation?

An in-depth understanding of the antimicrobial properties of these compounds is needed to answer these questions but such understanding has been lacking, despite the burgeoning literature on the subject. Methodological limitations (discussed in more detail below) in the evaluation of antimicrobial activity *in vitro* have led to many contradictory results. Moreover, there have been too few studies in real foods (these are considered laborious and often lead to negative outcomes). There is also a need to investigate the appropriate mode of application of an essential oil in a foodstuff. For instance, immersion, mixing, encapsulation, surface-spraying, and evaporating onto active packaging are some promising methods of adding these compounds to foods that have not been extensively investigated.

3.3 Measuring antimicrobial activity

The antimicrobial activity of plant-derived compounds against many different microorganisms, tested individually and *in vitro*, is well documented in the literature (Tables 3.1 and 3.2; Ippolito and Nigro, 2003). However, the results reported in different studies are difficult to compare directly. Indeed, contradictory data have been reported by different authors for the same antimicrobial compound (Mann and Markham, 1998; Manou *et al.*, 1998; Skandamis, 2001; Skandamis *et al.*, 2001b). Also, it is not always apparent whether the methods cited measure bacteriostatic or bactericidal activities, or a combination of both.

Antimicrobial assays described in the literature include measurement of:

- the radius or diameter of the zone of inhibition of bacterial growth around paper discs impregnated with (or wells containing) an antimicrobial compound on agar media;
- the inhibition of bacterial growth on an agar medium with the antimicrobial compound diffused in the agar;
- the minimum inhibitory concentration (MIC) of the antimicrobial compound in liquid media;
- the changes in optical density or impedance in a liquid growth medium containing the antimicrobial compound.

Three main factors may influence the outcome of the above methods when used with essential oils of plants: (i) the composition of the sample tested (type of plant, geographical location and time of the year), (ii) the microorganism (strain, conditions of growth, inoculum size, etc.), and (iii) the method used for growing and enumerating the surviving bacteria. Many studies have been based on subjective assessment of growth inhibition, as in the disc diffusion method, or on rapid techniques such as optical density (turbidimetry) without accounting for the limitations inherent in such methods. In the disc method, the inhibition area depends on the ability of the essential oil to diffuse uniformly through the agar as well as on the released oil vapours. Other factors that may influence results involve the presence

Table 3.1 Plant essential oils tested for antibacterial properties

Achiote,[14] Allspice,[16] Almond[1] (bitter, sweet), Aloe Vera,[14] Anethole,[11] Angelica,[1] Anise,[1,5,6] Asafoetida[14] (*Ferula* spp.)

Basil,[1,10,31] Bay,[1,20,28,31] Bergamot,[1] Birch[14]

Cajeput,[32] Calmus,[1] Camomile-German,[10] Cananga, Caraway,[1,3] Cardamon,[1] Carrot seed,[39] Cedarwood,[39] Celery,[39] Chilli,[39] Cinnamon casia,[1,19,16,18] Cinnamon (bark leaf),[28,33] Cinnamon,[28] Citronella,[1] Clove,[1,3,8,10,1,12,15,16,18,19,40] Coriander,[1,5,8] Cornmint,[5] Cortuk,[17] Cumin,[3,5,10] Cymbopogon,[38] Dill[1,5]

Elecampane, Estragon,[10] Eucalyptus,[24,35,38] Evening primrose[39]

Frankincense,[39] Fennel[1,5,10,23]

Gale (sweet), Gardenia,[39] Garlic,[10,16,18,22] Geranium,[1] Ginger,[1,10] Grapefruit[6]

Horseradish, Hassaku Fruit Peel[27]

Jasmine[14,32]

Laurel,[1,5,10] Lavender,[1] Lemon,[1,5,6,10] Lime,[1,6] Linden flower,[2] Liquorice, Lovage,[1] Lemongrass[24,31,36]

Mace,[20] Mandarin,[1,6,10] Marigold *Tagetes*,[39] Marjoram,[1,10,31] Mastich gum tree (*Pistachia lentiscus* var. *chia*),[14] Melissa,[1] Mint (apple),[1,29,30] Mugwort,[39] Musky bugle, Mustard,[16], Mountain tea (*Sideritis* spp.)[14]

Neroly,[10] Nutmeg[1,8,10,20]

Onion,[10,16,18,22] Orange,[1,5,6,10,21] Oregano,[4,9,10,16,18,31] Ocicum[38]

Palmarosa,[24] Paprika,[16] Parsley,[1,5,10] Patchouli,[39] Pennyroyal, Pepper, Peppermint,[1,10,24] Pettigrain,[10] Pimento[1,10,18]

Ravensara,[39] Rose,[1] Rosemary,[1,3,7,10,16] Rosewood[39]

Saffron,[10] Sage,[1,3,5,7,10,16] Sagebrush,[13] Savoury,[5] Sassafras,[1] Sideritis,[37] Senecio (chachacoma),[34] Spike,[1] Spearmint,[1] Star Anise,[1] St John's Wort[1]

Tangerine,[39] Tarragon,[4] Tea Thuja,[1] Thyme,[1,3,4,5,9,10,18,40] Tuberose,[39] Turmeric,[16] Teatree[25,26]

Valerian,[1] Verbena,[1] Vanilla[10]

Wintergreen,[39] Wormwood[39]

Data from: [1]Deans and Ritchie (1987); [2]Aktug and Karapinar (1987); [3]Farag *et al.* (1989); [4]Paster *et al.* (1990); [5]Akgul and Kivanc (1989); [6]Dabbah *et al.* (1970); [7]Shelef *et al.* (1980); [8]Stechini *et al.* (1993); [9]Salmeron *et al.* (1990); [10]Aureli *et al.* (1992); [11]Kubo and Himejima (1991); [12]Briozzo *et al.* (1989); [13]Nagy and Tengerdy (1967); [14]Nychas and Tassou (2000); [15]Al-Khayat and Blank (1985); [16]Azzouz and Bullerman (1982); [17]Kivanc and Akgul (1990); [18]Ismaiel and Pierson (1990); [19]Blank *et al.* (1987); [20]Hall and Maurer (1986); [21]Sankaran (1976); [22]Elnima *et al.* (1983); [23]Davidson and Branen (1993); [24]Pattnaik *et al.* (1995a,b,c); [25]Mann *et al.* (2000); [26]Nelson (2000); [27]Takahashi *et al.* (2002); [28]Smith-Palmer *et al.* (2001); [29]Iscan *et al.* (2002); [30]Tassou *et al.* (2000); [31]Mejlholm and Dalgaard (2002); [32]Skandamis (2001); [33]Chang *et al.* (2001); [34]Perez *et al* (1999); [35]Oyedeji *et al.* (1999); [36]Carlson-Castelan *et al.* (2001); [37]Ozcan *et al.* (2001); [38]Cimanga *et al.* (2002); [39]Nychas unpublished; [40]Smith-Palmer *et al.* (1998).

of multiple active components. These active compounds at low concentrations may interact antagonistically, additively or synergistically with each other. Some of the differences in the antimicrobial activity of oils observed in complex foods compared with their activity when used alone in laboratory media could be due to the partitioning of active components between lipid and aqueous phases in foods (Stechini *et al.*, 1993, 1998).

Turbidimetry is a rapid, non-destructive and inexpensive method that is easily automated but has low sensitivity. Turbidimetry detects only the upper part of growth curves, and requires calibration in order to correlate the results with viable counts obtained on agar media (Koch, 1981; Bloomfield, 1991; Cuppers and Smelt, 1993; McClure *et al.*, 1993; Dalgaard and Koutsoumanis, 2001; Skandamis *et al.*, 2001b). The changes in absorbance are only evident when population levels reach 10^6–10^7 CFU/ml, and are influenced by the size of the bacterial cells at different growth stages. The physiological state of the cells (injured or healthy), the state of oxidation of the essential oil as well as inadequate

Table 3.2 Naturally occurring antimicrobial compounds in plants

Apigenin-7-glucose, aureptan
Benzoic acid, berbamine, berberine, borneol
Caffeine, caffeic acid, 3-*o*-caffeylquinic acid, 4-*o*-caffeylquinic acid, 5-*o*-caffeylquinic acid, camphene camphor, carnosol, carnosic acid, carvacrol, caryophelene, catechin, 1,8 cineole, cinnamaldehyde, cinnamic acid, citral, chlorogenic acid, chicorin, columbamine, coumarine, *p*-coumaric acid, *o*-coumaric, *p*-cymene, cynarine
Dihydrocaffeic acid, dimethyloleuropein
Esculin, eugenol
Ferulic acid
Gallic acid, geraniol, gingerols
Humulone, hydroxytyrosol, 4-hydroxybenzoic acid, 4-hydroxycinnamic acid
Isovanillic, isoborneol
Linalool, lupulone, luteoline-5-glucoside, ligustroside, *S*-limonene
Myricetin, 3-methoxybenzoic acid, menthol, menthofurane
Oleuropein
Paradols, protocatechic acid, *o*-pyrocatechic, α-pinene, β-pinene, pulegone
Quercetin
Rutin, resocrylic
Salicylaldehyde, sesamol, shogoals, syringic acid, sinapic
Tannins, thymol, tyrosol, 3,4,5-trimethoxybenzoic acid, 3,4,5-thihydroxyphenylacetic acid
Verbascoside, vanillin, vanillic acid

Data from: Nychas and Tassou (2000); Iscan *et al.* (2002); Flamini *et al.* (2002); Mourey and Canillac (2002); Takahashi *et al.* (2002); Amakura *et al.* (2002); Gounaris *et al.* (2002), Hayes and Markovic (2002); Chang *et al.* (2001); Perez *et al.* (1999); Oyedeji *et al.* (1999); Carlson-Castelan *et al.* (2001); Ozcan *et al.* (2001); Cimanga *et al.* (2002)

dissolution of the compound tested may also affect absorbance measurements in growth media.

Unlike the plate counting technique, impedance-based methods can be used to monitor microbial metabolism in real time mode. The impedimetric method is recognized as an alternative rapid method not only for screening the biocide activity of novel antimicrobial agents but also for estimation of growth kinetics in mathematical modelling (Ayres *et al.*, 1993, 1998; Tranter *et al.*, 1993; Tassou *et al.*, 1995, 1997; Johansen *et al.*, 1995; Tassou and Nychas, 1995a,b,c; Koutsoumanis *et al.*, 1997, 1998; MacRae *et al.*, 1997; Lachowicz *et al.*, 1998). The technique depends on using a medium that offers a sharp detectable impedimetric change as the bacterial population grows and converts the low conductivity nutrients into highly charged products. As with turbidometry, calibration of impedimetric data with plate counts is necessary (Dumont and Slabyj, 1993; Koutsoumanis *et al.*, 1998).

Although time-consuming and laborious, the traditional microbiological method of determining viable numbers by plate counting remains the gold standard in antimicrobial studies. The latter method has a major advantage of requiring little capital investment; however, it is material-intensive, requires a long elapse time and may have poor reproducibility.

MICs are measured by serial dilution of the tested agents in broth media followed by growth determination by either absorbance reading or plate-counting (Carson *et al.*, 1995). The MIC technique has been miniaturized and automated using the bioscreen microbiological growth analyser (Lambert and Pearson, 2000) to allow a high throughput of compounds and microorganisms (Lambert *et al.*, 2001). The advantage of this method is the simultaneous examination of multiple concentrations of one or more preservatives and subsequent determination of MIC based on mathematical processing.

3.4 Studies *in vitro*

Almost all essential oils from spices and herbs inhibit microbial growth as well as toxin production. The antimicrobial effect is concentration dependent and may become strongly bacteriocidal at high concentrations. Gram-positive bacteria (spore- and non-spore-formers), Gram-negative bacteria, yeasts (Tables 3.1–3.3) and moulds (Ippolito and Nigro, 2003) are all affected by a wide range of essential oils. Well-known examples include the essential oils from allspice, almond, bay, black pepper, caraway, cinnamon, clove, coriander, cumin, garlic, grapefruit, lemon, mace, mandarin, onion, orange, oregano, rosemary, sage and thyme. The active compounds of some of these essential oils are shown in Tables 3.2 and 3.4.

Table 3.3 Some examples of microorganisms sensitive to the antimicrobial action of essential oils from herbs and spices

Gram-positive bacteria	Gram-negative bacteria	Yeasts/fungi
Arthobacter sp.	*Acetobacter* spp.	*Aspergillus niger*
Bacillus sp.	*Acinetobacter* sp.	*As. parasiticus*
B. subtilis	*A.calcoaceticus*	*As. flavus*
B. cereus	*Aeromonas hydrophila*	*As. ochraceus*
B. megaterium	*Alcaligenes* sp.	*Candida albicans*
Brevibacterium ammoniagenes	*A. faecalis*	*Candida tropicalis*
Brev. linens	*Campylobacter jejuni*	*Dekkera bruxellensis*
Brochothrix thermosphacta	*Citrobacter* sp.	*Fusarium oxysporum*
Clostridium botulinum	*C. freundii*	*F. culmorum*
Cl. perfrigenes	*Edwardsiella* sp.	*Mucor* sp.
Cl. sporogenes	*Enterobacter* sp.	*Pichia anomala*
Corynebacterium sp.	*En. aerogenes*	*Penicillium* sp.
Enterococcus feacalis	*Escherichia coli*	*Pen. chrysogenum*
Lactobacillus sp.	*E. coli* O157:H7	*Pen. patulum,*
Lac. plantarum,	*Erwinia carotovora*	*Pen. roquefortii*
Lac. minor	*Flavobacterium* sp.	*Pen. citrinum*
Leuconostoc sp.	*Fl. suaveolens*	*Rhizopus* sp.
Leuc. cremoris	*Klebsiella* sp.	*Saccharomyces cerevisiae*
Listeria monocytogenes	*K. pneumoniae*	*Trichophyton mentagrophytes*
L. inocua	*Moraxella* sp.	*Torulopsis holmii*
Micrococcus sp.	*Neisseria* sp.	*Pityrosporum ovale*
M. luteus	*N. sicca*	
M. roseus	*Mycobacterium smegmatis*	
Pediococcus spp.	*Pseudomonas* spp.	
Photobacterium phosphoreum	*P. aeruginosa, fluorescens, fragi* and *clavigerum*	
Propionibacterium acnes	*Proteus* spp.	
Sarcina spp.	*Pr. vulgaris*	
Staphylococcus spp.	*Salmonella* spp.	
Staph. aureus,	*Salmonella enteritidis, senftenberg, typhimurium, flexneri, pullorum*	
Staph. epidermidis	*Serratia* sp.	
Streptococcus faecalis	*S. marcecens*	
	Vibrio sp.	
	V. parahaemolyticus	
	Yersinia enterocolitica	

Based on: Nychas (1995); Mejlholm and Dalgaard (2002); Thangadural *et al.* (2002); Mangena and Muyima (1999); Karaman *et al.* (2001); Hayes and Markovic (2002); Chang *et al.* (2001); Cimagna *et al.* (2002).

Table 3.4 Examples of essential oils commonly used for food preservation and their main active constituents

Herb/spice	Active compound	Herb/spice	Active compound
Allspice	eugenol, methyl eugenol	Mint	α-, β-pinene, limonene, 1,8-cineole
Caraway	carvone	Onion	D-n-propyl disulphide, methyl-n-propyl disulphide
Cinnamon	cinnamaldehyde, eugenol	Oregano	thymol, carvacrol
Cloves	eugenol, eugenol acetate	Pepper	monoterpenes
Coriander	D-linalool, D-α-pinene β-pinene	Rosemary	borneol, 1,8-cineole, camphor, bornyl acetate
Cumin	cuminaldehyde	Sage	thujone, 1,8-cineol, borneol
Garlic	diallyl disulphide, diallyl tri-sulphide, allyl propyl disulphide	Thyme	thymol, carvacrol, menthol, menthone

Data from: Skandamis (2001); Iscan *et al.* (2002); Flamini *et al.* (2002); Karaman *et al.* (2001); Gounaris *et al.* (2002), Oyedeji *et al.* (1999); Ozcan *et al.* (2001); Cimanga *et al.* (2002).

and discussed in Adams and Smid, 2003). The antimicrobial activity of these compounds is influenced by the culture medium, the temperature of incubation and the inoculum size. In addition, a strong synergism with some membrane chelators acting as permeabilizing agents (e.g. ethylenediaminetetraacetic acid, EDTA) against Gram-negative bacteria has been reported (Tassou, 1993; Ayers *et al.*, 1998; Brul and Coote, 1999; Skandamis, 2001; Skandamis *et al.*, 2001b).

3.5 Applications in food systems

There have been relatively few studies of the antimicrobial action of essential oils in model food systems and in real foods (Table 3.5). The efficacy of essential oils *in vitro* is often much greater than *in vivo* or *in situ*, i.e. in foods (Nychas and Tassou, 2000; Davidson, 1997; Skandamis *et al.*, 1999b). For example, the essential oil of mint (*Mentha piperita*) has been shown to inhibit the growth of *Salmonella enteritidis* and *Listeria monocytogenes* in culture media for 2 days at 30°C. However, the effect of mint essential oil in the traditional Greek appetizers tzatziki (pH 4.5) and taramasalata (pH 5.0) and in paté (pH 6.8) at 4°C and 10°C was variable. *Salmonella enteritidis* died off in the appetizers under all conditions examined but not when inoculated in paté and maintained at 10°C. Similarly, *L. monocytogenes* numbers declined in the appetizers but increased in paté (Tassou *et al.*, 1995a,b, 2000).

Growth of *Escherichia coli*, *Salmonella* spp., *L. monocytogenes* and *Staphylococcus aureus* was inhibited by oregano essential oil (EO) in broth cultures. However, the antimicrobial action of this EO in an emulsion or pseudoemulsion type of food such as aubergine salad, taramasalata and mayonnaise depended on environmental factors such as pH, temperature and oil (vegetable or olive) used. Homemade aubergine salad and taramasalata were inoculated with *E. coli* O157:H7 and *Salmonella enteritidis*, respectively. The pH of these products was adjusted to 4–5.3. A range of concentrations (0–2.1%) of oregano essential oil was added and the foods were incubated at temperatures from 0 to 20°C. The survival curves for *E. coli* O157:H7 in aubergine salad at 0 and 15°C, modelled according to Baranyi, are shown in Figs 3.1 and 3.2. A reduction in viable counts for both pathogens in both foods tested was observed and their death rate depended on the pH, the storage temperature and the essential oil concentration (Koutsoumanis *et al.*, 1999; Skandamis and Nychas, 2000; Skandamis *et al.*, 1999a, 2002b) (see Fig. 3.3).

Table 3.5 Applications of essential oils in foods

Food	Microorganisms	Essential oil	References
Milk (fresh, skimmed)	*Staph. aureus* *Salmonella enteritidis* *P. fragi*	Mastic gum	Tassou and Nychas (1995c)
Dairy products: soft cheese, mozzarella	*L. monocytogenes* *Salmonella enteritidis*	Clove, cinnamon, thyme	Smith-Palmer *et al.* (2001); Menon and Garg (2001)
Fresh meat: block or minced	*Salmonella typhimurium* and *enteritidis* *Staph. aureus* *P. fragi* *L. monocytogenes* Lactic acid bacteria *Br. thermosphacta* Enterobacteriaceae Yeasts and indigenous flora	Oregano, clove, basil, sage	Tassou and Nychas (1995b); Menon and Garg (2001); Skandamis and Nychas (2001, 2002a,b); Tsigarida *et al.* (2000); Skandamis *et al.* (2002a,b); Stecchini *et al.* (1993)
Meat products: paté	*L. monocytogenes* *Salmonella enteritidis* Indigenous flora	Mint	Tassou *et al.* (1995)
sausage	*Br. thermosphacta, E. coli*	Mustard oil	Lemay *et al.* (2002)
Fish: Gilt-head bream	*Salmonella enteritidis* *Staph. aureus* Resident flora	Oregano	Tassou *et al.* (1996)
Cod fillets, salmon	*Photobacterium phosphoreum*	Basil, bay, cinnamon, clove, lemongrass, marjoram, oregano, sage, thyme	Mejlholm and Dalgaard (2002)
Salads and dressings: tuna, potato, aubergine (egg plant), tarama-salata, mayonnaise, tzatziki	*Staph. aureus* *Salmonella enteritidis* *P. fragi* *L. monocytogenes* *Sh. putrefaciens* *Br. thermosphacta* *E. coli* Indigenous flora	Carob Mint, oregano, basil, sage	Tassou *et al.* (1997) Tassou and Nychas (1995c); Tassou *et al.* (1995); Koutsoumanis *et al.* (1999); Skandamis and Nychas (2000); Skandamis *et al.* (1999a,b, 2001a, 2002c)
Sauces: meat gravy	*Salmonella enteritidis* and *typhimurium* *Staph. aureus* *P. fragi*	Basil, sage	Tassou and Nychas (1995c)

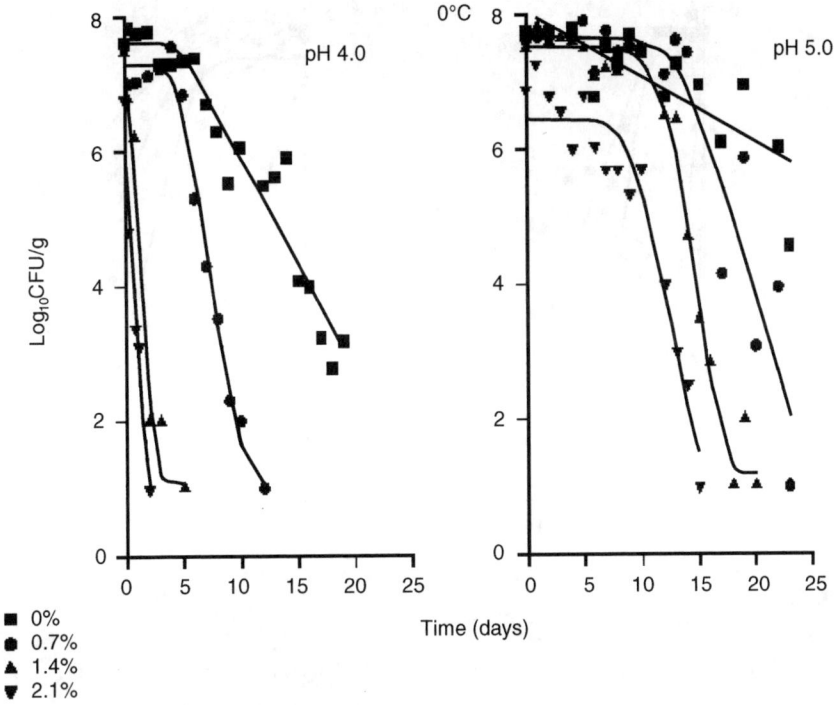

Fig. 3.1 Survival curves for *E. coli* O157:H7 in aubergine (egg plant) salad at 0°C, pH 4.0 and 5.0, in the presence of 0, 0.7, 1.4 and 2.1% oregano essential oil. (Data from Skandamis and Nychas, 2000.)

The type of oil or fat present in a food can affect the antimicrobial efficacy of essential oils. This was evident when the efficiency of four plant essential oils (bay, clove, cinnamon and thyme) was assessed in low-fat and full-fat soft cheese against *L. monocytogenes* and *Salmonella enteritidis* at 4°C and 10°C, respectively, over a 14-day period. In the low-fat cheese, all four oils at 1% reduced *L. monocytogenes* to below the detection limit of the plating method. In contrast, in the full-fat cheese, the oil of clove was the only substance to achieve such reduction. The oil of thyme was ineffective against *Salmonella enteritidis* in the full-fat cheese, despite the fact that this organism was completely inhibited in broth culture (Skandamis, 2001). Thyme oil was as effective as the other three oils in the low fat cheese, reducing *Salmonella* Enteritidis to less than 1 log CFU/g from day 4 onwards (Smith-Palmer *et al.*, 2001).

Table 3.5 summarizes some of the studies on the inhibitory action of essential oils in solid foods (e.g. fish and meat) stored under various packaging conditions (VP, MAP). For example, *L. monocytogenes* and *Salmonella typhimurium* were inhibited in meat treated with clove and oregano essential oil, respectively (Menon and Garg, 2001; Tsigarida *et al.*, 2000; Skandamis *et al.*, 2002a). *Salmonella typhimurium* survived in untreated meat, while the addition of oregano essential oil at a concentration of 0.8% v/w reduced viable numbers by 1–2 log CFU/g. The same level of oregano essential oil reduced the counts of *L. monocytogenes* by 2–3 log CFU/g on meat. A marked reduction of *Aeromonas hydrophila* was also reported in cooked, non-cured pork treated with clove or coriander oils and packaged either under vacuum or air and stored at 2°C and 10 °C. The lethal effect of these two oils was more pronounced under vacuum than in aerobic conditions (Stecchini *et al.*, 1993).

The availability of oxygen can affect the antimicrobial efficacy of essential oils. Paster *et*

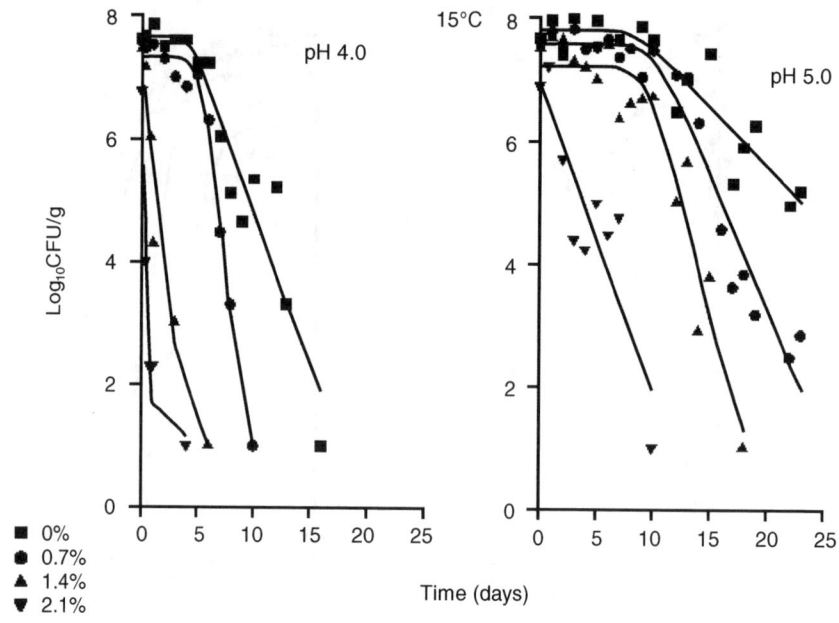

Fig. 3.2 Survival curves for *E. coli* O157:H7 in aubergine (egg plant) salad at 15°C, pH 4.0 and 5.0, in the presence of 0, 0.7, 1.4 and 2.1% oregano essential oil. (Data from Skandamis and Nychas, 2000.)

al. (1990, 1995) observed that the antimicrobial activity of the oregano essential oil on *Staph. aureus* and *Salmonella enteritidis* was enhanced when these organisms were incubated under microaerobic or anaerobic conditions. Under conditions of low oxygen tension, there are fewer oxidative changes in the essential oil (Paster *et al.*, 1990, 1995). Moreover, oregano essential oil was more effective under vacuum and a 40% CO_2 : 30% O_2 : 30% N_2 atmosphere when an impermeable film was used compared to aerobic incubation or packaging in bags that allowed O_2 to permeate the package (Tsigarida *et al.*, 2000; Skandamis *et al.*, 2002a).

Oregano EO has both bacteriostatic and bacteriocidal effects on raw fish (*Sparus aurata*) inoculated with *Staph. aureus* and *Salmonella enteritidis* and stored under MAP (40% CO_2, 30% O_2 and 30% N_2) or in air at 1°C. Growth of spoilage organisms such as *Shewanella putrefaciens* and *Photobacterium phosphereum* is also inhibited on gilt head seabream and cod treated with oregano EO (Tassou *et al.*, 1996; Mejlholm and Dalgaard, 2002). Similar reductions were also reported for many other meat and fish organisms, as shown in Table 3.5 (Greer *et al.*, 2000; Mejlholm and Dalgaard, 2002; Skandamis and Nychas, 2001, 2002a,b).

The studies reviewed above all show that antimicrobial activity demonstrated *in vitro* is not necessarily a good indication of practical value in food preservation. The active compounds of essential oils are often bound with food components (e.g. proteins, fats, sugars, salts). Therefore, only a proportion of the total dose of EO added to a food remains free to exert antibacterial activity. Extrinsic factors such as temperature also limit the antimicrobial action of essential oils (Davidson, 1997). Moreover, the spatial distribution of the different phases (solid/liquid) in a food and the lack of homogeneity of pH and water can also play a role in efficacy. Interactions between the different components in the food may create pH gradients in the final product as well as different bulk concentrations of the antimicrobial in the different phases. The local buffering capacity of the food ingredients

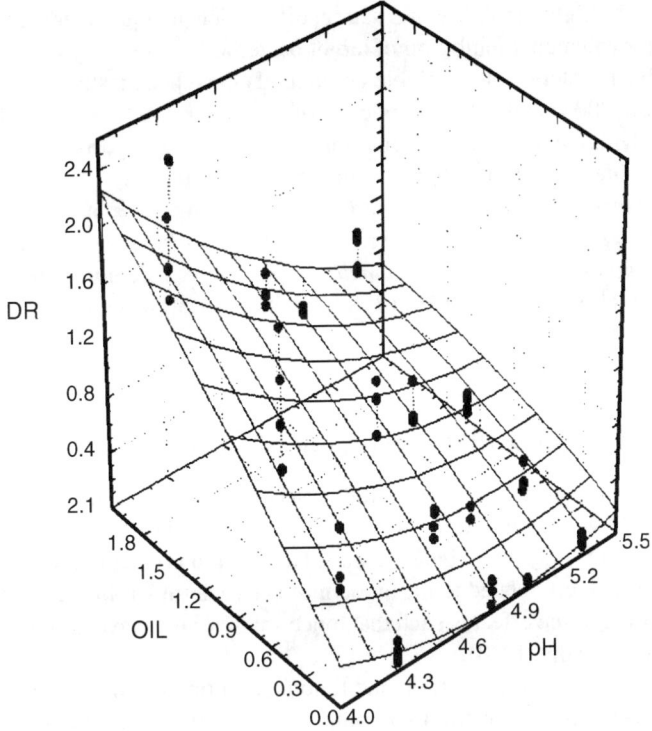

Fig. 3.3 Quadratic response surfaces predicting the death rate (DR) of *Salmonella typhimurium* in taramasalata as a function of pH and oregano essential oil (OIL%). (Data from Koutsoumanis *et al.*, 1999.)

determines the pH within specific regions of complex foods. Since the spatial distribution of microorganisms is not homogeneous, the antimicrobial activity could also depend on their population density, on the food structure *per se*, and on carbon source availability governed by diffusion factors. The microbial ecology of specific foodstuffs, buffering capacity, local pH and food structure should all be taken into account during the evaluation of an antimicrobial compound.

The growth of bacteria in liquids occurs planktonically, in contrast with the discrete colonies formed either on or within a solid matrix (Robins *et al.*, 1994; Wilson *et al.*, 2002). In the latter case, the cells are immobilized and localized in high densities in the food matrix (Skandamis *et al.*, 2000; Wilson *et al.*, 2002). Challenge tests have revealed that the physiological attributes of bacteria grown in model food matrices were significantly different from those of cells growing freely in liquid cultures (Brocklehurst *et al.*, 1997; Skandamis *et al.*, 2000; Wilson *et al.*, 2002). These differences can be accounted for by: (i) the population density *per se*, (ii) diffusivity and thus availability of major nutrients, (iii) oxygen availability, and (iv) accumulation of end products (Stecchini *et al.*, 1993, 1998; Thomas *et al.*, 1997; Skandamis *et al.*, 2000). Bacteria within solid matrices grow as submerged 'nests' (Thomas *et al.*, 1997). While the diffusivity of low molecular weight nutrients such as glucose may be very similar in liquids and gel matrices, that of antimicrobial agents may be very different and may strongly influence the efficacy of the agents in a solid matrix (Diaz *et al.*, 1993; Stecchini *et al.*, 1998). Oily substances within emulsions form droplets with diameters of 10–18 μm (Wilson *et al.*, 2002). The diffusion of such large droplets is very likely to be affected by the density, viscosity, tortuosity and other structure-related properties of the medium. Thus, the higher mobility of essential oil droplets in liquid media may be the most important factor enhancing inhibition of target bacteria.

To avoid this problem, water-soluble compounds have been tested *in vitro* and *in vivo* (Patsiouras *et al.*, 2003). Indeed, the extraction of essential oils by steam distillation of herbs and spices provides another useful fraction: the so-called 'hydrosols'. Hydrosols are currently not widely used and there is little published research on their antimicrobial activity. The hydrosol fractions of oregano, thyme, mint and rosemary have been tested in broth against *Listeria monocytogenes, Escherichia coli, Lactobacillus plantarum, Brochothrix thermosphacta* and *Salmonella enteridis*. All demonstrated antimicrobial activity, but their effectiveness in a sample food (minced meat) was limited (Patsiouras *et al.*, 2003).

3.6 Mode of action and development of resistance

In general, the mode of action of essential oils is concentration dependent (Prindle and Wright, 1977). Low concentrations inhibit enzymes associated with energy production while higher amounts may precipitate proteins. However, it is uncertain whether membrane damage is quantitatively related to the amount of active antimicrobial compound to which the cell is exposed, or the effect is such that, once small injuries are caused, the breakdown of the cell follows (Judis, 1963).

Essential oils damage the structural and functional properties of membranes and this is reflected in the dissipation of the two components of the proton motive force: the pH gradient (ΔpH) and the electrical potential ($\Delta\psi$) (Sikkema *et al.*, 1995, Davidson, 1997; Ultee *et al.*, 1999, 2000, 2002). Carvacrol, an active component of many essential oils, has been shown to destabilize the cytoplasmic and outer membranes and act as a 'proton exhanger', resulting in a reduction of the pH gradient across the cytoplasmic membrane (Helander *et al.*, 1998; Lambert *et al.*, 2001; Ultee *et al.*, 2002). The collapse of the proton motive force and depletion of the ATP pool eventually led to cell death (Ultee *et al.*, 2002). Like other many preservatives, the essential oils cause leakage of ions, ATP, nucleic acids and amino acids (Tranter *et al.*, 1993; Gonzalez *et al.*, 1996; Tahara *et al.*, 1996; Helander *et al.*, 1998; Cox *et al.*, 1998; Ultee *et al.*, 1999; Tassou *et al.*, 2000). Like carvacrol, the essential oils from tea and mint cause leakage of cellular material including potassium ions and 260 nm-absorbing substances (Cox *et al.*, 1998; Gustafson *et al.*, 1998; Ultee *et al.*, 1999). Nutrient uptake, nucleic acid synthesis and ATPase activity may also be affected, leading to further damage to the cell. Several reports have demonstrated that most essential oils (at approx. 100 mg/l) impair the respiratory activity of bacteria and yeasts (e.g. *Saccharomyces cerevisiae*) (Conner and Beuchat, 1984a,b; Denyer and Hugo, 1991; Tassou *et al.*, 2000).

Unlike many antibiotics, essential oils are capable of gaining access to the periplasm of Gram-negative bacteria through the porin proteins of the outer membrane (Helander *et al.*, 1998). The permeability of cell membranes is dependent on their composition and the hydrophobicity of the solutes that cross them (Sikkema *et al.*, 1995; Helander *et al.*, 1998; Ultee *et al.*, 2002). Low temperatures decrease the solubility of essential oils and hamper penetration of the lipid phase of the membrane (Wanda *et al.*, 1976). The partition coefficient of essential oils in cell membranes is a crucial determinant of antimicrobial efficacy.

The solubility of essential trace elements such as iron is negatively affected by essential oils. Consequently, reduced availability of iron could inhibit bacterial growth. Additionally,

Table 3.6 Lethal dose (LD_{50}) of some essential oils determined in rats

Plant/herb	LD_{50} (g/kg)	Plant/herb	LD_{50} (g/kg)
Prunus amygdalus	<1.0	*Juniperus communis*	>5
Angelica archangelica	2–>5	*Laurus nobilis*	2–5
Pimpinella anisum	2–5	*Lavandula angustifolia*	2–>5
Ocinum basilicum	1–2	*Citrus limonum*	>5
Pimenta racemosa	1–2	*Origanum marjorana*	2–5
Citrus bergamia	>5	*Pistacia lentiscus*	>5
Cinnamomum camphora	2–5	*Citrus aurantium*	>5
Anethum graveolens	2–5	*Origanum vulgare*	1–2
Allium sativum	>5	*Petroselinum sativum*	1–5
Anthemis nobilis	>5	*Piper nigrum*	>5
Cinnamomum zeylanicum	2–5	*Rosmarinus officinalis*	>5
Daucus carota	>5	*Menta viridis*	2–5
Cinnamomum cassia	2–5	*Salvia officinalis*	2–5
Syzygium aromaticum	1–5	*Thymus vulgaris*	2–5
Eucalyptus globulus	2–5	*Citrus reticulata*	>5
Foeniculum vulgare	2–5	*Coriandrum sativum*	2–5
Zingiber officinale	>5	*Camellia sinensis*	2–5

Data modified from: Skandamis (2001).

the reaction of ferrous ion with phenolic compounds can indirectly damage cells by causing oxidative stress (Friedman and Smith, 1984; Nagaraj, 2001). The highly reactive aldehyde groups of some plant-derived antimicrobial compounds (e.g. citral, salicylaldehyde) form Schiff's bases with membrane proteins and so prevent cell wall biosynthesis (Friedman, 1996, 1999; Patte, 1996). Phenolics, essential oils and phytoalexins generally cause static rather than outright toxic effects (Tokutake *et al.*, 1992); cell membranes that leak or function poorly would not necessarily lead to cell death but would most probably cause a deceleration of metabolic processes such as cell division (Darvill and Albersheim, 1984; Kubo *et al.*, 1985).

Antibiotics and related drugs have substantially reduced the threat posed by infectious diseases in the last century. However, the emergence and spread of antibiotic-resistant bacteria has, more recently, become a major concern. This concern has widened to include all microorganisms exposed to antimicrobial agents, including the so-called 'natural' compounds. However, there is relatively little information on the resistance mechanisms of microorganisms against plant-derived antimicrobial compounds.

Deans and Ritchie (1987), who studied the effect of 50 plant essential oils against 25 genera of bacteria, concluded that Gram-positive and Gram-negative organisms were equally susceptible to the antimicrobial action of essential oils. However, this conclusion is now under dispute. In general, Gram-positive bacteria are more sensitive than Gram-negative organisms to the antimicrobial compounds in spices (Dabbah *et al.*, 1970; Farag *et al.*, 1989; Shelef, 1983; Tassou and Nychas, 1995b,c, 1999). However, variation in the rate or extent of inhibition is also evident among the Gram-negative bacteria. For example, *E. coli* was less resistant than *Pseudomonas fluorescens* or *Serratia marcescens* to essential oils from sage, rosemary, cumin, caraway, clove and thyme (Farag *et al.*, 1989). *Salmonella enteritidis* and *typhimurium* were less sensitive than *P. fragi* to sage and mastic gum oils (Tassou and Nychas, unpublished) whereas *Salmonella typhimurium* was more sensitive than *P. aeruginosa* to the essential oils from oregano and thyme (Paster *et al.*, 1990). *Pseudomonas putida* and *P. aeruginosa* have been reported as relatively tolerant of essential oils, and efflux pumps in the outer membrane have been suggested as possible mechanisms for this resistance (Pattnaik et al., 1995a,b; Isken and de Bond, 1998; Mann et al., 2000). Mutants of *E. coli* and sub-populations of *Staph. aureus* resistant to pine and tea-tree oil, respectively, have also been reported (Moken et al., 1997; Nelson, 2000).

3.7 Legislation

Many essential oils from herbs and spices are used widely in the food, health and personal care industries and are classified as GRAS substances or are permitted food additives (Kabara, 1991). A large number of these compounds have been the subject of extensive toxicological scrutiny and an example of the data available is shown in Table 3.6. However, their principal function is to impart desirable flavours and aromas and not necessarily to act as antimicrobial agents. Therefore, it is possible that additional safety and toxicological data would be required before regulatory approval for their use as novel food preservatives would be granted.

3.8 Future prospects and multifactorial preservation

Given the high flavour and aroma impact of plant essential oils, the future for using these compounds as food preservatives lies in the careful selection and evaluation of their efficacy at low concentrations but in combination with other chemical preservatives or preservation processes. Synergistic combinations have been identified between garlic extract and nisin, carvacrol and nisin, vanillin or citral and sorbate, thyme oil and/or cinnamaldehyde in an edible coating, and low-dose gamma irradiation and extracts of rosemary or thyme.

3.9 References

ADAMS, M. and SMID, E. (2003), Nisin in multifactorial food preservation. In *Natural Antimocrobials for the Minimal Processing of Foods*, Roller, S. (ed.), Woodhead Publishing Limited, Cambridge.

AKGUL, A. and KIVANC, M. (1989), Sensitivity of four foodborne moulds to essential oils from Turkish spices, herbs and citrus peel. *Journal of Sciences & Food Agriculture* **47**: 129–132.

AKTUG, S.E. and KARAPINAR, M. (1987), Inhibition of foodborne pathogens by thymol, eugenol, menthol and anethole. *International Journal of Food Microbiology* **4**: 161–6.

AL-KHAYAT, M.A. and BLANK, G. (1985), Phenolic spice components sporostatic to Bacillus subtilis. *Journal of Food Science* **50**: 971–980.

AMAKURA, Y., UMINO, Y., TSUJI, S., ITO, H., HATANO, T., YOSHIDA, T. and TONOGAI, Y. (2002), Constituents and their antioxidative effects in eucalyptus leaf extract used as natural food additive. *Food Chemistry* **77**: 47–56.

AURELI, P., CONSTANTINI, A. and ZOLEA, S. (1992), Antimicrobial activity of some plant essential oils against *Listeria monocytogenes*. *Journal of Food Protection* **55**: 344–8.

AYRES, H.M., FURR, J.R. and RUSSELL, A.D. (1993), A rapid method of evaluating permeabilizing activity against *Pseudomonas aeruginosa*. *Letters in Applied Microbiology* **17**: 149–51.

AYRES H.M., PAYNE, D.N., FURR, J.R. and RUSSELL, A.D. (1998), Use of the Malthus-AT system to assess the efficacy of permeabilizing agents on the activity of antibacterial agents against *Pseudomonas aeruginosa*. *Letters in Applied Microbiology* **26**: 422.

AZZOUZ, M.A. and BULLERMAN, L.B. (1982), Comparative antimycotic effects of selected herbs, spices plant components and commercial antifungal agents. *Journal of Food Protection* **45**: 1298–301.

BLANK, G., AL-KHAYAT, M. and ISMOND, M.A.H. (1987), Germination and heat resistance of *Bacillus subtilis* spores produced on clove and eugenol based media. *Food Microbiology* **4**: 35–42.

BLOOMFIELD, S.F. (1991), Methods for assessing antimicrobial activity. In *Mechanisms of Action of*

Chemical Biocides; Their Study and Exploitation, Denyer, S.P. and Hugo, W.B. (eds), Society for Applied Bacteriology, Technical Series No. 27, Blackwell Scientific Publications, Oxford pp. 1–22.

BRIOZZO, J., NUNEZ, L., CHIRIFE, J. and HERSZAGE, L. (1989), Antimicrobial activity of clove oil dispersed in a concentrated sugar solution. *Journal of Applied Bacteriology* **66**: 69–75.

BROCKLEHURST, T.F, MITCHELL, G.A. and SMITH, A.C. (1997), A model experimental gel surface for the growth of bacteria on foods. *Food Microbiology* **14**: 303–11.

BRUL, S. and COOTE, P. (1999), Preservative agents in foods. Mode of action and microbial resistance mechanisms. *International Journal of Food Microbiology* **50**: 1–17.

CARLSON-CASTELAN, LH., MACHADO, R.A.F., SPRICIGO, C.B., PEREIRA, L. K. and BOLZAN, A. (2001), Extraction of lemongrass essential oil with dense carbon dioxide. *Journal of Supercritical Fluids* **21**: 33–9.

CARSON, C.F., HAMMER, K.A. and RILEY, T.V. (1995), Broth microdilution method for determining the susceptibility of *Escherichia coli* and *Stapylococcus aureus* to the essential oil of *Melaleuca alternifolia* (tea tree oil). *Microbios* **82**: 181–5.

CHANG, S.-T., CHEN, P.-F. and CHANG, S.-C. (2001), Antibacterial activity of leaf essential oils and their constituents from *Cinamomum osmophloeum*. *Journal of Ethnopharmacology* **77**: 123–7.

CIMANGA, K., KAMBU, K., TONA., L., APERS, S., DE BRUYEN, T., HERMANS, N., TOTTE, J., PIETERS, L. and VLIENTINCK, A.J. (2002), Correlation between chemical composition and antibacterial activity of essential oils of some aromatic medicinal plants growing in the Democratic Republic of Congo. *Journal of Ethnopharmacology* **79**: 213–20.

CONNER, D.E. and BEUCHAT, L.R. (1984a), Sensitivity of heat-stressed yeasts to essential oils of plants. *Applied Environmental Microbiology* **47**: 229–33.

CONNER, D.E. and BEUCHAT, L.R. (1984b), Effects of essential oils from plants on growth of food spoilage yeasts. *Journal of Food Science* **49**: 429–34.

COX, S.D., GUSTAFSON, J.E., MANN, C.M., MARKHAM, J.L., LIEW, Y.C., HARTLAND, R.P., BELL, H.C., WARMINGTON, J.R. and WYLLIE, S.G. (1998), Tea tree oil causes K$^+$ leakage and inhibits respiration in *Escherichia coli*. *Letters in Applied Microbiology* **26**: 355–8.

CUPPERS, H.G.A.M. and SMELT, J.P.P.M. (1993), Time to turbidity measurement as a tool for modeling spoilage by *Lactobacillus*. *Journal of Industrial Microbiolology* **12**: 168–71.

DABBAH, R., EDWARDS, V.M. and MOATS, W.A. (1970), Antimicrobial action of some citrus fruit oils on selected food-borne bacteria. *Applied Microbiology* **19**: 27–31.

DALGAARD, P. and KOUTSOUMANIS, K. (2001), Comparison of maximum specific growth rates and lag times estimated from absorbance and viable count data by different mathematical models. *Journal of Microbiological Methods* **43**: 183–96.

DARVILL, A.G. and ALBERSHEIM, P. (1984), Phytoalexins and their elicitors: a defence against microbial infection in plants. *Annual Review of Plant Physiology* **35**: 243–75.

DAVIDSON, P.M. (1997), Chemical preservatives and natural antimicrobial compounds. In *Food Microbiology Fundamentals and Frontiers*, Doyle, M.P., Beuchat, L.R., Montville, T.J. (eds), pp. 520–56, ASM Press, New York.

DAVIDSON, P.M. and BRANEN, A.L. (1993), *Antimicrobials in Foods*. Marcel Dekker, NY.

DEANS, S.G and RITCHIE, G. (1987), Antibacterial properties of plant essential oils. *International Journal of Food Microbiology* **5**: 165–80.

DENYER, S.P. and HUGO, W.B. (1991), Biocide induced damage to the bacterial cytoplasmic membrane. In *Mechanisms of Action of Chemical Biocides; Their Study and Exploitation*, Denyer, S.P. and Hugo, W.B. (eds), The Society for Applied Bacteriology, Technical Series No. 27. Blackwell Scientific Publications, Oxford.

DIAZ, G., WOLF, W., KOSTAROPOULOS, A.E. and SPIESS, W.E.L. (1993), Diffusion of low-molecular weight compounds in food model system. *Journal of Food Processing and Preservation* **17**: 437–54.

DUMONT, L.E. and SLABYJ, B.M. (1993), Impedimetric estimation of bacterial load in commercial carrageenan. *Food Microbiology* **11**; 375–83.

ELNIMA, N.I., SYED, A.A., MEKKAWI, A.G. and MOSSA, J.S. (1983), Antimicrobial activity of garlic and onion extracts. *Pharmazie* **38**: 743–8.

FARAG, R.S., DAW, Z.Y, HEWEDI, F.M. and EL-BAROTY, G.S.A (1989), Antimicrobial activity of some Egyptian spice essential oils. *Journal of Food Protection* **52**: 665–7.

FLAMINI, G., CIONI, P.L., MORELLI, I., MACCHIA, M. and CECCARINI, L. (2002), Main agronomic productive characteristics of two ecotypes of *Rosmarinus officinalis* L. and chemical composition of their essential oils. *Journal of Agricultural and Food Chemistry* **50**: 3512–17.

FRIEDMAN, M. (1996), Food browning and its prevention. *Journal of Agricultural and Food Chemistry* **47**: 1523–40.

FRIEDMAN, M. (1999), Chemistry nutrition and microbiology of D-amino acids. *Journal of Agricultural and Food Chemistry* **47**: 3457–79.

FRIEDMAN, M. and SMITH, G.A. (1984), Inactivation of quercetin mutagenicity. *Food Chemistry and Toxicology* **22**: 535–9.

GALLI, A., FRANZETTI, L. and BRIGUGLIO, D. (1985), Antimicrobial properties *in vitro* of essential oils and extract of spices used for food. *Industrial Alimentaries* **24**: 463–6.

GONZALEZ, B., GLAASKER, E., KUNJI, E.R.S., DRIESSEN, A.J.M., SUAREZ, J.E. and KONINGS, W.N. (1996), Bactericidal mode of action of plantaricin C. *Applied Environmental Microbiology* **62**: 2701–709.

GOUNARIS, Y., SKOULA, M., FOURNARAKI, C., DRAKAKAKI, G. and MAKRIS, A. (2002), Comparison of essential oils and genetic relationship of *Origanum* X *intercedens* to its parental taxa in the island of Crete. *Biochemical Systematics and Ecology* **30**: 249–58.

GREER, G.G., PAQUET, A. and DILTS, B.D. (2000), Inhibition of *Brochothrix thermosphacta* in broths and pork by myristoyl-L-methionine. *Food Microbiology* **17**: 177–83.

GUSTAFSON, J.E., LIEW, Y.C., CHEW, S., MARKHAM, J., BELL, H.C., WYLLIE, S.G. and WARMINGTON, J.R. (1998), Effects of tea tree oil on *Escherichia coli*. *Letters in Applied Microbiology* **26**: 194–8.

HALL, M.A. and MAURER, A.J. (1986), Spice extracts and propylene glycols as inhibitors of *Clostridium botulinum* in turkey frankfurter slurries. *Poultry Science* **65**: 1167–71.

HAYES, A.J. and MARKOVIC, B. (2002), Toxicity of Australian essential oil *Backhousia citriodora* (Lemon myrtle) Part 1. Antimicrobial activity and *in vitro* cytotoxicity. *Food and Chemical Toxicology* **40**: 535–43.

HELANDER, I.K., ALAKOMI, H.L., LATVA-KALA, K., MATTILA-SANDHOLM, T., POL, I., SMID, E.J. and VON WRIGHT, A. (1998), Characterization of the action of selected essential oil components on Gram negative bacteria. *Journal of Agricultural Chemistry* **46**: 3590–5.

IPPOLITO, A. and NIGRO, F. (2003), Natural antimicrobials in the postharvest storage of fresh fruits and vegetables. In *Natural Antimicrobials for the Minimal Processing of Foods*, Roller, S. (ed.), Woodhead Publishing Limited, Cambridge.

ISCAN, G., KIRIMER, N., KURKCUOGLU, M., CAN BASER, K.H. and DEMIRCI, F. (2002), Antimicrobial screening of *Menta piperita* essential oils. *Journal of Agricultural Food Chemistry* **50**: 3943–6.

ISKEN, S. and DE BOND, J.A.M. (1998), Bacterial tolerance to organic solvents. *Extremophiles* **2**: 29–238.

ISMAIEL, A.A. and PIERSON, M.D. (1990), Inhibition of germination, outgrowth and vegetative growth of *Clostridium botulinum* 67B by spice oils. *Journal of Food Protection* **53**: 755–8.

JOHANSEN, C., GILL, T. and GRAM, L. (1995), Antibacterial effect of protamine assayed by impedimetry. *Journal of Applied Bacteriology* **78**: 97–303.

JUDIS, J. (1963), Studies on the mechanism of action of phenolic disinfectants II. *Journal of Pharmacological Science* **52**(2): 126–31.

KABARA, J.J. (1991), Phenols and chelators. In *Food Preservatives*, Russell, N.J. and Gould, G.W. (eds), pp. 200–214. Blackie, Glasgow & London.

KARAMAN, S., DIGRAK, M., RAVID, U. and ILCIM, A. (2001), Antibacterial and antifugal activity of the essential oils of *Thymus revolutus* Celak from Turkey. *Journal of Ethnopharmacology* **76**: 183–6.

KIVANC, M. and AKGUL, A. (1990), Mould growth on black table olives and prevention by sorbic acid, methyl-eugenol and spice essential oil. *Die Nahrung* **34**: 369–73.

KOCH, A.L. (1981), Growth measurement. In *Manual of Methods for General Bacteriology*, Gerhardt, P., Murray, R.G.E and Costilow, R.N. (eds), pp. 179–207. American Society for Microbiology, Washington DC.

KOUTSOUMANIS, K., TAOUKIS, P.S., TASSOU, C.C. and NYCHAS G.-J.E. (1997), Predictive modelling of the growth of *Salmonella enteritidis*; the effect of temperature, initial pH and oleuropein concentration. *Proceedings of the International Conference on Predictive Microbiology Applied to Chilled Food Preservation*, 16–18 June, 1997 Quimper, France, pp. 113–19.

KOUTSOUMANIS, K., TASSOU, C.C., TAOUKIS, P.S. and NYCHAS, G.-J.E. (1998), Modelling the effectiveness of a natural antimicrobial on *Salmonella enteritidis* as a function of concentration, temperature and pH, using conductance measurements. *Journal of Applied Microbiology* **84**: 981–7.

KOUTSOUMANIS, K., LAMBROPOULOU, K. and NYCHAS, G.J.E. (1999), A predictive model for the non-thermal inactivation of *Salmonella enteritidis* in a food model system supplemented with a natural antimicrobial. *International Journal of Food Microbiology* **49**: 67–74.

KUBO, I. and HIMEJIMA, M. (1991), Anethole, a synergist of polygodial against filamentous microorganisms. *Journal of Agricultural Food Chemistry* **39**: 2290–2.

KUBO, I., MATSUMOTO, A. and TAKASE, I. (1985), A multichemical defense mechanism of bitter olive *Olea europaea* (Oleaceae). Is oleuropein a phytoalexin precursor? *Journal of Chemical Ecology* **11**(2): 251–63.

LACHOWICZ, K.J., JONES, G.P., BRIGGS, D.R., BIENVENU, F.E., WAN, J., WILCOCK, A. and COVENTRY, M.J. (1998), The synergistic preservative effects of the essential oils of sweet basil (*Ocimum basillicum* L.) against acid-tolerant food microflora. *Letters in Applied Microbiology* **26**: 209–14.

LAMBERT, R.J.W. and PEARSON, J. (2000), Susceptibility testing: accurate and reproducible minimum inhibitory concentration (MIC) and non-inhibitory concentration (NIC) values. *Journal of Applied Microbiology* **88**: 784–90.

LAMBERT, R.J.W., SKANDAMIS, P., COOTE, P.J. and NYCHAS, G.-J.E. (2001), A study of the minimum inhibitory concentration and mode of action or oregano essential oil, thymol and carvacrol. *Journal of Applied Microbiology* **91**: 453–62.

LEMAY, M-J., CHOQUETTE, J., DELAQUIS, P.J., GARIEPY, C., RODRIGUE, N. and SAUCIER L. (2002), Antimicrobial effect of natural preservatives in a cooked and acidified chicken meat model. *International Journal of Food Microbiology* **78**: 217–26.

MACRAE, M., REBATE, T., JOHNSTON, M. and OGDEN, I.D. (1997), The sensitivity of *Escherichia coli* O157 to some antimicrobials by conventional and conductance assays. *Letters in Applied Microbiology* **25**: 135–7.

MANGENA, T. and MUYIMA, N.Y.O. (1999), Comparative evaluation of the antimicrobial activities of essential oils of *Artemisis afra*, *Pteronia incana* and *Rosmarinus ofiicinalis* on selected bacteria and yeasts strains. *Letters in Applied Microbiology* **28**; 291–6.

MANN, C.M. and MARKHAM, J.L. (1998), A new method for determining the minimum inhibition concentration of essential oils. *Journal of Applied Microbiology* **84**: 538–44.

MANN, C.M., COX, S.D. and MARKHAM, J.L. (2000), The outer mebrane of *Pseudomonas aeruginosa* NCTC 6749 contributes to its tolerance to the essential oil of *Melaleuca alternifolia* (tea tree oil). *Letters in Applied Microbiology* **30**: 294–7.

MANOU, L., BOUILLARD, L., DEVLEESCHOUWER, M.J. and BAREL, A.O. (1998), Evaluation of the preservative properties of *Thymus vulgaris* essential oil in topically applied formulations under a challenge test. *Journal of Applied Microbiology* **84**: 368–76.

MCCLURE, P.J., COLE, M.B., DAVIES, K.W. and ANDERSON, W.A. (1993), The use of automated turbidimetric data for the construction of kinetic models. *Journal of Industrial Microbiology* **12**: 277–85, 4, 3.

MEJLHOLM, O. and DALGAARD, P. (2002), Antimicrobial effect of essential oils on the seafood spoilage micro-organism *Photobacterium phosphoreum* in liquid media and fish products. *Letters in Applied Microbiology* **34**: 27–31.

MENON, V.K. and GARG, S.R. (2001), Inhibitory effect of clove oil on *Listeria monocytogenes* in meat and cheese. *Food Microbiology* **18**: 647–50.

MOKEN, M.C., MCMURRY, L.M. and LEVY, S.B. (1997), Selection of multiple-antibiotic resistant (mar), mutants of *Escherichia coli* by using the disinfectant pine oil: roles of the mar and acrABloci. *Antimicrobial Agents and Chemotherapy* **41**: 2770–2.

MOUREY, A. and CANILLAC, N. (2002), Anti-*Listeria monocytogenes* activity of essential oils components of conifers. *Food Control* **13**: 289–92.

NAGARAJ, R. (2001), Glycation and oxidative stress. *7th Int. Symposium on the Mailard Reaction*, Kumamoto, Japan.

NAGY, J.G. and TENGERDY, R.P. (1967), Antibacterial action of essential oils of Artemis as an ecological factor. *Applied Microbiology* **15**: 819–21.

NELSON, R.R.S. (2000), Selection of resistance to the essential oil of *Melaleuca alternifolia* in *Staphylococcus aureus*. *Journal of Antimicrobial Chemotherapy* **45**; 549–50.

NYCHAS, G.J.E. (1995), Natural antimicrobials from plants. In *New Methods of Food Preservation*, Gould, G.W. (ed.), pp. 58–89. Blackie Academic Professional, London.

NYCHAS, G.J.E. and TASSOU, C.C. (2000), Preservatives: traditional preservatives – oils and spices. In *Encyclopedia of Food Microbiology*, Robinson, R., Batt, C. and Patel, P. (eds), pp. 1717–22. Academic Press, London.

OYEDEJI, A.O., EKUNDAYO, O., OLAWORE, O.N., ADENIYI, B.A. and KOENIG, W.A. (1999), Antimicrobial activity of the essential oils of five *Eucalyptus* species growing in Nigeria. *Filoterapia* **70**: 526–8.

OZCAN, M., CHALCHAT, J.C. and AKGUL, A. (2001), Essential oil composition of Turkish mountain tea (*Sideritis* spp.). *Food Chemistry* **75**: 459–63.

PASTER, N., JUVEN, B.J., SHAAYA, E., MENASHEROV, M., NITZAN, R., WEISSLOWICZ, H. and RAVID, U. (1990), Inhibitory effect of oregano and thyme essential oils on moulds and foodborne bacteria. *Letters in Applied Microbiology* **11**: 33–7.

PASTER, N., MENASHEROV, M., RAVID, U. and JUVEN, B. (1995), Antifungal activity of oregano and thyme essential oils applied as fumigants against fungi attacking stored grain. *Journal of Food Protection* **58**: 81–5.

PATSIOURIAS, E., KOUTSOUMANIS, K. and NYCHAS, G.J.E. (2003), Effect of hydrosols on pathogenic and spoilage bacteria. Unpublished data.

PATTE, J. (1996), Biosynthesis of threonine and lysine. In *Escherichia coli* and *Salmonella*, Frederci, M. and Neidhardt, F. (eds), 2nd Edition, pp. 528–41. ASM Press, Washington DC.

PATTNAIK, S., RATH, C. and SUBRAMANYAM, V.R. (1995a), Characterization of resistance to essential oils in a strain of *Pseudomonas aeruginosa* (VR-6). *Microbios* **81**(326): 29–31.

PATTNAIK, S. SUBRANANYAM, V.R. and RATH, C.C. (1995b), Effect of essential oils on the viability and morphology of *Escherichia coli* (SP-11) *Microbios* **84**: 195, 199.

PATTNAIK, S., SUBRAMANYAM, V.R., KOLE, C.R. and SAHOO, S. (1995c), Antibacterial activity of essential oils from Cymbopogon: inter- and intra-specific differences. *Microbios* **84**(341): 239–45.

PEREZ, C., AGNESE, A.M. and CABRERA, J.L. (1999), The essential oil of *Senecio graveolens* (Compositae): Chemical composition and antimicrobial activity tests. *Journal of Ethnopharmacology* **66**: 91–6.

PRINDLE, R.F. and WRIGHT, E.S. (1977), Phenolic compounds. In *Disinfection, Sterilisation and Preservation*, Block, S.S. (ed.), pp. 219–51. Lea & Febiger, Philadelphia.

RAMADAN, F.M., EL-ZANFALY, R.T., EL-WAKEIL, F.A. and ALLIAN, A.M. (1972), On the antibacterial effects of some essential oils I. Use of agar diffusion method. *Chem. Mikrobiol. Technol. Lebensm.* **2**: 51–5.

ROBINS, M.M., BROCKLEHURST, T.F. and WILSON, P.D.G. (1994), Food structure and the growth of pathogenic bacteria. *Food Technology International Europe*, 31–6.

ROLLER, S. (2003), *Natural Antimicrobials for the Minimal Processing of Foods*, Woodhead Publishing Limited, Cambridge.

SALMERON, J., JORDANO, R. and POZO, R. (1990), Antimycotic and antiaflatoxigenic activity of oregano (*Origanum vulgare*, L.) and thyme (*Thymus vulgaris*, L.). *Journal of Food Protection* **53**: 697–700.

SANKARAN, R. (1976), Comparative antimicrobial action of certain antioxidants and preservatives. *Journal of Food Science and Technology* **13**: 203–204.

SHELEF, L.A (1983), Antimicrobial effects of spices. *Journal of Food Safety* **6**: 29–44.

SHELEF, L.A., NAGLIK, O.A. and BOGEN, D.W. (1980), Sensitivity of some common food-borne bacteria to the spices sage, rosemary and allspice. *Journal of Food Science* **45**: 1042–4.

SIKKEMA, J., DE BONT, J.A.M. and POOLMAN, B. (1995), Mechanisms of membrane toxicity of hydrocarbons. *Microbiology Reviews* **59**: 201–222.

SKANDAMIS, P.N. (2001), Effect of oregano essential oil on spoilage and pathogenic microorganisms in foods. Ph.D. Thesis, Agricultural University of Athens.

SKANDAMIS, P.N. and NYCHAS, G.-J.E. (2000), Development and evaluation of a model predicting the survival of *Escherichia coli* O157:H7 NCTC 12900 in homemade eggplant salad at various temperatures, pHs, and oregano essential oil concentrations. *Applied and Environmental Microbiology* **66**: 1646–53.

SKANDAMIS, P. and NYCHAS, G.-J.E. (2001), Effect of oregano essential oil on microbiological and physicochemical attributes of mince meat stored in air and modified atmospheres *Journal of Applied Microbiology* **91**, 1011–22.

SKANDAMIS, P. and NYCHAS G.-J.E. (2002a), Essential oils; can be considered 'smart' enough for their potential to be used in active packaging system? In *Joint Meeting of the SFAM & DSM 'Frontiers in Microbial Fermentation and Preservation'*, Wageningen, The Netherlands, 9–11 January 2002.

SKANDAMIS, P.N. and NYCHAS, G.-J.E. (2002b), Preservation of fresh meat with active and modified atmosphere packaging conditions. *International Journal of Food Microbiology* **79**: 35–43.

SKANDAMIS, P.N., MICHAILIDOU, E. and NYCHAS G.-J.E. (1999a), Modelling the effect of oregano (*Origanum vulgare*) on the growth/survival of *Escherichia coli* O157:H7 in broth and traditional Mediteranean foods. *International Congress on 'Improved Traditional Foods for the next Century'*, Valencia 28–29/1999, Spain, pp. 270–3.

SKANDAMIS, P., TASSOU, C.C. and NYCHAS, G.-J.E. (1999b), Potential use of essential oils as food preservatives. In *17th International Symposium of the International Committee on Food Microbiology and Hygiene (ICFMH)*, Tuijtelaars, A.C.J., Samson, R.A., Rombouts, F.M. and Notermans, S. (eds), Veldhoven, The Netherlands, 13–17 September 1999, pp. 300–3.

SKANDAMIS, P., TSIGARIDA, E. and NYCHAS, G.-J.E. (2000), Ecophysiological attributes of *Salmonella typhimurium* in liquid culture and within gelatin gel with or without the addition of oregano essential oil. *World Journal of Microbiology and Biotechnology* **16**: 31–5.

SKANDAMIS, P., ELIOPOULOS, V. and NYCHAS, G.-J.E. (2001a), Effect of essential oils on survival of *Escherichia coli* O157:H7 NCTC 12900 and *Listeria monocytogenes* in traditional Mediterranean salads. Annual Conference of Society for Applied Microbiology, Swansea, UK, 1–17 July 2001.

SKANDAMIS, P., KOUTSOUMANIS, K., FASSEAS, K. and NYCHAS, G.-J.E. (2001b), Evaluation of the inhibitory effect of oregano essential oil on *Escherichia coli* O157:H7, in broth culture with or without EDTA, using viable counts, turbidity and impedance. *Italian Journal of Food Science and Technology* **13**: 65–75.

SKANDAMIS, P., TSIGARIDA, E. and NYCHAS, G.-J.E (2002a), The effect of oregano essential oil on survival/death of *Salmonella typhimurium* in meat stored at 5°C under aerobic, vp/map conditions. *Food Microbiology* **19**: 97–103.

SKANDAMIS, P., TSIGARIDA, E. and NYCHAS, G.-J.E. (2002b), Effect of conventional and natural preservatives on the death/survival of *Escherichia coli* O157:H7 NCTC 12900 in traditional Mediterranean salads. In *Joint Meeting of the SFAM & DSM 'Frontiers in Microbial fermentation and Preservation'*, Wageningen, The Netherlands, 9–11 January 2002.

SKANDAMIS, P.N., DAVIES, K.W., MCCLURE, P.J., KOUTSOUMANIS, K. and TASSOU, C. (2002c), A vitalistic approach or non-thermal inactivation of pathogens in traditional Greek salads. *Food Microbiology* **19**: 405–21.

SMITH-PALMER, A., STEWART, J. and FYFE, L. (1998), Antimicrobial properties of plant essential oils and essences against important food-.borne pathogens. *Letters in Applied Microbiology* **26**: 118–22.

SMITH-PALMER, A., STEWART, J. and FYFE, L. (2001), The potential application of plant essential oils as natural food preservatives in soft cheese. *Food Microbiology* **18**: 463–70.

STECCHINI, M.L., SARAIS, I. and GIAVEDONI, P. (1993), Effect of essential oils on *Aeromonas hydrophila* in a culture medium and in cooked pork. *Journal of Food Protection* **56**: 406–9.

STECCHINI, M.L., DEL TORRE, M., SARAIS, I., SARO, O., MESSINA, M. and MALTINI, E. (1998), Influence of structural properties and kinetic constraints on *Bacillus cereus* growth. *Applied and Environmental Microbiology* **64**: 1075–8.

TAHARA, T., OSHIMURA, M., UMEZAWA, C. and KANATANI, K. (1996), Isolation, partial characterization and mode of action of acidocin J1132, a two-component bacteriocin produced by *Lactobacillus acidophilus* JCM 1132. *Applied and Environmental Microbiology* **62**: 892–7.

TAKAHASHI, Y., INABA, N., KUWAHARA, S., KUKI, W., YAMANE, K. and MURAKAMI, A. (2002), Rapid and convenient method for preparing aurapten-enriched product from hassaku peel oil. Implication of cancer-preventive food additives. *Journal of Agricultural Food Chemistry* **50**: 3193–6.

TASSOU, C.C. (1993), Microbiology of olives with emphasis on the antimicrobial activity of phenolic compounds. Ph.D. Thesis, University of Bath, Bath, UK.

TASSOU, C.C. and NYCHAS G.-J.E. (1995a), Inhibition of *Salmonella enteritidis* by oleuropein in broth and in a model food system. *Letters in Applied Microbiology* **20**: 120–4.

TASSOU, C.C. and NYCHAS, G.-J.E. (1995b), The inhibitory effect of the essential oils from basil and sage in broth and in food model system. In *Developments in Food Science 37; Food Flavors: Generation, Analysis and Process Influence*, Charalambous, G. (ed.), pp. 1925–36. Elsevier, New York.

TASSOU, C.C. and NYCHAS, G.-J.E. (1995c), Antimicrobial activity of the essential oil of mastic gum (*Pistachia lentiscus* var.*chia*) on Gram-positive and Gram-negative bacteria in broth and in model food system. *International Biodeterioration and Biodegradation* **36**: 411–20.

TASSOU, C.C., DROSINOS, E.H. and NYCHAS, G.-J.E. (1995), Effects of essential oil from mint (*Mentha piperita*) on *Salmonella enteritidis* and *Listeria monocytogenes* in model food systems at 4 and 10°C. *Journal of Applied Bacteriology* **78**: 593–600.

TASSOU, C.C., DROSINOS, E.H. and NYCHAS, G.-J.E. (1996), Inhibition of the resident microbial flora and pathogen inocula on cold fresh fillets in olive oil, oregano and lemon juice under modified atmosphere or air. *Journal of Food Protection* **59**: 31–4.

TASSOU, C.C., DROSINOS, E.H. and NYCHAS, G.-J.E. (1997), Antimicrobial effect of carob (*Ceratonia siliqua*) extract against food related bacteria in culture media and model food systems. *World Journal of Microbiology and Biotechnology* **13**: 479–81.

TASSOU, C.C., KOUTSOUMANIS, K., SKANDAMIS, P. and NYCHAS, G.-J.E. (1999), Novel combinations of natural antimicrobial systems for the improvement of quality of agro-industrial products. Pp. 51–2 *International Congress on 'Improved Traditional Foods for the next Century'*, Valencia, Spain, 28–29 November 1999.

TASSOU, C.C., KOUTSOUMANIS, K. and NYCHAS, G.-J.E. (2000), Inhibition of *Salmonella enteritidis* and *Staphylococccus aureus* in nutrient broth by mint essential oil. *Food Research International* **33**: 273–80.

THANGADURAL, D., ANITHA, S., PULLAIAH, T., REDDY, P.N. and RAMACHAMDRAIAH, O.S. (2002), Essential oil constituents and *in vitro* antimicrobial activity of *Decalepis hamiltonii* roots against foodborne pathogens. *Journal of Agricultural and Food Chemistry* **50**: 3147–9.

THOMAS, L.V, WIMPENNY, W.T. and BARKER, G.C. (1997), Spatial interactions between subsurface bacteria colonies in a model system: a territory model describing the inhibition of *Listeria monocytogenes* by a nisin-producing lactic acid bacterium. *Microbiology* **143**: 2575–82.

TOKUTAKE, N., MIYOSHI, H. and IWAMURA, H. (1992), Effects of phenolic respiration inhibitors on cytochrome bc1 complex of rat-liver mitochondria. *Bioscience, Biotechnology and Biochemistry* **56**: 919–23.

TRANTER, H.S., TASSOU, C.C. and NYCHAS, G.-J.E. (1993), The effect of the olive phenolic compound, oleuropein, on growth and enterotoxin B production by *Staphylococcus aureus*. *Journal of Applied Bacteriology* **74**: 253–60.

TSIGARIDA, E., SKANDAMIS, P. and NYCHAS, G.-J.E. (2000), Behaviour of *Listeria monocytogenes* and autochthonous flora on meat stored under aerobic, vacuum and modified atmosphere packaging conditions with or without the presence of oregano essential oil at 5°C. *Journal of Applied Microbiology* **89**: 901–9.

ULTEE, A., KETS, E.P.W. and SMID, E.J. (1999), Mechanisms of action of carvacrol on the food-borne pathogen *Bacillus cereus*. *Applied and Environmental Microbiology* **65**: 4606–10.

ULTEE, A., SLUMP, R. A., STEGING, G. and SMID, E.J. (2000), Antimicrobial activity of carvacrol on rice. *Journal of Food Protection* **63**: 620–4.

ULTEE, A., BENNIK, M.H.J. and MOEZELAAR, R. (2002), The phenolic hydroxyl group of carvacrol is essential for action against the food-borne pathogen *Bacillus cereus*. *Applied Environmental Microbiology* **68**: 1561–8.

WANDA, P., CUPP, J., SNIPES, W., KEITH, A., RUCINSKY, T., POLISH, L. and SANDS, J. (1976), Inactivation of the enveloped bacteriophage O6 by butylated hydroxytoluene and butylated hydroxyanisole. *Antimicrobial Agents and Chemotherapy* **10**: 96.

WILKINS, K.M. and BOARD, R.G. (1989), Natural antimicrobial systems. In *Mechanisms of Action of Food Preservation Procedures*, Gould, G.W. (ed.), Chapter 11, pp. 285–362, Elsevier, London.

WILSON, P.D.G., BROCKLEHURST, T.F., ARINO, D., THUAULT, M., JAKOBSEN, M., LANGE, J.W.T., FARKAS, J., WIMPENNY, J.W.T. and VAN IMPE, J.F. (2002), Modelling microbial growth in structured foods: towards a unified approach. *International Journal of Food Microbiology* **75**: 273–89.

4

Screening for health effects of herbs

R. Rodenburg, TNO Pharma, The Netherlands

4.1 Introduction

It is estimated that 30–40% of all pharmaceutical preparations that are used nowadays are derived from or based on plant metabolites. Plants are therefore an important source of molecules that may be useful as a drug. Many plant-derived drugs are based on the knowledge of the medicinal effects of plants, for example from traditional medicine. Some well-known examples are quinine from the bark of the Cinchona tree, morphine and codeine from opium, and, more recently, taxol from the bark of the Pacific yew. In traditional and in herbal medicine, crude plant-derived preparations are often used, derived from the whole plant or parts of it, e.g. tinctures, syrups, or either dried plant parts or plant extracts. However, there is growing interest in identifying the active substances present in medicinal plants, in order to use these for pharmaceutical drug development.

What is the advantage of isolating the active compounds from medicinal plants? Knowing which compound or compounds are responsible for the pharmacological activity of a medicinal plant means safe therapeutic drugs can be produced, which can be administered in a controlled manner. Contrast this with using crude plant extracts, of which the composition, and thus the pharmacological potency, may be variable. Furthermore, the identified bioactive compounds may be chemically modified to further improve the properties of the compound as a drug. A well-known example of this is aspirin, which is based on salicin from willow or poplar bark. Finally, plant extracts typically contain thousands of different compounds, of which only a few are pharmacologically relevant. All other compounds are either inactive or may lead to toxicological problems.

Although the pharmaceutical industry has always had an interest in using plants as a source of new drugs, a consequence of the introduction of high-throughput screening (HTS) has been that the attention of the pharmaceutical industry has been focused on the screening of synthetic small molecules, for example those generated by combinatorial chemistry. These compounds are screened for specific bioactivities towards molecular targets, which can be a receptor, an enzyme or any other validated drug target.

However, the diversity generated chemically may be limited with respect to such properties as bioavailability and cytotoxicity. This may be the reason why the number of successful new drugs from this approach is relatively small. By contrast, nature provides a vast reservoir of bioactive molecules. In particular, plants are a rich source of bioactive metabolites, and provide enormous potential in the discovery of new drugs. Now that the modern screening technologies can be combined with the molecular diversity presented by plants, there is a growing interest in using natural compounds as a source of new drugs.

The initial experiments performed during the development of new drugs usually consist of *in vitro* screening experiments, designed to study the effects of bioactive molecules on a predefined, validated drug target. The starting material can either be extracts from a collection of plants, different extracts from a single plant, or individual compounds, e.g. a natural compound library. There is a fundamental difference between the screening of compound libraries and plant extracts. Compound libraries are a collection of usually large numbers of relatively pure compounds. Usually, the chemical identity and properties of the compounds present in the library are known. Often, the *in vitro* assays that are used for the screening of the compound libraries are implemented on (ultra) high-throughput robotized systems. The 'hits' are subsequently tested in secondary screening assays and *in vitro* toxicity assays, followed by *in vivo* experiments. In contrast to compound libraries, plant extracts typically are mixtures of thousands of different molecules, of which most have not yet been characterized. There are several ways to identify the bioactive molecules in these extracts. The common approach is to set up a fractionation scheme and to screen the fractions for the presence of the desired bioactive properties. Active fractions are subfractionated and tested, until the molecules responsible for the bioactivity can be identified. The assays that are used to test the bioactivity of plant extracts are not necessarily implemented in high-throughput systems, owing to the smaller sample numbers, and this also allows for assays with a somewhat higher complexity. An example of this is presented in Section 4.6. This chapter will focus on the screening of plant extracts and fractions thereof, although in some cases reference will be made to compound screening.

4.2 Types of assays

The ways in which the bioactivity of compounds or extracts can be analysed is almost limitless. An overview of types of assays that are often used is given below; however, this should not be regarded as a comprehensive list since for each scientific question several methods may be applied. The common denominator of screening assays is that they are performed *in vitro*. The complexity of the model system ranges from simple assays using a single molecule, up to assays using whole cells. For example, when screening for proteases, a single labelled polypeptide as protease substrate can be used for screening. By contrast, in case plant compounds have to be screened for effects on cell proliferation, a whole cell assay may be very useful.

For each type of biological effect that is the subject of analysis, different types of assays can be applied. The choice for a particular type of assay is something that has to be judged case by case.

4.2.1 Cell-based assays
Cell-based assays are predominantly developed from cultured cell lines, although primary cells can also be used. An often-used format is the reporter gene assay. In this case, the effects of compounds are tested at the level of transcription. For example, in case the drug target is a transcription factor, or a receptor that activates a transcription factor in a specific manner, a reporter gene assay can be used. For this purpose, cells are transfected with a

reporter gene plasmid, which contains a reporter gene encoding an enzyme that can be measured easily. One of the most frequently used reporter enzymes is firefly luciferase (de Wet et al., 1987), which upon ATP-dependent conversion of its product gives rise to the emission of light. Using a luminometer, the production of luciferase can be detected. The reporter gene is driven by a promoter that is activated by the transcription factor of interest. In this way, the reporter gene will be activated in the presence of a bioactive compound; thus the reporter enzyme will be produced and detected. As alternatives for luciferase, other reporter genes can be used, such as β-galactosidase (Flanagan and Wagner, 1987), green fluorescent protein (Chalfie et al., 1994) and secretory alkaline phosphatase (Berger et al., 1988). The latter has the advantage that it is secreted into the culture medium and therefore does not require cell lysis before it can be detected.

As an alternative to reporter gene assays, biological effects in cell lines can also be detected using molecular detectors. There are several examples of this, for example fluorescent probes that can be used to detect specific small molecules (e.g. reactive oxygen (Keston and Brandt, 1965), nitric oxide (Kojima et al., 1998), Ca^{2+} (Minta et al., 1989)), but also recombinant proteins that are introduced into the cell line by transfection. An example of this is aequorin, a protein derived from a jellyfish that fluoresces in the presence of Ca^{2+} (Sheu et al., 1993; Button and Brownstein, 1993).

One of the most straightforward cell-based assays is the screening for cell proliferation or cell death. Cells are simply grown for a certain amount of time in the presence of the compounds that are screened, after which the number of cells is quantified. There are several colorimetric methods available that can be used for this purpose (Denizot and Lang, 1986). Alternatively, cell death can be assessed by analysing the culture medium for the presence of cytoplasmic enzymes, such as lactate dehydrogenase (LDH).

4.2.2 Receptor binding assays

Perhaps one of the simplest ways to identify whether compounds interact with a receptor is by performing receptor binding assays. The receptor of interest is present in a cell membrane preparation from an organ/tissue/cell line that is known to express the receptor. The preparation is incubated with a radiolabelled ligand, usually a well-characterized reference compound that specifically interacts with the receptor. The unbound material is removed by washing and the remaining radiolabelled ligand is quantified using a radiodetector, although alternative detection methods are also possible. New binding partners are identified by testing their ability to compete with the radiolabelled ligand for binding to the target receptor.

4.2.3 Fluorescence assays

The detection of fluorescence is one of the most sensitive analytical techniques. It is now widely used as a detection method in so-called soluble assays, which are simple one-step assays. By contrast, insoluble assays involve the attachment of one of the assay reagents to a solid support (i.e. a bead or a well plate), and usually include a washing step before the assay can be read. Many fluorescence assays are based on the FRET principle, fluorescence resonance energy transfer. The assay principle is that a fluorescence donor is attached to one assay reagent (i.e. a receptor fragment) and a fluorescence acceptor to another assay reagent (i.e. a receptor ligand). When the two reagents come into close proximity (i.e. ligand binds the receptor), and the donor is excited, energy is transferred to the acceptor and light of a different wavelength is emitted, which can be measured. FRET is a very sensitive, non-radioactive method. However, for some applications problems may be encountered, such as interference by assay buffer constituents or scattering of the emitted light. The use of a time-resolved fluorescence measurement will result in a strong reduction of interference by background fluorescence.

4.2.4 Scintillation assays
The scintillation proximity assay (Hart and Greenwald, 1979) makes use of a radiolabelled ligand and a receptor that is attached to a bead or well-plate coated with scintillant. Upon binding of the ligand, the scintillant will be stimulated to emit light, which can be detected. Similar to fluorescence, this is also a sensitive technique, with the advantage that it is less sensitive to interference. One of the drawbacks of this technique is that it makes use of radiolabels.

4.2.5 Fluorescence polarization (FP) assays
Similar to fluorescence and scintillation assays, FP assays are used to screen for molecular proximity. FP is based on the observation that polarized light used to excite a fluorescent molecule will result in the emission of polarized light. The polarization of the emitted light will change if the fluorescent molecule physically interacts with another molecule. The change in fluorescence polarization can be measured using an FP detector. Advantages of FP assays are that the assays are soluble, non-radioactive and homogeneous. A drawback may be that an excess of binding partner has to be used in order to be able to detect a signal, and therefore in some cases suboptimal assay conditions have to be used.

These are some of the basic principles of screening assays. There are numerous methods derived from the principles described above that have been successfully developed. The techniques mentioned here have all been applied to HTS. In case throughput is not an issue, more complex assays can be developed depending on the target and compounds/extracts to be screened.

4.3 Throughput vs content assays

The drug discovery trajectory of the pharmaceutical industry in many cases makes use of screening assays with a tremendous throughput. In this way, hundreds of thousands of compounds can be screened in a matter of days, which can be achieved only by utilizing robotized screening facilities. The assays that are used in these HTS facilities are usually relatively simple and will provide a yes or no answer (e.g. this compound does or does not inhibit the activity of enzyme X). In other words, the throughput is maximized whereas the content of the information that is provided is minimized. Various reports have shown that the answers provided by these HTS-type assays may vary depending on the assay set-up. For example, if a receptor assay is developed using a FRET read-out, the answers may be different from a FP assay for the same receptor (Sills et al., 2002). This also implies that each assay suffers from false positives and false negatives. Although the false positives can be identified in secondary screening experiments, the false negatives obviously cannot be identified and are lost. At the other end of the spectrum, assays are developed that provide more detailed information on the mode of action of bioactive compounds. This will result in more informative data at the inevitable cost of a lower throughput. It should also be mentioned that there is a clear trend towards the development of high information content assays with a high throughput. An example of this is the use of microscopic techniques that allow the inspection of multi-well plate for cellular events, such as apoptosis, cell proliferation, and receptor translocation. Another emerging approach is the genomics-based screening, in which expression profiles are measured instead of a single variable.

4.4 Assay quality

Each screening assay can be regarded as an analytical method to determine the bioactivity of compounds or extracts. Therefore, screening assays have to fulfil a number of criteria that deal with assay quality, especially when used in a high-throughput environment. Important criteria are reproducibility and robustness. One of the most widely used criterion for reproducibility is the Z'-factor, which is given in equation 4.1:

$$Z' = 1 - \frac{3\sigma^{pos} + 3\sigma^{neg}}{|\mu^{pos} - \mu^{neg}|} \qquad [4.1]$$

In this equation, σ^{pos} is the standard deviation (SD) of the signal obtained from the positive control, σ^{neg} is the SD from the signal of the negative control, and μ^{pos} and μ^{neg} are the mean values obtained from the positive and negative controls, respectively (Zhang et al., 1999). The Z'-factor is more informative than the signal-to-noise ratio, since it takes into account the assay dynamic range as well as the data variation of both positive and negative control samples. Assays with a Z'-factor ≤ 0 can be regarded as not useful for screening. The closer the Z'-factor is to 1, the better the assay can discriminate between positive and negative controls. In case an assay has a low but positive Z' value, and therefore the assay conditions are suboptimal, the assay may still be very useful when multiple measurements per sample are performed.

Robustness of the assay is another important issue. The experiments that have to be performed in order to test the robustness of assays are dependent on the experimental set-up in which the assay will be implemented, and on the type of assay. Items that may be tested could be the effects of temperature, well-plate, solvents and end-of-run. During the actual screening, several control incubations are usually included to test the performance of the assay.

4.5 Screening bio-active compounds

An important consideration before starting a screening exercise aimed at identifying bioactive compounds from plant extracts is that plants often contain many different bioactive compounds. Thus, a well-characterized beneficial effect of a given plant may be the result of a combination of effects of different plant constituents. This may result in initial disappointing screening results, since fractionation of these extracts may either give inconclusive results, e.g. several fractions show activity, or show no activity from any fraction. In these cases, it may be very informative to include mixtures of fractions as well as the starting material in the same screening experiment as in which fractions are tested. Further, it is advisable to have at least some analytical data on the fractions to be tested at hand, such as high-performance liquid chromatography (HPLC) profiles. These may also be very helpful when interpreting screening results.

One of the more down-to-earth problems encountered when screening plant extracts or plant-derived compounds is that many plant metabolites are poorly soluble in water. An example of this is provided by the flavonoids, present in extracts of several different medicinal plant species. The screening assay has to be designed in such a manner that the solvents used for the plant extracts do not affect the outcome of the assay. Frequently used solvents such as methanol or *N*, *N*-dimethylformamide (DMF) affect assays at concentrations as low as 0.1% (v/v), depending on the type of assay. The effects of acetone and chloroform are even more dramatic and these solvents should not be used when preparing plant extracts for screening. Solvents such as dimethylsulphoxide (DMSO) and ethanol are usually more compatible with most screening assays, although it is still recommended to use concentrations well below 0.5% (v/v). In any case a solvent control should be included in each screening experiment.

4.6 Screening experiments for anti-inflammatory properties

This chapter describes some examples of screening experiments aimed at identifying anti-inflammatory constituents of plants. A large number of plants and herbs are known for their anti-inflammatory properties. Well-known examples are willow bark (contains salicin, from which aspirin is derived), *Boswellia serrata* (boswellic acids) and turmeric (curcumin). In addition to these, many other herbs have been suggested to be anti-inflammatory. Inflammation plays a role in many different clinical disorders. In addition to the obvious inflammatory diseases such as arthritis, asthma, Crohn's disease, psoriasis and so on, inflammation also plays an important role in diseases such as atherosclerosis, diabetes, Alzheimer's and many other diseases. In many of these, a disordered immune system contributes to the onset and/or progression of the disease.

4.6.1 Single target screening

What are important targets for anti-inflammatory therapies? A key regulatory factor in the inflammatory response is the transcription factor family NF-κB. This family of proteins is present in almost all human cells. In inflammatory cells such as macrophages and lymphocytes, NF-κB is activated after stimulation of the cells by a pro-inflammatory stimulus. Its activation leads to the transcription of many different genes involved in the inflammatory response, including cytokines such as TNFα. Inhibition of NF-κB attenuates the inflammatory response, and therefore it is a major target for the development of anti-inflammatory drugs. An example of a relatively simple cell-based assay to screen for NF-κB activation is given in Fig. 4.1. The assay is a reporter gene assay, in which a total of five NF-κB elements are placed in front of the reporter gene. Thus, the reporter gene is expressed only when NF-κB is activated. The reporter gene plasmid is transfected into a macrophage-like cell line, in this case the mouse cell line RAW264.7. The cells are incubated with a stimulus that activates NF-κB, and in the absence of an inhibitor a strong reporter gene response is detectable. When cells are incubated with a stimulus together with an inhibitor, the reporter gene response is attenuated. In this way, possible inhibitory compounds present in plant-derived extracts may be detected. Experiments such as these always include controls for cell viability, to check for possible interference of toxic compounds present in extracts that may give rise to false positives. Obviously, positive and negative controls for activation and inhibition (i.e. a known inhibitor such as dexamethasone) of NF-κB are always included.

There are many other important targets for anti-inflammatory therapies, in addition to transcription factors. An important class of molecules are the receptors, for example receptors for cytokines, chemokines and eicosanoids (Onuffer and Horak, 2002; Holgate

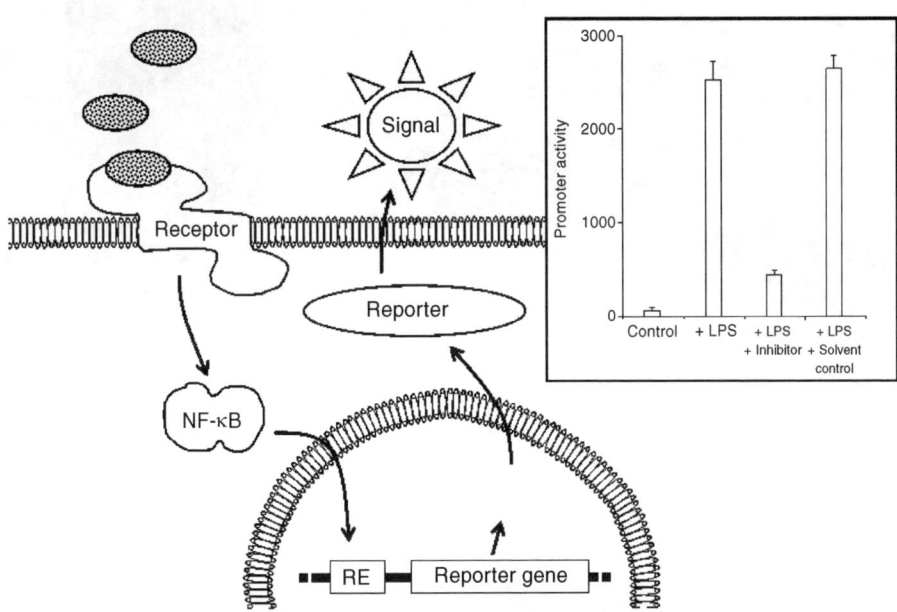

Fig. 4.1 Example of a reporter gene assay to monitor inflammatory responses. One of the main events in the pro-inflammatory response of macrophages is the activation of the transcription factor NF-κB. This cell-based assay is designed to monitor this process. The basis is formed by a macrophage-like cell line that was stably transfected with a reporter gene (luciferase) under the control of a promoter that contains NF-κB responsive elements (RE). When these cells are exposed to pro-inflammatory substances (e.g. LPS), this will lead to signal transduction from the receptor into the cell towards NF-κB, which in turn is activated and induces the transcription of luciferase. The expression of the luciferase protein is monitored by measuring the luciferase activity using luminometry. The inset shows an example of an anti-inflammatory inhibitor that strongly reduces luciferase activity even in the presence of LPS, whereas the solvent control does not have an effect. This system is useful to screen for anti-inflammatory plant-derived substances that target the NF-κB pathway.

et al., 2003). Adhesion molecules that are involved in the translocation of immune cells from the circulation into the sites of inflammation are also promising drug targets (Yusuf-Makagiansar *et al.*, 2002). Other important targets are the signalling molecules themselves, such as cytokines and chemokines. The best-known example of this is TNFα: therapies based on the specific inhibition of TNFα have proven to be very efficacious for Crohn's disease and rheumatoid arthritis (Elliot *et al.*, 1994; Targan *et al.*, 1997).

4.6.2 Genomics-based screening

A much more complex experiment to test whether plant-derived extracts possess anti-inflammatory activity is genomics-based screening. The basic principle is that, instead of just one or two targets being screened, the whole transcriptome, proteome or metabolome is analysed. As in 'normal' screening, an *in vitro* model system is usually used. This can be a macrophage, lymphocyte or any other type of relevant cell line, but it can also be applied to whole blood. The advantage of cell lines is that they provide a relatively stable background in which the experiments are performed.

In the example described here, a macrophage-like cell line was used. The cell line, U937 (Ralph *et al.*, 1976), can be cultured in such a manner that it adopts a macrophage-like phenotype. This can be verified using macrophage-specific markers: in our case we

48 Handbook of herbs and spices

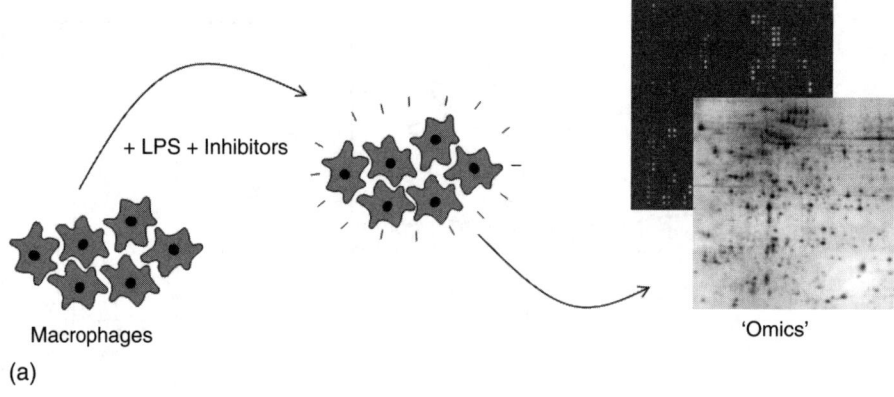

(a)

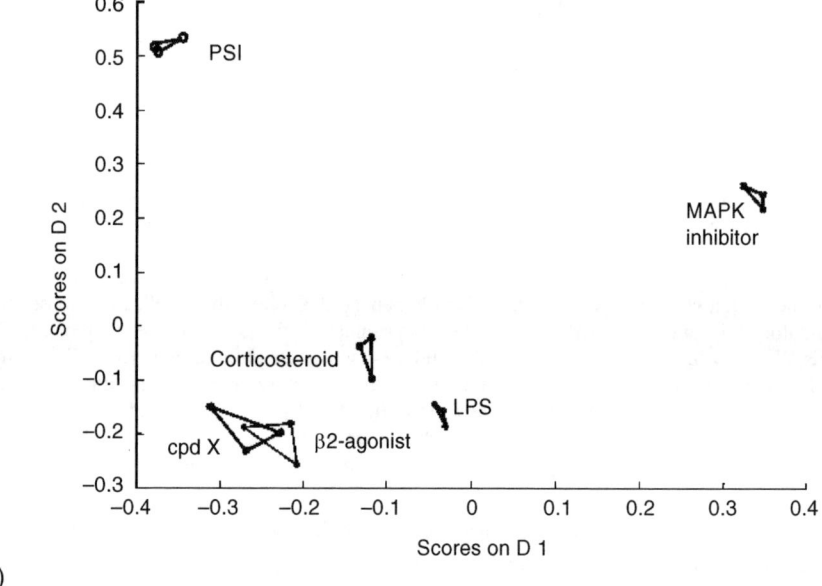

(b)

Fig. 4.2 Genomics-based screening. This experiment was performed to analyse the putative anti-inflammatory properties of an unknown plant-derived compound (cpd X). (a) In this example, macrophages were used as a model system. The macrophages were stimulated with LPS, a potent pro-inflammatory substance derived from *E. coli*. Stimulation was performed in the absence or presence of several different anti-inflammatory compounds, including the compound under investigation. In response to these treatments, the cells start producing a wide range of different mRNA and proteins. The complete set of proteins and mRNAs produced in response to the treatment can be regarded as characteristic for the treatment. To monitor this output, protein and mRNA expression patterns are analysed by proteomics (2D gel electrophoresis) and transcriptomics (cDNA arrays), respectively. (b) The enormous amount of data generated by proteomics and transcriptomics was analysed first by using software packages specifically designed to analyse data from both technologies. This resulted in normalized data sets from which outliers and other aberrant data were removed. Subsequently, the data were analysed by pattern recognition (multivariate analysis), which resulted in the two-dimensional presentation given here. Each dot represents a data set, for each treatment three independent data sets were generated. This representation clearly shows that different data sets can be grouped according to the anti-inflammatory compound that was added to the cells. The compound under investigation (cpd X) clearly overlaps with the data set derived from the cells treated with a beta-agonist; therefore, it is very likely that the anti-inflammatory mechanism of this compound bears resemblance to the mechanism of beta-agonists. (PSI = proteasome-specific inhibitor)

analysed the proteome and found several proteins that are up-regulated in differentiated macrophages, for example Cathepsin B (Krause *et al.*, 1996). The differentiated macrophages are subsequently activated with a pro-inflammatory compound. Usually lipopolysaccharide (LPS) is used, although other, less subtle, stimuli may also be used. Activation of the macrophages gives rise to a plethora of molecular responses, including the release of cytokines and prostaglandins, the induction of pro-inflammatory enzymes and many other proteins, mRNA molecules and metabolites. When testing an anti-inflammatory substance, such as a plant extract, this will be co-incubated together with the stimulus. An active substance will give rise to specific molecular effects, i.e. certain genes will no longer be activated, whereas others may be specifically up-regulated. The effects are monitored by determining the gene expression, protein and/or metabolite expression profiles (transcriptomics, proteomics and metabolomics, respectively). This system has been validated by testing the effects of several known anti-inflammatory compounds, including corticosteroids, beta-agonists and mitogen-activated protein kinase (MAPK) inhibitors. Although these different inhibitors have many molecular responses in common (e.g. inhibition of TNFα release), each inhibitor will give rise to a specific modulation of metabolite, mRNA and protein expression profile.

The very complex data provided by this type of assay are analysed by specialized data analysis tools, including multivariate analysis. The data analysis is a key element for this approach, since it will give information not only on whether a compound/extract is anti-inflammatory or not, but also on the molecular pathways that are affected by the compounds/extracts under study. It will also provide detailed information on the similarities and differences compared with the molecular responses of well-characterized anti-inflammatory compounds of which the molecular target is known. See Fig. 4.2 for a schematic overview of this system. It should be mentioned that the term 'screening' is not really apt here, since the throughput of this type of assay is very low. However, this set-up does not require relatively pure fractions to be tested: complex mixtures can also be used, thus reducing the number of samples to be tested. The amount of information that this approach provides is enormous. Not only will it tell us whether a plant extract is anti-inflammatory or not, but it will also give information on the possible mechanisms that underlie the effects. There are several applications for this technology. For example, it can be used to support claims of health-promoting products. Furthermore, it can be very useful to give information on the mode of action of new anti-inflammatory drugs. Finally, in cases where new anti-inflammatory compounds are being extracted from samples, it can be used to identify these by combining the screening results with plant metabolite profiles and analysing these using complex data-analysis tools.

4.7 Future trends

4.7.1 Technological: automation, miniaturization, novel detection methods
The trend in HTS has for long been driven by a demand to increase the speed and accuracy of screening assays. This has lead to the introduction of robotized screening facilities that operate virtually 24 hours a day, screening hundreds of thousands of compounds per day. Nowadays, the bottleneck is not so much the throughput of the screening but more the need for new compounds that can be screened for possible bioactivities, the development of new assays and the interpretation of the screening results. Moreover, experiments that classically were performed in the drug development stage, such as metabolic stability and other toxicity tests, are more and more transformed into rapid assays that can be performed at a very early stage in the drug discovery process.

4.7.2 Induced diversity

It is common knowledge that the vast number of different plant species in tropical rain forests are a rich source of plant metabolites that possess interesting pharmacological activities. The mining of this enormous storehouse of potentially interesting metabolites is a big challenge. Recently, at TNO Pharma we have developed a new technology platform that makes use of domesticated plants or crops, such as tobacco, arabidopsis, tomato and so on, to induce the production of a wide variety of metabolites in these plants. The principle of this so-called 'induced diversity' starts by the preparation of plant callus cultures. These can be cultivated in an unlimited variety of ways, in order to evoke the production of a wide variety of metabolites in the callus cultures. Culturing conditions such as light/dark, light wavelength, temperature, use of plant pathogen, and so on can all be used in any combination to create molecular diversity. Cultures can be prepared in 96-well plates, in which each well can be cultured in a different way, thus creating in principle 96 different sets of metabolites per plate. These cultures can subsequently be used to prepare extracts and screen for the presence of metabolites with a bioactivity of interest. This will result in a number of 'hits', that is plant callus cultures with the strongest effect in the screening assay. The culturing conditions of these callus cultures can subsequently be further refined to optimize the production of metabolites responsible for the bioactivity being screened for.

4.7.3 High information content screening/systems biology

Traditionally, screening methods were designed to provide information on whether a compound does or does not affect a drug target using just one output (e.g. optical density). The pressure to characterize drug candidates in the earliest stages of the drug discovery programme has, among others, led to the emerging of screening methods that provide more information than just a yes or no answer.

One of these recent trends is to use imaging technologies as a read-out of the screening. Instead of assaying the activity of a single enzyme or receptor, automated imaging-based screening is suitable to screen for biological processes, such as apoptosis, receptor distribution, transcription factor translocation, changes in cell morphology, and so on. In theory, assays can be developed for any biological process that can be visualized microscopically in cultured cells. Some biological processes can be analysed using time intervals, providing the possibility of screening for cell motility and other dynamic processes. A major advantage of this technology is that the screening is not limited to a single molecular target. Nevertheless, the throughput reached is fairly high, since the technology is available to perform imaging-based screening in 1536 well-plates, making it possible to perform this type of assay in a high-throughput format.

Another trend is to use arrays or profiles to read-out screens. This type of screen provides output on a large number of variables, such as a collection of mRNAs, proteins or metabolites. Examples are the use of cDNA arrays to screen for mRNA expression patterns and the use of liquid chromatography–mass spectrometry (LC–MS) to screen for changes in the production of particular classes of metabolites. An example of this type of screening is shown in Section 4.6.2. The throughput of this type of screening is low, although there is certainly a trend to speed up the processes by automation of the various steps in the screening process, not only the laboratory experiments, but also the data interpretation and/or database mining, which still is rather time-consuming.

4.8 Sources of further information

The Society for Biomolecular Screening is a useful source for more information on screening. The SBS produces the *Journal of Biomolecular Screening*, in which papers and editorial commentary are published that emphasize scientific and technical applications and advances in the field of HTS. A*ssay and Drug Development Technologies* publishes papers on early-stage screening techniques and tools that optimize the identification of novel drug leads and targets for new drug development. *Analytical Biochemistry* emphasizes analytical methods in the biological and biochemical sciences, and has a more broad perspective than the two journals mentioned above. In addition, pharmacological journals may also be of interest to the reader, such as *Biochemical Pharmacology*, *Current Opinion in Pharmacology*, *European Journal of Pharmacology*, *Journal of Pharmacology and Experimental Therapeutics*, and *Molecular Pharmacology*.

4.9 References

BERGER, J., HAUBER, J., HAUBER, R., GEIGER, R., and CULLEN, B.R. (1988), 'Secreted placental alkaline phosphatase: a powerful new quantitative indicator of gene expression in eukaryotic cells'. *Gene* **66**(1): 1–10.

BUTTON, D. and BROWNSTEIN, M. (1993), 'Aequorin-expressing mammalian cell lines used to report Ca^{2+} mobilization'. *Cell Calcium* **14**: 663–71.

CHALFIE, M., TU, Y., EUSKIRCHEN, G., WARD ,W.W. and PRASHER, D.C. (1994), 'Green fluorescent protein as a marker for gene expression'. *Science* **263**(5148): 802–5.

DENIZOT, F. and LANG, R. (1986), 'Rapid colorimetric assay for cell growth and survival. Modifications to the tetrazolium dye procedure giving improved sensitivity and reliability'. *J Immunol Methods* **89**(2): 271–7.

ELLIOTT, M.J., MAINI, R.N., FELDMANN, M., KALDEN, J.R., ANTONI, C., SMOLEN, J.S., LEEB, B., BREEDVELD, F.C., MACFARLANE, J.D., BIJL, H. and WOODY, J.N. (1994), 'Randomised double-blind comparison of chimeric monoclonal antibody to tumour necrosis factor alpha (cA2) versus placebo in rheumatoid arthritis'. *Lancet* **344**(8930): 1105–10.

FLANAGAN, W.M. and WAGNER, E.K. (1987), 'A bi-functional reporter plasmid for the simultaneous transient expression assay of two herpes simplex virus promoters'. *Virus Genes* **1**(1): 61–71.

HART, H.E. and GREENWALD, E.B. (1979), 'Scintillation proximity assay (SPA) – a new method of immunoassay. Direct and inhibition mode detection with human albumin and rabbit antihuman albumin'. *Mol Immunol* **16**(4): 265–7.

HOLGATE, S.T, PETERS-GOLDEN, M., PANETTIERI, R.A. and HENDERSON, W.R. JR (2003), 'Roles of cysteinyl leukotrienes in airway inflammation, smooth muscle function, and remodeling'. *J Allergy Clin Immunol* **111**(1 Suppl): S18–34.

KESTON, A.S. and BRANDT, R. (1965), 'The fluorometric analysis of ultramicro quantities of hydrogen peroxide'. *Anal. Biochem* **11**: 1–5.

KOJIMA, H., NAKATSUBO, N., KIKUCHI, K., KAWAHARA, S., KIRINO, Y., NAGOSHI, H., HIRATA, Y. and NAGANO, T. (1998), 'Detection and imaging of nitric oxide with novel fluorescent indicators: diaminofluoresceins'. *Anal Chem* **70**(13); 2446–53.

KRAUSE, S.W., REHLI, M., KREUTZ, M., SCHWARZFISCHER, L., PAULAUSKIS, J.D. and ANDREESEN, R. (1996), 'Differential screening identifies genetic markers of monocyte to macrophage maturation'. *J Leukoc Biol* **60**(4): 540–5.

MINTA, A., KAO, J.P. and TSIEN, R.Y. (1989), 'Fluorescent indicators for cytosolic calcium based on rhodamine and fluorescein chromophores'. *J Biol Chem* **264**(14): 8171–8.

ONUFFER, J.J. and HORUK, R. (2002), 'Chemokines, chemokine receptors and small-molecule antagonists: recent developments'. *Trends Pharmacol Sci* **23**(10): 459–67.

RALPH, P., MOORE, M.A. and NILSSON, K. (1976), 'Lysozyme synthesis by established human and murine histiocytic lymphoma cell lines'. *J Exp Med* **143**(6): 1528–33.

SHEU, Y.-A., KRICKA, L.J. and PRITCHETT, D.B. (1993), 'Measurement of intracellular calcium using bioluminescent aequorin expressed in human cells'. *Anal Biochem* **209**: 343–7

SILLS, M.A., WEISS, D., PHAM, Q., SCHWEITZER, R., WU. X. and WU, J.J. (2002), 'Comparison of assay technologies for a tyrosine kinase assay generates different results in high throughput screening'. *J Biomol Screen* **7**(3); 191–214.

TARGAN, S.R., HANAUER, S.B., VAN DEVENTER, S.J., MAYER, L., PRESENT, D.H., BRAAKMAN, T., DEWOODY, K.L., SCHAIBLE, T.F. and RUTGEERTS, P.J. (1997), 'A short-term study of chimeric monoclonal antibody cA2 to tumor necrosis factor alpha for Crohn's disease'. *N Engl J Med* **337**(15): 1029–35.

DE WET, J.R., WOOD, K.V., DELUCA, M., HELINSKI, D.R. and SUBRAMANI, S. (1987), 'Firefly luciferase gene: structure and expression in mammalian cells'. *Mol Cell Biol* **7**(2): 725–37.

YUSUF-MAKAGIANSAR, H., ANDERSON, M.E., YAKOVLEVA, T.V., MURRAY, J.S. and SIAHAAN, T.J. (2002), 'Inhibition of LFA-1/ICAM-1 and VLA-4/VCAM-1 as a therapeutic approach to inflammation and autoimmune diseases'. *Med Res Rev* **22**(2): 146–67.

ZHANG, J.H., CHUNG, T.D. and OLDENBURG, K.R. (1999), 'A simple statistical parameter for use in evaluation and validation of high throughput screening assays'. *J Biomol Screen* **4**(2): 67–73.

5
Under-utilized herbs and spices

P. N. Ravindran and Geetha S. Pillai, Centre for Medicinal Plants Research, India and K. Nirmal Babu, Indian Institute of Spices Research, India

5.1 Introduction

In ancient times spices and herbs were valued as basic ingredients of incense, embalming preservatives, ointments, perfumes, antidotes against poisons, cosmetics and medicines, and were used less in culinary preparations. A notable use of spices and herbs in ancient and medieval times was for the treatment of a variety of illnesses. Subsequently, spices and herbs came to be used to flavour food and beverages. In the course of time, spices and herbs were shown to be useful not only for making food palatable, but also in retarding or preventing rancidity and spoilage. This knowledge acted as a catalyst for the use of spices in a variety of processed foods. Based on use, herbs and spices are classified as culinary, cosmetic and pharmaceutical. In the modern world spices have wide affiliation in the culinary art of people around the world, and are used in the food industry for flavouring and seasoning, as well as in pharmaceutical preparations in the traditional systems of medicine and in beauty care. Spices and herbs are useful because of the chemical constituents contained in the form of essential oil, oleoresin, oleogum and resins, which impart flavour, pungency and colour to prepared dishes.

The International Organization for Standardization (ISO) lists 112 plant species that are used as spices and herbs. Among these, a few are very widely used and grown commercially in many countries, a few are less widely used but are well-known, while others are less known and are under-utilized. Such under-utilized herbs and spices are indeed valuable, not only as spices for flavouring dishes, but also as medicinal plants of great importance. A list of such under-utilized herbs and spices is given in Table 5.1. This chapter deals briefly with some of the more important under-utilized herbs and spices. A few have already been dealt with in this volume as well as in Volume 1.

5.2 Sweet flag

Sweet flag is the rhizome of *Acorus calamus* Linn. of the family Acoraceae and is highly valued as herbal medicine in India and other European countries. It is used as an ingredient

Table 5.1 List of some of the under-utilized herbs and spices

SI No.	Botanical name	Common name	Family	Part used
1.	*Acorus calamus* L.	Sweet flag	Acoraceae	Rhizome
2.	*Alpinia galanga* Willd.	Greater galangal	Zingiberaceae	Rhizome
3.	*Angelica archangelica* L.	Garden angelica	Apiaceae	Root
4.	*Armoracia rusticana* Gart.	Horseradish	Brassicaceae	Root
5.	*Bunium persicum* (Bosis) B Fedtsh.	Black caraway	Apiaceae	Seed, tuber
6.	*Capparis spinosa* L.	Caper	Capparidaceae	Unopened flower buds
7.	*Carum bulbocastanum* L	Black caraway	Apiaceae	Fruit, Bulb
8.	*Ferula asafoetida* L.	Asafoetida	Apiaceae	Oleogum
9.	*Garcinia gummi-gutta* (L.) N. Robson	Malabar tamarind	Clusiaceae	Pericarp lobes (fruit rind)
10.	*G. indica* Choisy	Kokum	Clusiaceae	Pericarp lobes (fruit rind)
11.	*Hyssopus communis* L. syn *H. officianalis*	Hyssop	Lamiaceae	Leaf
12.	*Kaempferia galanga* L.	Galangal	Zingiberaceae	Rhizome, tubers
13.	*Levisticum officianale*	Lovage	Apiaceae	Fruit, leaf
14.	*Marjorana hortensis* (*Origanum marjorana*)	Sweet marjoram	Lamiaceae	Leaf and flowering top
15.	*Murraya koenigii* L.	Curry leaf	Rutaceae	leaf
16.	*Nigella sativa* L.	Black cumin	Ranunculaceae	Seed
17.	*Pandanus amaryllifolius*	Pandan wangi	Pandanaceae	Leaf
18.	*Papaver somniferum* L.	Poppy seed	Papaveraceae	Seed
19.	*Piper betle* L.	Betel leaf	Piperaceae	Leaf
20.	*Punica granatum* L.	Pomegranate	Punicaceae	Seed dried with flesh
21.	*Rosmarinus officinalis*	Rosemary	Lamiaceae	Terminal shoot, leaf
22.	*Salvia officinalis*	Garden sage	Lamiaceae	Terminal shoot, leaf
23.	*Satureja hortensis* L.	Summer savory	Lamiaceae	Leaf, flowering top
24.	*Satureja montana*	Winter savory	Lamiaceae	Leaf, flowering top
25.	*Schinus terebenthifolius*	Brazilian pepper (Pink pepper)	Anacardiaceae	Fruit
24.	*Sinapis alba*	White mustard	Brassicaceae	Seed
25.	*Tamarindus indica* L.	Tamarind	Caesalpineaceae	Fruit pulp
26.	*Thymus vulgaris*	Thyme	Lamiaceae	Terminal shoot, leaf
27.	*Trachyspermum ammi.* (L) Sprague ex Tussil	Ajowan	Apiaceae	Fruit
28.	*Trigonella foenum-graecum*	Fenugreek	Fabaceae	Seed, leaf
29.	*Xylopia aethiopica*	Guinean pepper	Annonaceae	Fruit
30.	*Zanthoxylum piperitum*	Japanese pepper	Rutaceae	Fruit and rind

of several drugs of the *Unani, Ayurvedic* and modern systems of medicine. It is also well known for its insecticidal properties. The word *Acorus* is derived from *kore*, meaning pupil, and refers to the alleged ophthalmic virtues of the plant.

5.2.1 Origin and distribution

Acorus is native to most northern latitude countries around the world and may have been widely dispersed around the USA by Native Americans who planted the roots along their migratory paths to be harvested as needed. The species *A. calamus* is native to the southeastern USA, growing wild in wet areas in marshes and ditches.

Acorus is found wild or cultivated throughout India and Ceylon at up to 1800 m (6000 feet) height in the Himalayan region. It is a promising crop, especially for marshy land. In

India, *A. calamus* is grown abundantly in the marshy tracts of Kashmir, in certain areas of Manipur, the Naga Hills and Sikkim.

5.2.2 Botany and description

Acorus is an attractive, perennial, herbaceous, aquatic, marshy plant. This species inhabits perpetually wet areas such as the edges of streams and around ponds and lakes, in ditches and seeps. It is a grass-like, rhizome-forming, semi-aquatic perennial herb that can grow up to 2 m high. The plant has a creeping and much branched aromatic rhizome. The rhizome is cylindrical; light brown or white and spongy in colour.

The leaves are thick, erect and sword-shaped with crimped edges. The leaves, when bruised, emit a strong scent. *Acorus* produces small, yellow flowers arranged on a spike. Plants rarely flower or set seed.

Acorus is a rather remarkable plant in a number of respects. Until recently, it was just another member of the family Araceae, one of the larger and more complicated monocot families. Upon investigation of its morphology, anatomy and DNA sequences, it now appears that *Acorus* is the most primitive monocot and may represent an early stage in the evolution of the monocots (Albertazzi *et al.*, 1998, Duvall *et al.*, 1993; Duvall, 2001). Acorus is now included in a separate family, Acoraceae.

The rhizome is light brown with long internodes, root and leaf scars and a soothing aromatic odour. The transverse section shows narrow cortical and large stellar regions. The cortex consists of thin-walled parenchymatous cells arranged in chains, leaving large intercellular spaces, sheathed collateral vascular bundles and bundles of fibres. Endodermal cells are barrel shaped and possess abundant starch grains. Large oil cells with yellowish contents and cells containing dark brown oleoresin and starch grains are scattered in the ground tissue of both cortex and stele. Solitary polygonal crystals of calcium oxalate are present in each cell of the storied row of cells running parallel to the fibres (Sharma *et al.*, 2000). Calquist and Schneider (1997) demonstrated the vessels in the metaxylem of both roots and rhizomes by scanning electron microscopy (SEM), and the end walls of the vessel elements are characterized by perforations that retain porose pit membranes and are interpreted as a primitive character.

On microscopic examination the powder appears yellowish-white and consists of masses of whole or broken, oval-shaped parenchymatous cells. Some of these cells contain yellow-brown oleoresin, packed with small spherical starch grains. A few xylem elements in groups of vessels with annular thickening are also seen (Karnick, 1994a).

5.2.3 Cultivation and production

Acorus is propagated vegetatively. Sprouted rhizomes collected from the vigorously growing mother plants are used as planting material. About 80 000 propagules are required for one hectare of land. The planting time is June–July. The rhizome bits are planted in about 6 cm deep furrows with a spacing of 30 cm between the rows and 35 cm between the plants. Application of farmyard manure or compost, 8–10 tonnes/hectare supplemented with organic fertilizer is needed for good growth. For satisfactory cultivation and yield application of 100 kg/ha nitrogen is recommended (Tiwari *et al.*, 2000; Kumar *et al.*, 2000).

Propagation of sweet flag through tissue culture was reported by Hettiarachchi *et al.* (1997); Harikrishnan *et al.* (1999); Kulkarni and Rao (1999) and Rani *et al.* (2000). All the authors used Murashigue and Skoog basal medium supplemented with varying levels of BA (benzyl adenine) and NAA (α-naphthalene acetic acid). The cultures initiated from rhizome buds were multiplied, rooted and successfully established in soil. This method can be exploited for large-scale multiplication of quality planting material in this crop.

5.2.4 Chemistry

The root essential oil contains monoterpene hydrocarbons, sesquiterpine ketones (*trans* or α) asarone (2,4,5-trimethoxy-1-propenylbenzene) and β-asarone (*cis*-isomer). The American variety is consistently tested free of the carcinogenic β-asarone, whereas the Asian varieties contain varying amounts of β-asarone, and cause a more sedative feeling when ingested. European varieties of sweet flag have yielded various sesquiterpenoids with psychoactive or medicinal properties.

The volatile oil obtained by steam distillation of the rhizome was purified and subjected to liquid–gas chromatography. A total of 93 volatile components were detected from the Indian variety, of which β-asarone was found to be the major component. European calamus yielded 184 volatile components, including 67 hydrocarbons, 35 carbonyl compounds, 56 alcohols, 8 phenols, 2 furans and 4 oxido compounds. Its oil yield varies with temperature and method of storage. Sweet flag leaves, rhizome and roots contain 0.22–0.89, 3.58–7.80 and 1.77–3.15 ml/l00 g dry matter, of essential oil, respectively. Tannic substances are in the rangeof 1.22–1.85, 0.63–1.05 and 0% respectively in leaves, rhizome and roots. Leaves contain vitamin C (Kumar *et al.*, 2000).

The variation of essential oils and their major constituents in *Acorus*, with respect to season and geographical areas, was analysed. The major components of volatile oil obtained from the same part of the plants from different geographical areas exhibited no change in chemical structure and the best season for cropping was found to be June (Kumar *et al.*, 2000). The calamus root oil obtained from the plants grown in various geographical areas such as China, Japan (wild and cultivated types), Asian regions, Canada, Bangladesh and also the commercial sample from Germany were subjected to analysis by various researchers. It was found that there is variation in the presence and quantity of the components in those samples (Lawrence, 2002). The comparative percentage composition of the major components of various collections of Japanese and Asian calamus root oil is given in Tables 5.2 and 5.3.

Petrikova *et al.* (2000) studied the essential oil concentration in *Acorus* collected from 13 locations in the Czech Republic and found that the oil contents were higher in spring crops (0.8–2.6%) than in autumn crops (1–1.8%). The concentration of β-asarone ranged from 0.07 to 0.41%.

The volatile oil of accessions collected from Jammu and Kashmir had some common constituents, such as palmitic acid, isoeugenol, butyric ester, asarone and hydrocarbons. Both the oils differed in some components: eugenol, calamol and azulene are present in Jammu collections, whereas Kashmir collections have heptylic acid, 2-pinene, camphor, calamene and azulene. The presence of 124 mg% of choline per 100 g was reported in the *Acorus* plants (Chaudhary *et al.*, 1957). Sikkim collections of *Acorus* are reported to have a higher oil content and a higher percentage of other major constituents (Agarwal, 1987). Chowdhury *et al.* (1993) identified 1-(1-acetoxy-2-propenyl)-4-hydroxybenzene in Indian calamus roots. Essential oil isolated from calamus roots grown in India, in alkaline soil rich in exchangeable sodium, was analysed by GC–MS (gas chromatography–mass spectroscopy) for its constituents and more than 25 compounds were detected. The major components detected were (*E*)-asarone (58%), (*Z*)-asarone (2%), asaronaldehyde (8%), α-terpineol (2%), calamol (2%), etc. (Chowdhary *et al.*, 1997). The accuracy of this study is doubtful as the authors have reported presence of high amount of (*E*)-asarone, but Indian calamus is rich

Table 5.2 Comparative percentage composition of the major components of Japanese calamus root oils

Compound	Chemotype 1 (6)*	Chemotype 2 (12)*	Chemotype 3 (2)*
(Z)-methyl isoeuginol	0–2.9	0–11.3	1.6–7.6
Epi-shyobunone	0.6–5.1	4.1–10.3	10.7–12.1
Shyobunone	1.0–4.8	7.1–15.2	14.3–22.1
Elemicin	1.4–1.7	0–3.7	0.7–2.2
Preisocalaminidiol	1.2–6.2	9.4–28.9	22.8–34.9
(Z)-asarone	64.7–92.1	23.5–48.7	6.0–8.1
(E)-asarone	1.8–10.0	1.4–13.8	2.6–4.1
Other constituents	0.4–10.9	7.7–34.4	24.1–26.1

*Number of samples.
Source: Sugimoto et al. (1997a).

Table 5.3 Comparative percentage composition of the major components calamus root oils of Asian origin

Compound	Source*							
	1 (1)†	2 (1)†	3 (1)†	4 (8)†	5 (3)†	6 (2)†	7 (1)†	8 (2)†
Methyl euginol	–	–	–	–	–	–	–	0–1.6
(Z)-methyl isoeuginol	1.2	1.6	0	0.3–9	0–0.7	0–2.9	87.3	0–1.3
Epi-shyobunone	1.8	0.7	3.4	0–1.1	0–3..5	2.2–3.0	–	–
(E)-methyl isoeuginol	–	–	–	–	–	–	–	0–1.6
Shyobunone	2.2	14.3	4.5	0–1.3	0–5.3	2.8–4.0	–	–
Elemicin	1.7	2.2	0.6	0–1.6	0–0.7	0.5–2.3	–	42.7–72.7
Preisocalaminidiol	3.0	34.9	1.4	0–4.5	5.0–12.0	4.4–7.7	–	–
(Z)-asarone	85.7	8.1	20.9	73.8–96.3	10.2–17.9	43.6–53.7	12.7	25.7–37.3
(E)-asarone	2.6	4.1	3.5	1.8–15.0	0–6.0	3.9–4.5	–	1.6–2.4
Other constituents	1.8	24.1	65.7	0–13.6	55.9–85.1	29.0–35.5	–	0–13.0

*1. Cultivated Osaka, Japan 2. Cultivated Hokkldo, Japan. 3. Cultivated Chongging, China. 4. Various Asian sources. 5. Three areas in China. 6. Henen, China. 7. Jllin, China. 8. Hubei, China.
† Number of samples.
Source: Sugimoto et al. (1997b).

in β-asarone (=Z-asarone). Wu *et al.* (1993) characterized a new compound from *Acorus* roots of Chinese origin using X-ray diffraction analysis and named it as calamensesquiterpinol. The same group of workers (Wu *et al.*, 1994) identified calamenidiol, isocalamenidiol and calamenene in *A. calamus* roots.

Sweet flag oil also contains acoradin, 2,4,5-trimethoxy-benzaldehyde; 2,5-dimethoxybenzoquinone; galangin; sitosterol; the phenylpropane derivatives isoeugenol methyl ether, γ-asarone, *cis*-asarone, *trans*-asarone and acoramone (Patra and Mitra, 1981). The compounds Z-3-(2,4,5-trimethoxyphenyl)-2-propenal and 2,3-dihydro-4,5,7-trimethoxy-l-ethyl-2-methyl-3-(2,4,5-trimethoxyphenyl) indene were also isolated from *Acorus* plants. A phenyl propane derivative 1-(*p*-hydroxyphenol)-*l*-(*o*-acetyl)prop-2-ene, was isolated from the rhizome of *Acorus*.

The *Acorus* oil of Dutch origin was subjected to fractionation after separation of acidic and phenolic substances. The lower boiling fraction contains terpenes. The sesquiterpenes diol-isocalamenidiol and three monocyclic sesquiterpenes were isolated from essential oil of Japanese plant rhizomes. Two new selinane-type sesquiterpenes, two new sesquiterpenic ketones and a new tropane were isolated from sweet flag oil, and their structures were also elucidated. The structure of the sesquiterpenic hydrocarbon, (–)cadala-1,4,9-triene, was determined from chemical and spectral data. Kumar *et al.* (2000) have made a detailed review on the chemistry of *Acorus*.

Various compounds such as calacone, telekin, isotelekin, calarene, monocyclic ketones – shyobunone, epishyobunone and isoshyobunone – were also isolated and their structure elucidated during the period from 1966 to 1968 (Rastogi and Mehrotra, 1990). (Z,Z)4,7-dicadienal was isolated from oil and synthesized; two new compounds: (Z)3-(2,4,5-trimethoxyphenyl)-2-propenal and 2,3-dihydro-4,5,7-trimethoxy-1-ethyl-2-methyl-3-(2,4,5-trimethoxyphenyl) indene, were isolated from rhizomes and their structures elucidated and confirmed by synthesis (Rastogi and Mehrotra, 1995).

Sugimoto *et al.* (1999) analysed the phylogenetic relationship of *A. gramineus* and three types of *A. calamus* by comparing the 700 bp sequence of a 5S-rRNA gene spacer region. *A. calamus* was classified into two chemotypes: chemotype A in which Z-asarone is the major essential oil constituent and chemotype B which contained mainly sesquiterpenoids. An intermediate type (M) of these two chemotypes in various ratios was also observed. The results revealed that the phylogenetic relationship predicted by the spacer region data correlated well with the essential oil chemotype pattern of *A. calamus*.

5.2.5 Functional properties and toxicology

The biological properties of calamus were reviewed by Kumar *et al.* (2000). Both the dried, pleasant smelling rhizome and its essential oil are used as aromatic, bitter carminative compositions and in bronchial troubles. The sedative-potentiating principle was found in the petroleum ether extract of rhizomes of *Acorus*. The same active principle was also obtained during steam distillation of its rhizomes. This fraction showed depressant action on normotensive dogs, inhibited the rate of contraction of frog and dog hearts, relaxed the tone of isolated intestine, uterus and bronchi and antagonized acetylcholine and histamine-induced spasm. The former action was not modified by atropinization or the latter by pre-treatment with antihistamines. Root extract exhibited antimicrobial activity against *Staphylococcus aureus*, *Escherichia coli* and *Aspergillus niger* (Rastogi and Mehrotra, 1995). The complete extract of *Acorus* rhizomes exhibited significant anti-bacterial and anti-inflammatory effects in experimental animals. The dichloromethane extract of rhizome recorded the highest aphidicidal activity. The extract was found to have fumigative toxicity to aphids and β-asarone was found to be an active ingredient in this extract. An ethanolic extract of rhizome was screened for central nervous system (CNS) activity against mice and rats. The extract exhibited a large number of actions similar to β-asarone but differed in several other respects, including the response to electroshock, apomorphine and isolation-induced aggressive behaviour, amphetamine toxicity in aggregated mice, behavioural despair syndrome in forced swimming, etc. The *Acorus* plant extract was screened for anti-implantation activity in female rats.

The insecticidal properties of alcohol extracts of rhizomes of sweet flag against the fully grown larvae of *Trogoderma granarium* was reported (Pal *et al.*, 1996). The sterilizing effect of *Acorus* oil on the female of *Trogoderma granarium* was attributed to the absorption of the terminal oocytes of the ovaries and disturbance to follicular epithelium. Oral administration of 2 ml of boiled coconut oil extract of the rhizome of *Acorus* showed anti-inflammatory activity in rats.

The rhizome extract of *Acorus* plant has produced a hypotensive response in dogs. The ethanolic extract of *Acorus* rhizome inhibited gastric secretion and protected gastroduodenal mucosa against the injuries caused by pyloric ligation, indomethacin, reserpine and cystamine administration. The extract had a highly protective effect against cytodestructive agents. Mutagenetic and DNA damaging activity was found in calamus oil, β-asarone and commercial calamus drugs (Ramos-Ocampo and Hsia, 1987). The essential oil obtained from *Acorus* showed spasmolytic activity in isolated organs of certain experimental animals. It was found that chromosome damage in the human lymphocytes is induced by β-asarone. Because of the genotoxicity potency of β-asarone, *Acorus* should be used at a very low concentration. Feeding studies in rat using Indian calamus oil (high β-asarone) had shown death, growth depression, hepatic and heart abnormalities and serious effusion in abdominal and/or peritoneal cavities.

The essential oil of *Acorus* has shown marked nematicidal activity against larvae of root knot nematode, *Meloidogyne. incognita*, the most menacing pest of Indian soils. The minimal lethal concentration of (Z/E) asarone mixture against the second stage larvae of dog roundworm *Toxocara canis* was 1.2 mM. The calamus oil (0.1%) effectively controlled the population of *Diplodia natalensis* and *Penicillium digitatum*. The insect growth-inhibiting activity was observed in the extract of *Acorus* plant. Bioactivation of *p*-asarone was mediated via insect mixed function oxidases.

There are various reports on the toxic effect of *Acorus* on stored pests and insects. Toxicity of *A. calamus* oil vapours and rhizome powder collected from different altitudes at different temperatures to stored pests and insects such as *Sitophilus granarius*, *S. oryzae*, *Callosobruchus maculatus*, *C. phaseoli*, *Lasioderma serricorne* and *Tribolium confusum* were reported (Schmidt, *et al.*, 1991; Su, 1991; Paneru *et al.*, 1997; Rahman and Schmidt, 1999). Reddy and Reddy (2000) described a treatment based on an *Acorus calamus* powder solution for the control of *Oryzaephilus surinamensis*, *Lasioderma serricorne*, *Araecerus fasciculatus* and *Tribolium castaneum* in stored turmeric rhizomes.

A-asarone produces stimulating effects and also sedative effects similar to reserpine and chlorpromazine. High antioxidant activity was reported from the ethanolic extracts of rhizomes of *Acorus* (Acuna *et al.*, 2002).

Acorus is considered unsafe for human consumption by the Food and Drug Administration (FDA) because massive doses given to laboratory rats over extended time periods proved to be carcinogenic (JECFA, 1981). Increased incidence of hepatomas was also observed in rats treated with β-asarone (Wiseman *et al.*, 1987).

FDA studies have shown that only calamus native to India contains the carcinogenic β-asarone. The North American variety contains only asarone. Calamus has been banned by the FDA as a food additive and within the last few years many herbal shops have stopped recommending or dispensing it. The presence of β-asarone in flavourings and other food ingredients with flavouring properties have been reviewed by the Scientific Committee on Food. The committee analysed the implications of the presence of β-asarone in food on human health in the light of the available information and recommended limitations in exposure and use levels (SCF, 2002). The use of *Acorus* is now more medicinal than as a spice.

5.2.6 Uses

Attempts have been made to utilize this plant's important biological activities in the development of herbal formulations for different kinds of ailments and uses. For skin care, a bath preparation has been prepared in which the ingredients are lactose and plant extracts. This preparation provides a moisturizing effect on the skin. A number of hair-care preparations have been formulated in which *Acorus* is one of the major constituents. The decoction of *Acorus* rhizome in combination with 0.25% solution of anesthesisn (1 : 1) is suggested for alvealitis prophylaxis.

Table 5.4 Some of the herbal tonic/medicinal drugs in which *Acorus* is one of the ingredients

SI no.	Drugs	Uses
1.	Krumina syrup	For intestinal diseases
2.	Antospray powder	Skin care for softness
3.	Automer drops	To check ear infections
4.	Cybil Tab	As tranquillizer (sedative)
5.	Libobel drops	In liver disorders
6.	Galachol	To enhance lactation
7.	Sepra Tab	As mental restorative
8.	Traquinil	As tranquillizer
9.	Ciladin	To cure pschychosometic disorders and use as tranquillizer
10.	Phortage	For sex tonic
11.	Suktin	Acidity, gastrocordiac syndrome flatulence, dyspepsia, gastroentritis and peptic ulcer, etc.
12.	Cilpazyme syrup	For constipation
13.	Remanil Tab	In swelling of bones, joints andrheumatic pains
14.	Nab Tab	In epilepsy

Source: Kumar *et al.* (2000).

The extract of *Acorus* rhizome is one of the major constituents in the formulation of many general tonics (Table 5.4). *Acorus* rhizome powder has been used in the preparation of antacid tablets with purgative properties. P-tab was found to be most effective drug for patients suffering from insomnia and irritability. This is composed of eight herbal drugs, of which *Acorus* is one. The water-ethanolic extract of *Acorus* plant was found to exhibit antioxidant property and can be used in the food industry as fat oxidant. In the formulations of mosquito repellent preparations, *Acorus* is a major constituent. *Acorus* is also used in waste-water treatment such as inactivated sludge treatment in the aeration tank; secondary setting and subsequent post-treatment for increased N and P compound removal and disinfection. *Acorus* cultivation in the treatment areas enhances N and P removal and disinfection. Treatment of clarified sewage leads to the removal of microorganisms.

5.3 Greater galangal

Alpinia galanga (L.) Willd belongs to the family Zingiberaceae, and is commonly known by such names as galangal, galanga, greater galangal and Java galangal. The related species *A. officinarum* is the lesser galangal.

5.3.1 Production and international trade

Data on production, consumption and trade are scarce and not reliable because traders make no distinction between *A. galanga* and *A. officinarum;* both are used as the source plant for the Ayurvedic raw drug *rasna*. India is a major supplier, along with Thailand and Indonesia (Scheffer and Jansen, 1999). However, its volatile oil attracts more international interest because of its high medicinal value (http://www.indianspices.com).

5.3.2 Origin and distribution

Greater galangal is native to Indonesia, but has become naturalized in many parts of South and South-East Asian countries. The oldest reports about its use and existence are from southern China and Java. It occurs frequently in the sub-Himalayan region of Bihar, West Bengal and Assam. It is currently cultivated in all South-east Asian countries, India, Bangladesh, China and Surinam (Scheffer and Jansen, 1999). It shows exuberant growth along the eastern Himalayas and in southwest India, cultivated throughout the Western Ghats (Warrier et al., 1994). India exports galanga in different forms (http://www.indian spices.com).

5.3.3 Botany and description

Greater galanagal is a perennial, robust, tillering, rhizomatous herb, which grows up to 3.5 m tall, with a subterranean, creeping, copiously branched aromatic rhizome. The rhizomes are 2.5–10.0 cm thick, reddish brown externally, and light orange brown internally. The aerial leafy stem (pseudostem) formed by the rolled leaf sheaths is erect. The leaves are 23–45 by 3.8–11.5 cm, oblong-lanceolate, acute and glabrous. The inflorescence is terminal, erect, many flowered, racemose, 10–30 × 5–7 cm, pubescent; the bracts are ovate, up to 2 cm long, each subtending a cincinnus of two to six greenish white flowers; the bracteoles are similar to the bracts but smaller; the flowers are fragrant, 3–4 cm long, yellow-white. The fruit is a globose to ellipsoidal capsule, 1–1.5 cm in diameter, orange-red to wine red.

The anatomy of the rhizome shows a central stele surrounded by an outer cortical zone. Fibrovascular bundles are distributed throughout the cortex and steel. Numerous resin canals are also present.

5.3.4 Chemistry

The composition of galangal rhizomes per 100 g dry matter is: moisture – 14 g, total ash – 9 g, matter soluble in 80% ethanol – 49 g, matter soluble in water – 19 g, total sugar–9 g, total nitrogen – 3 g, total protein – 16 g, essential oil content – 0.2–1.5% (dry wt). Fresh rhizomes yield about 0.1% of oil on steam distillation with a peculiar strong and spicy odour. Earlier investigations indicated camphor, 1,8-cineole (20–30%), methyl cinnamate (48%) and probably D-pinene as the oil components. 1'-Acetoxychavicol acetate, a component of newly dried rhizomes, is active against dermatophytes, and together with another compound, 1'-acetoxyeugenol acetate, exhibits anti-tumour activity in mice. The same compounds isolated from roots showed anti-ulcer activity in rats. Oil shows potential insecticide property. Galangal root, root oil and root oleoresin are given the regulatory status 'generally regarded as safe' (GRAS) in the USA (Scheffer and Jansen, 1999). The root contains a volatile oil (0.5–1.0%), resin, galangol, kaempferid, galangin, alpinin, etc. The active principles are the volatile oil and acrid resin. Galangin has been obtained synthetically (http://www.naturedirect2u.com).

Compounds such as 1'-acetoxy chavicol acetate, 1'-acetoxyeuginol acetate and 1'-hydroxychavicol acetate and two diterpenes – galanal A and galanal B – were isolated from seeds and their structures were elucidated. Galanolactones (E)-8-(17),12-labdadien-15,16-dial and (E)-8-(17)-epoxylabd-12-en-15,16-dial were isolated from seeds and characterized. A compound, di(p-hydroxy-cis-styryl) methane, was isolated along with p-hydroxy cinnamaldehyde from rhizomes (Rastogi and Mehrotra, 1995). Haraguchi et al. (1996) have isolated an antimicrobial diterpene (diterpene 1) from A. galanga and the structure was elucidated by spectral data and identified as (E)-8-beta,17-epoxylabd-12-ene-15,16-dial.

The volatile constituents of the rhizomes and leaves of A. galanga from the lower Himalayan region of India were analysed by GC and GC–MS. The main constituents identified in the rhizome were 1,8-ciniole, fenchyl acetate and β-pinene. The leaf oil contained 1,8-ciniole, β-pinene and camphor as the major constituents (Raina et al., 2002).

Jirovetz et al. (2003) investigated the essential oils of the leaves, stems, rhizomes and roots of *A. galanga* from southern India by GC–FID (flame iuonization detection), GC–MS and olfactometry. Mono- and sesquiterpenes and (*E*)-methyl cinnamate could be identified in all the four samples and these are responsible for the characteristic odour and the reported use in (folk) medicine as well as in food products. They found that the essential oil of *A. galanga* leaf is rich in 1,8-cineole (28.3%), camphor (15.6%), β-pinene (5.0%), (*E*)-methyl cinnamate (4.6%), bornyl acetate (4.3%) and guaiol (3.5%). The stem essential oil contains 1,8-cineole (31.1%), camphor (11.0%), (*E*)-methyl cinnamate (7.4%), guaiol (4.9%), bornyl acetate (3.6%), β-pinene (3.3%) and alpha-terpineol (3.3%). 1,8-Cineole (28.4%), α-fenchyl acetate (18.4%), camphor (7.7%), (*E*)-methyl cinnamate (4.2%) and guaiol (3.3%) are the main constituents of the rhizome essential oil. The root essential oil contains α-fenchyl acetate (40.9%), 1,8-cineole (9.4%), borneol (6.3%), bornyl acetate (5.4%) and elemol (3.1%).

5.3.5 Cultivation and production

A. galanga is found in wild/semi-wild and cultivated states. The plant requires sunny or moderately shady locations. Soils should be fertile, moist but not swampy. Sandy or clayey soils rich in organic matter and with a good drainage are preferred. Wild or semi-wild types occur in old clearings, thickets and forests. In the tropics, galangal occurs up to an altitude of 1200 m.

Long tips of rhizomes are used for propagation. Soil should be well tilled before planting. Alternatively, holes, 35 cm × 35 cm and 15–20 cm deep, are dug, filled with manure mixed with soil, inorganic fertilizers and lime (for acid soils). One piece of rhizome is planted per hole, and covered with mulch. Shoots from pieces of galanga rhizome emerge about one week after planting. About four weeks after planting, three or four leaves develop. Rhizomes develop quickly and reach their best harvest quality in three months after planting. If left too long, they become fibrous and large clumps will hamper harvesting. Seeds rarely reach maturity.

Often trenches are dug to drain the field after rainfall, as rhizomes do not develop under waterlogged conditions. Usually planted along the borders of gardens, in rows at distances of 0.5–1 m square. Weeding and subsequent earthing up are carried out, respectively, one or two months after planting.

5.3.6 Harvesting and processing

Harvesting is done usually three months after planting (during late summer or early autumn) for market purposes. Whole plants are pulled out, shoots cut off and rhizomes washed and cleaned. Rhizomes more than four months old turn woody, fibrous and spongy and lose their value as spice. For local use, plants are left in the field almost permanently and small quantities of good quality rhizomes can always be harvested. For production of essential oil, rhizomes are harvested when plants are more than seven months old. No reliable data are present on the yield (Scheffer and Jansen, 1999).

Harvested rhizomes are washed, trimmed, dried and marketed fresh or dried after packing (Scheffer and Jansen, 1999). The dried product is ground before use. Ground rhizomes are not traded in bulk as they may be adulterated. Essential oil is also a product.

5.3.7 End uses

Rhizomes are bitter, acrid, thermogenic, aromatic, nervine tonic, stimulant, repulsive, carminative, stomachic, disinfectant, aphrodisiac, expectorant, bronchodilator, febrifuge, anti-inflammatory and tonic (Warrier *et al.*, 1994). They have many applications in traditional medicine such as for skin diseases, indigestion, colic, dysentery, enlarged spleen, respiratory diseases, cancer of mouth and stomach, treatment for systemic infections and cholera, as an expectorant, after childbirth (Scheffer and Jansen, 1999), in vitiated conditions of *vāta* (all the body phenomena controlled by the central and autonomic nervous systems) and *kapha* (for the function of heat regulation, and formation of various preservative fluids such as mucus, synovia, etc.). The main function of *kapha* is to provide coordination of the body system and regularization of all biological activities. Galanga is also reported to be useful in the treatment of rheumatoid arthritis, inflammation, stomatopathy, pharyngopathy, cough, bronchitis, asthma, hiccough, dyspepsia, stomachalgia, obesity, cephalagia, diabetes, tubercular glands and intermittent fevers (Warrier *et al.*, 1994).

Rhizomes show antibacterial, antifungal, antiprotozoal and expectorant activities (Scheffer and Jansen, 1999). Galangal's antibacterial effect acts against germs, such as *Streptococci, Staphylococci* and coliform bacteria. This plant is used to treat loss of appetite, upper abdominal pain, and sluggish digestion. It relieves spasms, combats inflammation and has stress-reducing properties.

In Asia, galangal is also used for arthritis, diabetes, stomach problems and difficulty in swallowing. It is especially useful in flatulence, dyspepsia, nausea, vomiting and sickness of the stomach, being recommended as a remedy for seasickness. It tones the tissues and is sometimes prescribed in fever. Galangal is used in cattle medicine, and the Arabs use it to make their horses fiery. It is included in several compound preparations. The reddish-brown powder is used as a snuff for catarrh (http://www.naturedirect2u.com). The young rhizome is a spice and is used to flavour various dishes in Malaysia, Thailand, Indonesia and China.

5.3.8 Functional properties and toxicology

Antifungal activity of *A. galanga* and the competition for incorporation of unsaturated fatty acids in cell growth was reported by Haraguchi *et al.* (1996). They isolated an antimicrobial diterpene (diterpene 1) and found that this compound synergistically enhanced the antifungal activity of quercetin and chalcone against *Candida albicans*. Its antifungal activity was reversed by unsaturated fatty acids. Protoplasts of *C. albicans* were lysed by diterpene 1. These results suggest that the antifungal activity of this compound is due to a change of membrane permeability arising from membrane lipid alternation. The ethanolic extract of *A. galanga* rhizome exhibited hypolipidaemic activity *in vivo*. The oral administration of the extracts (20 mg/day) effectively lowered the serum and tissue levels of total cholesterol, triglycerides and phospholipids and significantly increased the serum levels of high-density lipoproteins (HDL) in high-cholesterol fed white wistar rats over a period of four weeks. The study suggests that galanga is useful in various lipid disorders especially atherosclerosis (Achuthan and Padikkala, 1997).

A novel composition of aromatic and terpinoid compounds present in *A. galanga* showed synergistic effects with respect to immunomodulation, and effectively suppressed hypersensitivity reactions. These compounds are used for preparing medicaments for the treatment or prevention of allergic reactions and conditions, such as asthma, allergic rhinitis, anaphylaxis and autoimmune disorders, such as ulcerative colitis, rheumatoid arthritis, as well as for the alleviation of pain (Weidner *et al.*, 2002). The constituents isolated from the seeds of *A. galanga* are reported to exhibit anti-ulcer activities (Mitsui *et al.*, 1976).

5.3.9 Quality issues

Dried powdered rhizome is sometimes adulterated with lesser galangal (*A. officinarum*) (Scheffer and Jansen, 1999). Other species of *Alpinia* such as *A. calcarata*, *A. conchigera*, *A. mutica*, *A. nigra*, *A. rafflesiana* and *A. scabra* are sometimes substituted for the genuine drug. Inferior ginger and rhizomes of *Acorus calamus* are also used as adulterants (Anon., 1985).

The fruits of *A. galanga* are used in traditional Chinese medicine, but the dry fruits are easy to adulterate with other species and so adulterated substances may be used as a medicine in local areas. The dry fruits of the adulterants are very similar in odour, morphology, chemical constituents and anatomical characters and are difficult to distinguish. Zhao *et al.* (2001) characterized *A. galanga* and the species used as adulterants using the nuclear ribosomal DNA internal transcribed spacer (nrDNA ITS) region sequences: the molecular markers are used to distinguish the drug at DNA level.

5.4 Angelica

The genus *Angelica* is a unique member of the family Apiaceae (formerly Umbelliferae), for its pervading aromatic odour, entirely different from other members such as fennel, parsley, anise and caraway: here even the roots are aromatic. There are more than 40 species of *Angelica*, but *A. archangelica* (syn. *A. officinalis* Moench; *Archangelica officinalis* (Moench) Hoffm.) is the only one officially used in medicine and as a spice. As the name indicates, the folklore of North European countries and nations affirms to its merits as a protection against communicable disease, for purifying the blood, and for curing every conceivable malady. According to one Western legend, *Angelica* was revealed in a dream by an angel as a gift of Mother Angel to cure the plague. Another explanation of the name of this plant is that it blooms on the day of Michael the Archangel, and is on that account an additive against evil spirits and witchcraft. It was valued so much that it was called 'The Root of the Holy Ghost' (Grieve, 1931).

The fruit, young stem and roots are used as food additives and for flavouring (Anon., 2001), for human consumption as a beverage base such as herbal tea and liquors, in medicines (Duke, 1985) and also as an ornamental.

5.4.1 Origin and distribution

The crop is indigenous to Northern Europe and distributed in Temperate Asia – in regions such as Georgia, the Russian Federation (Ciscaucasia, Western Siberia) and also European countries such as Belarus, the Czech and Slovak republics, Denmark, Estonia, Finland, Germany, Iceland, Latvia, Lithuania, the Netherlands; Norway, Sweden, Ukraine, and also the European part of the Russian Federation and New Zealand. The crop seems to be widely naturalized elsewhere (Wiersema and Leon, 1999). In several London squares and parks, angelica has continued to grow, self-sown, for several generations as a garden escape; in some cases it is appreciated as a useful foliage plant, in others, it is treated rather as an intruding weed. It was exceedingly common on the slopes bordering the Tower of London on the north and west sides and the inhabitants held the plant in high repute, both for its culinary and medicinal use.

Angelica grows in temperate regions at altitude 1000–4000 m and is commercially grown in Belgium, Hungary and Germany. There are 30 or more varieties of angelicas growing around the world. China alone boasts at least ten varieties. In India, angelica is found in a natural state in Kashmir (near water channels) at altitudes of 1000–3900 m, in Himachal Pradesh, Uttar Pradesh, at altitudes of 1800–3700 m and Sikkim at 3000–3300 m. It has also been reported from Rajasthan at an attitude of 1200 m and from Bihar. Both *A. archangelica* and its related species *A. glauca* are aromatic and used as herbal spices.

5.4.2 Botany and description

Angelica is a stout, aromatic perennial herbaceous plant that flowers every two years. Its habit is confusing, and it is a biennial in the botanical sense of that term. The seedlings attain maturity within 12 months. The plants usually set seed in their third year of growth and the plants die off after seeding once. Rarely, plants flower in their second year.

Angelica is glabrous and it grows to a height of 2–3 m. The stems are hollow, round, jointed, channelled, smooth and purplish; the leaves are ovate, 30–90 cm, 2–3 pinnate, ultimate pinna toothed, leaflets few, ovate or lanceolate. The root is tuberous, aromatic warm, pungent and of bitter sweet taste. The roots of the common angelica are long and spindle-shaped, thick and fleshy and with many long, descending rootlets.

Angelica blossoms in July. The flowers are small and numerous, yellowish or greenish-white in colour, and are grouped into large, globular umbels. Fruits are pale yellow, oblong, 4–6 mm (1/6 to 1/4 inch) in length when ripe, with membranous edges, flattened on one side and convex on the other, which bear three prominent ribs. Both the odour and taste of the fruits are pleasantly aromatic. The schizocarps are oblong or sub-quadrate, somewhat corky, 13 mm × 6 mm. The seeds are dorsally much compressed.

5.4.3 Chemistry

Angelica contains an essential oil, 0.1–0.4% in fresh roots and 0.5–1% in dried root. Fruits contain 1.2–1.3% oil. Oil is extracted by prolonged steam distillation. The major constituents of the root oil are α-pinene, β-pinene, *p*-cymene, dihydrocarvone, terebangelene and other terpenes, sesquiterpene ketones, angelic acid, valeric resin, alcohols and various acids such as aconitic acid, malic acid, quinic acid, citric acid and oxalic acid. The roots contain five furanocoumarins. The medicinal properties are attributed to these compounds, namely archangelin, prangolarin, oxypeucedanin hydrate, ostsathol and osthol. These are reported to have effect in curing leucoderma. The root oil also contains angelicin, archangelicin, umbelliferone, tiglic acid, etc. Angelicin and archangelin are reported to have spasmolytic activity (Harborne and Baxter, 1993). The phellopterin from fruit is identified as 4-methoxy-7-(γ,γ-dimethylallyloxy)-psoralen by degradation and synthesis. Seed oil typically is quite a bit higher in β-phellandrene (35–65%) and lower in the musk components (pentadecanolide and tridecanolide) than the root oil. Root oil contain between 10 and 30% β-phellandrene. The seed oil also contains in addition methyl-ethylacetic acid and hydroxymyristic acid umbelliprenin, isoimperatorin, bergapten, prangolarin, ostruthol and oxypeucedanin hydrate (Harborne and Baxter, 1993; Rastogi and Mehrotra, 1990, 1993; Anon., 2001; http://www.naturedirect2u.com).

Essential oils isolated by hydrodistillation and supercritical CO_2 extraction on analysis revealed that the two oils had a widely different percentage composition and the one extracted through supercritical fluid extraction (SFE) exhibited a higher number and concentration of oxygenated compounds (Paroul *et al.*, 2002).

The effect of potential precursors (cinnamic acid, phenylalanine, tyrosine) in three concentrations (0.01; 0.1; 1 mmol/l) on the growth of the culture and coumarin production was investigated in the suspension culture of *A. archangelica*. The cultures were cultivated under constant illumination (3500 lux) and in the dark. Under constant illumination, coumarin production was decreased by the action of cinnamic acid, but not by tyrosine and phenylalanine (0.01 and 0.1 mmol/l), increased it in comparison with the culture without precursors (Siatka and Kasparova, 2002). Angelica balsam is obtained by extracting the roots with alcohol, and evaporating and extracting the residue with ether. It is a dark brown colour and contains angelica oil, angelica wax and angelicin.

Twenty solvents were tested in the extraction of compounds from the roots of angelica and the calcium-antagonistic activity of the extracts was investigated. Chloroform was found to be the best solvent for the extraction of non-polar, biologically active compounds from the roots of *A. archangelica* (Harmala *et al.*, 1992).

5.4.4 Functional properties and toxicology

The furanocoumarin content in the leaves is reported to be phytotoxic (Ojala, 1999). Salikhova and Poroshenko (1995) reported that the antimutagenic effect of angelica against thio-TEPA (triethylenethiophosphoramide) mutagenicity in murine bone marrow cells is greater with pretreatment (two hours before) than simultaneous treatment. A commercial preparation, STW 5, consisting of angelica extract along with eight other plant extracts, was tested for its potential anti-ulcerogenic activity against indometacin-induced gastric ulcers of the rat and found beneficial. The cytoprotective activity of the extract was assigned to its flavanoid content and free radical scavenging properties (Khayyal *et al.*, 2001).

5.4.5 Cultivars and varieties

One variety, Dong Quai, is used in China (http://www.naturedirect2u.com). Lundqvist and Andersson (2001) studied the genetic diversity of Angelica along the free-flowing Vindel River in northern Sweden, using starch gel electrophoresis. The diversity was found to increase downstream. Angelica is an insect-pollinated out breeder and the seeds may float for over a year. Dispersal appears to be related to the floating ability of propagules.

5.4.6 Cultivation and production

Propagation is through seed and root propagules. The plant is cultivated in ordinary deep, moist loam, in a shady position, as it thrives well in a damp soil and loves to grow near running water. Seeds should be sown as soon as possible after removing them from the plant; however, they can be stored in a plastic container under refrigeration. Fresh seeds are sown outdoors in autumn for exposure to frost or prechilled in a refrigerator for a few weeks before sowing in spring. Four to six-leaved seedlings are transplanted to a moist shady position, before the roots become immovable. Mulching and irrigation must be provided as required. Angelica needs plenty of fertilizer and moisture.

Red spider mites attack angelica when conditions are dry, so spraying the underside of leaves daily during dry spells is recommended. Application of sulphur on the infested plants early in the morning when the plants are damp are also practised, as the powder will stick better.

Offshoots, produced after harvest of stems, can be transplanted to 60 cm apart and provides a quick method of propagation. This method is considered inferior to that of raising by seed, which as a rule will not need protection during winter.

5.4.7 Harvesting and processing

Leaves are harvested in the spring before the plant blooms. The leaves, stems, seeds and roots are edible and used in cooking, candying, tisanes, teas and liqueurs. Flower stalks and leaf stalks are best when harvested in April–May while leaves are best for flavouring when harvested in June, just before flowering. Roots are dug up just before flowering and dried slowly (Westland, 1987; Clevely and Richmond, 1999).

The seeds are gathered when ripe and dried. The seed-heads should be harvested on a fine day and dry in shade. When dry they are beaten with a rod to remove seeds. Seeds are further dried and stored.

Oil of angelica, which is very expensive, is obtained from seeds by distillation with steam, the vapour being condensed and the oil separated by gravity. A mass of 100 kg of angelica seeds yield 1 kl of oil, and the fresh leaves a little less, the roots yielding only 0.15–0.3 kg.

5.4.8 Quality issues

Leaves, leaf stalks, flower stalks and root oil are the products. Oil is extracted from the root, fruit or seed of the plant. Fresh roots yield oils of lighter colour and more pronounced terpene content. Oil distilled from older roots is darker, more viscous and has a characteristic musk-like odour. Oil from young roots (or from the seed) exhibits a light, somewhat peppery top note missing in oils from older (2–3 yr) roots. Seed oil is colourless or very pale yellow with a strong, fresh, light peppery odour. It is sometimes used to adulterate the root oil and can be difficult to detect (http://www.naturedirect2u.com).

5.4.9 End uses

In food flavouring

Angelica is a favourite flavouring herb in Western culinary art. Leaves are used dried or fresh as a tisane, which helps in reducing fever and cold. Because of its lovely colour and scent it is often used to decorate cakes and pastry and for flavouring jams. Angelica jams and jellies are favourites. Leaf stalks are employed in confectionery. Young leaves and shoots are used to flavour wine and liquors, while the stout stems are candied as a cake decoration or cooked like rhubarb. Essential oil is used in the perfume and flavour industries. Angelica root is the main flavouring ingredient of gin. It is widely used in liqueurs such as benedictine, chartreuse, cointreau and vermouth (http://www.naturedirect2u.com).

Angelica is largely used in the grocery trade, as well as for medicine, and is a popular flavouring for confectionery and liqueurs. The appreciation of its unique flavour was established from ancient times. The preparation of angelica is a small but important industry in the south of France, its cultivation being centralized in Clermont Ferrand. The stem is largely used in the preparation of preserved fruits and '*confitures*' generally, and is also used as an aromatic garnish by confectioners. The seeds, which are aromatic and bitterish in taste, are employed in alcoholic distillates, especially in the preparation of vermouth and similar preparations, as well as in other liqueurs. From ancient times, angelica has been one of the chief flavouring ingredients of beverages and liqueurs. The seed oil is used as flavouring for beverages and also medicinally. Chopped leaves of angelica may be added to fruit salads, fish dishes and cottage cheese in small amounts. Leaves are added to sour fruit such as rhubarb to neutralize acidity. Stems are boiled with jams to improve the flavour. Young stems can be used as a substitute for celery. All parts promote perspiration, stimulate appetite and digestion, and are used to treat ailments of the chest. Young leaves and shoots are used to flavour wines and liqueurs, while the stout stems are candied as a cake decoration. Fresh or preserved roots have been added to snuff and used by Laplanders and North American Indians as tobacco (Clevley and Richmond, 1999).

In order to retain their medicinal virtues for many years, angelica roots are dried rapidly and placed in airtight containers. Fresh root has a yellowish-grey epidermis, and yields a honey-coloured juice, when bruised, having all the aromatic properties of the plant. If an incision is made in the bark of the stems and the crown of the root at the commencement of spring, this resinous gum will exude. It has a special aromatic flavour of musk or benzoin, and can be used as a substitute for either of these. The dried root, as it appears in commerce, is greyish brown and much wrinkled externally, whitish and spongy within and breaks with a starchy fracture, exhibiting shining, resinous spots. The odour is strong and fragrant, and the taste at first sweetish, afterwards warm, aromatic, bitterish and somewhat musky. These properties are extracted by alcohol.

In medicine

The roots, leaves and seeds are used for medicinal purposes. The whole plant is aromatic, but the root is official only in the Swiss, Austrian and German pharmacopoeias. For medicinal use, the whole herb is collected in June and cut shortly above the root. If the stems are too thick, the leaves may be stripped off separately and dried on wire or netting trays. The stem, which is in great demand when trimmed and candied, should be cut in about June or early July.

Properties of the herb (and extract) are: antispasmodic, aphrodisiac, anticoagulant, bactericidal, carminative, depurative, diaphoretic, digestive, diuretic, emmenagogue, expectorant, febrifuge, hepatic, nervine, stimulant, stomachic and tonic. Powdered root is administered to children in warm water for stomach complaints to check vomiting. It is also used in leucoderma. All parts promote perspiration, stimulate appetite, and are used to treat ailments of the chest and digestion (Westland, 1987). It is an alternative to artificial hormones during the menopause, a remedy for menstrual problems, a tonic for anemia, and a treatment for heart disease and high blood pressure. Medieval and Renaissance herbalists noted the blood-purifying powers of angelica. It was used as a remedy for poisons, agues and all infectious maladies. The fleshy roots were chewed and burnt to ward off infection during the 14th- and 15th-century plagues. It stimulates production of digestive juices, improves the flow of bile into the digestive tract, and combats digestive spasms. The oil has been recommended for treating a weak stomach or digestive system, lack of appetite, anorexia, flatulence, chronic gastritis and chronic enteritis. It is also used to reduce accumulation of toxins, arthritis, gout and rheumatism and water retention. In the traditional Chinese medicine, angelica is used for damp, cold intestinal conditions with underlying spleen *Qi* deficiency, as well as chronic lung, phlegm, cold syndromes with painful wheezing. In aromatherapy, it is a germ killer, excellent for coughs and colds, flu, muscular aches, fatigue, migraine, nervous tension, stress and rheumatism. It has a calming effect on the digestion and is relaxing (http://www.naturedirect2u.com).

The yellow juice from the stem and root, when dry, is a valuable medicine in chronic rheumatism and gout. Taken in medicinal form, angelica is said to cause disgust for alcoholic spirits. It is a good vehicle for nauseous medicines and forms one of the ingredients in compound spirit of aniseed.

5.4.10 Recipes and formulations

To preserve angelica
Cut into 10 cm (4 inch) long pieces and steep for 12 hours in salt and water. Put a layer of cabbage or cauliflower leaves in a clean brass pan, then a layer of angelica, then another layer of leaves and so on, finishing with a layer of leaves on the top. Cover with water and vinegar. Boil slowly until the angelica becomes quite green, then strain and weigh the stems. Use 1 kg loafsugar to each kg of stems. Put the sugar in a clean pan with water to cover; boil for ten minutes and pour this syrup over the angelica and allow to stand for 12 hours. Pour off the syrup, boil it up for five minutes and pour it again over the angelica. Repeat the process, and after the angelica has stood in the syrup for 12 hours, put in a brass pan and boil until tender. Then take out the angelica pieces, put them in a jar and pour the syrup over them, or dry them on a sieve and sprinkle them with sugar: they then form candy. Confectioners have evolved their own methods of candying angelica.

Angelica liqueur
A delicious liqueur, which is also a digestive, preserving all the virtues of the plant, is made in this way: 28 g (1 ounce) of the freshly gathered stem of angelica is chopped up and steeped in 1.2 l (two pints) of good brandy for five days together with 28 g (1 ounce) of skinned bitter almonds reduced to a pulp added. The liquid is then strained through fine muslin and 0.6 l (1 pint) of liquid sugar added to it.

Angelica is used in the preparation of vermouth and chartreuse. Though the tender leaflets of the blades of the leaves have sometimes been recommended as a substitute for spinach, they are too bitter for the general taste, but the blanched mid-ribs of the leaf, boiled and used as celery, are delicious, and Icelanders eat both the stem and the roots raw, with butter. In Lapland, the inhabitants regard the stalks of angelica as a great delicacy. The Finns eat the young stems baked in hot ashes, and an infusion of the dried herb is drunk either hot or cold. Angelica makes a drink much in use in Continental Europe for typhus fever: pour 1.2 l (2 pints) of boiling water on 170 g (6 oz) of angelica root sliced thin, infuse for half an hour, strain and add juice of two lemons, 110 g (4 oz) of honey and 70 ml (1/8 pint) of brandy. The Norwegians use the roots for making breads.

5.5 Horseradish

Horseradish (*Armoracia rusticana* P. Gaertn., B. Mey. and Scherb) belongs to the family Brassicaceae (Cruciferae) and contains the distinctive mustard oils that are common to this family. The plant is known by various common names such as horseradish, red cole, creole mustard, German mustard, horse-radish root (archaic) and red horseradish.

Horseradish is a pungent herb, with leaves that are used in salads and sandwiches, and roots that are used for sauces that are added to meat. It is also used for various medical complaints. Both the leaves and roots were used extensively as medicine in Europe during the Middle Ages.

5.5.1 Origin and distribution
Horseradish is native to Europe and Asia (southern Russia, eastern Ukraine), but has become naturalized in North America and New Zealand, where it can be found growing along roadsides. Cultivation dates back only to about Roman and Greek times, about 2000

years ago (Simon *et al.,* 1984; Phillips and Rix, 1993; Brown, 2002). The crop was introduced into Western Europe in the 13th century.

It is grown in the USA and about 7×10^6 kg of horseradish are processed annually for consumption as food. The crop is cultivated over 600 ha (1500 acres) in the USA. In India it is found growing to a small extent in gardens in North India and hill stations of South India.

For proper growth, horseradish needs a temperature between 5 and 19°C with an annual precipitation of 50–170 cm and a soil pH of 5.0–7.5. The hardy horseradish thrives in moist, semi-shaded environments of the north-temperate regions of North America. Although the plant will grow on any soil type, best growth is in deep, rich loam soil, high in organic matter (Simon *et al.,* 1984). Principal production areas are located in the USA and, to a lesser extent, Europe.

5.5.2 Botany and description

Horseradish is a hardy perennial root crop, grown for the very pungent roots, which contain oil with a strong pungent odour and hot, biting taste. The plant attains a height of 0.6–0.9 m (2–3 ft) when in flower. Propagation is by planting pieces of side roots, which are taken from the main root when the latter is harvested. The roots develop entirely underground and grow to a metre (3 ft) in length. The top of the plant consists of a rosette of large paddle-shaped leaves and a flower stalk; it rarely produces seeds. White flowers, with a sweet honey scent, are produced on terminal panicles in late spring. Horseradish may be an interspecific hybrid and is reported to be generally sterile (Anon., 2001).

There are two types of horseradish: (i) 'common' type with broad crinkled leaves and roots of high quality and (ii) 'Bohemian' type with narrow smooth leaves and poor-quality roots, but which is more disease resistant (Anon., 2001).

5.5.3 Chemistry

The root contains a pungent, acrid and vesicating volatile oil. Distillation of the dried and powdered root gives about 0.05–0.2% volatile oil. The intense pungency and aroma of horseradish are the result of isothiocyanates released from the glucosinolate sinigrin and 2-phenylethylglucosinolate by the naturally occurring enzyme myrosinase, in the presence of water. The active constituents are sinigrin (a glycoside, combined with water yields mustard oils), asparagine and resin (Karnick, 1994b). The root is a rich source of vitamin C; the fresh root contains an average of 302 mg per 100 g. Harborne and Baxter (1993) reported the presence of glucoberteroin, glucobrassicanapin, glucocapparin, glucocheirolin, glucochlearin, glucoiberin, glucoiberverin, glucolepidiin, gluconapin, glucotropaeolin and sinigrin from horseradish. Though the undisturbed root has little odour, pungency develops upon crushing or grinding the tissue. The roots are usually processed under refrigeration immediately after dicing, because of the high volatility of the oil.

5.5.4 Cultivation and production

Horseradish is planted with root crowns and root cuttings. Traditionally grown as a perennial in Eastern Europe, the plant is cultivated as an annual in the USA. The originally planted root cuttings are harvested for market and the newly developed lateral roots are broken off and stored in the dark for planting during the following season. The planted roots increase in diameter, but not length, by the end of the growing season (October or November). Horseradish prefers deep, fertile soil with good moisture retention. However, it is tolerant to most soils and grows in full sun or semi-shade. The ground should be prepared in the spring before planting and well-rotted manure and garden compost added. Crown cuttings can be taken in the spring by carefully lifting a healthy section of the plant and gently teasing out a portion of the root, with a section of the crown and at least one fresh crown bud. This should be placed in a prepared site and watered well. Root cuttings can be taken in the spring, autumn or early winter. Pieces of older roots, the thickness of a pencil, should be cut into 13–21 cm long pieces and planted at 30 cm spacing in a trench 10–13 cm deep.

5.5.5 Harvesting and processing

The flavour of the root is reported to be improving in cold weather. The roots, approx. 20–35 cm long are dug between October and December. Large roots should be used for flavouring sauces whereas the thinner roots can be used for propagation. Roots harvested in the spring produce a milder flavour. Leaves are picked when young in the spring and early summer and added to salads. Leaves can also be dried and stored in an airtight container.

5.5.6 Quality issues

The processed horseradish is sometimes adulterated with a mixture of turnips or parsnips. It is sometimes fortified with allylisothiocyanate (synthetic mustard oil) to get the desired pungency. Chemical analysis and infrared spectrum of volatile oil can detect the nature and extent of this type of adulteration, but to a very limited extent.

5.5.7 End uses

Culinary uses
Horseradish is used as an appetizing spice. The high vitamin C content present in it is attributed for its digestive and anti-scorbutic properties. Leaves are used in salads and sandwiches. Grated roots are used alone, or in combination with apple, as a spice for fish. Horseradish is made into a sauce with vinegar and cream that is used with roast beef, cold chicken or hard-boiled eggs. In Eastern Europe, it is used as a condiment in combination with beets. Leaves and roots are used as food in Germany. As a spice the horseradish root is usually grated or minced and mixed with vinegar, salt, or other flavourings to make sauce or relish. These are often used with fish or other seafood or as an appetizer with meats. The plant material is also employed as an ingredient in some ketchups and mustards. Horseradish is available in a dehydrated form.

Medicinal uses
The herb evidently controls bacterial infection, this effect being attributed to allylisothiocyanate. It lowers fever by increasing perspiration, acts as a diuretic, stimulant, diaphoretic, digestive and also stimulates circulation. Excess internal consumption can lead to vomiting or the development of an allergic response. It is claimed to be used in the treatment for general debility; arthritis; gout; respiratory infections; urinary infections; and fevers. It is applied externally as a poultice for infected wounds; inflammation of the pleura; arthritis and inflammation of the pericardium (Phillips and Rix, 1993; Brown, 2002). The fresh root of horseradish has been considered to be an antiseptic, diaphoretic, diuretic, rubefacient, stimulant, stomachic and vermifuge. The material has also been used as a remedy for

asthma, coughs, colic, scurvy, toothache, ulcers, venereal diseases and cancer. The roots are also used as a digestive stimulant, diuretic, to increase blood flow and also in rheumatism (Karnick, 1994b). Peroxidase enzyme is extracted from the plant root and used as an oxidizer in chemical tests, such as blood glucose determinations. Horseradish has strong irritant activity and ingestion of large amounts can cause bloody vomiting and diarrhea. Livestock feeding on tops or roots of horseradish may be poisoned. The volatiles of horseradish root are reported to have herbicidal and microbial activity.

5.5.8 Functional properties and toxicology
Horseradish is generally recognized as safe for human consumption as a natural seasoning and flavouring. The root and leaves are said to contain oils with antibiotic qualities. The pungency of the volatile oil has been known to clear sinuses. The root also contains useful minerals including calcium, sodium, magnesium and vitamin C.

5.5.9 Recipes and formulations

Horseradish sauce
Ingredients: full cup of thick cream (or Greek yogurt); two tablespoons of freshly grated horseradish, one teaspoon of freshly chopped parsley, two tablespoons of white wine vinegar and pinch of salt. Place the cream and horseradish in a bowl and mix gently. Add parsley and other ingredients and mix well. Keep at room temperature or store in an airtight tub. Serve with various meats or fish.

5.6 Black caraway

Black caraway (*Bunium persicum* Boiss Fed. (syn. *Carum bulbocastanum*)) is a perennial aromatic spice belonging to the family Apiaceae. It is a temperate plant, naturally occurring in the dry temperate regions of the northwest Himalayas, where the winter is severe, and the ground is under snow in winter. A long chilling period is essential for germination of seeds. In India the plant occurs in the alpine areas of Himachal Pradesh and Kashmir and Utharanchal. Black caraway is often confused with black cumin (*Niglella sativa*) and caraway (*Carum carvi*).

5.6.1 Production and international trade
The production and export figures of black caraway are not available. The area under the crop is estimated to be about 300 ha and the annual yield is around 400–600 tonnes.

5.6.2 Botany and description
Black caraway is a temperate perennial; the plant habit is dwarf or tall, spreading or compact, the height ranging from 30 to 80 cm. The plant is branched, tuberous; leaves 2–3 pinnate, finely dissected, flowers white, borne on compound umbels, fruit vicid, ridged, vittae 3–5 mm long, brown to dark brown. The crop is naturally cross-pollinated. The plant has $2n = 14$ as its chromosome number. The crop is not subjected to any vigorous crop improvement work.

5.6.3 Chemistry
The chemical composition has not been worked out in detail. The principal constituents of the essential oil are cuminaldehyde (45.4%) and *p*-cymene (35%). Carvone, limonene, α-pinene, β-pinene, cymene and terpinene are the minor constituents (Kaith, 1981).

5.6.4 Cultivars and varieties
There are no approved varieties or improved cultivars. However, four distinct morphotypes are available (dwarf compact, dwarf spreading, tall compact, tall spreading).

5.6.5 Cultivation and production
The propagation is both vegetative (through bulbs) or through seeds. In vegetative propagation bulbs that are three or four years old and of 3–4 cm diameter are used: About 2.5×10^5–3×10^5 bulbs are needed for a hectare (Munshi *et al.*, 1989). When seed is used, 1–1.5 kg seeds/ha is sown in the first year, and in the second year re-seeding at the rate of 200 g/ha is practised to maintain the required population. Sowing is in September–October in rows spaced at 15–20 cm, in raised beds. Germination takes place after the winter in April. During the growing period, growth and development of aerial shoot and underground tubers takes place, and in the ensuing winter the aerial portion dies out and the tubers remain dormant in the soil (Panwar 2000). A fertilizer dose of 20–25 kg farmyard manure (FYM), 60 kg of nitrogen, 30 kg of phosphorus and 30 kg potash per hectare is recommended for good yield (Panwar *et al.*, 1993, Panwar, 2000). Irrigation is recommended at peak flowering and seed formation stage (Badiyala and Panwar, 1992).

Black caraway is attacked by blight caused by *Alternaria*; rust caused by *Puccinia bulbocastanii*, powdery mildew caused by *Erysiphae polygoni* and bulb rot caused by *Fusarium solani*. However, the growers do not adopt any fungicide application. The major insect pests are white grub, hairy caterpillar, armyworms and semi-loopers (Sharma *et al.*, 1993).

5.6.6 Harvesting and processing
The plant takes four years from seed to seed, but when grown from 3–4-year-old bulbs, the flowering takes place in the next season itself. The seeds are ready to harvest in July. Harvesting is when the fruits turn light brown and before full ripening, to avoid shattering. Plants are cut and stacked for drying and then threshed by beating with sticks. Seeds are winnowed, dried, cleaned and stored in airtight containers.

Black caraway oil is extracted by steam distillation of crushed seeds. The oil content is about 5–14% in fresh seeds and 3–6% in dried seeds. The straw contains black caraway herb oil to the extent of 1.25%. The commercial products are seed, seed oil and solvent extracted oleoresin.

5.6.7 End uses
Seeds are widely used as a spice, for flavouring dishes, especially in north Indian, Persian and Mughalai dishes. The hill tribes eat the tubers either raw or after cooking. The essential oil is used in processed food industry and in perfumery. Oleoresin is used in processed foods.

Black caraway is also important medicinally and used in *Ayurvedic* medicinal formulations.

Seeds are stimulants and carminative and are used in treating diarrhoea, dyspepsia, fever, flatulence, stomachic, haemorrhoids and hiccoughs.

5.6.8 Quality issues
The bazaar product is always adulterated with fruits of *Bupleurum falcatum* L., coloured with walnut bark decoction and sometimes with the seeds of *Daucus carota*.

5.7 Capers
Capers (also known as caperberry or caperbush) are immature flower buds of *Capparis spinosa* L. (syn. *Capparis rupestris*), also known as *Capparis ovata* Desf. belonging to the family Capparidaceae. The flower buds are pickled in vinegar or preserved in granular salt. Semi-mature fruits (caperberries) and young shoots with small leaves may also be pickled for use as a spice. Two types of capers occur: *C. spinosa* (spiny in nature) and *C. inermis* (no spines). Use of this plant has been known since Biblical times (Morris and Mackley, 1999).

5.7.1 Origin and distribution
There is a strong association between the caper bush and oceans and seas. *Capparis spinosa* is said to be native to the Mediterranean basin, but its range stretches from the Atlantic coasts of the Canary Islands and Morocco to the Black Sea to the Crimea and Armenia, and eastward to the Caspian Sea and into Iran. Capers probably originated in the dry regions in west or central Asia (Jacobs, 1965; Zohary, 1969). Known and used for millennia, capers were mentioned by Dioscorides as being a marketable product of the ancient Greeks. Capers are also mentioned by the Roman scholar Pliny the Elder.

Dry heat and intense sunlight provide the preferred environment for caper plants. Plants are productive in zones having 350 mm annual precipitation (falling mostly in winter and spring months) and survive summertime temperatures of 40°C. However, caper is a tender plant of the cold and has a temperature hardiness range similar to the olive tree (–8°C).

Plants grow well in nutrient-poor sharply drained gravelly soils. Mature plants develop extensive root systems that penetrate deeply into the earth. Capers are salt-tolerant and flourish along shores within sea-spray zones.

5.7.2 Production and international trade
Locally, capers are collected from wild plants within their natural range. European sources are Spain, Greece, Dalmatia, Grenada and Balearic Islands, France and Italy (especially Sicily and the Aeolian island of Salina and the Mediterranean island of Pantelleria). Capers are also cultivated in Armenia, Algeria, Egypt, Morocco, Tunisia, Asia Minor, Cyprus and the Levant, the coastal areas of the Black Sea, and Iran. Areas with intensive caper cultivation and production are Spain (2600 ha) and Italy (1000 ha).

5.7.3 Botany and description
Caper plants are small shrubs, and may reach about 1 m in height. However, uncultivated caper plants are more often seen hanging, draped and sprawling as they scramble over soil and rocks. The caper's vegetative canopy covers the soil surface, which helps to conserve

soil water reserves. Leaf stipules are transformed into spines. Flowers are borne on first-year branches. The flowers are pink with long tassels of purple stamens. The flowers open in the morning and close by noon.

5.7.4 Chemistry

Flower buds contain a glycoside, rutin, which on acid hydrolysis yields rhamnose, dextrose and quercetin. Flower buds also contain about 4% pentosans on a dry weight basis, rutic acid, pectic acid and saponin. Caper seeds yield about 35% pale yellow oil containing palmitic, stearic, oleic and linoleic acids. The root bark contains rutic acid and a volatile substance with a garlic odour. A series of isomers of the compound cappaprenols have been isolated from *Capparis*. Glucobrassicin, neoglucobrassicin and 4-methoxyglucobrassiin were identified in the roots by high-performance liquid chromatography (HPLC) (Rastogi and Mehrotra, 1995).

Two (6*S*)-hydroxy-3-oxo-alpha-ionol glucosides, together with corchoionoside C ((6*S*, 9*S*)-roseoside) and a prenyl glucoside, were isolated from mature fruits of *C. spinosa* (Calis *et al.*, 2002).

5.7.5 Cultivars and varieties

Varieties have been developed for characters such as spinelessness, round, firm buds, and flavour, through selection. High-yielding caper plants and types with short and uniform flowering periods have not been developed. Some of the varieties are:

- 'enza spina' – Italian selection, form without stipular spines;
- 'spinosa comune' – Italian form with stipular spines;
- 'inermis' – without stipular spines;
- 'josephine' – one of the better Mediterranean selections;
- 'aculeata'; 'dolce di Filicudi e Alicudi' – from the Aeolian Archipelago;
- 'nuciddara' or 'nucidda' 'nocellana' – spineless, with globose buds, mustard-green colour, and strong aroma;
- 'testa di lucertola'; 'tondino' – grown on the island of Pantelleria.

5.7.6 Cultivation and production

Plants are grown from seeds as well as through vegetative cuttings. Caper seeds are very small, and germinate readily – but only in low percentages (Barbera and Di Lorenzo, 1984; Bond, 1990). Various factors such as unit fruit weight, fruit position on mother plant, maturity of the fruit, etc are reported to influence caper seed germination (Pascual *et al.*, 2003). Dried seeds should be initially immersed in warm water (40°C) and then allow to soak for one day. Seeds are then wrapped in a moist cloth, placed in a sealed glass jar and kept in the refrigerator for two or three months. After refrigeration, seeds should be soaked again in warm water overnight and then sown about 1 cm deep in a loose, well-drained soil medium. Young caper plants can be grown in a greenhouse (preferable minimum temperature of 10°C).

Vegetative propagation by stem cuttings is easy. Cuttings from the basal portions are collected in February, March or April. A loose, well-drained medium with heat from below is used for rooting. A dip in an indole 3-butyric acid (IBA) solution of 1.5–3.0 ppm is recommended (15 seconds). Transplanting is carried out during the wet winter and spring periods, and first-year plants are mulched with stones. In Italy, plants are spaced 2–2.5 m apart (depending on the roughness of the terrain; about 2000 plants per hectare). A full yield is expected in three or four years. Plants are pruned back in winter to remove dead wood and water sprouts. Pruning is crucial to high production: heavy branch pruning is necessary, as flower buds arise on 1-year-old branches. Three-year-old plants will yield 1–3 kg of caper flower buds per plant. Grown from seed, California caper bushes reportedly begin to flower in the fourth year; however, Italian sources report some flowering from first year transplants. Caper plantings will live for 20–30 years. Propagation by cutting as well as seeds present serious problems which affects cultivation (Barbera and Di Lorenzo, 1984). Propagation of caper through tissue culture was also reported (Ancora and Cuozzo, 1984).

Two viruses, namely, Caper Latent Carla virus and Caper Vein Yellowing Virus, have been reported in Puglia, Italy. Viruses are transmitted by mechanical inoculation, by grafting and by vegetative propagation of cultivated varieties. Certain insect pests may also be vectors. Various fungi that infect the crop are *Albugo capparidis* De Bary, *Aschochyta capparidis* (Cast.) Sacc., *Botrytis* – grey mould, *Camarosporium suseganense* Sacc. and Speg., *Cercospora capparidis* Sacc., *Cloeosporium hians* Peck and Sacc., *Hendersonia rupestris* Sacc. and Speg., *Leptosphaeria capparidis* Pass., *Phoma capparidina* Pass., *Phoma capparidis* Pass., *Phyllosticta capparidis* Sacc. and Speg., *Septoria capparidis* Sacc., etc.

Various insect pests of capers are *Acalles barbarus* Lucas – a weevil that attacks roots; *Asphondylia capparis* Rubs. – a dipterian (Cecidomyiidae) that disfigures flower buds; *Calocoris memoralis* Sacc., *Cydia capparidana* Zeller – a lepidopteran that disfigures flower buds and *Eurydema ventralis* Kolen.

5.7.7 Harvesting and processing
The unopened flower buds should be picked on a dry day. Harvesting is carried out regularly throughout the growing season. The bushes are checked every morning for small, hard buds that are just at the right stage for harvesting. These buds are to be picked by hand and this labour-intensive harvesting makes this herb an expensive one. In southern Italy, caper flower buds are collected by hand about every 8–12 days, resulting in 9–12 harvests per season.

The capers are washed and allowed to wilt for a day in the sun. The wilted buds are stored in jars and covered with salted wine vinegar, brine, olive oil or in salt alone (Morris and Mackley, 1999). Capers should always be submerged in their pickling medium to prevent them from developing an off-odour. Capers are preserved either in vinegar or under layers of salt in a jar. Raw capers are bland flavoured and need to be cured to develop their piquant flavour. In Italy, capers are graded on a scale from 7 to 16, which indicates their size in millimetres. Mechanized screens are used to sort the various sized capers after being handpicked from the hillsides.

In French-speaking countries, capers are graded using the terms '*nonpareilles*' and '*surfines*'. Capers under a centimetre in diameter are considered more valuable than the larger *capucines* and *communes* (up to 1.5 cm diameter). Capers in vinegar are traditionally packaged in tall narrow glass bottles. Caper fruits (caperberry, capperone or taperone) may be used in making caper-flavoured sauces, or sometimes pickled for eating, like small gherkins.

5.7.8 Biological activity
Cappaprenol 13 from roots inhibited carrageenin-induced and oxyphenbutazone induced paw oedema in rats (Rastogi and Mehrotra, 1995).

5.7.9 End uses

Capers have a sharp, piquant flavour and add pungency, a peculiar aroma and saltiness to pasta sauces, pizza, fish, meats and salads. The flavour of caper may be described as being similar to that of a combination of mustard and black pepper. In fact, the caper's strong flavour comes from mustard oil: methyl isothiocyanate (released from glucocapparin molecules) arising from crushed plant tissues.

Capers make an important contribution to the pantheon of classic Mediterranean flavours that include: olives, rucola (argula, or garden rocket), anchovies and artichokes. Tender young shoots including immature small leaves may also be eaten as a vegetable, or pickled. More rarely, mature and semi-mature fruits are eaten as a cooked vegetable. Additionally, ash from burned caper roots has been used as a source of salt.

Capers are said to reduce flatulence and have an anti-rheumatic effect. In *Ayurvedic* medicine capers are recorded as hepatic stimulants and protectors, improving liver function. Capers have reported uses for arteriosclerosis, as diuretics, kidney disinfectants, vermifuges and tonics. Infusions and decoctions from caper root bark have been traditionally used for dropsy, anaemia, arthritis and gout. Capers contain considerable amounts of the antioxidant bioflavinoid, rutin. Caper extracts and pulps have been used in cosmetics, but there has been reported contact dermatitis and sensitivity from their use (Mitchell, 1974; Schmidt, 1979).

Chopped capers are an ingredient of a wide range of classic sauces, such as tartare, remoulade and ravigote sauces. They are also used in Italian tomato sauce, and in the famous dish of cold braised veal, *vitello tonnato*. In Britain, hot caper sauce is traditionally served with boiled mutton, salmon or pan-fried or grilled fish with the addition of a little grated lemon rind to complement the distinctive flavour. Capers are also used in other areas of Italian cooking, as flavouring in antipasti salads and as a topping on pizza. They are also used with fish and vegetable dishes in Northern and Eastern Europe (Morris and Mackley, 1999).

5.8 Asafoetida

Asafoetida is the dried latex (oleogum) obtained from the rootstocks (or taproots) of certain species of *Ferula* such as *F. asafoetida* L., *F. foetida* Regel, *F. alliacea* Boiss, *F. rubricaulis* Boiss, Linn. and *F. narthex* Boiss. *Ferula* belongs to the family Apiaceae. It is also known as Devil's dung, food of gods, asafetida, etc. Early records state that Alexander the Great carried this 'stink finger' to the West in 4 BC. It is also used as a flavouring agent in the kitchens of ancient Rome. This pungent, resinous gum is used widely in Indian vegetarian cooking (Morris and Mackley, 1999).

The whole plant exudes a strong characteristic smell. Several species of *Ferula* yield asafoetida. The bulk of the product comes from the official plant, *F. asafoetida*, which grows from 600 to 1200 m (2000–4000 feet) above sea level in Iran and Afghanistan. These high plains are arid in winter but are thickly covered in summer with a luxuriant growth of these plants. The cabbage-like folded heads are eaten raw by the local people.

5.8.1 Origin and distribution

The genus *Ferula* is indigenous to Iran and Afghanistan and in the Kashmir region of India. Major areas of occurrence are Iran and Afghanistan, followed by Turkey and Northern Kashmir. Commercially asafoetida is produced only in Iran and Afghanistan. *Ferula narthex* occurs in Northern Kashmir.

5.8.2 Botany and description

Asafoetida is a herbaceous perennial with fleshy, massive, carrot-shaped fleshy root covered with bristly fibres, root with one or more forks. The stem is 1.8–3 m high, solid, clothed with membraneous leaf sheaths; the leaves are radical, *ca* 45 cm long, shiny, coriaceous with pinnatifid segments and channelled petiole; there are pale green yellow flowers, about 10–20 in the main umbels and 5–6 in the partial umbels; fruits are thin, flat, foliaceous, reddish brown with pronounced vittae. *Ferula* is reported to be dioecious; the male plant producing only flowers without oleogum. It is the female plants that produce asafoetida.

5.8.3 Cultivars and varieties

Based on the relative flavour and quality there are various commercial varieties available. *Irani Ras*, *Irani Khada* and *Irani No. 1* are '*Irani*' varieties, whereas *Naya Chal*, *Hadda*, *Naya Zamin*, *Charas*, *Galmin*, *Khawlal*, *Kabuli* and *Shanbundi* are '*Pathani*' varieties.

5.8.4 Chemistry

Asafoetida contains about 62% of resin, 25% of gum and 7% oil, together with free ferulic acid, water, and small quantities of various impurities. In its raw state, the resin or powder has an unpleasant smell, but this completely disappears when the spice is added to preparations. The odour of asafoetida is stronger and more tenacious than that of the onion, the taste is bitter and acrid; the odour of the gum resin depends on the volatile oil.

The resin portion consists of asaresino tannol combined with ferulic acid, the other di- and trisulphides, and traces of various other compounds. The disagreeable odour of the oil is due to the disulphides. The volatile oil (6–17%) consists of sulphated turpenes, resin (40–60%), saresinatannol, ferulic acid and gum (25%) (Martindale,1996).

A sequeterpinoid coumarin (foetidin) and two coumarins (asafoetidin and ferocolicin) were isolated from roots and gum resin, respectively. Three new compounds (asadisulphide, asacoumarin and asacoumarin B) were isolated from resin prepared from roots and their structures elucidated usising ^{13}C-NMR (Rastogi and Mehrotra, 1995). Six new sulphide derivatives (foetisulphide A, foetisulphide B, foetisulphide C, foetisulphide D, foetithiphene A and foetithiphene B) along with six known compounds were isolated and identified from ethyl acetate soluble fraction from a methanol extract of *F. foetida* (Duan *et al.*, 2002).

Luteolin exhibited antipolio virus activity, which was comparable to that of ascorbate-stabilized quercetin.

5.8.5 Cultivation and production

Nothing much is known about the agronomy of *Ferula*. At present it is grown probably as a poor man's crop in Iran and Afghanistan and little is known about the crop requirements.

5.8.6 Harvesting and processing

Asafoetida is an exudate obtained by tapping the rootstock or the thick carrot-shaped taproot of the plant. The process of tapping and asafoetida production involves the following operations.

Tapping is done in March–April following the winter. The upper part of the taproot is exposed by removing the surrounding soil and debris. Leaves are removed, leaving only a tuft of brush-like leaves at the top. The plant is allowed to remain like this for a week or so. Then the top of the rhizome along with the tuft of leaves is cut off with a sharp knife. The cut surface is then covered with leaves and earthed up to form a dome-shaped structure, probably to make the inside warm enough for the easy flow of resin from the cut end during the cool season and keep the area cool during the hot summer months. A milky juice exudes from the surface of the cut end. After four or five days the first collection of resin is given. Then a small portion of the root is again chopped off so that a fresh surface is exposed. After collecting the sap from the second cut a third cut is given. This process is repeated. Every time after the collection of resin, a fresh cut is given until the exudation stops, which takes about three months.

The resin collected is stored in pits dug in the ground. Usually the sides and bottom of the pits are plastered with mud and the top covered with leaves and twigs, leaving a small window. In the beginning asafoetida will be in the form of a sticky paste. Maturing takes place in the pits, and then the asafoetida is packed in jute bags for marketing.

A very fine variety of asafoetida is obtained from the leaf bud in the centre, but this does not come onto world market, and is only used in India, where it is known as *Kandaharre Hing*. It appears in reddish-yellow flakes and when squeezed gives out an oil.

5.8.7 Produce and products

There are two main types of asafoetida: '*Hing Kabuli Safaid*' is the milky white asafoetida obtained from *F. rubicaulis* and the '*Hinglal*' is the brown or red asafoetida (Irani and Pathan types). There is only one type of milky white asafoetida and two types of red asafoetida. Asafoetida is marketed in three forms – tears, mass and paste. Tears are the purest form, they are round or flat, about 15–30 mm diameter and have greyish or dull yellow colour. Mass is agglutinated tears mixed with extraneous matters. Paste is semi-solid and contains extraneous matter. Asafoetida is often adulterated with gum arabic, other gum resins, barley and wheat flour, red clay, gypsum, chalk, etc.

All forms of asafoetida are produced in Iran. The tears produced in Iran are called '*Irani Ras*' and mass is called '*Irani Hing*'. Afghanistan produces white and red varieties. The Irani and Pathani (from Afghanistan) products have the following properties.

Irani
Dry, blackish brown, reddish brown, or yellow in early stage, changing to deeper shades. Sweet fetid odour, sweet taste, contain wood chips except in 'Irani Ras', Alcohol solublity: soluble in 10–30%, ash insoluble in HCl: 0.5–7.75%, volatile oil: 5–10%, resin portion: 40%.

Pathani
Agglutinated and wet, blackish brown, reddish brown, yellow or white. Bitter fetid odour and bitter taste. Alcohol solubility: soluble in 25–50%, ash insoluble in HCl: 0.7–1.90%, volatile oil: 10–20%, resin portion: 40–60%.

Compound asafoetida
Natural asafoetida is very strong and as such cannot be used for cooking. For commercial uses natural asafoetida is hence blended with gum arabic and flour. This is the compound asafetida available for consumers in the market. The blending formula differs from manufacturer to manufacturer and is a trade secret.

5.8.8 Related products

Galabanum
This is known in the trade, as '*Jawashir*' or '*Gaoshir*' and is the oleoresin derived from *F. galbaniflua,* a tall herb occurring in Iran. It is obtained in a similar way to asafoetida. It initially occurs as yellowish or brownish tears and later forms lumps or masses.

The resin in galabanum contains umbelliferone combined with galbaresinotannol, galbaresinic acid and essential oil that contains D-α-pinene, β-pinene, myrecine, cadinene, L-cadinol and traces of other compounds.

Sumbul (musk root)
This is a product obtained from *F. sumbul* and *F. suaveolens,* both growing in Central Asia. The commercial product is the dried, sliced rhizomes that are about 10 cm long and 7 cm in diameter, dark brown externally and yellow inside. It has a bitter taste. *Sumbul* contains 17–18% resinous matter, the main constituent of which is umbellic acid, phytosterol, umbellifernone, betaine, angelic acid and valerianic acid. The essential oil (0.2–1.4%) possesses a characteristic odour and contains a sesquiterpene sumbulene, a mixture of various esters and alcohols.

Sagapenum
In the trade sagapenum is known as *sagbinaj*. It is an oleogum derived from *F. persica* and *F. szowitziana,* indigenous to Iran and neighbouring areas. The oleogum obtained as in the case of asafoetida, and resembles galabanum tears. Its uses are similar to those of *galabanum*.

5.8.9 End uses
Asafoetida is mostly used in Indian vegetarian cooking, in which the strong onion–garlic smell enhances the flavour, especially those of the Brahmin and Jain castes where onions and garlic are prohibited. It is much used in Persian cuisine also, in spite of its offensive odour, as a spice and is thought to exercise a stimulant action on the brain. It is a local stimulant to the mucous membrane, especially to the alimentary tract, and therefore is a remedy of great value as a carminative in flatulent colic and a useful addition to laxative medicine. There is evidence that the volatile oil is eliminated through the lungs, therefore it is excellent for asthma, bronchitis, whooping cough, etc. and even hysteria (Morris and Mackley, 1999). Owing to its vile taste it is usually taken in pill form, but is often given to infants through the rectum in the form of an emulsion. The powdered gum resin is not advocated as a medicine, the volatile oil being quickly dissipated. In India the fruit is also used for medicinal purposes. In traditional medicine, asafoetida is also used in hysterical afflictions and epilepsy as well as in cholera. White asafoetida is believed to be a panacea for many stomach troubles and diarrhoea.

Certain species of *Ferula* yield oleogum, related to asafoetida and used mainly for pharmaceutical purposes. They are *Galabanum, Sumbul* and *Sagapenum* (or *sagbinaj*). *Galbanum* has a characteristic aromatic odour and a bitter acidic taste. It is considered a stimulant, carminative, expectorant and antispasmodic. In indigenous medicine it is used as a uterine tonic and is effective as an anti-inflammatory agent. *Sumbul* is used as a sedative in hysteria and other nervous disorders and is also used as a mild gastro-intestinal stimulant.

Asafoetida oleoresin is bitter, acrid, carminative, antispasmodic, expectorant, anthelmintic, diuretic, laxative, nervine tonic, digestive, sedative and emmenagogue. It is used in flatulent colic, dyspepsia, asthma, hysteria, constipation, chronic bronchitis, whooping cough, epilepsy, psychopathy, hepatopathy, splenopathy and vitiated conditions of *kapha* and *vāta* (Warrier *et al.*, 1995).

Asafoetida is admittedly the most adulterated drug in the market. Besides being largely admixed with inferior qualities of asafoetida, it often has red clay, sand, stones and gypsum added to it to increase its weight.

5.8.10 Other species
Various species of the genus *Ferula* are: *F. narthex* – found in Kashmir, *F. galbaniflua* Boiss and Bulise – a tall herb occurring mainly in Iran, *F. sumbulferula* and *F. suaveolens* Aitch and Henosel – both occurring in central Asia, *F. persica* – Willd, *F. zowitziana* DC – indigenous to Iran, *F. foetida* Regel., *F. alliacea* Boiss and *F. rubricaulis* Boiss, Linn.

The Tibetan asafoetida (*Narthex asafetida/Ferula narthax*) is closely allied to *Ferula. Ferula narthex,* found in Kashmir, grows to 1.5–3 m (5–10 ft) high, possesses large leaf sheaths; upper leaves much reduced, flowers small, yellow, in single or scarcely branched compound umbels arising from within the leaf sheaths. The umbels have no involucre, the limb of the calyx is suppressed, and the stylopods are depressed and cup-shaped, styles recurved, fruit compressed at the back, dilated at the margin. The tap roots are thick, carrot-like and branched. This variety produces some of the asafoetida used in commerce.

Scorodosma foetida, another gigantic umbelliferous plant found on the sandy steppes of the Caspian, also is a source of commercial asafoetida. The Persian *sagapenum,* or *serapinum*, a species of *Ferula,* that was formerly imported to Bombay, is in appearance very similar to asafoetida, but does not go pink when freshly fractured, and in smell is less disagreeable than asafoetida. This species is an ingredient of Confection Rutea (*British Pharmacopoeia Codex*).

5.9 Hyssop

Hyssop is the flowering top of the evergreen perennial shrub *Hyssopus officinalis*, which is a valuable expectorant.

5.9.1 Origin and distribution
Hyssop is native to southern Europe and the temperate zones of Asia. It grows wild in countries bordering the Mediterranean Sea. It is cultivated in Europe, especially in southern France, mainly for its essential oil. In India it is found in the Himalayas from Kashmir to Kumaon at altitudes of 2435–3335 m and is cultivated in Baramullah, in Kashmir.

5.9.2 Botany and description
The plant grows to a height of 60 cm, branches are erect or diffuse; leaves linear–oblong or lanceolate, obtuse, entire, narrow, sessile, green and fragrant, hairy and dotted with oil-bearing glands. The plant flowers in autumn. Whorls of blueish-purple flowers are produced on long narrow spikes.

5.9.3 Chemistry
The herb contains volatile oil, fat, sugar, choline, tannins, carotene and xanthophyll. The

flower tops contain ursolic acid (0.49%) and a glucoside diosmin, which on hydrolysis yields rhamnose and glucose. The fresh herb contains iodine in a concentration of 14 mcg/kg. The aerial part on steam distillation yields a volatile oil, 0.15–0.30% and 0.3–0.8%, from fresh and dried materials, respectively.

Hyssop oil is colourless or greenish yellow with an aromatic, camphoraceous odour and slightly bitter taste. The major component of the volatile oil is ketone 1-pinocamphone. The content of essential oil is rather low (0.3–0.9%); it is mostly composed of cineol, β-pinene and a variety of bicyclic monoterpene derivatives (L-pinocamphene, isopinocamphone, pinocarvone). Hyssop contains large amounts of bitter and antioxidative tannins: phenols with a diterpenoid skeleton (carnosol, carnosolic acid), depsides of coffeic acid (= 3,4-dihydroxycinnamic acid) and several triterpenoid acids (ursolic and oleanolic acid) (Galambosi, 1993; Kerrola, 1994; Dordevic, 2000).

5.9.4 Cultivation and production
The herb is cultivated mostly in the Mediterranean region and is propagated through seeds and cuttings. It grows in hot, arid conditions in full sun, in well-drained, near-neutral sandy soil. The seeds are sown in spring season in indoors and the mature seedlings (six to eight weeks old) are transplanted to the field. The cuttings are taken in early summer and planted 20 cm apart. Irrigation during initial establishment and moderate fertilization is preferred. Over-fertilization of the crop results in more foliage with reduced flavour.

5.9.5 Harvesting and processing
The time of harvest depends on the type of uses. The herb is harvested fresh for cooking purposes. The herbs for processing and distilling purpose are harvested just before flowering; when the leaves contain the highest concentration of essential oil. The leaves are harvested in the morning for higher yield of oil. The leaves are dried in a dry, dark room with adequate ventilation for one or two weeks. The dried herb is stored in airtight containers in the dark. Seeds are harvested when they turn brown. Roots are harvested after the aerial parts die down.

5.9.6 End uses
Hyssop is used as a condiment and also in medicines. The leaves and flowering tops of hyssop are employed in flavouring for salads and soups. It is also used in the preparation of liquor and perfumes. It is also used as a pot herb.

Hyssop is considered a stimulant, carminative and expectorant and is used in colds, coughs, congestion and lung complaints. A tea made from the herb is effective in nervous disorders and toothache. It is also effective in pulmonary, digestive, uterine and urinary troubles and asthma and coughs. Leaves are stimulating, stomachic, carminative and colic and leaf juice is used for the treatment of roundworms.

Hyssop oil is used as a flavouring agent in bitters and tonics and also in perfumery. In small quantities it promotes expectoration in bronchial catarrh and asthma.

5.9.7 Functional properties and toxicology
Antimicrobial activity of essential oil of hyssops was investigated and the property was attributed to the linalool and 1,8-ciniole components of the essential oil (Mazzanti *et al.*, 1998).

5.9.8 Quality issues
Hyssop oil is occasionally adulterated with lavender or rosemary oils. Sometimes it is also mixed with camphor oil fractions.

5.10 Galangal

Galanga or galangal (*Kaempferia galanga* L.) which should not be confused with the greater galangal (*Alpinia galanga*), is a perennial aromatic rhizomatous herbaceous plant belonging to the family Zingiberaceae. This genus comprises about 70 species, among which *K. galanga* and *K. rotunda* are of economic value. These plants are used for flavouring food and also in medicine. Rhizome and roots are aromatic and are used as spice.

5.10.1 Origin and distribution
The genus is presumably native to tropical Asia and is distributed in the tropics and subtropics of Asia and Africa, but is now rarely found growing wild. It is cultivated in home gardens in India, Sri Lanka, Malaysia, Moluccas (Indonesia), Philippine Islands and South-East Asia.

5.10.2 Botany and description
Plants attain a height of 30 cm, but often much shorter, and have fleshy, cylindrical aromatic root tubers. The plant possesses two to a few broad, round leaves that are usually spread horizontally; leaves are sessile, ovate, deltoid-acuminate, thin deep green, and the petioles are short channelled; flowers are irregular, bisexual, white, 6–12 from the centre of the plant between the leaves, fugacious, fragrant and opening successively, bracts are lanceolate, green, short, calyx is as long as the outer bracts, short cylindrical, there are three petals, corolla tube 2.5 cm long, lobes are equal, usually spreading, lanceolate, pure white, one stamen, perfect, filament is short, arcuate, anther is two-celled, cells discrete. Flowering starts in June and ends in September, with peak flowering during July to August.

Macroscopic analysis of its powder revealed that it is has a camphoraceous odour, bitter aromatic taste and is brown in colour. The cross-section of the rhizome and root showed thin-walled parenchyma cells, fragments of thick walls of tracheids, with irregular-shaped parenchyma cells and their parts and a number of starch granules that have come out from the cells. Cytological studies showed that the somatic chromosome number of *K. galanga* is $2n = 54$ (Ramachandran, 1969).

5.10.3 Chemistry
Kempferia galanga rhizome contains about 2.5–4% essential oil. The main components of the oil are ethyl cinnamate (25%), ethyl-*p*-methoxycinnamate (30%) and *p*-methoxycinnamic acid and a monoterpene ketone compound, 3-carene-5-one (Kiuchi *et al.*, 1987). The other constituents are camphene, δ-3-carene, *p*-methoxy styrene, γ-pinene, β-myrcene, *p*-cymene, 1,8-cineole, isomyrcene, camphor, α-terpineol, *p*-cymene-8-ol, eucarvone, δ-cadinene, etc. The leaves contain kaempferol, quercetin, cyanidin and delphinidin. The root contains camphene, 1,8-cineole, camphor, borneol, cinnamaldehyde, ethyl cinnamate, quinozoline, ethyl *p*-methoxy cinnamate and quinazoline-4-phenyl-3-oxide.

The rhizome is also reported to display cytotoxic properties. The essential oil is used in flavouring curries, in perfumery and also for medicinal purposes (Bhattacharjee, 2000).

Deoxypodophyllotoxin and ethyl *p*-methoxy-*trans*-cinnamate were isolated from rhizomes; monoterpeneketone -car-3-en-5-one was isolated from rhizomes and characterized (Harborne and Baxter, 1993). Deoxypodophyllotoxin exhibited cytotoxic activity by inhibiting HeLa cells.

5.10.4 Cultivars and varieties

Not much work has been undertaken to identify the extent of variability in the crop. In an attempt to induce mutation, bushy mutants were induced with 7.5 krad gamma rays. Irradiation at lower doses (below 1.0 krad) stimulated the germination of rhizomes. In an evaluation of five geographical races of *K. galanga* from Kerala, significant variation was observed in rhizome and oil yields but there was little variation in oil quality. One high-yielding, cultivar Kasturi, has been released from Kerala Agricultural University (KAU). Another high yielding cultivar, Rajani, has also been identified from the germplasm collection at KAU.

5.10.5 Cultivation and production

Galanga requires fertile sandy soils and a warm humid climate. It thrives well up to an elevation of about 1500 m above mean sea level. A well-distributed annual rainfall of 1500–3000 mm is required during the growing period and dry spells during land preparation and harvesting.

The species is propagated by rhizome fragments. Mother rhizomes are superior to finger rhizomes. The rhizome bits are planted on beds of 1–2 m width and 25 cm height at a spacing of 40–60 cm^2 (IBPGR, 1981; Bhattacharjee, 2000). About 750 kg of seed rhizomes per hectare is required. Planting during the third week of May gives significantly higher rhizome and oil yields.

The mean nutrient uptake of the crop is 22.8 kg N, 28 kg P_2O_5 and 36.9 kg K_2O per hectare. Application of 50–75 kg N, 60 kg P_2O_5 and 50–75 kg K_2O is found to be beneficial for increased rhizome and oil yields. Application of farmyard manure at 30 tonnes/ha is superior to the application of nutrients through inorganic form of fertilizers and it increased the yield by 60%. A well-managed plantation yields about 4–6 tonnes of fresh rhizomes per hectare. The dry recovery varies from 23 to 28%. Leaf rot disease may occur during the rainy season and can be controlled by trenching the beds with 1% Bordeaux mixture.

In Kerala, cultivation of *K. galanga* is restricted to some localized tracts and the productivity of the crop is low, ranging from 2–5 tonnes of fresh rhizomes per hectare. There is an acute shortage of planting material and the absence of seed set limits the scope for breeding (Kurian *et al.*, 1993).

Propagation of this crop through tissue culture was reported by Vincent *et al.* (1992) and Geetha *et al.* (1997). The tissue cultured plants could not be used directly for field planting as it takes two crop seasons to produce enough rhizomes. However, these plants can be used for planting material production through high-density planting. Propagation by *in vitro* rhizomes is possible, and is a method that can be commercialized.

5.10.6 Harvesting and processing

The crop matures about six or seven months after planting. The aerial portion dries off on maturity. The rhizomes are dug out, cleaned and washed to remove soil and they are dried in sun.

The essential oil is extracted by steam distillation of sliced and dried rhizomes. The oil yield varies with season and maturity stage of the rhizome.

5.10.7 End uses

Kampferia galanga is cultivated for its aromatic rhizomes and also as an ornamental. It is used extensively as a spice throughout tropical Asia and has a long history of medicinal use. The rhizome is chewed and ingested. It is used as flavouring for rice. The rhizomes are considered stimulatory, expectorant, carminative and diuretic. They are used in the preparation of gargles and administered with honey for coughs and chest inflictions. In the Philippines, a decoction of the rhizomes is used for dyspepsia, headache and malaria. The juice of the plant is an ingredient in the preparation of some tonic preparations. The rhizomes and roots are used for flavouring food and also in medicine in South-East Asia (CSIR, 1959, 1992).

The rhizome mixed with oil is used externally for healing wounds and may be applied to rheumatic regions. A lotion prepared from the rhizome is used to remove dandruff or scales from the head. The powdered rhizome mixed with honey is given as an expectorant. The leaves are used in lotions and poultices for sore eyes, rheumatism and fever. In Thailand the dried rhizome of this plant is used as a cardiotonic (CSIR, 1959). In India the dried rhizomes, along with some other plants, are used for heart disease. It is also used for treatment of abdominal pain, vomiting, diarrhoea, and toothache with the functions of promoting vital energy circulation and alleviating pain.

The herb is used as food flavouring in Malaysia and also in perfumery. The rhizomes are used to protect clothes against insects. It is also used as a masticatory along with betel leaf and arecanut. Slices of the dried rhizome may be cooked with vegetable or meat dishes, but mostly the spice is used fresh and grated or crushed. It is essential for Javanese cooking (*Rijstafel*) and is especially used in the Indonesian island of Bali.

5.10.8 Functional properties and toxicology

Essential oil from the root induced glutathione-s-transferase activities in the stomach, liver and small intestine of mouse. An ethanol extract of dried rhizome showed antispasmodic activity vs histamine-induced contraction and barium-induced contraction in Guinea pigs. An ethanol–water extract indicated smooth muscle stimulant activity. Water extracts of dried rhizomes exhibited antitumour activity. Rhizome and root oils showed antibacterial activity against *Escherichia coli*, *Staphylococcus aureus* and antifungal activity against *Cladosporium* sp. Nematocidal activity was observed in the rhizome of *K. galanga*.

The hypolepidaemic action of the ethanolic extract of *K. galanga* was observed *in vivo*. The oral administration of the extract was effective in lowering the total cholesterol, triglycerides and phospholipid levels in serum and tissues (Achuthan and Padikkala, 1997).

5.11 Betel vine

The betel vine (*Piper betle* L.) is a perennial climber belonging to the family Piperaceae. Betelvine has been widely used in various parts of India and other South Asian countries for centuries. The name *Piper* is probably originated from the Sanskrit term '*Pippali*' (meaning long pepper) and the name '*betle*' might have come from the Malayalam word '*Vetila*'. It is cultivated, as a commercial crop, for its leaf, which is used as a masticatory. As a masticatory, the betel leaf (pan) is credited with aromatic, digestive, stimulant and carminative properties. Betel chewing imparts a pleasant odour to the oral cavity and also warmth and a feeling of well-being. Hydrated lime is spread over the betel leaf and is chewed with a few pieces of arecanut (*Areca catechu*). Other spices and masticatories such as cardamom, fennel, nutmeg, clove and tobacco are also added to the betel (leaf roll) quid preparation. Betel chewing is prevalent in all south Asian countries, Indonesia, many Pacific Ocean Islands and Middle East and South-East Asia.

5.11.1 Production and international trade

Betel vine is commercially cultivated in countries such as India, Bangladesh, Pakistan, Malaysia, Indonesia, Sri Lanka, Thailand, Papua New Guinea, Madagascar, Bourbon and the West Indies. India is the major producer, where it is cultivated in an area of 43 000 ha, with an annual production worth Rs. 7000 millions. Bangladesh is the second largest producer. Sri Lanka is also a major producer, which exports most of its produce to Pakistan.

In India the crop is extensively cultivated in the states of Andhra Pradesh, Bihar, Gujarat, Karnataka, Kerala, Maharashtra, Madhya Pradesh, Assam, Orissa, Rajastan, Tamil Nadu, Uttar Pradesh and West Bengal. Countries such as Oman, Kuwait, Quatar, Saudi Arabia, the UAE, the UK, the USA and Nepal are the main importers. Improving the quality of the leaves, proper pre- and post-harvest handling, and improved methods of packing, storage and transportation can make manifold increase in the export of betel leaves.

5.11.2 Origin and distribution

According to De Candolle (1884), *P. betle* might have originated in the Malaya Archipelago. Burkill (1966) described the native place as Central and Eastern Malaysia where the crop was cultivated and spread through tropical Asia and Malaysia. Later on it reached Africa and then the West Indies. Cultivation of betel vine started in Southern Asia, but there is doubt about the exact place of domestication.

5.11.3 Botany and description

Piper betle is a perennial, dioecious climber that belongs to the dicot family Piperaceae. Male vines are cultivated and grow vigorously up to a height of 20 m, with a stem diameter of 15–20 cm. Stems are semi-woody, green or pinkish green, cylindrical or bilaterally pressed with dimorphic branching. The plant grows creeping on earth or climbing up on the trees with orthotropic vegetative branches by means of adventitious roots arising from the nodes. Roots are few, sparingly branched and short. The nodes are conspicuously swollen and the internodes are elongated. Leaves are simple, alternate, stipulate, bifarious, petiolate with 5–20 cm long, broad, cordate to obliquely ovate, thick and often unequal. Lamina is oblique at the base, slightly acuminate, acute, entire with undulate margin, glabrous, bright or dark green with reticulate venation. The leaves are aromatic and the taste varies from sweet to pungent. The petiole is usually 2–15 cm long.

Flowering is rare, mainly because the plants are replanted in every four or five years under cultivation. Plants flower when they are 8–10 years old.. The inflorescence is cylindrical, pendulous spike and flowers are naked, unisexual, dioecious, fairly long, peduncled (3–10 cm long) and oppositifolius. Female spikes are 3.5–6 cm long. Male spikes are dense, cylindrical, 8–10 cm long, sub-pendulous, consisting of numerous unisexual bracteate flowers. Fruit is a drupe, seen very rarely, often sunken in fleshy spike. There are

10–20 seeds in each fruit, but they are poor in germination. There is much variation in the chromosome numbers reported in betelvine: figures $2n = 26, 32, 52, 58, 62, 78$ and 195 have been given. The most frequent number is $2n = 78$ for the majority of cultivars and varieties (Jose and Sharma, 1984).

Cultivated betelvines are mostly male plants selected and multiplied over a period by the growers for vigorous growth and leaf production.

5.11.4 Chemistry

The varying taste of betel vine ranging from sweet to pungent is due to the presence of essential oils. The chief constituent of leaves is a volatile oil, known as betel leaf oil, and its amount varies in leaves from different varieties. The oil is of a clear yellow colour and is obtained from fresh leaves. The essential oil consists of euginol, cadinene, chavicol, chavibetol, cineole, sesquiterpene, allylpyrochavicol, caryophyllene, methyl euginol, hydroxy-chavicol, sitosterols, stigma sterols, etc. (Balasubrahmanyam *et al.*, 1994).

Chemical analysis of betel vine using modern analytical tools has revealed that the presence or absence and the quantity of any of these chemical constituents vary with the variety. The betel vines in the Indian subcontinent were grouped in one or other of the five varieties, *Bangla, Desasvari, Kapoori, Meetha* and *Sanchi*. The percentage of essential oil in each of the varieties varies from 0.10 to 1.0%. About 52 compounds have been identified in the betel leaf oil and the composition of these varies with varieties. The major constituents are monoterpenes, sesqueterpenes, oxygenated compounds, aledehydes, acids, oxides, phenols, phenolic esters and esters (Balasubramanyam and Rawat, 1990; Balasubrahmanyam *et al.*, 1994; Ravindran, 2000). The quantity of essential oil increases with maturity and also depends on the external environment. The presence of eugenol and lusitanicoside have also been reported (Harborne and Baxter, 1993).

The volatile oil of *P. betle* cultivated in the Hue area of Vietnam, obtained by steam distillation of the fresh leaves, was analysed and found to contain isoeuginol (72%) and isoeugenyl acetate (12.2%). These constituents indicated the existence of a new isoeugenol chemotype of *P. betle* (Thahn *et al.*, 2002).

5.11.5 Cultivars and varieties

More than 150 types and cultivars are grown by cultivators and recognized by traders in India. The most important cultivars common in India are given in Table 5.5.

Rawat *et al.* (1988) identified five distinct varieties, from the germplasm, that differ in their morphology and chemistry. They are *Bangla, Desawari, Kapoori, Meetha* and *Sanchi*. A few disease-tolerant or resistant lines were identified through the screening of cultivars. Crop improvement work is limited to germplasm evaluation and selection (Ravindran, 2000).

5.11.6 Cultivation and production

The suitable environment for betel vine cultivation is a cool shady area in the humid tropics with plenty of moisture in the soil. It also thrives well in areas with well-distributed high rainfall and in areas with high humidity, moderate temperature and copious rainfall. Such natural conditions are available in certain parts of Western Ghats, Assam, Meghalaya, Tripura, Kerala, and the uplands of North Kanara. The crop is grown under artificial conditions in the hot arid zones of north India under shade and irrigation. Betel vine is very

Table 5.5 Important cultivars of betel leaf in India

SI No.	Cultivars	Pungency*
1.	Bangla (Madhya Pradesh)	P
2.	Bangla (Uttar Pradesh)	P
3.	Bangla Nagaram (Uttar Pradesh)	P
4.	Calcutta (West Bengal)	P
5.	Calcutta Bengal (West Bengal)	P
6.	Deshi Calcutta (west Bengal)	P
7.	Desvar Mahoba (Uttar Pradesh)	MP
8.	Ghanghatte (West Bengal)	P
9.	Godi Bangla (Orissa)	P
10.	Halisahar Sanchi (West Bengal)	P
11.	Kakir (Bihar)	P
12.	Kalipatti (Maharashtra)	P
13.	Kappori (Bihar)	NP
14.	Kappori (Orissa)	NP
15.	Karapaku (Andra Pradesh)	P
16.	Karpuri (Tamil Nadu)	NP
17.	Kaljedu (Andra Pradesh)	NP
18.	Maghai (Bihar)	P
19.	Meetha Pan (West Bengal)	Sweet
20.	Nov Bangla (Orissa)	P
21.	Ramtek Bangla (Maharashtra)	P
22.	S. G. M. 1 (Tamil Nadu)	MP
23.	Sachi Pan (Assam)	P
24.	Sangli Kapoori	NP
25.	Tellaku (Andra Pradesh)	NP
26.	Vellai Kodi (Tamil Nadu)	NP

*MP = mildly pungent; P = pungent; NP = non-pungent.
Source: Ravindran (2000).

sensitive to sudden temperature changes and is a shade-loving plant. The crop thrives in well-drained, fertile, humus-rich soil. It grows in a wide range of soil types, ranging from clay-loam to sandy, provided there is good drainage and moisture-holding capacity.

There are four main types of betel vine cultivation practice in India. It is cultivated as an inter-crop in arecanut and coconut plantations, in open conservatories with wind-breaks and live standards (in a bed system or in a trench system), in closed conservatories or as an open system in the backyards of houses. Betelvine is propagated vegetatively by hard stem cuttings. Cutting from the middle portion of a vine is the ideal planting material, as the tender as well as over-matured portions take longer time for sprouting. The planting time is determined according to the availability of suitable conditions and also on the availability of standards. The planting season in various regions is spread throughout the year. Close spacing of about 30×30 cm^2 between the vines is advisable for better leaf yield. The inter-row spacing is usually 10 to 20 cm. Micropropagation protocols for betel vine have been standardized by Nirmal Babu *et al.* (1992) and by Aminudin *et al.* (1993).

5.11.7 Harvesting and processing

Fresh betel leaves are usually used for chewing. Leaves are ready for harvest in four to six months. The harvested leaves are packed moist in different types of well-aerated baskets and marketed. In some places the leaves are blanched or bleached and marketed. This process leads to some changes in the chemical composition of the essential oils (Table 5.6).

Table 5.6 Essential oil composition (%) of bleached leaves of two *Bangla* lines

Compound	Jganathi *Bangla*		Tamluk *Bangla*	
	Bleached	Control	Bleached	Control
Linalool	0.12	0.08	0.44	0.23
Chavicol	–	0.09	0.31	–
Safrol	0.86	0.18	–	–
Eugenol	64.30	64.00	46.14	63.66
Methyl eugenol	0.23	0.07	–	0.11
β-caryophyllene	3.34	3.37	4.70	1.19
L-lunalene	1.05	1.23	1.33	1.12
Germacrene D	5.93	6.15	5.96	–
γ-elemene	3.80	–	–	4.98
Eugenyl acetate	4.12	3.84	5.25	5.03
ƒ-cadinene	1.89	1.81	3.08	3.05
Sesquiterpene alcohol	Traces	–	–	Traces
Phytol	–	–	–	0.18
Essential oil	0.01	0.01	0.01	0.01

Source: Ravindran (2000).

(Ravindran, 2000). The impact of various drying methods on the quality of betel leaf has been analysed and the results reveal that the solar-dried leaves, followed by shade and sun-dried, maintained the best quality (Ramalakshmi, 2002).

5.11.8 End uses

The most extensive use of betel leaves is for chewing. Leaves are chewed with arecanut and lime, and with or without tobacco. Betel leaf chewing is an ancient practice in India and other countries of South-East Asia. In India it is associated with many religious and social practices. As a masticatory, it is aromatic, digestive, stimulant and carminative. However, excessive indulgence in chewing produces various afflictions of the mouth including carcinoma, mainly because tobacco is used as an accompaniment.

The leaves are stimulant, antiseptic and sialogogue. Leaf juice is used in eye afflictions. Aqueous extract is useful in throat inflammation and in alleviating coughs and indigestion. The essential oil from leaves is used in respiratory catarrh and also as an antiseptic. The oil also possesses antibacterial and antifungal activities. The oil is an active local stimulant used in the local application or gargle, also as an inhalant in diphtheria. In India the leaves are used as a counter-irritant to suppress the secretion of milk in mammary abscesses. The juice of four leaves is equivalent in power to one drop of the oil. Betel leaves possess anti-oxidant action, because of the phenols such as hydroxy chavicol present in it.

5.12 Pomegranate

Pomegranate, used as spice, constitutes the dried seed with the pulp of *Punica granatum*. The tree has been placed by various authorities in different orders, but is now included in the family Punicaceae. The pomegranate is mentioned in the *Papyrus of Ebers*. It is still used by the Jews in some ceremonies, and as a design has been used in architecture and needlework from the earliest times. It formed part of the decoration of the pillars of King Solomon's Temple, and was embroidered on the hem of the High Priest's ephod.

There are three kinds of pomegranate: one very sour, the juice of which is used instead of unripe grape juice; the other two moderately sweet or very sweet. These are eaten as dessert after being cut open, the seeds, strewn with sugar and sometimes sprinkled with rosewater. A wine is extracted from the fruits, and the seeds are used in syrups and preserves. For medicinal and spice purposes the sour variety is used. It is said to have originated in Western Asia and now grows widely in Mediterranean countries, China, Japan, India and in many other tropical and subtropical countries.

5.12.1 Botany and description
Pomegranate is a glabrous and deciduous shrub or small tree with dark grey bark. Leaves are opposite or subopposite, often fascicled on short petiole, oblong or obovate and 2.5–6.0 cm long. The flowers are terminal or axillary, solitary, large and showy and orange-red coloured. The calyx is coriaceous and persistent, prolonged above the ovary and the distal end and campanulate in shape. Petals are 1.2–2.5 cm long, thin and wrinkled. The ovary is inferior. The fruits are large, globose, 5.0–8.0 cm across, crowned by somewhat tubular limb of the calyx and indehiscent with red pulp and juicy. The seeds are angular with coriaceous testa. Flowering is in April–May and fruiting during June–August, but flowering and fruiting both also may occur at different seasons.

The dried seed is used as a spice, while the dried root is used in traditional medicine. It is marketed as quills 7–10 cm (3 to 4 inches) long. It is yellowish-grey and wrinkled outside, the inner bark being smooth and yellow, having little odour and a slightly astringent taste.

5.12.2 Chemistry
Various parts of the plant contain malvidin, pentose, glucosides, tannin and ursolic acid. The stem yields carbohydrates, carotene and D-mannitol. The chief constituent of the bark (about 22%) is called punicotannic acid. It also contains gallic acid, betulic acid, mannite, friedelin and four alkaloids, pelletierine, methyl-pelletierine, pseudo-pelletierine and isopelletierine. The liquid pelletierine boils at 125°C, and is soluble in water, alcohol, ether and chloroform. The drug probably deteriorates with age.

The fruits contain nicotinic acid, pectin, protein, riboflavin, thiamine, vitamin C, delphinidin diglycoside, aspartic, citric, ellagic, gallic and malic acids, glutamine and isoquercetin. The rind contains tannic acid, sugar and gum. Pelletierine tannate is a mixture of the tannates of the alkaloids obtained from the bark of the root and stem, and represents the taenicidal properties. The seeds contain asiatic and maslinic acids, pelargonidin-3, 5-diglucoside, sitosterol and β-D-glucoside. Betulic acid, granatins A and B and punicatolin are found in leaves (Chatterjee and Pakrashi, 1994). Oestrone with oestrogenic activity is isolated from the seeds of pomegranate (Harborne and Baxter, 1993).

Rastogi and Mehrotra (1995) reported the isolation of cyanidin-3-glucoside and 3,5-diglucoside, delphinidin-3-glucoside and 3,5-diglucoside from seed coat; isolation of punicafolin from leaves and its characterization as 1,2,4-o-galloyl 3,6(R) hexahydroxy-diphenoyl-β-D-glucose, granatin B corilagin, strictinin, 1,2,4,6-tetra-o-galloyl-β-D-glucose and 1,2,3,4,6-penta-o-galloyl 3,6 (R)-hexahydroxy-diphenoyl-β-D-glucose. Isolation of a new hydrolysable tannin-2-O-galloyl-4, 6-(S,S)-galloyl-D-glucose and its characterization; structures of punicalin, punicalagin (revised); determination of punicic (33.3%), nonadecanoic (5.9%), heneicosanoic (5.0%), tricosanoic (4.9%) and 13-methylstearic (1.5%), 4-methyllauric (0.5%) acids in seed oil by GC were also reported.

5.12.3 Cultivars and varieties

Pinana is a dwarf variety naturalized in the West Indies. Many horticultural varieties have been developed for culinary purposes.

5.12.4 End uses

Use as spice

The rind of the fruit is in curved, brittle fragments, rough and yellowish-brown outside, paler and pitted within. It is called Malicorium. The fruit is used for dessert, and in the East the juice is included in cool drinks. The seed dried with the pulp is used as a spice in many dishes.

Medicinal uses

According to Chatterjee and Pakrashi (1994), the green leaves are made into a paste and applied on eyes for conjunctivitis, and leaf juice is given in dysentery. The bark of the root and stem is considered astringent and anthelmintic and are specially used against tapeworm (Chopra, 1982). The fruit juice is cooling and refrigerant. A decoction of fruit-rind is useful in chronic dysentery and diarrhoea and this decoction, together with that of the bark of *Holarrhena antidysenterica*, is an effective remedy for dysentery (Chatterjee and Pakrashi, 1994). The pulp and seeds are stomachic (Chopra, 1982) and are also used as laxative. The flower buds are used in bronchitis. Chatterjee and Pakrashi (1994) stated that the flower buds are dried and powdered to a snuff, which is applied to epitaxis and internally used as an effective remedy in infantile diarrhoea and dysentery. The flowers are also used to stop nose bleeds. An extract of leathery pericarp is taken orally at bedtime to cure pinworm disease. The flower buds are powdered and used in dysentery and diarrhoea (Singh *et al.*, 2000). In southern Italy, a decoction of the pericarp is prepared by boiling 30 g in 1 l of water, with lemon or orange juice added. It is taken two cups a day as an astringent and to treat helminthiasis and dysentery. In Turkey, the pericarp of the fruit is dried, powdered and mixed with honey to prepare pills; three to six pills are taken internally to stop bleeding from piles. It is non-toxic and can be used for a long time.

The seeds are demulcent. The fruit is a mild astringent and refrigerant in some fevers, and especially in biliousness, and the bark is used to remove tapeworm. In India the rind is used in diarrhoea and chronic dysentery, often combined with opium. It is used as an injection in leucorrhoea, as a gargle in sore throat in its early stages, and in powder for intermittent fevers. The flowers have similar properties. The rind often causes nausea and vomiting, and possibly purging. Use of it should be preceded by strict dieting and followed by an enema or castor oil if required. It may be necessary to repeat the dose for several days. A hypodermic injection of the alkaloid may produce vertigo, muscular weakness and sometimes double vision. The root bark was recommended as a vermifuge. It may be used fresh or dried (Singh *et al.*, 2000).

The flowers yield a red dye, and with the leaves and seeds were used by the Ancients as astringent medicines and to remove worms. The bark is used in tanning and dyeing giving the yellow hue to Morocco leather. The barks of three wild pomegranates are said to be used in Java: the red-flowered *merah*, the white-flowered *poetih* and the black-flowered *hitam*.

5.13 Summer savory

The genus *Satureja* Linn. (Lamiaceae) comprises about 14 species of highly aromatic, hardy annual or perennial herbs or under-shrubs. Two important species of this genus are

S. hortensis (summer savory) and *S. montana* (winter savory) (CSIR, 1972). Summer savory (*Satureja hortensis*) is a hairy aromatic annual and is grown as a popular garden herb. The savory of commerce is the dried leaves and flowering tops, but the best class comprises only leaves (CSIR, 1972).

5.13.1 Production and international trade
France, the former Yugoslavia and Albania are the major producers (Anon., 2002). Savory is also cultivated in Spain, Germany and other parts of continental Europe, Canada, the UK and the USA. In India it is cultivated in Kashmir (CSIR, 1972). The Yugoslavian variety is recognized as the premier grade (McCormick – *Spice Encyclopedia*, http://www.mccormick.com/content).

5.13.2 Origin and distribution
The crop is indigenous to southern Europe and the Mediterranean area. It is distributed in the warmer regions of both the hemispheres. Several species have been introduced into England, but only two, the annual summer savory and the perennial winter savory, are generally grown. It grows wild in dry, light soils and on rocky hillsides on chalk and is locally cultivated for commercial use. The plant is cultivated in several areas of Iran.

5.13.3 Botany and description
Summer savory is an annual herbaceous plant with small erect stems, grows about 30 cm in height. The branches are pinkish, leaves dark green, petiolate, leathery, elliptical, about 1 cm long and often fascicled. The hairs on the stem are short and decurved. Lilac, pink or white flowers appear in small spikes in the leaf axils, during late summer (Rosengarten, 1969; CSIR, 1972; Tainter and Grains, 1993).

5.13.4 Chemistry
The herb has a thyme-like flavour. The fresh leaves contain moisture (72%), protein (4.2%), fat (1.65%), sugar (4.45%), fibre (8.60%) and ash (2.11%). The leaves contain 11.95% (dry weight basis), pentosans and also labiatic acid, ursolic acid, β-sitosterol and volatile oil (CSIR, 1972).

There are many reports on the composition of essential oil of the aerial parts and leaves of savory from different parts of the world (Ghannadi *et al.*, 2000; Opdyke, 1976; Thieme and Nguyen, 1972a,b; Hajhashemi *et al.*, 2000; Gora *et al.*, 1996). The essential oil obtained from the full flowering spice is between 0.1 and 0.15%. Savory oil is described as light yellow to dark brown liquid and comprises carvacrol, *p*-cymene, pinene, dipentane, ursolic acid, etheral oil, phenolic substances, resins, tannins and mucilage (Prakash, 1990; Karnick, 1994b).

Lawrence (1981) compared the chemical composition of savory oils from Europe, Canada and North Africa. The oil exhibited differences in *p*-cymene, myrcene and γ-terpinene contents. Prakash (1990) made a comprehensive literature survey on the chemical composition of savory oil. The seed contains fixed oil (45%) and protein (24%) on a dry basis. Ghannadi (2002) analysed the seed oil of savory collected from Iran using GC and GC–MS. The seeds yielded 0.3% of a pale yellowish oil with a pleasant spicy odour. Forty-two components were characterized, representing 96.7% of the total oil. The oil was rich in

Table 5.7 Percentage composition of the seed oil of *Satureja hortensis* from Iran

Compound	Percentage
Hexanol	0.2
Heptanal	0.1
α-thujene	0.2
α-pinene	0.7
Camphene	0.1
β-pinene	0.5
p-menth-3-ene	trace
Myrcene	1.1
α-phyllandrine	0.2
α-terpinine	2.1
p-cymene	9.3
β-phellandrene	0.5
γ-tepinene	12.8
Terpinolene	0.2
Methyl benzoate	0.2
Linalool	0.2
Cis-thujone	0.1
Borneol	0.1
Terpinene-4-ol	1.1
α-terpineol	0.1
Myrtenol	0.2
Cuminaldehyde	0.3
Methyl carvacrol	0.5
Bornyl acetate	0.1
Thymol	0.3
Perillyl alcohol	0.1
Carvacrol	59.7
Eugenol	1.7
Carvacrol acetate	0.2
α-copaene	0.1
β-caryophyllene	1.2
Aromadendrene	0.1
α-humulene	0.1
Germacrene D	trace
β-bisabolene	1.1
δ-cadinene	trace
Elemol	0.1
Germacrene B	0.1
Ledol	trace
Spathulenol	0.2
Caryophyllene oxide	0.4
Humulene epoxide	0.2

Source: Ghannadi (2002).

monoterpenes. The major components were carvacrol (59.7%), γ-terpene (12.8%). p-cym-ene (9.3%, and α-terpinine (2.1%) (Table 5.7). Many of these compounds are also common in the oil from the vegetative parts.

5.13.5 Cultivation and production
Savory grows wild, propagated vegetatively and also through seeds. The most preferred method of propagation is through seeds. The species is cold sensitive. Seeds are sown in

well-drained soil during spring in rows 30 cm apart. Temperate climate, full sun and rich and light soil are preferred. The seedlings need thinning out, when large enough, to 15 cm apart.

The seeds may also be sown scattered, when they must be thinned out, the thinned-out seedlings being planted in another bed at 15 cm distance from each other and well watered. The seeds are very slow in germinating.

5.13.6 Harvesting and processing

Harvesting takes place 75–120 days after seed sowing. The harvest is dried in the shade or at 35°C and stored in closed containers. The dried leaves are brownish green in colour. It is marketed both as whole leaf, dried and ground form.

5.13.7 End uses

The use of savory as a culinary herb dates back to the early Romans. The leaves are gathered before flowering and the flowering shoots used fresh or dried. It is used sparingly in meat dishes and stuffings, with peas, beans and cabbage to improve their digestibility, and in liqueurs (Verghese, 2003).

Savory, which has a distinctive taste, though it somewhat recalls that of marjoram, is not only added to stuffings, pork pies and sausages as a wholesome seasoning, but fresh sprigs of it are boiled with broad beans and green peas, in the same manner as mint. It is also boiled with dried peas in making pea soup. For garnishing it has been used as a substitute for parsley and chervil (McCormick – *Spice Encyclopedia*, http://www.mccormick.com/content).

An infusion of leaves treats gastric upsets, indigestion and loss of appetite. The tea made out of this is used as a tonic. Savory has aromatic and carminative properties, and though chiefly used as a culinary herb, may be added to medicines for its aromatic and warming qualities. It was formerly deemed a sovereign remedy for the colic and a cure for flatulence, and was also considered a good expectorant (Karnick, 1994b). Flowering stalks are used as a moth repellent for cloths.

5.13.8 Quality issues

There are different definitions for savory, the spice of commerce, such as the plant cut down at flowering time and dried (Parry, 1969); plant freshly harvested during the flowering season (Guenther, 1974); the leaf harvested before the plant blooms or before flowering (Lewis, 1984; Prakash, 1990); the whole ground dried leaves and flowering tops (Farrell, 1990); the dried leaves of the herb (McCormick – *Spice Encyclopedia*, http://www.mccormick. com/content); whole dried plant (FCC, 1996) and dried leaves and flowering tops (CSIR, 1972).

Farrell (1985) described the US specifications for savory: savory shall be the whole or ground dried leaves and flowering tops of *S. hortensis* L. The brownish-green leaves are fragrantly aromatic with a warm, slightly sharp taste. The produce should contain about 10% total ash, 2% acid insoluble ash, 10% moisture, 25 ml volatile oil per 100 g and granulation 95% (95% of the ground product should pass through a US standard sieve No. 40).

5.14 Winter savory

Winter savory (*S. montana*) is a semi-evergreen bushy and woody perennial shrub, with smaller pink or white flowers and a stronger flavour. Essential oil is extracted commercially from this species and other uses are similar to summer savory.

The stems are woody at the base, diffuse, much branched. The leaves are oblong, linear and acute, or the lower ones spatulate or wedge-shaped and obtuse. Flowering is in June; the flowers are very pale purple, the cymes shortly pedunculate. It is propagated either from seeds, sown at a similar period and in the same manner as summer savory, or from cuttings and divisions of root.

Winter savory is dried and powdered and mixed with grated breadcrumbs, 'to bread their meat, be it fish or flesh, to give it a quicker relish'. It is recommended by old writers, together with other herbs, in the dressing of trout. When dried, it is used as seasoning in the same manner as summer savory, but is not employed medicinally.

Satureja thymbra, which is used in Spain as a spice, is closely allied to the savories grown in English kitchen gardens, yields oil containing about 19% of thymol. Other species of *Satureia* contain carvacrol. The oil from wild plants of winter savory contains 30 or 40% of carvacrol, and that from cultivated plants still more.

5.15 Other

5.15.1 Mango ginger

Mango ginger (*Curcuma amada*) is a rhizomatous aromatic herb of the family Zingiberaceae and is cultivated throughout India, Sri Lanka, Bangladesh and in many South-East Asian countries for its rhizomes that are used as flavouring for pickles and other dishes and also valued for their medicinal properties. The fresh as well as dried rhizomes are used for flavouring curries. The fresh cut rhizomes have the flavour and the colour of mango, hence the name mango ginger. The herb attains 60–90 cm height, leaves are long, petiolate, oblong-lanceolate, tapering at both ends, glabrous, green on both sides; flowers are white or pale yellow in spikes that occur in the centre of the leaves, lip is semi-elliptic, yellow, three-lobed, the middle lobe emarginated. The ethanol extract of rhizome showed the presence of hydroxyl, carbonyl, ester and olefin functional groups in it and also methyl, methylene, methionine proteins and olefinic proteins (Jain and Mishra, 1964; Gholap and Bandyopadhyay, 1984; Rao *et al.*, 1989; Mujumdar *et al.*, 2000).

High-frequency microrhizome production from the *in vitro* shoot cultures in liquid Murashigue and Skoog medium with 5 mg l^{-1} BA and 8% sucrose was reported by Nayak (2002).

The rhizomes are bitter, sweet, sour aromatic (a mixture of tastes, starting from bitter initially, turning to a sweet and then sour aromatic sensation), and cooling; used as an appetizer, carminative, digestive, stomachic, demulcent, febrifuge, alexeteric, aphrodisiac, laxative, diuretic, expectorant, anti-inflammatory and antipyretic and used in the treatment of anorexia, dyspepsia, flatulence, colic, bruises, wounds, chronic ulcers, skin diseases, pruritus, fever, constipation, hiccough, cough, bronchitis, sprains, gout, halitosis, otalgia and inflammations (Hussain *et al.*, 1992; Warrier *et al.*, 1994).

There is only very limited literature available on the pharmacological activity of the extract (Bhakuni *et al.*, 1969; Rao *et al.*, 1989). The rhizome extract of the plant exhibited an hyper-cholesteremic effect in rabbits (Pachuri and Mukherjee, 1970). The extract showed presence of an antibiotic principle with strong inhibitory activity on *Aspergillus niger* and *Trichophyton rubrum* (Gupta and Banerjee, 1972).

The rhizome is a favourite spice and vegetable owing to the rich flavour of raw mango.

The essential oils in the rhizome make it useful as a carminative and stomachic. The pulped rhizome is also used on concussions and sprains. An improved cultivar (Amba) has been developed at the high altitude research station at Pottangi, Orissa (India).

5.15.2 Lovage

Lovage (*Levisticum officinale* Koth.) is a perennial plant that belongs to the family Apiaceae, and is a native of Europe. Centres of lovage cultivation are located principally in central Europe. It is also found cultivated in some areas in New England, USA. It has been grown over the centuries for its aromatic fragrance, its fine ornamental qualities and, to a lesser extent, its medicinal values. All parts of the plant, including the roots, are strongly aromatic and contain extractable essential oils.

It is a pungent, clump-forming herb with rhizomatous roots and stout hollow-ridged stems up to 2.4 m. Leaves are broad and glossy; a tall flower stalk that grows 2 m high with greenish-yellow flowers in large, dense umbels are produced in summer. The fruits are ridged and golden brown in colour (Clevely and Richmond, 1999).

Chemical constituents of lovage oil are mainly phthalides and terpenoids, including *n*-butylidene phthalide, *n*-butyl-phthalide, sedanonic anhydride, D-terpineol, carvacrol, eugenol and volatile oil. The principal components of volatile oil are angelic acid and β-terpenol, coumarins, furocoumarins including psoralins, rotoside, sitosterols, resins, pinene, phellandrene, terpinine, carvacol, terpineol, isovaleric acid, umbelliferone and bergapten. Fresh leaves contain a maximum 0.5% essential oil; the most important aroma components are phthalides (ligustilide, butylphthalide and a partially hydrogenated derivative thereof called sedanolide). Terpenoids (terpineol, carvacrol) and eugenol are less important (Simon *et al.*, 1984; Karnick, 1994b).

Najda *et al.* (2003) studied the composition of various compounds in various plant parts of lovage. The phenolic acids in various plant parts were as follows: roots 0.12–0.16%, herb 0.88–1.03%, stems 0.30–0.39%, leaf 1.11–1.23% and fruits 1.32–1.41%. The quantity of tannins in various plant parts was: roots 6.6%, herb 5.3%, stems 7.4%, leaf 2.7%, and fruits 1.8%. Free phenolic acids such as chlorogenic, caffeic, *p*-coumaric and *m*-coumaric were detected using HPLC.

The crop is propagated either through seeds or through root divisions. It prefers a well-drained, fertile soil. The seeds are sown outdoors during spring in a seedbed. The roots are divided in spring or autumn and planted. Mature plants require wider space, as they are large and bulky. Deep, rich moist soil and full sun or partial shade are required for better growth. The plants need to be cut back during summer to get a continuous supply of tender leaves. Fertilization with balanced organic fertilizer is required in spring and mulching is done in summer. Young flower stalks are removed to keep the foliage fresh for longer.

Harvesting is done in the second or third year of the crop and is usually in October. Young leaves, hollow main stems before flowering, sliced dried roots of 2–3-year-old plants and ripe seeds are the useful parts. The fresh roots are generally first harvested from 2–3-year-old plants. Subsequent harvests take place every third year. The fresh roots are washed, cut into approximately 13 mm thick pieces and dried.

Leaves are used in flavouring soups, salads, casseroles and stews because of their pungent, celery-like flavour. The stems are used for candied products. Roots are peeled and cooked as a vegetable. Powdered root is sometimes used as a spice. The volatile oil extracted from the roots is highly valued for use in perfumery, soaps and creams, and it has been used for flavouring tobacco products. The seeds and seed oil are used for flavouring agents in confectionery and liqueurs.

As a medicinal plant, lovage has been used as a digestive, carminative, diaphoretic, diuretic, emmenagogue, anti-dyspeptic, expectorant, stimulant and stomachic; and also as a treatment for jaundice. Current medicinal applications include use as a diuretic and for regulation of menstrual cycle. Lovage is generally recognized as safe for human consumption as a natural seasoning and flavouring agent (Karnick, 1994b).

5.15.3 *Zanthoxylum* spp.

The term Szechuan pepper or Japanese pepper refers to a spice obtained from a group of closely related plants of the genus *Zanthoxylum*, belonging to the family Rutaceae and consisting of approximately 200 species with a pan-tropical distribution. It is a large genus of aromatic, prickly trees or shrubs and is mostly distributed in the Himalayan region, furthermore in Central, South, South-East and East Asia. The most important species are *Z. piperitum* DC, *Z. simulans* Hance, *Z. bungeanum* Max., *Z. schinifolium* Sieb. and Zucc, *Z. nitidum* Roxb, *Z. ovalifolium* Wight., *Z. rhetsa* Pierre., *Z. alatum* Roxb. and *Z. acanthopodium* DC. All these species are widely distributed over Asia, but are not used as a spice throughout the region. All species mentioned have their place in local cuisine. The literature often gives contradicting information on the genuine species of the spice used. *Zanthoxylum* is a confusing genus and the information available is very scanty.

Szechuan pepper or Japanese pepper is very important in the cuisine of central China and Japan, but it is also known in parts of India, especially in the Himalayan region, and in certain regions of South-East Asia. The fruit of *Z. piperitum* (Japanese pepper) is the genuine source of the spice. It is a small tree and often wrongly assumed to be part of the pepper family. The spice, which is the ground husks of the berries, is common in the Szechuan region of China, and the leaves of the plant are also used in Japan as spice. The ripe fruits of the tree open out in a similar way to star anise. This spice is also known by various common names such as anise pepper, *fagara*, Chinese brown pepper, *poivre anise*, *anispfeffer*, *pimenta de anis*, *pepe d'anis* and Szechuan pepper.

Most *Zanthoxylum* species produce pungent alkamides derived from polyunsaturated carboxylic acids, stored in the pericarp. The commonly found alkamides are α-, β- and γ-sanshool and hydroxy sanshools. Total amide content in *Z. piperitum* is as high as 3%. Non-volatile constituents such as flavonoids, terpene alkaloids, benzophenthredine alkaloids, pyranoquinoline alkaloids, etc. were also identified. The composition of leaf oil of *Z. piperitum* from Japan has been reported (Kusumoto *et al.*, 1968; Shimoda *et al.*,1997). The volatile compounds in the leaves were isolated by steam distillation and the aroma components were evaluated by an aroma extraction dilution analysis. The main components responsible for the aroma are glycosides such as (Z)-3-hexenol, C-6 compounds, citronellal, citronellol, geraniol and 2-phenylethanol (Kojima *et al.*, 1997).

Xanthoxylin and (–)-sesamin are isolated from *Z. piperitum* (Harborne and Baxtor, 1993). β-Sanshool and γ-sanshool, unsaturated aliphatic acid amides isolated from the pericarp, were found to relax the circular muscle of the gastric body, as well as contract the longitudinal muscle of the ileum and distal colon in an experimental system using the gastrointestinal tract isolated from a guinea pig (Hashimoto *et al.*, 2001). Epple *et al.* (2001) investigated the effects of a total extract from *Z. piperitum* fruit on food intake in rats and found that they failed to habituate to the stimuli.

The rust-red berries contain bitter, black seeds that are usually removed before the spice is sold. This spice is used whole or ground and is much used in Chinese cookery, especially with chicken and duck. It is one of the spices in the Chinese five-spice powder and is used in Japanese seven-spice seasoning mix. The leaves are dried and ground to make *sansho*, a Japanese spice. In the Goa and Konkan region of India the dried immature fruits of Z. *rhesta* are used for flavouring fish and chicken preparations.

In the past the ground bark was used as a remedy for toothache in the USA. Both bark and berries are used in traditional medicines and herbal cures to purify the blood, promote digestion and as an anti-rheumatic.

5.16 References

ACHUTHAN, C.R. and PADIKKALA, J. (1997), Hypolipidemic effect of *Alpinia galanga* (Rasna) and *Kaempferia galanga* (Kachoori). *Indian J. of Clini. Biochem.* **12**(1): 55–8.
ACUNA, U.M., ATHA, D.E., MA, J., NEE, M.H. and KENNELLY, E.J. (2002), Antioxidant capacities of ten edible North American Plants. *Phytotherapy Res.* **16**: 63–5.
AGARWAL, S.O. (1987), Some industrially important aromatic plants of Sikkim. *Indian Perfumer* **31**: 113–15.
ALBERTAZZI, F.J., KUDLA, J. and BOCK, R. (1998), The cox2 locus of the primitive angiosperm plant *Acorus calamus*: molecular structure, transcript processing and RNA editing. *Molecular and General Genetics* **259**(6): 591–600.
AMINUDDIN, JOHRI, J.K., ANIS, M. and BALASUBRAHMANYAM, V.R. (1993), Regeneration of *Piper betle* from callus tissue. *Curr. Sci.* **63**: 793–6.
ANCORA, G. and CUOZZO, L. (1984), *In vitro* propagation of caper (*Capparis spinosa* L.) In: XXVIII *Conv. Ann. Ital. Gen Agr.* Bracciano, 82–3.
ANON. (1985), *The Wealth of India*, Vol. I. Publications and Informations Directorate, New Delhi. pp. 196–7.
ANON. (2001), *Herbs Cultivation and their Utilization*. NIIR Board, National Institute of Industrial Research, Asia Pacific Business Press Inc., New Delhi, India.
ANON. (2002), *Handbook on Spices*. NIIR Board, National Institute of Industrial Research, Asia Pacific Business Press Inc., New Delhi, India.
ANU, A., NIRMAL BABU, K., JOHN, C.Z. and PETER, K.V. (2001), *In vitro* clonal multiplication of *Acorus calamus* L. *J. Plant Biochem. Biotech.* **10**: 53–5.
BADIYALA, D. and PANWAR, K.S. (1992), Effect of bulb size and row spacing on the performance of Kala-Zira. *Indian Perfumer* **36**; 34–6.
BALASUBRAHMANYAM, V.R. and RAWAT, A.K.S. (1990), Notes on economic plants – Betelvine. *Econ. Bot.* **44**(3): 540–3.
BALASUBRAHMANYAM, V.R., JOHRI, J.K, RAWAT, A.K.S., TRIPATI, R.D. and CHAURASIA, R.S. (1994), Betelvine *(Piper betle L.)*. Economic Botany Information Service, National Botanical Research Institute, Council of Scientific and Industrial Research (CSIR), Lucknow, India.
BARBERA, G. and DI LORENZO, R. (1984), The caper culture in Italy. *Acta Horticulturae* **144**: 167–71.
BHAKUNI, D.S., DHAR, M.L., DHAR, M.M., DHAVAN, B.N. and MEHROTRA, B.N. (1969), Screening of Indian plants for biological activity. *Indian J. Exp. Biol.* **7**: 250–62.
BHATTACHARJEE, S.K. (2000), *Handbook of Aromatic Plants*. Pointer Publishers, Jaipur, India.
BOND, R.E. (1990), The caper bush. *The Herbalist* **56**: 77–85.
BROWN, D. (2002), *The Royal Horticultural Society New Encyclopedia of Herbs and their Uses*. Dorling Kindersley, London.
BURKILL, I.H. (1966), *A Dictionary of Economic Products of the Malayan Peninsula*, Vol. II. Ministry of Agriculture, Malaysia, 1767–72 (Reprint).
CALIS, I., KURUUZUM-UZ, A., LORENZETTO, P.A. and RUEDI, P. (2002), (6S)-hydroxy-3-oxo-alpha-ionol glucosides from *Capparis spinosa* fruits. *Phytochemistry* **59**(4): 451–7.
CARLQUIST, S. and SCHNEIDER, E.L. (1997), Origins and nature of vessels in monocotyledons. 1. *Acorus*. *Int. J. Plant. Sci.* **158**(1): 51–6.
CHATTERJEE, A. and PAKRASHI, S.C. (1994), *The Treatise on Indian Medicinal Plants*, Vol. 1–5, PID, CSIR, New Delhi.
CHAUDHARY, S.S., GAUTAM, S.K. and HANDA, K.L. (1957), Components of calamus oil from calamus roots growing in Jammu and Kashmir. *Ind. J. Pharm.* **19**: 183–6.
CHOPRA, R.N. (1982), *Indigenous Drugs of India*, 2nd Ed., Academic Publishers, Calcutta.

CHOWDHURY, A.K.A., ARA, T., HASHIM, M.F. and AHMED, M. (1993), A new phenyl propene derivative from *A corus calamus*. *Pharmazie* **48**: 786–7.
CHOWDHURY, A.R., GUPTA, R.C. and SHARMA, M.L. (1997), Essential oil from the rhizomes of *Acorus calamus* Linn. raised on alkaline soil . *Indian Perfumer* **41**(4): 154–6.
CLEVELY, A. and RICHMOND, K. (1999), *Growing and Using Herbs*. Sebastian Kelly, Oxford.
CSIR (1972), *The Wealth of India, Raw Materials*, Vol. IX. Council of Scientific and Industrial Research (CSIR), New Delhi, 273–238.
CSIR (1959), *The Wealth of India, Raw Materials*, Vol. V. Council of Scientific and Industrial Research, New Delhi.
CSIR (1992), *The Useful Plants of India*. Publications and Information Directorate, Council of Scientific and Industrial Research, New Delhi, p. 307.
DE CANDOLLE, A. (1884), *Origin of Cultivated Plants*. Kegan Paul, London.
DUAN, H., TAKAISHI, Y., TORI, M., TAKAOKA, S., HONDA, G., HO, M. and ASHURMETOV, O. (2002), Polysulfide derivatives from *Ferula foetida*. *J. Nat. Prod.* **65**(11):1667–9.
DUKE, J.A. (1985), *CRC Handbook of Medicinal Herbs*. CRC Press Inc., Boca Raton, Florida, pp. 43–4, 120–21.
DUVALL, M.R. (2001), An anatomical study of anther development in *Acorus* L.: phylogenetic implications. *Plant Syst. Evol.* **228**(3–4): 143–52.
DUVALL, M.R., LEARN JR., G.H., EGUIARATE, L.E. and CLEGG, M.T. (1993), Phylogenetic analysis of rbcL sequences identifies *Acorus calamus* as the primal extant monocotyledon. *Proc. Natl. Acad. Sci., USA* **90**: 4641–4.
EPPLE, G., BRYANT, B.P., MEZINE, I. and LEWIS, S. (2001), *Zanthoxylum piperitum*, an Asian spice, inhibits food intake in rats. *J. Chem. Ecol.* **27**(8): 1627–40.
FARRELL, K. (1985), *Spices, Condiments and Seasonings*, AVI Publ., Westport, CT.
FARRELL, K. (1990), *Spices, Condiments and Seasonings*, 2nd Edition, Van Nostard Reinhold, New York, 189–192.
GALAMBOSI, B., SVOBODA, K.P., DEANS, S.G. and HETHELYI, E. (1993), Agronomical and phytochemical investigation of *Hyssopus officinalis*. *Agric. Sci. Finland* **2**: 293–302.
GEETHA, S.P., MANJULA C., JOHN, C.Z., MINOO, D., NIRMAL BABU, K and RAVINDRAN P.N. (1997), Micropropagation of *Kaempferia galanga* L. and *K. rotunda* L. *J. Spices and Aromatic Crops* **6**(2): 129–35.
GHANNADI. A. (2002), Composition of the essential oil of *Satureja hortensis* L. seeds from Iran. *J. Essent. Oil Res.* **14**; 35–6.
GHANNADI, A., HAJHASHEMI, V., SADRAEI, H. and MOHSENI, M. (2000), Composition of the essential oil of *Satureja hortensis* L. grown in Kashan. *Iranian J. Basic Med. Sci.* **2**(3): 166–9.
GHOLAP, A.S. and BANDYOPADHYAY, C. (1984), Characterization of mango like aroma in *Curcuma. amada* Roxb. *J. Agric. Food Chem.* **32**: 7–9.
GORA, J., LIS, A. and LEWANDOWSKI, A. (1996), Chemical composition of the essential oil of cultivated summer savory (*Satureja hortensis* L. cv Saturn). *J. Ess. Oil Res.* **8**(4): 427–8.
GRIEVE, M. (1931), *A Modern Herbal*, Leyel, C.F. (ed.), Jonathan Cape, London.
GUENTHER, E. (1974), *The Essential Oils* Vol. 3, Van Nostand, New York, 739–741.
GUPTA, S.K. and BANERJEE, A.B. (1972), Screening of selected West Bengal plants for antifungal activity. *Econ. Bot.* **26**: 255–7.
HAJHASHEMI, V., SADRAEI, H., GANNADI, A. and MOHSENI, M. (2000), Antispasmodic and antidiarrhoeal effect of *Satureja hortensis* L. essential oil. *J. Ethnopharmacol.* **71**: 187–92.
HARAGUCHI, H., KUWATA, Y., INADA, K., SHINGU, K., MIYAHARA, K., NAGAO, M. and YAGI, A. (1996), Antifungal activity from *Alpinia galanga* and the competition for incorporation of unsaturated fatty acids in cell growth. *Planta Med.* **62**(4): 308–313.
HARBORNE, J.B. and BAXTER, H. (1993), (eds) *Phytochemical Dictionary – A Handbook of Bioactive Compounds from Plants*. Taylor and Francis, UK.
HARIKRISHNAN, K.N., MARTIN, K.P., ANAND, P.H.M. and HARIHARAN, M. (1997), Micropropagation of sweetflag (*Acorus calamus*) – a medicinal plant. *J. Medicinal and Aromatic Plant Sciences* **19**(2): 427–9.
HARMALA, P., VUORELA, H., TORNQUIST, K. and HILTUNEN, R. (1992), Choice of solvent in the extraction of *Angelica archangelica* roots with reference to calcium blocking activity. *Planta Med.* **58**(2): 176–83.
HASHIMOTO, K., SATOH, K., KASE, Y., ISHIGE, A., KUBO, M., SASAKI, H., NISHIKAWA, S., KUROSAWA, S., YAKABI, K. and NAKAMURA, T. (2001), Modulatory effect of aliphatic acid amides from *Zanthoxylum piperatum* on isolated gastrointestinal tract. *Planta Med.* **67**(2): 179–81.

HETTIARACHCHI, A., FERNANDO, K.K.S. and JAYASURIYA, A.H.M. (1997), *In vitro* propagation of wadakaha (*Acorus calamus* L.) *J. Natl. Sci. Council of Sri Lanka* **25**(3): 151–7.
HTTP://WWW.INDIANSPICES.COM
HTTP://WWW.NATUREDIRECT2U.COM
HUSSAIN, A., VIRMANI, O.P. POPLI, S.P. MISRA, L.N. and GUPTA, M.M. (1992), *Dictionary of Indian medicinal plants*. Central Institute of Medicinal and Aromatic Plants, Lucknow, 161–162.
HUXLEY, A., ED. (1992), *The New Royal Horticultural Society Dictionary of Gardening*. RHS Publications, London.
IBPGR (1981), *Root and Tuber Crops*. International Bureau of Plant Genetic Resources Secretariat, Rome, 72–5.
JACOBS, M. (1965), The genus *Capparis* (Capparaceae) from the Indus to the Pacific. *Blumea* **12**(3): 385–541.
JAIN, M.K. and MISRA, R.K. (1964), Chemical examination of *Curcuma amada* Roxb. *Indian J. Chem.* **2**: 39.
JECFA (1981), Joint FAO/WHO Expert Committee on Food Additives. *Monograph on* β*-asarone*. WHO Food Additive Series No. 16.
JIROVETZ, L., BUCHBAUER, G., SHAFI, M.P. and LEELA, N.K. (2003), Analysis of the essential oils of the leaves, stems, rhizomes and roots of the medicinal plant *Alpinia galanga* from southern India. *Acta Pharm.* **53**: 73–81.
JOSE, J. and SHARMA, A.K. (1984), Chromosome studies in the genus *Piper* L. *J. Indian Bot. Soc.* **63**: 313–19.
KAITH, D.S. (1981), Effect of harvest management on the yield of essential oil content and flavour of Kalazira. *Indian Perfumer* **25**: 97–98.
KARNICK, C.R. (1994a), *Pharmacopoeial Standards of Herbal Products*. Vol. I. Indian Medicinal Science Series No. 37. Sri Satguru Book Centre, Indian Book Centre, New Delhi, India.
KARNICK, C.R. (1994b), *Pharmacopoeial Standards of Herbal Products*. Vol. II. Indian Medicinal Science Series No. 36. Sri Satguru Book Centre, Indian Book centre, New Delhi, India pp. 151.
KERROLA, T., GALAMBOSI, B. and KALLIO, H. (1994), Volatile components and odour intensity of four phenotypes of Hyssop (*Hyssopus officinalis* L.), *J. Agr. Food Chem.* **42**: 776–81.
KHAYYAL, M.T., EL-GHAZALY, M.A., KENAWY, S.A., SEIF-EL-NASR, M., MAHRAN, L. G., KAFAFI, Y.A. and OKPANYI, S.N. (2001), Antiulcerogenic effect of some gastrointestinally acting plant extracts and their combination. *Arzneimittelforschung* **51**(7): 545–53.
KIUCHI, F., NAKAMURA, N. and TSUDA, Y. (1987), 3-Caren-5-One from *Kaempferia galanga*. *Phytochemistry* **26**(12): 3350–1.
KOJIMA, H., KATO, A., KUBOTA K. and KOBAYASHI, A. (1997), Aroma compounds in the leaves of Japanese Pepper (*Zanthoxylum piperitum* DC) and their formation from glycosides. *Bioscience Biotech. Biochem.* **61**(3): 491–4.
KULKARNI, V.M. and RAO, P.S. (1999), *In vitro* propagation of sweet flag (*Acorus calamus*, Araceae). *J. Med. Arom Plant Sci.* **21**(2): 325–30.
KUMAR, V.S., SRIVASTAVA, R.K., KRISHNA, A., TOMAR, V.K.S., SINGH, A.K. and KUMAR, S. (2000), Cultivation, chemistry, biology and ulilization of bach (*Acorus calamus*): a review. *J. Med. Arom. Plant Sci.* **22**: 338–48.
KURIAN, A., PREMALETHA, T. and NAIR, G.S. (1993), Effect of gamma irradiation in Katcholam (*Kaempferia galanga* L.). *Ind. Cocoa Arecanut Spices J.* **16**: 125–6.
KUSUMOTO, S., OHSUKA, A., KOTAKE, M. and SAKAI, T. (1968), Constituents of leaf oil from Japanese pepper. *Bull. Chem. Soc. Japan* **41**: 1950–3.
LAWRENCE, B.M. (1981), Savory oil, Progress in essential oils. *Perf. Flavorist* **3**(6): 57–8.
LAWRENCE, B.M. (2002), Progress in essential oils. *Perf. Flavorist* **27**: 84–7.
LEWIS, Y.S. (1984), *Spices and Herbs for the food industry*. Food Trade Press, Orpington, 193–194.
LUNDQVIST, E. and ANDERSSON, E. (2001), Genetic diversity in populations of plants with different breeding and dispersal strategies in a free-flowing boreal river system. *Hereditas* **135**(1): 75–83.
MAZZANTI, G., BATTINELLI, L. and SALVATORE, G. (1998), Antimicrobial properties of the linalol-rich essential oil of *Hyssopus officinalis* L. var. decumbens (Lamiaceae). *Flav. Frag. J.* **13**(5): 289–94.
MITCHELL, J.C. (1974), Contact dermatitis from plants of the caper family, Capparidaceae. *Br. J. Dermatol.* **91**: 13–20.
MITIC, V. and DORDEVIC, S. (2000), Essential oil composition of *Hyssopus officinalis* L. cultivated in Serbia. *Facta Universitatis Series: Phys., Chem. Technol.* **2**(2); 105–8.
MITSUI, S., KOBAYASHI, S., NAGAHORI, H. and OGISO, A. (1976), Constituents from the seeds of *Alpinia galanga* Wild. and their anti-ulcer activities. *Chem. Pharm. Bull. (Tokyo)* **24**(10): 2377–82.

MORRIS, S. and MACKLEY, L. (1999), *Choosing and Using Spices*. Sebastian Kelley, Oxford, 51.
MUJUMDAR, A.M., NAIK, D.G., DANDGE, C.N. and PUNTAMBEKAR, H.M. (2000), Antiinflammatory activity of *Curcuma amada* Roxb. in albino rats. *Indian J. Pharmcol.* **32**: 375–7.
MUNSHI, A.M., ZARGAR, G.H., BABA, G.H. and BHATT, G.N. (1989), Effect of plant density and fertilizer levels on the growth and seed yield of black zira under rainfed conditions. *Indian Cocoa, Arecanut and Spices J.* **13**(4): 134–6.
NAJDA. A., WOLSKI, T., DYDUCH, J. and BAJ, T. (2003), Determination of quantitative composition of poliphenolic compounds occur in anatomically different parts of *Levisticum officinale* Koch. *Electronic Journal of Polish Agricultural Universities, Series Horticulture* **6**(1): http://www.ejpau.media.pl/series/volume6/issue1/horticulture/art-02.html
NAYAK, S. (2002), High frequency *in vitro* production of microrhizomes of *Curcuma amada*. *Indian J. Exp. Biol.* **40**(2): 230–2. http://www.ejpau.media.pl/series/volume6/issue1/horticulture/art-02.html.
NIRMAL BABU, K., REMA, J., GEETHA, S.P. MINOO, D., RAVINDRAN, P.N. and PETER, K.V. (1992), Micropropagation of betel vine (*Piper betle* L.). *J. Spices and Aromatic Crops* **1**(2): 160–2.
OJALA T, VUORELA P, KIVIRANTA J, VUORELA H, HILTUNEN R. (1999), Bioassay using *Artemia salina* for detecting phototoxicity of plant coumarins. *Planta Med.* **65**(8): 715–8.
OPDYKE, D.L.J. (1976), Savory oil (summer variety) *Food Cosmet. Toxicol.* **14**: 859–60.
PACHURI, S.P. and MUKHERJEE, S.K. (1970), Effect of *Curcuma amada* on the cholestertol level in experimental hypercholesterolemia in rabbits. *J. Res. Indian Med.* **5**: 27–31.
PAL, R.K., TRIPATHI, R.A. and PRASAD, R. (1996), Relative toxicity of certain plant extracts to khapra beetle, *Trogoderma granarium*. *Ann. Plant Protection Sci.*, **4**(1): 35–7.
PANERU, R.B., KENNEDY, S.H., LE-PATOUREL, G.N.J. (1997), Toxicity of *Acorus calamus* rhizome powder from Eastern Nepal to *Sitophilus granarius* (L.) and *Sitophilus oryzae* (L.) (Coleoptera, Curculionidae). *Crop Protection* **16**(8): 759–63.
PANWAR, K.S. (2000), Black caraway – Kala zira. In Arya, P. (ed.) *Spice Crops of India*. Kalyani Publ., New Delhi, 171–8.
PANWAR, K.S. SAGWAL, J.C., SHARMA, S.K. and SAROCH, K. (1993), Economic viability of Kala zira cultivation in high altitude dry temperate region of Himachal Pradesh. *Agric. Situation in India* **48**: 151–4.
PAROUL, N., ROTA, L., FRIZZO, C., ATTI DOS SANTOS, A.C., MOYNA, P., GOWER, A.E., SERAFINI, L.A. and CASSEL, E. (2002), Chemical composition of the volatiles of *Angelica* root obtained by hydridistillation and supercritical CO_2 extraction. *J. Ess. Oil. Res.* **14**(4): 282–5.
PARRY, J.W. (1969), Spices: Morphology, Histology and Chemistry Vol. II. Chemical Publ. Co., New York.
PASCUAL, B., BAUTISTA, A.S., FERREROS, N., LOPEZ-GALARZA, S. and MAROTO, J.V. (2003), Analysis of germination of caper seeds as influenced by the position of fruit on the mother plant, fruit maturation stage and fruit weight. *J. Hort. Sci. Biotech.* **78**(1): 73–8.
PATRA, A. and MITRA, K. (1981), Constituents of *Acorus calamus*: structure of acoramone. Carbon-13 NMR spectra of cis-8 trans-asarone. *J Nat. Proc.* **44**: 668–9.
PETRIKOVA, K., OPRAVILIVA, J. and SCHUBERTOVA, V. (2000), Essential oil and beta asarone contents in the sweet flag (*Acorus calamus* L.) collected at various locations in the Czech Republic, *Zahradnictvi Horticultural Sci.* **27**(1): 23–7.
PHILLIPS, R. and RIX, M. (1993), *Vegetables*. Pan Books, London.
PRAKASH, V. (1990), *Leafy Spices*. CRC Press Inc., Boca Raton, USA, 89–94.
RAHMAN, M.M. and SCHMIDT, G.H. (1999), Effect of *Acorus calamus* (L.) (Araceae) essential oil vapours from various origins on *Callosobruchus phaseoli* (Gyllenhal) (Coleoptera: Bruchidae). *J. Stored Products Research* **35**(3): 285–95.
RAINA, V.K., SRIVASTAVA, S.K. and SYAMASUNDER, K.V. (2002), The essential oil of the 'greater galangal' (*Alpinia galanga* (L.) Willd.) from the lower Himalayan region of India. *Flavor Fragrance J.* **17**(5): 358–60.
RAMACHANDRAN, K. (1969), Chromosome numbers in Zingiberaceae. *Cytologia* **34**: 213–21.
RAMALAKSHMI, K., SULOCHANAMMA, G., RAO, J.M.L., BORSE, B.B. and RAGHAVAN, B. (2002), Impact of drying on quality of betel leaf (*Piper betle* L.). *J. Food Sci. Technol.* **39**(6): 619–22.
RAMOS-OCAMPO, V.E. and HSIA, M.T.S. (1987), Mutagenic and DNA-damaging activity of calamus oil, asarone isomers and dimethoxypropenylbenzenes analogues. *Phill. Ent.* **7**: 275–91.
RANI, A S., SUBHADRA, V.V. and REDDY, V.D. (2000), *In vitro* propagation of *Acorus calamus* Linn. – A medicinal plant. *Indian J Exp. Biol.* **38**(7): 730–2.
RAO, A.S., RAJANIKANTH, B. and SESHADRI, R. (1989), Volatile aroma components of *Curcuma. amada*. *J. Agric. Food Chem.* **37**: 740–3.

RASTOGI, R.P. and MEHROTRA, B.N. (1990), *Compendium of Indian Medicinal Plants* Vol. 1 (1960–1969). Central Drug Research Institute, Lucknow and National Institute of Science Communication, New Delhi.

RASTOGI, R.P. and MEHROTRA, B.N. (1993), *Compendium of Indian Medicinal Plants* Vol. 3 (1980–1984). Central Drug Research Institute, Lucknow and National Institute of Science Communication, New Delhi.

RASTOGI, R.P. and MEHROTRA, B.N. (1995), *Compendium of Indian Medicinal Plants* Vol. 4 (1985–1989). Central Drug Research Institute, Lucknow and National Institute of Science Communication, New Delhi.

RAVINDRAN, P.N. (2000), Other economically important species of *Piper*. Ravindran, P.N. (ed.) *Black Pepper* – Piper nigrum, *Medicinal and Aromatic Plants – Industrial Profiles*. Harwood Academic Publishers, 497–509.

RAWAT, A.K.S., TRIPATHI, R.D., KHAN, A.J. and BALASUBRAHMANYAM, V.R. (1988), Essential oil as marker for identification of *Piper betle* L. cultivars. *Biochemical Syst. and Ecol.* **17**: 35–8.

REDDY, M.R.S. and REDDY, P.V.R M. (2000), Vasa (*Acorus calamus*) – a botanical pesticide against turmeric beetles. *Insect Environment* **6**(1): 8–9.

ROSENGARTNER, F. JR. (1969), *The Book of Spices*. Churchill Livingstone, London.

SALIKHOVA, R.A. and POROSHENKO, G.G. (1995), Antimutagenic properties of *Angelica archangelica* L. *Vestn Ross Akad Med Nauk* (**1**): 58–61.

SCF (2002), Scientific Committee on Food, European Commission. Opinion of the Scientific Committee on Food on the presence of β-asarone in flavourings and other food ingredients with flavouring properties, 1–15.

SCHEFFER, J.J.C and JANSEN, P.C.M. (1999), *Alpinia galanga* (L.) Willd. de Guzman, C.C. and Siemonsma, J.S. (eds): *Plant Resources of South-East Asia* No. 13. Spices. Backhyus Publishers, Leiden, the Netherlands. 65–8.

SCHMIDT, R.J. (ed.) (1979), Capparidaceae – Botanical Dermatology Database (BoDD). Electronic version of Botanical Dermatology by J. Mitchell and A. Rook, Greengrass, Vancouver, 1979. http://www.sfc.ucdavis.edu/pubs/SFNews/DecJan97–98/capers.html

SCHMIDT, G.H., RISHA, E.M. and EL-NAHAL, A.K.M. (1991), Reduction of progeny of some stored-product Coleoptera by vapours of *Acorus calamus* oil. *Jour. Stored Products Research* **27**(2): 121–7.

SHARMA, P.C. YELNE, M.B. and DENNIS, T.J. (2000), *Database on medicinal plants used in Ayurveda* Vol. 1. *Acorus calamus*. Central Council for Research in Ayurveda and Siddha, ISM & H, Ministry of Health & Family Welfare, Govt. of India, New Delhi, 469–95.

SHARMA, P.N., PANAWR, K.S., KAPOOR, A.S. and SAGWAL, J.C. (1993), Diseases of Kalazira in Himachal Pradesh. *Indian Cocoa, Arecanut and Spices J.* **17**: 9–12.

SHIMODA, M., WU, Y., NONAKA, S. and OSAJIMA, Y. (1997), Cluster analysis of GC data on oxygenated terpenes of young leaf and green fruit samples of Japanese pepper (*Xanthoxylum piperitum* DC). *J. Agric. Food Chem.* **45**: 1325–8.

SIATKA, T. and KASPAROVA, M. (2002), Effect of precursors on the production of coumarins in a suspension culture of *Angelica archangelica* L. *Ceska Slov Farm* **51**(1): 47–50.

SIMON, J.E., CHADWICK, A.F. and CRAKER, L.E. (1984), *Herbs: An Indexed Bibliography*. 1971–1980. Archon Books, Hamden, CT.

SINGH, S., PANDEY, P. and KUMAR, S. (2000), *Traditional Knowledge on the Medicinal Plants of Ayurveda*. Central Institute of Medicinal and Aromatic Plants (CIMAP), Lucknow, India, 340–349.

SU, H.C.F. (1991), Laboratory evaluation of toxicity of calamus oil against four species of stored-product insects. *Jour. Entomological Science* **26**(1): 76–80.

SUGIMOTO, N., MIKAGE, M., OHTSUBO, H., KIUCHI, F. and TSUDA, Y. (1997a), Pharmacognostical investigations of *Acori* rhizomes (1) Histological and chemical studies of rhizomes of *Acorus calamus* and *A. gramineous* distributed in Japan. *Natural Medicines* **51**: 259–64.

SUGIMOTO, N., OHTSUBO, H., MIKAGE, M., KIUCHI, F., LIU, H.M. and TSUDA, Y. (1997b), Pharmacognostical investigations of *Acori* rhizomes In Asian markets. *Natural Medicines* **54**: 316–24.

SUGIMOTO, N., KIUCHI, F., MIKAGE, M., MORI, M., MIZUKAMI, H. and TSUDA, Y. (1999), DNA profiling of *Acorus calamus* chemotypes differing in essential oil composition. *Biol. Pharm. Bull.* **22**(5): 481–5.

TAINTER, D.R. and GRENIS, A.T. (1993), Spices and Seasonings. VCH Publ., New York.

THAHN, L., DUNG, N.X., LUU, H.V. and LEELERCQ, P.A. (2002), Chemical composition of the leaf oil from *Piper betle* L. cultivated in Vietnam. *J. Ess. Oil Bearing Plants* **5**(1): 38–42.

THIEME, H. and NGUYEN, T.T. (1972a), Studies on the accumulation of volatile oils in *Satureja hortensis* L., *Satureja montana* L. and *Artemisia dracunculus* L. during ontogenesis. 1 – Review of literature, thin layer chromatography and gas chromatographic studies. *Pharmazie* **27**: 255–65.

THIEME, H. and NGUYEN, T.T. (1972b), Studies on the accumulation of volatile oils in *Satureja hortensis* L., *Satureja montana* L. and *Artemisia dracunculus* L. during ontogenesis. 2 – Changes in the content and composition of the volatile oil. *Pharmazie* **27**: 324–31.

TIWARI, J.P., TIWARI, A.B., TIWARI, G. and KUMAR, S. (2000), Effect of nitrogen application on growth and yield of *Acorus calamus*, (Proc. Natl. Sem. on the Res Dev. In Aromatic Plants). *J. Med. Arom. Plant Sci.* **22**(IB): 636–8.

VERGHESE, J. (2003), Savory. *Spice India* **16**(3): 12–15.

VINCENT, K.A., MARY, M. and MOLLY, H. (1992), Micropropagation of *Kaempferia galanga* L. – a medicinal plant. *Plant Cell Tissue and Organ Culture* **28**: 229–30.

WARRIER, P.K., NAMBIAR, V.P.K., RAMANKUTTY, C. (eds) (1994), *Indian Medicinal Plants – a Compendium of 500 Species*, Vol. 1. Orient Longman Pvt. Ltd., Chennai, 106.

WARRIER, P.K., NAMBIAR, V.P.K. and RAMANKUTTY, C. (eds) (1995), *Indian Medicinal Plants – a Compendium of 500 Species*. Vol. 3. Orient Longman Pvt. Ltd., Chennai, 13.

WEIDNER, M.S., PETERSON, M.J. and JACOBSON, N. (2002), Novel synergistic compositions containing aromatic compounds and terpenoides present in *Alpinia galanga*. *United Nations Patent Application – 20020086906 (A1)*. US Patent and Trademark Office.

WESTLAND, P. (1987), Angelica. *The Encyclopedia of Herbs and Spices*. Marshall Cavendish Ltd, London, 79.

WIERSEMA, J.H. and LEON, B. (1999), *World Economic Plants – A Standard Reference*. CRC Press LLC, Boca Raton, Florida.

WISEMAN, R.W., MILLER, E.C., MILLER, J.A. and LIEM, A. (1987) Structure–activity studies of the hepatocarcinogenicities of alkenylbenzene derivatives related to estragole and safrole on administration to preweaning male C57BL/6J×C3H/HeJ F1 mice. *Cancer Res.* **47**: 2275–83.

WU, L.J., SUN, L.L., LI, M.X., JIANG, Z.R., LU, Y., TIAN, Z.Y., ZHENG, Q.T. and ZHAO, H.B. (1993), The structural analysis of a new crystal sequiterpinenol in *Acorus calamus*. *Chin. J. Med. Chem.* **3**: 201–2.

WU, L.J., SUN, L.L., LI, M.X., YANG, H., JIANG, Z.R., LU, Y., TIAN, Z.Y., ZHENG, Q.T., MIYASE, T. and UENO, A. (1994), Studies on the constituents of the roots of *Acorus calamus* L. *Yakugaku Zasshi* **114**: 182–5.

ZHAO, Z.L., ZHUO, K.Y., DONG, H. and XU, L.S. (2001), Characters of nrDNA ITS region sequences of fruits of *Alpinia galanga* and their adulterants. *Planta Med.* **67**(4): 381–3.

ZOHARY, M. (1969), The species of *Capparis* in the Mediterranean and the near eastern countries. *Bull. Res. Counc. Israel* **8D**: 49–64.

Part II

Particular herbs and spices

6
Ajowan

S. K. Malhotra and O. P. Vijay, National Research Centre on Seed Spices, India

6.1 Introduction and description

The ajowan, *Trachyspermum ammi* (L.) Sprague ex Turrill belonging to the family Apiaceae is an important seed spice. It is known as bishop's weed, carum seed or carum ajowan. The common synonyms are *Trachyspermum copticum* Linn, *Carum copticum* Benth and Hook, *Ammi copticum* Linn., *Ptychotis coptica* DC and *Lingusticum ajowain*, Roxb. The correct generic position of this spice is very uncertain. Boissier considers it to belong to the genus *Ammi*, where Linnaeus originally put it, and as per *Genera Plantarum* it has been referred to *Carum*. In the recent past it was placed in the section *Trachyspermum*, which includes about 14 species (Bentley and Trimen, 1999). Ajowan is indigenous to India and Egypt (Sayre, 2001).

Ajowan is an annual, aromatic and herbaceous plant. It is profusely branched with a height of 60–90 cm small, erect with soft fine hair. It has many branched leafy stems, feather-like leaves 2–3 pinnately divided, segments linear with flowers terminal and compound. The fruits are small, ovoid, muricate, around cremocarps, 2–3 mm long, with greyish-brown compressed mericarps with distinct five ridges and tubercular surface. The fruits are the size and shape of parsley. The fruits have a very pungent aromatic taste and, when rubbed, they evolve a strong aromatic odour resembling that of thyme (*Thymus vulgaris*).

The crop belongs to family Apiaceae and order Apiales. As per the conventional classification of spices, out of the five types, ajowan is classified as aromatic spice, mostly dried fruits of which are used as spices. *Trachyspermum* is a cross-pollinated crop and has a somatic chromosome number of $2n = 18$. The flowers are self-fertile but cross-pollination occurs through insects.

6.2 Production

Ajowan is cultivated in the Mediterranean region and South-West Asian countries: Iran, Iraq, Afghanistan, Egypt and predominantly in India. In India it is grown in large areas in the

Table 6.1 Export of ajowan seed from India during 1996–97 to 2000–2001 (quantity in tonnes and value in Rs lakhs*)

	1996–97 Quantity	Value	1997–98 Quantity	Value	1998–99 Quantity	Value	1999–2000 Quantity	Value	2000–2001 Quantity	Value
Pakistan	–	–	–	–	–	–	–	–	335	76
Saudi Arabia	401	158	207	87	283	138	236	149	159	65
USA	41	19	21	11	33	25	39	165	55	30
UAE	46	17	15	5	28	11	5	2	79	29
Malaysia	35	10	29	13	20	5	–	–	62	25
Indonesia	40	13	35	10	–	–	–	–	45	16
Nepal	2	0.3	35	6	4	1	1	0.7	31	13
South Africa	29	14	11	6	6	5	13	10	14	10
Kenya	22	8	0.7	0.3	44	17	9	4	25	10
Bangladesh	–	–	–	–	–	–	–	–	44	9
Canada	16	8	3	2	29	9	9	7	22	9
UK	60	24	42	18	63	32	50	31	20	8
Other countries	212	62	99	32	150	67	53	34	71	23
Total	904	333	498	190	660	310	465	403	962	323

*Lakh = 100 000.

states of Rajasthan and Gujarat, and at a smaller scale in Uttar Pradesh, Bihar, Madhya Pradesh, Punjab, Tamil Nadu, West Bengal and Andhra Pradesh. It is cultivated almost throughout India.

In India, during 2000–2001, about 2900 tonnes of ajowan seed was produced from 13 600 ha, whereas 962 tonnes of ajowan seed worth Rs 32.3 million and 200 kg of ajowan oil with a value of Rs 1.61×10^5 was exported. India is the largest producer and exporter of the ajowan seed in the world exporting to around 46 countries. The major importing countries (Table 6.1) are Pakistan, Saudi Arabia, the USA, the UAE, Malaysia, Indonesia, Nepal, South Africa, Kenya, Bangladesh, Canada and the UK (Vijay and Malhotra, 2002; Selvan, 2002).

6.3 Cultivation

Ajowan is cultivated extensively as a cold season crop in the plains and as a summer crop in the hills. *Trachyspermum ammi* requires a warm and long frost-free growing season. It requires warm weather during seed development. The crop has moderate tolerance to drought and possesses wider climatic adaptability. It can be grown on any soil type: loamy to sandy loam and even in black soils. In India, it is grown under both rain-fed and irrigated cultivation system. The ripe fruits germinate relatively quickly and the germination time is 12 days. The seed from previous years harvest germinate well, but long storage quickly reduces germination vigour. The seeds are sown, broadcasted or drilled in rows 45 cm apart during September–October at a depth of 1 cm. The seed rate is 4 kg per hectare. The plant to plant spacing should be maintained at 20–30 cm (Malhotra, 2002).

The seeds are sown in well-prepared seed beds because seeds are small and have a low germination of 60–70%. An organic manure of 10–15 tonnes, 80–100 kg N, 30–50 kg P_2O_5, 30–50 kg K_2O and 50 kg sulphur per ha is recommended. The half dose of N, full quantity of P_2O_5, K_2O and sulphur should be added as a basal dose whereas the remaining dose of N should be applied in two equal parts at an interval of 30 and 60 days after sowing (Thomas et al., 2000; Krishna De, 2000; Malhotra, 2004b). A total of about five or six irrigations are required depending on the climate and soil type. Krishnamoorthy *et al.* (2000) have reported a significantly higher seed yield of 2050 kg/ha, when phosphorus 50 kg and nitrogen 100 kg/ha were applied.

The small white flowers bloom in November and December in the plains and mid summer in the hills. The harvesting is usually done from February to May. Flower production ceases when the seeds start maturing and become greyish-brown in colour. The yield is 400–600 kg/ha under a rain-fed farming system and 1200–2000 kg/ha under irrigated conditions. Collar root rot and powdery mildew are the major diseases and are controlled by spraying mancozeb (0.2%) and wettable sulphur (0.2%), respectively. Insects do not cause much damage to the crop.

Ajowan is cultivated in Iran, Iraq, Afghanistan, Egypt and Pakistan on a smaller scale and information related to cultivars is not available. The growing countries mostly grow regional cultivars. India predominantly produces a considerable amount of ajowan seed over large areas and farmers grow mostly local cultivars. Some of the regional cultivars developed through selections in India are GA-1 for Gujarat ; BEN-1 for Karnataka; RPA180 for Bihar; Sel-1 and Sel-2 for Andhra Pradesh. The NRC Seed Spices have developed AA 19 and AA 61 high-yielding cultivars suitable for cultivation under both rainfed and irrigated production systems (Malhotra and Vijay, 2003).

6.4 Chemical structure

The ajowan seed has the following chemical composition. The composition varies with variety, region and stage of harvest. The chemical composition of seeds is moisture 8.9%, protein 15.4%, fat (ether extract) 18.1%, crude fibre 11.9%, carbohydrates 38.6%, mineral matter 7.1%, calcium 1.42%, phosphorus 0.30%, iron 14.6 mg/100 g and a calorific value of 379.4 per 100 g (Pruthi, 2001). The detailed chemical composition is given in Table 6.2.

The volatile oil present in the seeds of ajowan is one of the principal constituents responsible for providing a typical flavour owing to the presence of thymol. Constituents of the seed are an aromatic volatile essential oil and a crystalline substance, stearoptene. It also contains cumene and terpene-'thymene'. One of the famous Indian *Ayurvedic* products,

Table 6.2 Chemical composition of ajowan ground spice per 100 g

Composition	Content
Carbohydrate (g)	24.6
Protein (g)	17.1
Fibre (g)	21.2
Water (g)	7.4
Food energy (calory)	363
Minerals (g)	7.9
Ca (g)	1.525
P (g)	0.443
Na (mg)	56
K (mg)	1.38
Fe (mg)	27.7
Thiamine (mg)	0.21
Riboflavin (mg)	0.28
Niacin (mg)	2.1

Source: Agarwal *et al.* (2000).

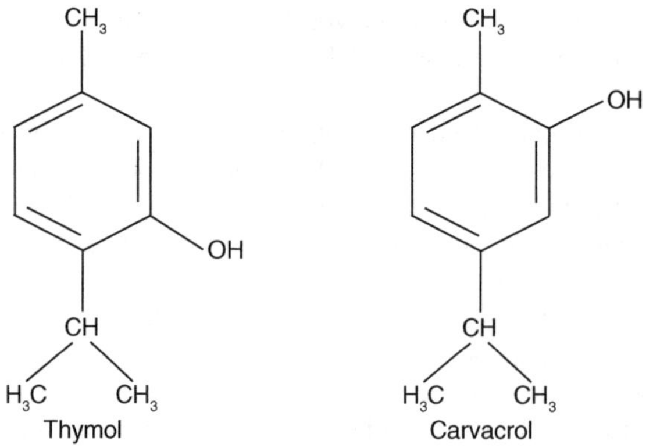

Fig. 6.1 Chemical structures of thymol and carvacrol.

ajowan-ka-phul contains stearoptene. A phenolic glucoside has been isolated and identified as 2-methyl-3-glucosyoxy-5-isopropylophenol. The fruits of ajowan yield 2–4% of essential oil, containing thymol (35–60%) as the major ingredient. Thymol crystallizes easily from the oil on cooling. The remainder of the oil consists of *p*-cymene, β-pinene, dipentene, β-terpinene and carvacrol (Prajapati *et al.*, 2003).

The chemical structures of thymol and carvacrol are shown in Fig. 6.1.

Ajowan oil has been reported to contain 27 compounds, of which thymol (61%) is the largest, with paracymene (15.6%), terpinene (11.9%), β-pinene (4–5%), dipentene (4–6%), comphene and myrcene in trace (Krishnamoorthy *et al.*, 2000). The water distilled oil from

Table 6.3 Essential oil composition of ajowan seed

Component	Essential oil (%)
Phenolic part	
Safrole	0.10
Thymol	87.75
Carvacrol	11.17
Non-phenolic part	
α-thujene	0.27
α-pinene	0.28
β-pinene	2.38
Myrcene	0.81
ρ-cymene	60.78
Limonene	8.36
γ-terpinene	22.26
Terpinolene	0.13
Linalool	0.27
Camphor	0.28
(Z) β-terpineol	0.19
(E) β-terpineol	1.35
Borneol	0.49
Terpinen-4-ol	0.12
α-teropineol	0.22
Carvone	0.15
Safrole	0.16

Source: Bhattacharya *et al.* (1998).

aerial parts and fruits of ajowan contain thymol (42.7% and 46.2%), γ-terpinene (38.5% and 38.9%) and *p*-cymene (14.1% and 13.9%) as main compounds (Masoudi *et al.*, 2002). Bhattacharya *et al.* (1998) reported the chemical composition of seed oil of ajowan. The phenolic components of the oil contained 87.75% thymol and 11.17% carvacrol as major constituents, and major non-phenolic components are *p*-cymene (60.78%) and γ-terpinene (22.26%) as given in Table 6.3.

The iodine content of the fruit from Patna was 0.45 mg/kg. Ajowan owes its characteristic odour and taste to the presence of an essential oil (2–4%). Other constituents in the fruits include sugars, tannins and glycosides. The alcoholic extract was found to contain a highly hygroscopic saponin, with a haemolytic index of 500. Yellow crystalline (mp 91–94°C) and steroidal substances (mp 140–150°C) called stearoptenes have also been isolated from ajowan fruits (Pruthi, 2001).

6.5 Main uses in food processing

The ajowan seed has been popular from ancient times for its use in folk medicines. In addition it has many uses for flavouring, culinary, household and cosmetic purposes. The entire plant has its herbal value in medicinal industry but commercially it is valued for its seed. Ajowan seeds have an aromatic smell and a warm pungent taste. They are used both as spices and condiment in many countries. They are used in India as a traditional spice in many foods, including curries. The major processed products are ajowan oil, oleoresin, thymol, thymol crystals, dethymolized oil (thymene) and fatty oils.

Ajowan oil is extracted from the seed by the steam distillation method. The two kinds of oils, i.e. essential oil (volatile oil) and non-volatile fatty oils, are extracted. Two integrated methods have been developed by Ramachandraih *et al.* (1988) to recover both these oils from the crushed seeds: the sequential method and the combined extraction method. Ajowan seeds contain 3–4% essential oil and 26% fatty oils. The yields of essential oil and fatty oils obtained through different methods are given in Table 6.4.

The ajowan oleoresin prepared from seeds gives a warm, aromatic and pleasing flavour to food products. The ajowan oleoresins are used in processed foods, snacks, sauces and various vegetable preparations. According to Pruthi (2001), by treating ajowan oil with aqueous alkaline solution, thymol can be extracted from it with ether or steam distillation. Both are used in the medicine/pharmaceutical industries. Indian standards have been laid by the Bureau of Indian Standards (BIS) for thymol, dethymolized oil or thymene production for industrial purposes. Fatty oils produced from ajowan seed, have their use in various pharmaceutical and cosmetic industries. Fatty oils are mainly used in soap industry for flavouring and as deodorant. They are also used for perfuming disinfectant soaps and as an insecticide. Thymol isolated from the oil is a powerful antiseptic and an ingredient in a

Table 6.4 Yield of essential oils and fatty oils from ajowan seeds by sequential and integrated methods *vis-à-vis* traditional method

Extraction method	Essential oil (%)	Fatty oil (%)	Meal (%)
1. Traditional	3.0	0.0	90.0
2. Sequential	3.0	27.0	64.0
3. Combination of above two methods	3.3	26.4	71.0

Source: Ramachandraiah *et al.* (1988).

number of skin ointments/powders, deodorant, mouthwashes, toothpastes and gargles. A thymol-free fraction of the oil, known as 'thymene', finds application in soap perfumes.

The Oil Technological Research Institute, Anantpur (AP) in India has developed an integrated method of developing oil, and fatty oil. As reviewed by Bentley and Trimen (1999), thymol may be obtained in large tubular crystals an inch or more (2.54 cm) in length on cooling. The oil also contains, in addition to thymol, a liquid hydrocarbon, which is called as cymol or cymene. These crystals from the oil are sold as *ajowan-ka-phool* (crystals) or *sat-ajowan* (water of ajowan) in the Indian market and is valued as medicine. It is probable also that the oil contains another hydrocarbon, which is isomeric with oil of turpentine. According to Stenhouse, as cited by Bentley and Trimen (1999), the liquid portion of the oil may be separated by rectification from the stearopten, as it boils at about 341°C while the latter begins to boil at only 424°C and is thus left behind in the still as a crystalline mass. Neither the thymol nor the liquid constituent (cymol) of the oil of ajowan has any rotatory power.

Ajowan powder is produced by grinding dried seeds. The pre-chilling and reduced temperature grinding can be used to overcome the loss of volatile oils. The finer powder product is mostly used for seasoning of foods whereas the coarse product is used for the purpose of extraction of oils, oleoresin and other extractives (Malhotra, 2000; Gopinath and Poonacha, 1992). Among other products, ajowan salt is commercially prepared by mixing finely ground rock salt and is mostly used for folk remedies of digestive problems.

The whole ajowan seed, powder and oil are used as adjuncts for flavouring foods, as antioxidants and as a preservative in confection, beverages and pan mixtures. Ajowan oil is also used in the preparation of lotions and ointments in the cosmetic industries (Malhotra, 2004a).

6.6 Functional properties and toxicity

The pharmaceutical data mentioned in the literature mainly refer to ajowan seed, oil and thymol. Ajowan oil and thymol are known to possess a number of functional properties:

- antimicrobial;
- antiflatulent and antispasmodic;
- antirheumatic;
- diuretic;
- stimulant, carminative and expectorant.

The antimicrobial activities of the essential oil distilled from ajowan seed was tested against a range of micro-organisms such as *Lactobacillus acidophilus*, *Bacillus cereus*, *Saccharomyces cerevisiae*, *Mycoderma* sp. and *Aspergillus niger*. Meena and Sethi (1994) reported various degrees of inhibition against test organisms. *Mycoderma* sp. was the most susceptible and *Bacillus cereus* was the most resistant. The susceptibility followed the order of *B. cereus*, *L. acidophilus*, *S. cerevisiae*, *A. niger* and *Mycoderma* sp.; the greater antimicrobial activity was observed in oil of ajowan both at ambient temperature and 37°C. Alcoholic extracts of ajowan also exhibited potent antimicrobial effects inhibiting the growth of *B. subtilis*, *Escherichia coli* and *S. cerevisiae*. Ajowan seeds are reported to be useful in flatulence, colic, atonic dyspepsia, diarrhoea, cholera, hysteria and spasmodic affections of bowels (reviewed by Pruthi, 2001; Latif and Rahman, 1999). The seed produces a feeling of warmth and relieves sinking and fainting feelings which accompany bowel disorders. Ajowan seed in conjunction with asafoetida, myrobalan and rock salt proved beneficial in stomach ache problems. A teaspoonful of seeds with a little salt is a common domestic remedy for indigestion from irregular diet. For stomach ache, cough and digestion, the seeds are masticated and swallowed, and this is followed by intake of a glass of hot water. A hot poultice of seed is used as a dry fomentation to the chest in asthma and expectoration from bronchitis.

The methanolic extracts of ajowan seed possess natural antioxidant properties. Thymol is also a powerful antiseptic and has agreeable odour. Thus it is also useful in controlling a variety of fungal infections of the skin. According to Krishnamoorthy *et al.* (2000), the aqueous portion left after the separation of essential oil from ajowan is known as *omum*-water (ajowan water), which is used against flatulence and in gripe water preparation for children. The oil is mainly carminative and antiflatulent. It is also applied to relieve rheumatic and neuralgic pain. It is also used to eradicate worms and in urticaria (Prajapati *et al.*, 2003). Traditionally, the seeds have been used in India as a folk remedy for arthritis, asthma, coughs, diarrohea, indigestion, intestinal gas, influenza and rheumatism (Sayre, 2001).

The folk Indian remedies as reviewed by Nadkarni (2001) are as follows:

- Ajowan seed with little rock salt mixture with the dose of a teaspoonful daily after meals improves indigestion and irregular diet.
- Compound powder of ajowan seed, rock salt, *sonchal* salt, *yavakshdra*, asafoetida, *myrobalan* equal part, to a dose of a teaspoonful daily after meals for a week relieves colic or pain in bowel.

Ajowan seed and its extract do not appear to have any significant toxicity. The amount of ajowan normally used in food are non-toxic. Normally, the concentrations of compounds in ajowan do not pose a health threat for consumption or to field workers handling the plants.

6.7 Quality issues

6.7.1 Specification for whole seed

The quality of ajowan seed depends mainly on the following:

- External appearance, which provides visual perception of quality such as colour, uniformity of size, shape and texture. Ajowan fruits are ovoid in shape and greyish-brown in colour and measure 1.7–3.0 mm long, 1.5–2.4 mm broad and 0.5–1.4 mm thick. Each mericarp has five ridges and the odour is similar to thyme (Chopra, 1998).
- Agmark of India provides three grades of ajowan seed, viz. special, good and fair (see Table 6.5).

Table 6.5 Agmark grade specification of ajowan seed

Grade designation	Special characteristics			
	Inorganic foreign matter (% by weight maximum)	Organic foreign matter (% by weight maximum)	Shrivelled, immature, damaged, discoloured and weevilled (% by weight maximum)	Moisture (%)
Special	0.24	0.50	1.0	11
Good	0.50	0.75	2.0	11
Fair	1.00	1.00	3.0	11

Source: Anon. (1997).

General characteristics of ajowan seed:

- seed shall be the dried ripe fruits of plant botanically known as *Trachyspermum ammi* (Linn);
- seed shall have the characteristic size, shape, colour, taste and aroma normal to the variety;
- seed shall be free from visible mould or insect, living or dead;
- seed shall be free from musty odour.

Ajowan has not received a place in the American Spice Trade Association (ASTA), the European Spices Association (ESA) and other ISO specifications list presumably because it has been considered of little importance because of the availability of thymol from *Thymus vulgaris*. The minimum specific quality indices for ajowan seed as per Pruthi (2001) are:

- seed moisture = not more than 12% by weight.
- total ash = not more than 7% by weight.
- ash insoluble in dilute HCl = not more than 1.5% by weight.
- organic extraneous matter = not more than 3% by weight.
- inorganic extraneous matter = not more than 2% by weight.
- volatile oil = not less than 1% (v/w).
- insect damaged matter = not more than 5% by weight.

Ajowan powder is produced by grinding dried, cleaned and sterilized seeds. After sieving through the required mesh size, the powder is packed in airtight containers. Some flavour may be lost by heat development during grinding. This can be reduced by using freeze-grinding techniques. Ajowan powder is yellowish brown with an aroma similar to thyme. The whole seed specification should be strictly followed in addition to seed powder quality specifications.

6.7.2 Volatile oil and oleoresins

The volatile oil content of ajowan seed averages 2–4% and it contains primarily 35–60% thymol, *p*-cymene (10–16%), α-terpinene (10–12%), β-pinene (4–5%) and dipenene (4–6%). The aroma of ajowan oil is warm, spicy, slightly fatty persisting and with a burning sensation. It is a colourless to brownish yellow liquid with the characteristic odour of thymol. The physiological properties of ajowan oil (from Singhal *et al.*, 1997) are:

- specific gravity = 0.910–0.930.
- refractive index = 1.498–1.504.
- optical rotation = up to 5°.
- soluble in 1–2 vols and more of 80% alcohol.
- phenols, 45–57%.

Ajowan oleoresin represents the overall flavour profile of the spices. It consists of the volatile essential oil and non-volatile resinous fraction comprising taste components. The ajowan oleoresin should be prepared with recommended organic solvents followed by the subsequent removal of the solvent as per specifications of importing countries. Oleoresin of ajowan is a pale green oily liquid with characteristic aroma and sharp taste attributable to the essential oil. The non-volatile fraction of the oleoresin contains essentially the fixed oils of the seed.

6.7.3 Adulteration

Ajowan seed is available both as whole or in ground form. It is subject to adulteration by addition of exhausted or spent seed (from which oil or oleoresin has been extracted) excess stems, chaff and earth or dust, etc. The oil is also adulterated with ajowan chaff oil. The range of essential oil is 2–4% and it should contain thymol ranging from 35 to 60%. If chaff oil is added, the thymol content will reduce to below 35%. The oleoresin may be adulterated by added synthetic saturated acid. Detection of these adulterants can be done by gas chromatography or by thin layer chromatography coupled with high-performance liquid chromatography. The adulteration at any level can be detected by using the specifications as explained separately for whole seed, powdered seed, volatile oil and oleoresin.

6.8 References

AGARWAL, S., SASTRY, E.V.D and SHARMA, R.K. (2000), *Seed Spices – Production, Quality and Export*. Pointer Publisher, Jaipur, India.

ANON.(1997), Indian Standards 'Ajowan', National Agmark Standards for Spices, Directorate of Marketing and Inspection, Ministry of Agriculture, Govt of India, Bureau of Indian Standards, New Delhi. IS: 4403–1979.

BENTLEY, R. and TRIMEN, H. (1999), *Medicinal Plants*. Asiatic Publishing House, Delhi, India.

BHATTACHARYA, A.K., KAUL, P.N and RAJESHWARA, R.B.R. (1998), Essential oil composition of ajowan seed production in Andhra Pradesh. *Indian Perfumer* **42**: 65–7.

CHOPRA, G.L. (1998), *Angiosperms*. Pradeep Publications, Jalandhar, India, pp. 55–6.

GOPINATH, G. and POONACHA, N.M. (1992), Bishop's weed. *Spice India* **5**(9): 9–14.

KRISHNA, DE A. (2000), The wonders of ajowan. *Spice India* **13**(1): 14–15.

KRISHNAMOORTHY, V., MADALGERI, M.B. and KANAN, C. (2000), Effect of interaction of N and P on seed and essential oil yield of ajowan genotypes. *J.Spices Arom.Crops* **9**(2): 137–9.

LATIF, A. and RAHMAN, S.Z. (1999), Medicinal use of spices for skin care in Unani medicine. In. Proc. Golden Jubilee *National Symposium on Spices, Medicinal and Aromatic Plants – biochemistry conservation and utilization*. IISR Calicut, 10–12 August 1998, pp. 274–81.

MALHOTRA, S.K. (2000), Value added spices products. In *Spices Crops of India* (ed.) P.S. Arya. Kalyani Publishers, New Delhi,. pp. 73–7.

MALHOTRA, S.K. (2002), *Ajowan Cultivation Practices* (in Hindi). NRCSS, Ajmer. Extension Folder No. 5, pp. 1–4.

MALHOTRA, S.K. (2004a), Underexploited seed spices. In *Spices, Medicinal and Aromatic Crops* (ed.) J. Singh. University Press, Hyderabad, India (in press).

MALHOTRA, S.K. (2004b), Minor seed spices 1 – Ajowan, dill, celery and anise. In *Fifty Years of Spices Research in India* (ed.) P.N. Ravindran. IISR, Calicut, India (in press).

MALHOTRA, S.K. and VIJAY, O.P. (2003), Plant genetic resources of seed spices in India. *Seed Spices Newsletter* **3**(1): 1–4.

MASOUDI, S., RUSTAIYAN, A., AMERI, N., MONFARED, A., KOMEILIZADEH, H., KAMALINEJED, M. and JANU ROODI, J. (2002), Volatile oils of *Carum copticum*. *J. Essential Oil Res.* **14**(4): 288–9.

MEENA, M.R. and SETHI, V. (1994), Anti-microbial activity of essential oils from spices. *J. Food. Sci. Tech.* **31**(1): 68–78.

NADKARNI, KM. (2001), *Indian Plants and Drugs with their Medicinal Properieties and Uses*. Asiatic Pub. House, Delhi, India, pp. 259–60.

PRAJAPATI, N.D., PUROHIT, S.S., SHARMA, A. and KUMAR, T. (2003), *A Handbook of Medicinal Plants*. Agribios India, Jodhpur, India, pp. 362–3.

PRUTHI, J.S. (2001), *Minor Spices and Condiments*. ICAR, New Delhi, pp. 124–33, 659–60.

RAMACHANDRAIAH, O.S., AZEEMODDIN, G. and THIRUMALA RAO, S.D. (1988), Integrated methods of obtaining essential and fatty oils from umbelliferous seeds. *Indian Perf.* **32**(1): 55–60.

SAYRE, J.K. (2001), *Ancient Herbs and Modern Herbs*. Bottlebrush Press, San Carlos, CA.

SELVEN, T.M. (2002), *Arecanut and Spices Database*. Directorate of Arecanut and Spices Development, Calicut, Kerala, India, pp. 1–105.

SINGHAL, R.S., KULKARNI, P.R. and REGE, D.V. (1997), In *Handbook of Indices of Food Quality and Authenticity*, Woodhead Publishing Limited, Abington, pp. 386–456.

THOMAS, J, JOY, P.P., MATHE, S., SKARIA, B.P., DUETHI, P.P. and JOSEPH, T.S. (2000), *Agronomic Practices for Aromatic and Medicinal Plants*. Directorate of Arecanut and Spices Development, Calicut, Kerala, India.

VIJAY, O.P. and MALHOTRA, S.K. (2002), Seed Spices in India and World. *Seed Spices Newsletter* **2**(1): 1–4.

7
Allspice

B. Krishnamoorthy and J. Rema, Indian Institute of Spices Research, India

7.1 Introduction and description

Allspice, *Pimenta dioica* (L.) Merr. (syn: *P. officinalis* Lindl., *Myrtus pimenta* L., *M. dioica* L. and *Eugenia pimenta* DC (Merrill, 1947) is a polygamodioecious evergreen tree, the dried unripe fruits of which provide the culinary spice pimento of commerce. It belongs to the family Myrtaceae and is known in English as allspice or pimento, in French as *piment jamaique* or *toute-epice,* in Portuguese as *pimenta da Jamaica* and in Spanish as *pimienta gorda*. The vernacular names of allspice are given in Table 7.1. The name allspice was coined by John Ray (1627–1705), an English botanist, who identified the flavour to a combination of clove, cinnamon and nutmeg (www.fragrant.demon.co.uk/allspice.html). The family Myrtaceae consists of about 3000 woody species, most of which grow in the tropics. The genus *Pimenta* Lindl. consists of about 18 species of aromatic shrubs and trees native to tropical America (Willis, 1966). The genus is closely related to *Myrtus* L. and *Eugenia* L. The commercially important *Pimenta* spp. is *Pimenta dioica* (L.) Merr. providing the spice pimento (allspice) and *P. racemosa* (Mill) Moore, bay or bay rum tree providing oil of bay. The basic chromosome number for the genus is $x = 11$ and allspice is a diploid with $2n = 22$ (Purseglove *et al.*, 1981).

Allspice is a small, functionally dioecious evergreen tree, 7–10 m tall, slender trunk profusely branched at its extremities. The bark is smooth and shiny, pale silvery brown, shedding in strips of 25–75 cm long at intervals. Leaves are borne in clusters at the ends of the branches, simple, opposite, entire, thinly, coriaceous, punctate with pellucid glands, aromatic when crushed. The petiole is 1–1.5 cm long, lamina elliptic to elliptic–oblong, 6–15 cm long and 3–6 cm wide, rounded at the apex and tapering at the base, dark green above, paler beneath and pinnately veined with the midrib impressed on the upper surface and prominent beneath, lateral veins not very prominent. Inflorescence axillary, compound, paniculate, separately branched, 5–15 cm long, composed of many flowered cymes. Flowers structurally hermaphrodite, but functionally male or female, white, aromatic, 8–10 mm diameter. Pedicels are about 1 cm long, pale green and pubescent, with small brownish pubescent bracteoles. The receptacle has four, cream-coloured, thick rounded calyx lobes, 1.5–2 mm long, wide spreading at anthesis and persistent in the fruit. There are four petals, reflexed, rounded, white, about 3–4 mm long, quickly deciduous. Stamens are free, numerous,

Table 7.1 Vernacular names of allspice (*Pimenta dioica*)

Language	Vernacular name
Arabic	*Bahar, Bhar hub wa na'im*
Danish	*Allehande*
Dutch	Jamaica pepper, piment
English	Jamaica pepper, myrtle pepper, pimento, newspice
Estonian	*Harilik pimendipuu, Vurts*
Finnish	*Maustepippuri*
French	*Piment, piment Jamaique, poivre aromatique, toute-epice, poivre de la Jamaique*
German	*Piment Neugewurz, Allgewurz, Nelkenpfeffer, Jamaicapfeffer, Englisches Gewurz*
Hungarian	*Jamaikai szegfubors, Szegfubors, Pimento, Amomummag*
Icelandic	*Allrahanda*
Italian	*Pimento, pepe di Giamaica*
Norwegian	*Allehande*
Polish	*Ziele angielskie*
Portuguese	*Pimenta da Jamaica*
Russian	*Yamaiskiy pjerets*
Spanish	*Pimienta de Jamaica, pimienta gorda*
Swedish	*Kryddpeppar*
Turkish	*Yeni bahar*

Source: http://www.GernotKatzer's Spice Directory_files\Prim_dio.htm.

5 mm long, about 100 in functionally male and 50 in functionally female. The anther is cream-coloured, filament white, slender, small, basifixed, bilocular, dehiscing by longitudinal slits. The style is white, shortly pubescent, about 5 mm long, stigma yellow. The ovary is inferior, two-celled, usually with one ovule in each cell, attached to the apex of the inner angle. The fruit is a sub-globose berry, 4–6 mm in diameter, green when unripe, deep purple to glossy black when ripe, aromatic on drying, dried unripe fruits dark brown. The embryo is involute–spiral in 2–2.5 coils, with very short cotyledons and a thick, long radicle.

Variants are rarely reported in allspice. Two seedling variant types with dwarf/semi-dwarf habit and short internodes and bushy nature possessing a large number of branches are being conserved in the field germplasm repository of Indian Institute of Spices Research, Calicut, Kerala, India (Krishnamoorthy *et al.*, 1997). The leaves are smaller (about one-third the size) than the leaves of normal trees. The variants were multiplied clonally through approach grafting and all the clones exhibited the parental character (Mathew *et al.*, 1999). This dwarf/semi-dwarf plant type in allspice with a large number of branches offers great potential in crop improvement programmes.

7.1.1. Etymology

The word pimento is derived from the Spanish word *pimienta* for black pepper, as allspice resembles peppercorns. It is known as pepper in many languages. In Russian it is known as *Yamaiskiy pjerets* – Jamaica pepper; in French *poivre aromatique* – aromatic pepper; in German *Nelkenpfeffer* – clove pepper; and in Swedish as *kryddpeppar* – condiment pepper. Newspice (German *Neugewurz*), also refers to its origin from the New World and the French *toute-epice*, reflects the complex aroma of this spice. However, the berries were widely known as *pimienta*, later anglicized as pimento. The genus name *Pimenta* was derived from the Spanish *pimiento* for black pepper. Since the Spaniards initially called allspice pimiento, the name was also introduced to many European countries along with the spice when the spice was introduced to Europe in the 16th century. The species name *dioica* (Greek *di-* from

Table 7.2 Average price of allspice in New York

Country	Year	Price (dollars/pound)
Guatemala/Honduras	1997	0.961
	1998	1.041
	1999	1.920
	2000	2.452
Jamaica	1997	1.180
	1998	1.353
	1999	2.268
	2000	3.579
Mexico	1997	0.583
	1998	1.071
	1999	1.912
	2000	2.360

Source: http://www.fas.usda.gov/htp/tropical/2001/03-01/spcavg.pdf.

dyo 'two', *oikos* 'house') indicates that the functional male and female flowers grow on different plants.

7.2 Production and trade

Jamaica is the largest producer and exporter of pimento, accounting for 70% of the world trade. The remaining 30% is produced by Honduras, Guatemala, Mexico, Brazil and Belize. The dried mature but unripe berries, berry oleoresin, berry oil and leaf oil are the products of commercial importance obtained from *P. dioica* and they find varied uses in the food, medicine and perfume industries. Among the pimentos from various geographical locations, Jamaican pimentos are considered to be of high quality because of their flavour, appearance and size and receive a premium price in the market (Table 7.2). The major importing countries are USA (Table 7.3), Germany, the UK, Finland, Sweden and Canada. Leaf oil is mainly exported to the USA and the UK. Pimento is generally classified with capsicum in the import statistics of most countries and hence analysis of the market situation is difficult.

7.2.1 Origin and distribution

The tree is indigenous to West Indies (Jamaica). The trees are also found in Central America (Mexico, Honduras, Guatemala, Costa Rica and Cuba) and in the neighbouring Caribbean islands, although its original home is in dispute. Christopher Columbus discovered allspice in the Caribbean islands in about 1494. Spanish explorers and later settlers in Jamaica harvested and used the leaves and berries. Reports indicate that, there has been continuous production of berries in Jamaica from about 1509 to the present day. The berries reached London in 1601 as described by Clusius in his *Liber Exoticorum* and the plants were first cultivated in England in a hot house in 1732 (Weiss, 2002). Before World War II, allspice was more widely used than today; however, during the war many trees were cut down and there was a shortage of the spice. Though cultivation was taken up after the war, production never fully recovered.

Allspice was introduced into West Indian Islands (Grenada, Barbados, Trinidad and Puerto Rico) from its place of origin. Attempts to introduce it into countries in tropical

Table 7.3 Imports of allspice by the USA

	1999		2000	
Country	Quantity (kg)	Value (dollars)	Quantity (kg)	Value (dollars)
China	99 850	126 658	94 320	69 458
Guatemala	185 349	540 039	269 339	1 077 571
Honduras	424 028	1 044 247	297 831	1 137 345
India	167 379	118 132	62 809	62 136
Jamaica	367 384	1 218 489	359 881	1 789 828
Lebanon	750	3 000	0	0
Mexico	80 849	201 311	387 081	1 205 363
Pakistan	13 230	31 832	28 592	39 483
Portugal	5 000	2 955	0	0
Spain	12 668	52 486	19 800	13 766
Taiwan	1 081	3 400	0	0
Thailand	0	0	4 830	16 325
Turkey	1 326	2 323	2 852	4 808
Other	39 529	91 470	40 164	145 499
Total	1 398 423	3 436 342	1 567 499	5 561 582

Source: http://www.fas.usda.gov/htp/tropical/2001/03-01/tropic.htm.

regions namely, India, Sri Lanka, Fiji, Malaysia, Singapore and Indonesia (Java, Sumatra), have, for various reasons, not succeeded fully. In India, there are a few trees in Maharashtra, Tamil Nadu, Karnataka and Kerala.

7.3 Chemical composition

7.3.1 Berry

The dried, mature but not ripe, berries are the pimento spice of commerce. Pimento is also sold as ground spice. The berries of international standard should be between 6.5 and 9.5 mm in diameter, medium to dark brown in colour, with an uneven surface and with a pleasant odour, characteristic of the spice and with approximately 13 fruits/g. The dried berry contains aromatic steam volatile oil, fixed (fatty) oil, resin, protein, starch, pigments, minerals, vitamins (Table 7.4), etc. The constituents present in the oil influence the quality and aroma of the spice. The phenolic compound eugenol and isoeugenol and the sesquiterpene hydrocarbon, β-caryophyllene are the major compounds present in allspice (Table 7.5). Several other compounds have been identified in allspice, which is present in lesser quantities (Table 7.6). The geographical variation, cultivar differences, stage of maturity, etc. also influence the quality of the berry. The quality of the berries from Jamaica are superior to those from other islands and are preferred for trade. Prolonged storage of allspice is detrimental to both oil content and flavour of the spice.

7.3.2 Berry oil

Extraction of berry oil can be carried out by different methods. Berry oil is generally obtained by hydrodistillation or steam distillation of dried immature berries. When supercritical CO_2 extraction techniques are employed for extraction of berry oil, the oil obtained is of superior quality and flavour, compared with steam distilled or hydrodistilled oil. The composition of berry oil extracted by steam distillation, hydrodistillation and

Table 7.4 Nutrient composition of allspice (per 100 g)

Composition	Quantity
Proximates	
Water	8.5 g
Food energy	262.6 kcal
Carbohydrates	72.1 g
Protein	6.1 g
Fat	8.7 g
Dietary fibre	21.6 g
Ash	4.6 g
Minerals	
Calcium	660.6 mg
Iron	7.1 mg
Magnesium	134.1 mg
Phosphorus	113.3 mg
Potassium	77.0 mg
Sodium	77.0 mg
Zinc	1.0 mg
Copper	0.6 mg
Manganese	2.9 mg
Vitamins	
Vitamin C	39.2 mg
Thiamin B1	0.1 mg
Riboflavin B2	0.1 mg
Niacin	2.9 mg
Vitamin B6	0.3 mg
Folate	36.0 µg
Vitamin E	1.0 mg

Source: USDA Nutrient Databases: http//www.organic.planet.com/products/g_allspice.html.

supercritical CO_2 extraction techniques are compared in Table 7.7 (Garcia-Fajardo *et al.*, 1997) The berry oils extracted by supercritical CO_2 method and steam distillation have been characterized based on their physicochemical properties (Table 7.8).

The yield of berry oil ranges from 3.0 to 4.5%. The oil is yellow to brownish yellow with a warm spicy sweet odour and fresh and sweet top-note, and is placed in the warm, sweet spicy group (Arctander, 1960). About 60 constituents have been detected, including phenols, monoterpene hydrocarbons, oxygenated hydrocarbons, sesquiterpene hydrocarbons and oxygenated sesquiterpenes, and about 34 constituents were reported in steam-distilled berry oil using gas chromatography (Nabney and Robinson, 1972). The oil from green berries is similar in composition to that from dried berries, but has a higher monoterpene content (Ashurst *et al.*, 1972). The principal components are usually eugenol, methyl eugenol, β-caryophyllene, humulene, terpinen-4-ol and 4,5-cineole (Fig. 7.1) (Tables 7.9, 7.10 and 7.11) (Nabney and Robinson, 1972; Purseglove *et al.*, 1981; Lawrence; 1999). The main constituents affecting taste and flavour are the abundance and ratio of 1,8-cineole (Fig. 7.1) and α-phellandrene.

Allspice contains various essential oils (Pino *et al.*, 1989), phenolic acids (Schulz and Herrmann, 1980), flavanoids (Vosgen *et al.*, 1980), catechins and phenyl propanoids (Kikuzaki *et al.*, 1999). The flavonol content of allspice is low and consists mainly of quercetin glycosides (Vosgen *et al.*, 1980). Three new galloylglucosides, (4*S*)-alpha-terpineol 8-*o*-beta-D-(6-*o*-galloyl) glucopyranoside; (4*R*)-alpha-terpineol 8-*o*-beta-D-(6-*o*-galloyl) glucopyranoside and 3-(4-hydroxy-3-methoxyphenyl) propane-1, 2-diol 2-*o*-beta-

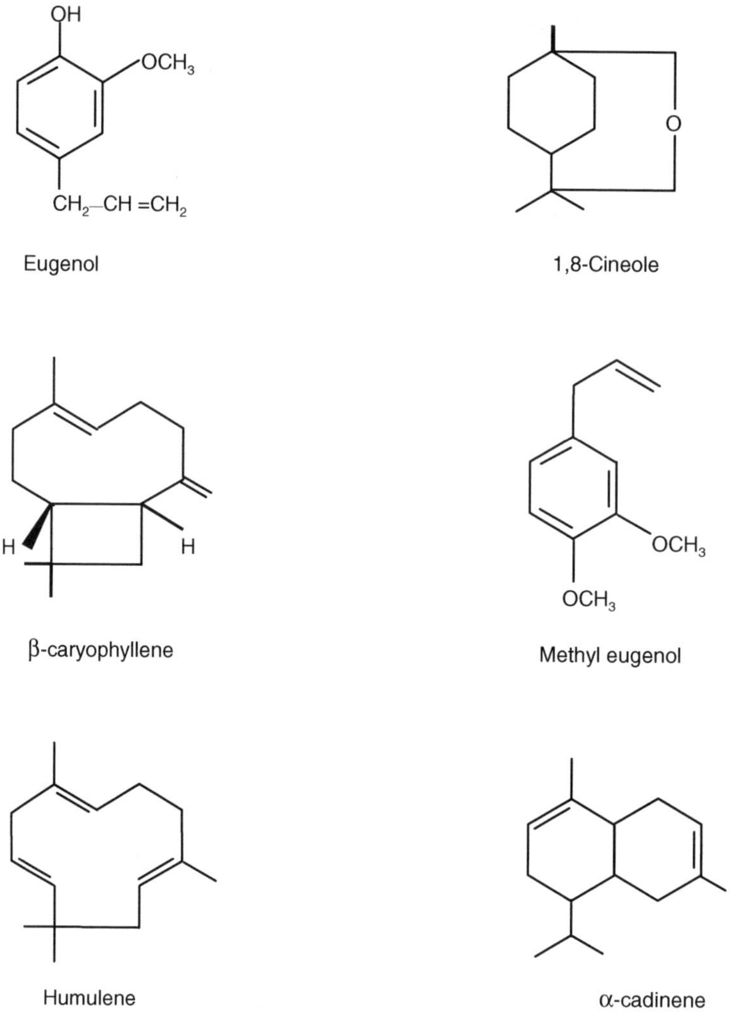

Fig. 7.1 Structures of some of the compounds in allspice.

D-(6-*o*-galloyl) glucopyranoside were isolated from berries of *P. dioica* (from Jamaica) together with three known compounds, gallic acid, pimentol and eugenol 4-*o*-beta-D-(6-*o*-galloyl) glucopyranoside (Kikuzaki *et al.*, 2000).

Allspice berry oil extracted by supercritical CO_2 extraction procedure is light red brown with the full sweetness and fresh natural odour and flavour of the freshly ground spice. The sensory character of the pimento berry oil obtained by steam distillation and liquid CO_2 extraction is represented in Fig 7.2 (Charalambous, 1994).

The initial impact of the liquid CO_2 extracted oil is sweet, spicy with a distinctly heavy fruity and floral dianthus character. After 6 h the profile becomes warmer, more fruity and peppery, less phenolic and spicy. These notes are still prominent after 24 h and continue for several days. The initial profile of steam distilled oil, although strong, is more phenolic, medicinal and less fruity. After 6 h the profile becomes warmer with increased fruitiness but not attaining the richness of fruit notes of the CO_2 extract. The floral character is hardly noticeable at any stage of evaporation. All these notes are still prominent after 24 h (Charalambous, 1994).

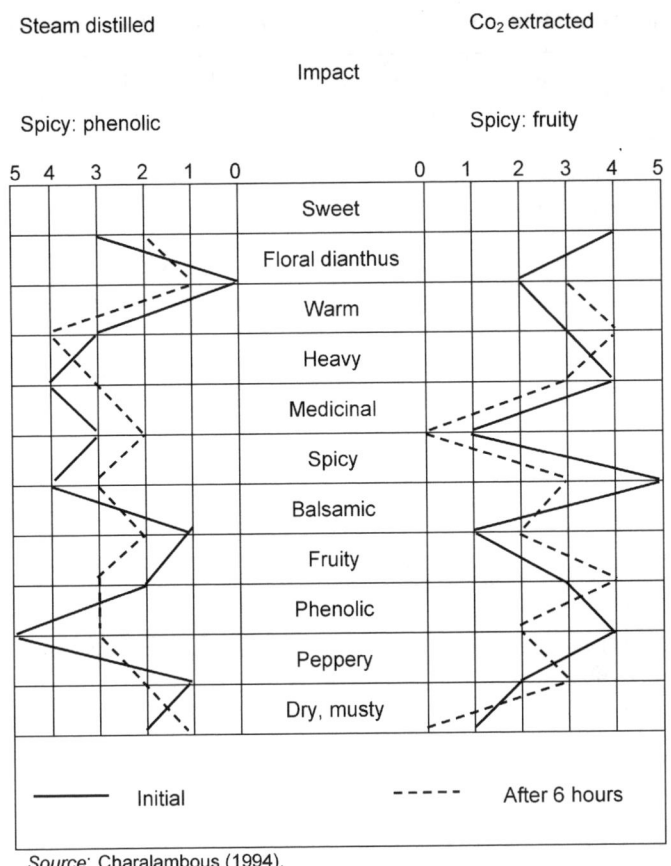

Fig. 7.2 Comparative odour profiles of steam distilled and CO_2 extracted pimento berry oil.

7.3.3 Oleoresin

Oleoresin is prepared by extraction of the crushed spice with organic solvents followed by evaporation of the solvent. The composition of the oleoresin depends upon the raw materials and the solvents used for extraction of oleoresin. The oleoresin is a brownish to dark green oily liquid and two grades are normally available, based on the volatile oil content namely, 40–50 and 60–66 ml per 100 g. An US specification requires a minimum of 60 ml per 100 g. A small quantity is sufficient to get the required flavour and aroma in food.

Kollmannsberger and Nitz (1993) compared the extracts made by supercritical CO_2 extraction procedure at various pressures and temperatures (150 bar and 350 bar at 50°C, 350 bar at 70°C); direct diethyl ether extract and simultaneous distillation and extraction of pimento berries. Supercritical CO_2 extracted at 350 bar pressure and 50°C temperature was found to be the best (Table 7.5).

7.3.4 Leaf oil

Pimento leaf oil is produced by distilling fresh or dry leaves. Leaves used for distillation may be fresh, withered or dried and stored for two or three months prior to distilling. Yield from dried and fresh leaves is 0.5–3.0% and 0.3–1.25%, respectively. The leaf oil is a brownish

Table 7.5 Constituents identified in allspice extracts (ppm) using different extraction procedures

Compound	CO_2 extracts			SDE	DDE
	150 bar/50°C	350 bar/50°C	350 bar/70°C		
α-pinene	40	60	39	50	46
β-pinene	39	55	37	56	54
Myrcene	38	48	37	79	72
(e)-β-ocimene	23	29	23	48	46
α-thujene	23	31	21	31	27
Sabinene	24	36	24	26	45
δ-3-carene	55	74	52	102	95
α-phellandrene	107	138	101	188	138
Limonene + β-phellandrene	138	188	119	298	233
p-cymene	85	111	79	193	171
α-terpinene	16	19	13	38	11
γ-terpinene	87	111	83	183	150
Terpinolene	146	199	136	261	170
1,8-cineole	272	355	249	472	403
Linalool	32	43	30	48	37
Terpinen-4-ol	110	160	107	198	123
p-cymen-8-ol	21	31	22	41	20
α-terpineol	47	70	47	90	44
Trans-p-menth-2-en-1-ol- + cis-p-menth-2-en-1-ol	16	21	15	31	15
β-caryophyllene	1 749	2 534	1 595	1 915	1 838
α-humulene	423	610	380	452	414
α-selinene	267	383	236	262	237
β-selinene	173	243	153	173	161
δ-cadinene	210	307	188	216	186
β-elemene	105	150	95	113	113
Allo-aromadendrene	81	118	73	83	82
Germacrene d	82	121	77	43	77
Spathulenol	42	54	39	47	34
Caryophyllene oxide + viridiflorol	187	225	147	168	142
Humulene oxide ii	39	55	35	47	34
t-cadinol + t-muurolol	76	103	66	79	55
α-muurolol	29	38	25	29	21
α-cadinol	60	90	56	78	51
Selin-11-en-4-ol	131	170	113	120	78
Caryophylla-2(12),6(7)-dien-5-ol	30	37	25	28	29
Eugenol	18 176	29 976	18 178	22 240	11 135
Methyl eugenol	1 822	2 670	1 661	2 025	1 424
Chavicol	57	73	58	60	25
Myristicin	33	44	26	31	23
Elemicin	13	19	12	15	10

SDE: simultaneous distillation and extraction using diethyl ether.
DDE: direct diethyl ether extract.
Source: Lawrence (1999).

yellow liquid with dry-woody, warm-spicy aromatic odour. The main composition of the leaf oil of allspice is eugenol. Eugenol content of leaf oil (65–96%) is somewhat higher than that in berry oil (Pino and Rosado, 1996; Pino *et al.* 1997). Leaf oil composition of allspice extracted by supercritical CO_2 method is given in Table 7.12. The chemical composition of the oil is also influenced by the geographical origin of the spice (Tables 7.9 and 7.13) (Pino *et al.*, 1997).

Table 7.6 Minor compounds in allspice berries

Camphene (3 ppm)	cis-sabinene hydrate (2 ppm)
(Z)-β-ocimene (1 ppm)	Linalool oxide-furanoid (1 ppm)
α-p-dimethylstyrene (1 ppm)	β-phellandren-6-ol (11 ppm)
δ-elemene (1 ppm)	trans-piperitol (1 ppm)
α-cubebene (10 ppm)	cis-piperitol (3 ppm)
α-ylangene (6 ppm)	Hexanal (<1 ppm)
α-copaene (35 ppm)	Benzaldehyde (<1 ppm)
β-cubebene (1 ppm)	Cinnamaldehyde (1–10 ppm)
α-gurjunene (43 ppm)	Vanillin (1–10 ppm)
α-bulnesene (27 ppm)	Methyl salicylate (3 ppm)
Aromadendrene (31 ppm)	Guaiacol (<1 ppm)
Selina-4,11-diene (35 ppm)	4-vinylguaiacol[†] (1 ppm)
γ-muurolene (57 ppm)	Methyl chavicol (6 ppm)
Ar-curcumene (10 ppm)	Safrole (5 ppm)
Zingiberene (25 ppm)	(E)-isoeugenol (1 ppm)
α-muurolene (42 ppm)	Methyl (E)-isoeugenol (1 ppm)
Germacrene a (11 ppm)	6-methyloxyeugenol (1–10 ppm)
β-bisabolene (4 ppm)	Palustrol (1 ppm)
cis-calamene (10 ppm)	Caryophyll-5-en-12-ol[†] (1 ppm)
β-sesquiphellandrene (5 ppm)	Isocaryophyllene oxide (1–10 ppm)
Cadina-4,11-diene (11 ppm)	Salvial-4(14)-en-1-one (1 ppm)
α-cadinene (11 ppm)	Globulol[†] (12 ppm)
cis-calacorene (3 ppm)	Humulene oxide (4 ppm)
trans-calacorene (1 ppm)	Ledol[†] (1–10 ppm)
Camphor (1 ppm)	Eudesmol* (1–10 ppm)
Ascaridole* (3 ppm)	Selineol*[†] (1–10 ppm)
Carvone (11 ppm)	Eudesmol* (1–10 ppm)
Geranial (<1 ppm)	Epi-cubenol (21 ppm)
Linalyl acetate (3 ppm)	Caryophylla-2(12),6(13)-dien-5-ol (12 ppm)
α-terpinyl acetate (10 ppm)	Isospathulenol (7 ppm)
Neryl acetate (2 ppm)	Cubenol[†] (1–10 ppm)
Geranyl acetate (6 ppm)	trans-sabinene hydrate (2 ppm)

*Correct isomer not identified; [†]Tentative identification.
Source: Lawrence (1999).

7.4 Cultivation

7.4.1 Propagation

Seeds

Allspice is traditionally propagated through seeds, but vegetative propagation is also adopted to get true to type plants. High-yielding trees that fruit regularly and have well-formed fruit bunches are selected as mother trees. Fresh ripe fruits from such high-yielding trees are collected and seeds are extracted from the ripe fruits after soaking them overnight in water and rubbing them in a sieve to remove the pericarp. Allspice seeds lose their viability soon after harvest and hence seeds are planted without delay after extraction. If seeds are to be transported or kept for a few days it is advisable not to extract the seeds from the fruits. It is reported that the viability of the seeds can be maintained at 50% for nine weeks by storing them at 21–30°C (Devadas and Manomohandas, 1988). Seeds are sown in beds of 15–20 cm height, 1 m width and convenient length made of loose soil-sand mixture over which a layer of sand (about 5–8 cm thick) is spread. Seeds are sown at 2–3 cm spacing and

Table 7.7 Percentage composition of a steam distilled oil, a hydrodistilled oil and a supercritical CO_2 extract of Mexican allspice

Compound	Steam distilled oil	Hydrodistilled oil	Supercritical CO_2 extract
α-pinene	trace	0.1	trace
β-pinene	trace	0.2	trace
Sabinene	0.3	0.3	0.2
Myrcene	17.7	16.5	6.0
δ-3-carene	trace	trace	–
α-terpinene	trace	0.1	trace
p-cymene	0.2	trace	trace
Limonene	0.7	trace	trace
1,8-cineole	1.9	4.1	1.3
(Z)-β-ocimene	trace	1.2	0.9
γ-terpinene	1.1	0.2	trace
Terpineolene	trace	0.6	0.4
Linalool	0.4	trace	trace
Terpinen-4-ol	0.3	0.5	0.3
Methyl salicylate	trace	trace	–
α-terpineol	0.7	0.7	0.4
Eugenol	17.3	8.3	14.9
Methyl eugenol	48.3	62.7	67.9
β-caryophyllene	6.2	2.7	5.2
α-humulene	1.1	0.2	0.2
γ-cadinene	0.6	0.1	0.2
β-selinene	trace	trace	trace
α-selinene	0.4	trace	trace
δ-cadinene	trace	trace	trace

Source: Lawrence (1999).

Table 7.8 Physicochemical comparison of liquid CO_2 extracted and steam distilled allspice berry oils

	Extraction procedure	
	Supercritical CO_2	Steam distillation
Specific gravity at 20°C	0.98 to 1.03	1.027 to 1.048
Refractive index at 20°C	1.505 to 1.525	1.525 to 1.54
Optical rotation at 20°C	–5 to 0	–5 to 0
Solubility in 70% v/v ethanol at 20°C	1 to 2	1 to 2
Total phenols v/v, minimum	75%	65%

Source: Charalambous (1994).

depth of about 2 cm. The seed bed has to be protected from direct sunlight. If only a small quantity of seeds is available for sowing, they can be sown directly in polybags filled with a soil–sand–cowdung mixture and kept in the shade. The beds may be mulched with dried leaves or straw to hasten germination. Watering should be done regularly. Germination commences in about 9–10 days and continues over a month. All the mulchings on the seed bed must be removed as the seeds start germinating. The seedlings are transplanted into polythene bags (25 cm × 15 cm) containing a mixture of soil, sand and well-decomposed cowdung (3 : 3 : 1) at the three- to four-leaf stage. The seedlings are ready for transplanting to the field at 9–10 months old, when they are 25–40 cm high.

Table 7.9 Constituents identified in Jamaican allspice berry and leaf oils

	Berry oil	Leaf oil
Phenolics	Eugenol	Eugenol
	Methyl eugenol	Methyl eugenol
	Chavicol	Isoeugenol
Monoterpene hydrocarbons	Δ-3-carene	Limonene
	p-cymene	cis-β-ocimene
	Limonene	trans-β-ocimene
	Myrcene	α-pinene
	α-pinene	α-phellandrene
	β-pinene	Sabinene
	α-phellandrene	γ-terpinene
	α-terpinene	Terpinolene
	γ-terpinene	
	Terpinolene	
	Thujene	
Oxygenated monoterpenes	1,8-cineole	1,8-cineole
	Linalool	Linalool
	α-terpineol	Terpinen-4-ol
	Terpinen-4-ol	
	Terpinen-4,8-oxide	
Sesquiterpene hydrocarbons	Alloaromadendrene	Alloaromadendrene
	γ-cadinene	δ-cadinene
	Calamene	β-caryophyllene
	β-caryophyllene	α-copaene
	Ar-curcumene	α-gurgunene
	β-elemene	α-humulene
	α-humulene	α-muurolene
	β-humulene	α-selinene
	Isocaryophyllene	
	γ-muurolene	
	α-selinene	
	β-selinene	
Oxygenated sesquiterpenes	β-caryophyllene alcohol	
	Caryophyllene oxide	
	Caryophyllene aldehyde	
	Humulene epoxide ii	

Source: Purseglove *et al.* (1981).

Table 7.10 Chemical composition of allspice berry oil

Eugenol (80.1%)
Methyl eugenol (5.0%)
β-caryophyllene (4.5%)
α-muurolene (1.1%)
α-selinene (1.1%)
Ledene (0.8%)
Allo-aromadendrene (0.7%)
Calamenene (0.3%)
p-cymene (0.3%)
10-α-cadinol (0.2%)
Methyl chavicol (0.2%)
Spathulenol (0.2%)
δ-cadinene (0.2%)
γ-cadinene (0.2%)
1,8-cineole (0.2%)
Myrcene (0.2%)
α-gurjunene (0.1%)
Linalool (0.1%)
Terpinolene (0.1%)
(E)-β-ocimene (0.1%)
Globulol (0.1%)
γ-terpinene (0.1%)
δ-3-carene (0.1%)
p-cymen-8-ol (0.1%)
Copaene (unknown isomer) (0.1%)
α,p-dimethylstyrene (0.1%)
Limonene (0.1%)
α-pinene (0.1%)
α-thujene (0.1%)
α-phellandrene trace
2-methylbutyl acetate trace
α-terpinene trace

Source: Guzman and Siemonsma (1999).

Table 7.11 Chemical composition of allspice berry oil from Cuba

Eugenol (87.0%)	β-selinene (0.2%)
1,8-cineole (3.3%)	γ-terpinene (0.2%)
β-caryophyllene (2.5%)	α-terpineol (0.2%)
α-humulene (1.6%)	Calamenene (0.1%)
p-cymene (0.7%)	Caryophyllene oxide (0.1%)
Terpinen-4-ol (0.5%)	α-copaene (0.1%)
Terpinolene (0.5%)	γ-muurolene (0.1%)
δ-cadinene (0.4%)	β-phellandrene (0.1%)
Guaiene (unknown isomer) (0.4%)	β-pinene (0.1%)
Limonene (0.4%)	α-terpinene (0.1%)
α-phellandrene (0.4%)	γ-cardinene (0.1%)
Camphene (0.2%)	α,p-dimethylstyrene (0.1%)
β-elemene (0.2%)	Humulene oxide (0.1%)
Myrcene (0.2%)	
α-pinene (0.2%)	Total 100%

Source: Guzman and Siemonsma (1999).

Table 7.12 Leaf oil composition of allspice extracted by supercritical CO_2

Methyl chavicol (0.31%)	α-muurolene (0.05%)
Thymol (1.82%)	Calamenene* + γ-cadinene (0.05%)
Carvacrol (1.08%)	Caryophyllene oxide (0.07%)
Eugenol (93.87%)	t-cadinol (0.17%)
β-caryophyllene (1.79%)	α-cadinol (0.17%)
α-humulene (0.35%)	α-amorphene (0.37%)

*Correct isomer not identified.
Source: Lawrence (1999).

Table 7.13 Chemical composition of leaf oil of allspice of Cuban origin

α-pinene (0.56%)	γ-terpinene (0.56%)
Myrcene (0.19%)	Terpinolene (1.38%)
α-phellandrene (1.12%)	Menthol (0.56%)
p-cymene (1.87%)	Methyl chavicol (0.09%)
1,8-cineole (14.50%)	Carvone (0.10%)
Limonene (0.10%)	Thymol (1.00%)
Carvacrol (1.00%)	δ-cadinene (5.49%)
Eugenol (28.04%)	Cadina-1,4-diene (0.49%)
β-caryophyllene (1.00%))	α-calacorene (1.23%)
α-humulene (10.12%)	Caryophyllene oxide (2.69%)
Allo-aromadendrene (2.13%)	α-eudesmol (0.52%)
α-amorphene (2.77%)	β-eudesmol (0.82%)
α-muurolene (1.76%)	t-cadinol (6.64%)
Calamenene* + γ-cadinene (11.12%)	α-cadinol (4.94%)

*Correct isomer not identified.
Source: Lawrence (1999).

Vegetative propagation

Allspice is polygamodioecious and it is difficult to identify the functional male and female trees until they flower. Hence clonal propagation is necessary to obtain uniformly high-yielding trees. Cuttings of allspice could be rooted in seven to eight months with hormonal application. Air layering of softwood and semi-hardwood shoots with hormone application (indolebutyric acid 4000 ppm + naphthalene acetic acid 4000 ppm) aided in rooting of allspice (Rema *et al.*, 1997). Studies on air layering in Maharashtra indicated that rooting is a slow process, taking 18–28 months and that January is the best season for rooting (Haldanker *et al.*, 1995). Propagation of allspice through chip budding is also possible though the percentage of success is low (30%). Approach grafting of allspice was reported with 90% success in Jamaica (Chapman, 1965). Approach grafting on its own rootstock was also successful in India.

7.4.2 Climate and soil

The natural habitat of allspice in Jamaica is limestone forest. Although allspice is planted on a wide range of soils, a well-drained, fertile, loam limestone soil with a pH of 6–8 suits the crop best. Pimento grows well in semitropical lowland forests with a mean temperature of 18–24°C, a low of 15°C and a maximum of 32°C. Allspice flourishes well up to 1000 m above sea level. An annual rainfall of 150–170 cm evenly spread throughout the year is desirable, but allspice grows well with a rainfall of 120–250 cm.

7.4.3 Planting and after-care

The spacing recommended for allspice is 6 m × 6 m. Pits of about 60 cm deep and 30 cm wide are dug and are filled with topsoil to which well-rotted manure or compost are incorporated. Although permanent shade trees are not considered necessary for allspice, they may be required in very exposed conditions. Transplanting should be done at the beginning of the rainy season. For vegetatively propagated trees, one male tree should be planted for every ten females for adequate pollination. When trees are grown for leaves to produce oil, the sex of the tree is not important.

The base of the young seedlings should be kept free of weeds. After three to four years of growth, slashing once or twice annually around the tree would be sufficient. The larger weeds in the plantation may be controlled from time to time by slashing. The branches cut from the trees during harvesting can be used as mulch. Allspice has to be irrigated until it is two or three years old. Generally, fully grown trees of allspice are not irrigated. However, in a severe summer, irrigating trees on alternate days at 10 l/tree is recommended.

Very little is known about the manurial requirements of pimento. As the tree is found mainly on soils derived from limestone, it is generally assumed that it is a lime-demanding plant and there are indications that the crop requires a soil relatively high in potash (Ward, 1961). A fertilizer dose of 20 g N, 18 g P_2O_5 and 50 g K_2O/tree in the first year after planting is progressively increased to 300 : 250 : 750/year for a grown tree of 15 years or more. The fertilizers are to be applied in two equal doses (May and September), in shallow trenches dug around the plant about 1.0–1.5 m away from the tree. The Department of Agriculture, Jamaica, recommends 1 kg of 10 : 10 : 10 or 15 : 15 : 15 NPK mixture applied during February and September at 0.4 kg/tree/application.

Young plantations can be intercropped for one to three years with crops such as banana or any other low-growing plants such as pulses.

7.4.4 Harvesting, processing and storage

The clonally propagated plants start flowering in three years and seedlings in five to six years under well-managed conditions. Seedling trees take 18–20 years to come into full bearing. The berries are harvested when fully grown, but still green, about three to four months after flowering. The time and extent of flowering are affected by the local conditions and climate, particularly the time of onset of the spring rains, so that the time of harvesting varies between seasons and places. It normally occurs from August–September in Jamaica, July–August in Guatemala and Honduras and September–October in Mexico.

Allspice does not set fruits in the plains. Spraying paclobutrazol was reported to induce flowering in allspice and further spraying of indoleacetic acid + benzylaminopurine induces fruit set in allspice (Krishnamoorthy et al., 1995).

A healthy, well-managed tree would produce on average 10 kg green berries/tree annually after 10 years. Allspice gives a good crop once every three years. Care must be given while harvesting berries to be used as spice as the quality of the berry is assessed mainly on appearance, colour, flavour and essential oil content. Berries for distillation require less care. The harvested berries are taken to the drying shed and left in heaps up to five days to ferment. Berries are then spread in drying yards and turned frequently to ensure uniform drying. It takes about 5–10 days for drying (12–14% moisture content) depending upon the weather. Well-dried fruits should be brownish black in colour and rattle when a handful is shaken. About 55–65 kg berries is obtained from 100 kg green. The dried berries are cleaned and stored in a clean dry place. In Guatemala, berries are blanched in boiling water for 10 min. This process reduces contamination and produces an attractive colour in the dried spice. Because of frequent shifting of the berries in and out of the sheds during rainy days, many berries break and hence, mechanical drying is preferred. Artificial drying is adopted in places where the berries mature during the rainy season. Solar energy dryers and many other simple dryers using firewood and forced air dryers are available for drying allspice. A small-scale unit of hot air drying can dry 250 kg (550 lb) of green pimento in 8 h (Breag et al., 1973). A maximum temperature of 75°C is recommended for obtaining good quality allspice without any loss in essential oil content. Microbial contamination is also reported to be minimum in artificial drying.

The dried fruits should be stored in poly-lined corrugated cardboard containers or in airtight containers and kept in a cool, dry area with a maximum temperature of 21°C and maximum humidity of 70%. Excessive heat volatilizes and dissipates aromatic essential oils and high humidity tends to cake them. Dried fruits should be stored off the floor and away from outside walls to minimize the chances of dampness. The product has to be kept away from heavy aromatic materials. The essential oil is stored in sealed opaque containers. The industry standard has recommended a shelf-life of 24 months.

7.4.5 Diseases

Leaf rust
The most serious disease of pimento in Jamaica is the leaf rust, caused by *Puccinia psidii* Wint. The young leaves, shoots, inflorescence and young fruits are covered by a bright yellow powdery mass of urediospores in the infested trees. Severe infection results in defoliation of the young leaves, with successive attacks culminating in the death of the tree. Leaf rust has also been reported in Florida. The variety of *P. psidii* reported on allspice in Jamaica is different from that found in south Florida (Marlatt and Kimbrough, 1979). The disease is severe during late winter and early spring on flushes of new growth in Florida. The symptoms are observed on both upper and lower surfaces of the leaves. Mature leaves bear circular, brown, necrotic lesions covered with urediospores.

Die back
The tree is also affected by die back or canker, caused by *Ceratocystis fimbriata* Ell. and

Halst. The disease usually is localized and spreads to other parts of the tree. Bark canker and dark streaking of the wood with drying of the leaves is observed in infected trees. When primary infection occurs below a fork in the tree, death occurs within few months. The disease can be controlled by pruning and removal of all dead and infected branches and application of 1% Bordeaux mixture.

Leaf rot
A leaf rot disease caused by *Cylindrocladium quinqueseptatum* was reported in India. The disease is severe during June–September. The disease can be controlled by a prophylactic spraying of 1% Bordeaux mixture in June (Anandaraj and Sarma, 1992).

7.4.6 Pests

Borer
The larvae of red borer *Zeuzera coffeae,* Nietner (*Cossidae lepidoptera*) damage allspice by tunnelling into the collar region (Abraham and Skaria, 1995). The branches wither and wilt. Swabbing the main stem with a suspension of 0.25% carbaryl was found to be effective against the pest.

Tea mosquito
The tea mosquito *Helopeltis antonii* has been reported to attack allspice in Kerala (Devasahayam *et al.*, 1986). The bug causes necrotic lesions on young shoots of allspice. The pest can be controlled by spraying quinalphos 0.05% on tender flushes.

Leaf-damaging pests
Caterpillars of the bagworm *Oeceticus abboti* and related species feed on young leaves and shoots of allspice. Young leaves are also damaged by whiteflies, *Aleyrodidae,* and the red-banded thrips, *Selenothrips rubrocintus.* Adults of the weevils *Prepodes* spp. and *Pachnaeus* spp. also feed on leaves and their larvae damage roots. Scale insects, soft and hard are frequently present on trees but normally do little damage (Purseglove *et al.*, 1981).

Fruit fly
The fruit fly *Anastrepha suspensa* is reported to occur on allspice in Jamaica (Van Whervin, 1974) and cause damage to the berries.

7.5 Uses

Whole spice, ground spice, berry oil, leaf oil and oleoresin are the major products obtained from pimento. In olden days Mayans used allspice to embalm and preserve the bodies of their leaders. Allspice was more popular in the early 20th century than it is today. It is reported that during World War II a shortage of the spice occurred in Europe and its popularity was never regained (Tainter and Grenis, 1993). The major use of allspice is in the food industry (65–70%). A small quantity is used for domestic use (5–10%), for production of pimento berry oil (20–25%), for extraction of oleoresin (1–2%) and in pharmaceutical and perfume industry.

7.5.1 Food industry
Allspice is mostly used in Western cooking and is less suitable for Eastern cooking. It is most used in British, American and German cooking.

Whole spice
The dried mature fruits are mainly used as a flavouring and curing agent in processed meats and bakery products and as a flavouring ingredient for domestic and culinary purposes. Whole fruits are preferred in prepared soups, gravies and sauces. Whole ripe berries are an essential component of the local Jamaican drink *Pimento dram* and as an ingredient of the liqueurs Chartreuse and Benedictine.

Ground spice
The major use of allspice in the ground form is for flavouring processed meats, baking products, desserts, fruit cakes, pies, desserts, pickles, sauces, salads, vegetables, soups, fish, poultry, sausages, meats, marinades, mulled wine and preserves. For domestic culinary use, pimento is often mixed with other ground spices.

Oleoresin
Oleoresin is also used in the meat processing and canning industries in the same way as ground spice is used. Allspice oleoresin is prepared in very small quantities and has not become a substitute for ground spice in the food industry. However, it has an advantage over ground spice in that it avoids the risk of bacterial contamination and its strength and quality are more consistent.

Essential oil
The berry oil contains all the odour principles of the ground spice and oleoresin but lacks some of the flavour principles. Essential oils from leaf oil and berry oil are used as a flavouring agent in meat products and confectioneries. The maximum permitted level of berry oil in food products is about 0.025%.

7.5.2 Perfumery
The oil is used in perfumery, notably for oriental fragrances. It is used as a fragrance component in perfumes, cosmetics, soaps and after-shaves.

7.6 Functional properties
Allspice is not only valued as a spice to add flavour to food but has medicinal, antimicrobial, insecticidal, nematicidal, antioxidant and deodorizing properties.

7.6.1 Medicine
The powdered fruit of allspice is used in traditional medicine to treat flatulence, dyspepsia, diarrhoea and as a remedy for depression, nervous exhaustion, tension, neuralgia and stress. In small doses it can also help to cure rheumatism, arthritis, stiffness, chills, congested coughs, bronchitis, neuralgia and rheumatism. It has anaesthetic, analgesic, antioxidant, antiseptic, carminative, muscle relaxant, rubefacient, stimulant and purgative properties (Rema and Krishnamoorthy, 1989). It is also useful for oral hygiene and in cases of halitosis. An aqueous suspension of allspice is reported to have anti-ulcer and cytoprotective activity by protecting gastric mucosa against indomethacin and various other necrotizing agents in rats (Rehaily *et al.*, 2002).

7.6.2 Fungicide
The antifungal potential of extracts of allspice was tested *in vitro* against the field fungus (*Fusarium oxysporum*) and six storage fungi (*Aspergillus candidus, A. versicolor, Penicillium*

Table 7.14 Minimum inhibitory concentration of hexane extracts of allspice for pathogenic bacteria

Bacteria	Minimum inhibitory concentration (%)
Escherichia coli	10
Salmonella sp.	>10
Staphylococcus aureus	10
Bacillus cereus	10
Camphytobacter	10

Source: Hirasa and Takemasa (1998).

aurantiogriseum, P. brevicompactum, P. citrinum and *P. griseofulvum*) and *in situ* against the initial mycoflora of wheat grains after harvest (mainly *Fusarium* spp., *Alternaria* spp. and *Cladosporium* spp.). Allspice suppressed the growth of all the above fungus *in vitro* (Scholz *et al.*, 1999).

7.6.3 Bactericide
Allspice had a strong bactericidal effect against *Yersinia enterocolitica* (Bara and Vanetti, 1995). The minimum inhibitory concentrations (%) of hexane extracts of allspice for several pathogenic bacteria are given in Table 7.14. (Hirasa and Takemasa, 1998). A study testing thymol (thyme and oregano), eugenol (clove, pimento and cinnamon), menthol and anathole (anise and fennel) on three pathogenic bacteria, *Salmonella typhimurium, Staphylococcus aureus* and *Vibrio parahaemolyticus*, showed that all these spice components inhibited the bacteria to different extents. Eugenol was more active than thymol, which was more active than anethole. Eugenol is also sporostatic to *Bacillus subtilis* at 0.05–0.06% level (Tainter and Grenis, 1993). Allspice was also reported to suppress *Escherichia coli, Salmonella enterica* and *Listeria monocytogenes* (Friedman *et al.*, 2002).

7.6.4 Insecticide
Allspice is reported to have insecticidal properties. The effect of 103 plant powders on the mortality and emergence of adults of *Sitophilus zeamais* and *Zabrotes subfasciatus* was evaluated in the laboratory. Powdered allspice caused >20% mortality of *S. zeamais*. Allspice oils at all concentrations inhibited egg hatch of *Corcyra cephalonica* compared with the control (Bhargava and Meena, 2001).

7.6.5 Nematicide
The nematicidal activity of the essential oil of allspice (*Pimenta dioica* L. Merr.) leaves and its major constituent eugenol was tested against *Meloidogyne incognita*. The essential oil and eugenol exhibited promising nematicidal activity at 660 µg/ml (Leela and Ramana, 2000).

7.6.6 Antioxidant
Antioxidants help to preserve food from oxidation and deterioration and to increase their shelf life. They can also be used as a natural preservative. Spices and herbs are recognized as sources of natural antioxidants and thus play an important role in the chemoprevention of diseases resulting from lipid peroxidation (Chung *et al.*, 1997). Allspice has a strong hydroxyl radical-scavenging activity (Nakatani, 2000). Compounds that markedly inhibit the formation of malondialdehyde from 2-deoxyribose and the hydroxylation of benzoate

with the hydroxyl radical were isolated from methanol extracts of allspice. These compounds were identified as pimentol and had a strong antioxidant activity as hydroxyl radical scavengers at 2.0 μm (Oya *et al.*, 1997). A phenylpropanoid, threo-3-chloro-1-(4-hydroxyl-3-methoxyphenyl)propane-1,2-diol isolated from berries of *P. dioica* inhibited autoxidation of linoleic acid in a water–alcohol system (Kikuzaki *et al.*, 1999).

The effect of different allspice extracts (ethanol, chloroform, diethylether, benzene and hexane) on the stability of rapeseed oil was examined. The ethanol extract exhibited a remarkable antioxidant effect and the antioxidant effectiveness of various extracts was in the order ethanol extract > chloroform extract > diethylether extract > benzene extract > hexane extract (Vinh *et al.*, 2000).

7.6.7 Deodorizing effect
The major function of allspice is to flavour food but it has a subfunction of deodorizing or masking unpleasant odours. The concentration of methyl mercaptan is a major cause of bad breath and it was observed that allspice has a deodorizing rate of 61% (deodorizing rate is the percentage of methyl mercaptan (500 ng) captured by methanol extract).

7.6.8 Toxicity
Allspice oil should only be used in low dilutions since it is found to irritate the mucous membrane, owing to the presence of eugenol in allspice oil. It is also reported to cause dermal irritation. At low doses it is non-toxic, non-irritant, non-sensitizing and non-phototoxic.

7.7 Quality issues and adulteration

7.7.1 Specifications
Cleanliness, safety issues (microbes and moisture levels) and economic parameters (aroma, flavour and granulation) are the main quality aspects dealing with spice. The cleanliness specifications have been set out in laws such as Food and Drug Administration Defect Action Levels (FDADALs) (USA) or in trade practices such as the American Spice Trade Association (ASTA), European Spices Association (ESA), etc.

Description
As per the ISO specifications, allspice is described as the dried, fully mature but unripe, whole berry of *Pimenta dioica* (L.) Merrill, 6.5–9.5 mm in diameter, a dark brown colour, the surface somewhat rough and bearing a small annulus formed by the remains of the four sepals of the calyx. Allspice may also be in the pure ground form.

Odour and taste
The odour and taste of pimento, either whole or ground, shall be fresh, aromatic and pungent. It shall be free from any foreign taste or odour, including rancidity or mustiness.

Freedom from moulds, insects, etc.
Allspice, whole or ground, shall be free from living insects and moulds and shall be practically free from dead insects, insect fragments and rodent contamination visible to the naked eye with such magnification as may be necessary in any particular case. In case of

dispute, the contamination of ground pimento shall be determined by the method specified in ISO 1208.

Extraneous matter
All that does not belong to the fruits of allspice and all other extraneous matter of animal, vegetable and mineral origin shall be considered as extraneous matter. Broken berries are not considered as extraneous matter. The total percentage of extraneous matter in whole dried allspice shall not be more than 1% (m/m) when determined by the method described in ISO 927.

Product and cleanliness specification
The standard specifications of various countries for berries, leaf oils and berry oil allspice is given in (Tables 7.15–7.20) and the cleanliness specifications are given in Tables 7.21 and 7.22.

Table 7.15 Chemical requirements of allspice

Characteristics	Requirement		Methods of test
	Whole	Ground	
Moisture content, % (m/m), max.	12	12	ISO 939
Total ash, % (m/m) on dry basis, max.	4.5	4.5	ISO 928
Acid insoluble ash, % (m/m) on dry basis, max.	0.4	0.4	ISO 930
Volatile oil, % (ml/100 g) on dry basis, min.			ISO 6571
Group A, more than	3	2	
Group B, min.	2	1	
max.	3	2	
Non-volatile ether extract, % (m/m) on dry basis, max.	–	8.5	ISO 1108
Crude fibre, % (m/m) on dry basis, max.	–	27.5	ISO 5498

Source: Purseglove *et al.* (1981).

Table 7.16 US Government standard specifications for allspice

Moisture, not more than	10%
Total ash, not more than	5%
Acid-insoluble ash, not more than	0.3%
Volatile oil, ml per 100 g, not less than	3
Sieve test	
US standard sieve size	No. 25
Percentage required to pass through, not less than	95

Source: Purseglove *et al.* (1981).

Table 7.17 Canadian Government standard specifications for allspice

Total ash, %, not more than	6.0
Ash insoluble in HCl, %, not more than	0.4
Crude fibre, %, not more than	25
Quercitannic acid, calculated from the total oxygen absorbed by the aqueous extract, % not less than	8

Source: Purseglove *et al.* (1981).

Table 7.18 European Spice Association (ESA) product specification for allspice

Product	Ash % w/w (max.)	Acid insoluble ash % w (max.)	Moisture % w/w (max.)	Volatile oil % v/w (min.)
Jamaica	5 (ESA)	0.4 (ISO)	12 (ISO)	3.5 (ISO)
Other origins	5 (ESA)	1 (ESA)	12 (ISO)	2 (ESA)

ISO = International Organization for Standardization.
Source: Sivadasan and Kurup (1998).

Table 7.19 British Standards Institute specifications for allspice

	Berry oil	Leaf oil
Apparent density, g/ml at 20°C	1.025 to 1.045	1.037 to 1.050
Optical rotation at 20°C	0°C to –5°C	–
Refractive index at 20°C	1.526 to 1.536	1.531 to 1.536
Phenolic* % volume, minimum	65	80
Solubility in ethanol at 20°C (70% v/v)	2 volumes	2 volumes

*Determined by absorption with 5% KOH.
Source: Purseglove et al. (1981).

Table 7.20 Essential oil association of the USA specification

	Berry oil EOA No. 255	Leaf oil EOA No. 73
Specific gravity at 25°C	1.018 to 1.048	1.018 to 1.048
Optical rotation at 20°C	0° to –4°C	–0°30' to –2°
Refractive index at 20°C	1.527 to 1.540	1.5319 to 1.5360
Phenols*, % by volume, minimum.	65	50 to 91
Solubility in 70% alcohol at 25°C	2 volumes	2 volumes

* Determined by absorption with 1N KOH.
Source: Purseglove et al. (1981).

Table 7.21 American Spice Trade Association (ASTA) cleanliness specifications for allspice

Total extraneous matter, determined by sifting and by hand picking, % by weight	0.5
Mammalian excreta, mg/lb	2.0
Other excreta, mg/lb	5.0
Whole insects, dead (by count) per lb	2.0
Insect-bored or otherwise defiled berries, % by weight	1.0
Mouldy berries, % by weight	2.0

Source: Sivadasan and Kurup (1998).

Table 7.22 Dutch regulations regarding cleanliness for allspice

Ash content (max %)	6.0
Sand content (max %)	1.5

Source: Sivadasan and Kurup (1998).

7.7.2 Sampling
Sampling shall be carried out in accordance with the method specified in ISO 948.

7.7.3 Packing
Allspice, whole or ground, shall be packed in clean and sound containers made of a material that does not affect the product but that protects it from the increase or loss of moisture and volatile matter. The packaging shall also comply with any national legislation relating to environmental protection.

7.7.4 Marking
The following particulars shall be marked directly on each package or on label attached to the package: name of the product (type: whole or ground) and trade name; name and address of the producer or packer and trademark, if any; code or batch number; net mass; grade; producing country; any other information requested by the purchaser, such as the year of harvest and date of packing (if known).

7.7.5 Pesticide residues
The limits for pesticide residue prescribed for other agricultural products are generally followed for spices. Maximum permitted limits of trace metals in allspice are given in Table 7.23.

Table 7.23 Maximum permissible limits of trace metals in allspice

Metal	Concentration (in ppm)
Aluminium	73
Arsenic	0
Barium	4.8
Beryllium	0.037
Bismuth	0
Boron	8.6
Cadmium	0
Copper	5.1
Lead	0
Lithium	0
Magnesium	1300
Manganese	11
Molybdenum	0.4
Nickel	0.57
Selenium	0.16
Silicon	18
Strontium	2.4
Tin	7.8
Titanium	1.6
Zinc	9.4

Source: Sivadasan and Kurup (1998).

7.7.6 Adulteration

Ground pimento is sometimes adulterated with powdered clove stem or with the aromatic berries of the Mexican tree *Myrtus tobasco*, known as 'pimienta de tobasco'. The powdered berries of the aromatic shrub *Lindera benzoin* (called wild allspice) has a strong spicy flavour in bark and berries and is used as a substitute for allspice by the Americans. A mixture of pimento leaf oil and clove stem and leaf oils can serve as a relatively inexpensive substitute for berry oil. Pimento berry oil is sometimes adulterated with eugenol from cheaper sources. Samples are also considered adulterated or of poor quality for trade if they contain an average of 30 or more insect fragments per 10 g, or an average of one or more rodent hairs per 10 g or an average of 5% or more mouldy berries by weight.

7.8 References

ABRAHAM, C.C and SKARIA, B.P (1995), 'New record of the red borer *Zeuzera coffeae* Nietner (Cossidae: Lepidoptera) as a pest of allspice (*Pimenta dioica* Linn.) Merril'. *Insect Environment* **1**(1): 6.

ANANDARAJ, M. and SARMA, Y.R. (1992), 'A new leaf rot in *Pimenta dioica*'. *Indian Phytopathology* **45**(2): 276–7.

ARCTANDER, S. (1960), *Perfumer and Flavour Materials of Natural Origin*. Elizabeth, New Jersey.

ASHURST, P.R., FIRTH, A.R. and LEWIS, O.M. (1972), 'A new approach to spice processing'. *Proceedings of the Conference on Spices*. Tropical Products Institute, London.

BARA, M.T.F and VANETTI, M.C.D. (1995), 'Antimicrobial effect of spices on the growth of *Yersinia enterocolitica*'. *Journal of Herbs, Spices and Medicinal Plants* **3**(4): 51–8.

BHARGAVA, M.C. and MEENA, B.L. (2001), 'Effect of some spice oils on the eggs of *Corcyra cephalonica* Stainton'. *Insect Environment* **7**(1): 43–4.

BREAG, G.R., CROWD, L.P.G., NAHNEY, J. and ROBINSON, F.V. (1973), 'Artificial drying of pimento'. *Proc. Int. Conf. Spices. Trop. Prod. Inst, London*.

CHAPMAN, G.P. (1965), 'A new development in the agronomy of pimento'. *Caribbean Quarterly* **2**: 1–12.

CHARALAMBOUS, G. (1994), *Spices, Herbs and Edible Fungi*. Elsevier, Amsterdam.

CHUNG, S.K., OSAWA, T. and KAWAKISHI, S. (1997), 'Hydroxyl radical scavenging effects of spices and scavengers from brown mustard (*Brassica nigra*)'. *Bioscience, Biotechnology and Biochemistry* **61**(1): 118–23.

DEVADAS, V.S. and MANOMOHANDAS, T.P. (1988), 'Studies on the variability of allspice seeds'. *Indian Cocoa Arecanut Spices Journal* **11**: 99.

DEVASAHAYAM, S., ABDULLA KOYA, K.M. and KUMAR, T.P. (1986), 'Infestation of tea mosquito bug *Helopeltis antonii* Signoret (Heteroptera: Miridae) on black pepper and allspice in Kerala'. *Entomon* **11**(4): 239–41.

FRIEDMAN, M., HENIKA, P.R. and MANDRELL, R.E. (2002), 'Bactericidal activities of plant essential oils and some of their isolated constituents against *Campylobacter jejuni, Escherichia coli, Listeria monocytogenes* and *Salmonella enterica*'. *Journal of Food Protection* **65**(10): 1545–60.

GARCIA-FAJARDO, J., MARTINEZ SOSA, M., ESTARRON ESPINOSA, M., VILAREM, G., GASET, A. and DE SANTOS, J.M. (1997), Comparative study of the oil and supercritical CO_2 extract of Mexican pimento (*Pimenta dioica* Merrill)'. *Journal of Essential Oil Research* **9**: 181–5.

GUZMAN, DE C.C. and SIEMONSMA, J.S. (1999), *Plant Resources of South-East Asia, No. 13*. Backhuys Publishers, Leiden.

HALDANKAR, P.M., NAGWEKAR, D.D., DESAI, A.G., PATIL, J.L. and GUNJATA, R.T. (1995), 'Air layering in allspice'. *Indian Cocoa Arecanut Spices J.* **19**(2): 56–7.

HIRASA, K. and TAKEMASA, M. (1998), *Spice Science and Technology*. Marcel Dekker Inc, New York.

KIKUZAKI, H., HARA, S., KAWAI, Y. and NAKATANI, N. (1999), 'Antioxidative phenylpropanoids from berries of *Pimenta dioica*', *Phytochemistry* **52**(7): 1307–12.

KIKUZAKI, H., SATO, A., MAYAHARA, Y. and NAKATANI, N. (2000), 'Galloylglucosides from berries of *Pimenta dioica*'. *Journal of Natural Products* **63**(6): 749–52.

KOLLMANNSBERGER, H. and NITZ, S. (1993), 'Uber die Aromastoffzusammensetzung von hoch-druck-extrakten: 2 Piment (Pimenta *dioica*)'. *Chem. Microbiol. Technol. Lebensmitt.* **15**(3/4): 116–26.

KRISHNAMOORTHY, B., KRISHNAMURTY, K.S. and REMA, J. (1995), 'Increase in fruit and seed size and seed seeds per fruit in allspice (*Pimenta dioica* L. Merr.) by hormonal application'. *Journal Spices Aromatic Crops* **4**(2): 162–3.

KRISHNAMOORTHY, B., SASIKUMAR, B., REMA, J. and GEORGE, J.K. (1997), 'Genetic resources of tree spices and their conservation in India'. *Plant Genetic Resources Newsletter* **111**: 53–8.

LAWRENCE, B.M. (1999), 'Progress in essential oils'. *Perfumer and Flavorist* **24**(2): 35–47.

LEELA, N.K. and RAMANA, K.V. (2000), 'Nematicidal activity of the essential oil of allspice (*Pimenta dioica* L. Merr.)'. *Journal of Plant Biology* **27**(1): 75–6.

MARLATT, R.B. and KIMBROUGH, J.W. (1979), '*Puccinia psidii* on *Pimenta dioica* in South Florida'. *Plant Disease Reporter* **63**(6): 510–12.

MATHEW, P.A., KRISHNAMOORTHY, B. and REMA, J. (1999), 'Seedling variants in allspice (*Pimenta dioica* L. Merr.)'. *Journal of Spices and Aromatics Crops* **8**(1): 93–4.

MERRILL, E.D. (1947), 'The technical name of allspice'. *Contr. Gray Herb* **165**: 30–8.

NABNEY, J. and ROBINSON, E.V. (1972), 'Constituents of pimento berry oil (*Pimenta doica*)'. *Flav. Ind.* **3**: 50–1.

NAKATANI, N. (2000), 'Phenolic antioxidants from herbs and spices'. *Bio factors* **13**(1–4): 141–6.

OYA, T., OSAWA, T. and KAWAKISHI, S. (1997), 'Spice constituents scavenging free radicals and inhibiting pentosidine formation in a model system'. *Bioscience, Biotechnology and Biochemistry* **61**(2): 263–6.

PINO, A., GARCIA, J. and MARTINEZ, M.A. (1997), 'Solvent extraction and supercritical carbon dioizide extraction of *Pimenta dioica* Merrill leaf'. *Journal of Essential Oil Research* **9**: 689–91.

PINO, J.A. and ROSADO, A. (1996), 'Chemical composition of the leaf oil of *Pimenta dioica* L. from Cuba'. *Journal of Essential oil Research* **8**(3): 331–2.

PINO, J., ROSADO, A. and GONZALEZ, A. (1989), 'Analysis of the essential oil of pimento berry (*Pimenta dioica*)'. *Nahrung* **33**: 717–20.

PURSEGLOVE, J.W., BROWN, E.G., GREEN, C.L. and ROBBINS, S.R.J. (1981), *Spices* Volume I. Longman, London and New York.

REHAILY AL, A.J., SAID AL, M.S., YAHYA AL, M.A., MOSSA, J.S. and RAFATULLAH, S. (2002), 'Ethnopharmacological studies on allspice (*Pimenta dioica*) in laboratory animals'. *Pharmaceutical Biology* **40**(3): 200–5.

REMA, J. and KRISHNAMOORTHY, B. (1989), 'Economic uses of tree spices'. *Indian Cocoa, Arecanut and Spices Journal* **12**(4): 120–1.

REMA, J., KRISHNAMOORTHY, B. and MATHEW, P.A. (1997), 'Vegetative propagation of major tree spices – a review'. *Journal of Spices and Aromatic Crops* **6**(2): 87–105.

SCHOLZ, K., VOGT, M., KUNZ, B., LYR, H., RUSSELL, P.E., DEHNE, H.W. and SISLER, H.D. (1999), 'Application of plant extracts for controlling fungal infestation of grains and seeds during storage'. Modern fungicides and antifungal compounds II. *12th International Reinhardshbrunn Symposium*, Friedrichroda, Thuringia, Germany.

SCHULZ, J.M. and HERRMANN, K.Z. (1980), *Lebensm. Unters Forsch.* **171**: 278–80.

SIVADASAN, C.R. and KURUP, M.P. (1998), *Quality Requirements of Spices for Export*. Spices Board, Kochi.

TAINTER, D.R. and GRENIS, A.T. (1993), *Spices and Seasonings*. VCH Publishers, Inc., New York.

VAN WHERVIN, L.W. (1974), 'Some fruitflies (Tephritidae) in Jamaica' *PANS* **20**(1): 11–19.

VINH, N.D., TAKACSOVA, M., NHAT, D.M. and KRISTIANOVA, K. (2000), 'Antioxidant activities of allspice extracts in rape seed oil'. *Czech Journal of Food Sciences* **18**(2): 49–51.

VOSGEN, B., HERRMANN, K. and KIOK, B. (1980), 'Flavonol glycosides of pepper (*Piper nigrum* L.), clove (*Syzygium aromaticum*) and allspice (*Pimenta dioica*) 3 phenolics of spices'. *Zeitschrift für Lebensmittel Untersuchung und Forschung* **170**(3): 204–7.

WARD, J.F. (1961), *Pimento*. Kingston Govt. Printer.

WEISS, E.A. (2002), *Spice Crops*. CABI Publishing, New York.

WILLIS, J.C. (1966), *A Dictionary of the Flowering Plants and Ferns,* 7th Edition (rev. by H. K. Airey Shaw). Cambridge University Press, Cambridge.

8
Chervil

A. A. Farooqi and K. N. Srinivasappa, University of Agricultural Sciences, India

8.1 Introduction and description

Chervil (*Anthriscus cerefolium* L. Hosffm.) is a warmth-giving herb belonging to the family Apiaceae (Umbilliferae). Its taste and fragrance fill the senses the way warmth does, slowly and subtly. Chervil was once called *myrhis* for its volatile oil, which has an aroma similar to the resinous substance of myrrh. One of the traditional fine aromas with a hint of myrrh, chervil is noticed even when in the background, because of its warm and cheering flavour and fragrance. This herb is called many different names in different countries: *Maqdunis afranji* in Arbaic, *San lo po* in Chinese, *Korvel* in Danish, *Kervel* in Dutch, garden chervil or French parsley in English, *Cerefolio* in Esperanto, *Maustekirveli* in Finnish, *Cerfenil* in French, *Aed-harakputk, Harakputk* in Estonian, *Kerbel*, Gartenkerbel, *Franzosiche or Petersilic* in German, *Tamcha* in Hebrew, *Turboloya or zamatos turbolya* in Hungarian, *Kerfill* in Icelandic, *Cerfoglio* in Italian, *Kjorvel* or *Hagekjorvel* in Norwegian, *Trybula Ogrodowda* in Polish, *Certolho* in Portuguese, *Kervel* in Russian, *Perifollo* or *Certafolia* in Spanish and *Korvel*, *Dansk Korvel* or *Tradgardskorvel* in Swedish.

Leaves of the chervil are nearly always used fresh, but can be preserved by deep freezing or by making a pesto-like preparation. The plant contains only a small amount of essential oil (0.3% in the fresh herb, 0.9% in the seeds) with methyl chavicol (estragol) as the main constituent and is popular in Central and Western Europe. Because of its resemblance to the myrrh given to Jesus and as well as the way it symbolized new life, it became traditional to serve chervil soup on Holy Thursday.

Chervil is a hardy annual, grows to a height of 25–70 cm and width of 30 cm. The lacy, light green leaves are opposite, compound and bipinnate, they are sub-divided again into opposite and deeply cut leaf lets. The lower leaves are pointed and the upper leaves are sessile with stem sheaths. The stems are finely grooved, round, much branched, light green and hairy. The white flowers are arranged in tiny umbels and grow into compound umbels. The whole plant smells of anise and tastes a little of pepper and of anise; it blooms during May to August. Chervil has a white, thin and single tapering root. The oblong fruit is 0.5–0.75 cm long, segmented and beaked. The seeds are long, pointed with a conspicuous furrow from end to end.

Chervil is available in two distinct types, salad chervil and turnip-rooted chervil. Salad chervil is grown in a similar way to parsley. Both have a fern-like leaf structure as delicate and dainty as the flower. The stems are branched and finely grooved and the root is thin and white.

8.2. Cultivation and production technology

Chervil probably originated in southern Europe or the Caucasus region. It is found in Europe and Asia. It has been cultivated in England since 1597 and in America since 1806. However, it can also be found growing in other places where the right conditions prevail.

8.2.1 Soil

Chervil grows well in any good garden soil with high fertility. However, moist, humus-rich soils with good drainage are most suitable. It can be successfully grown in soils with a pH of 6.5, especially turnip-rooted chervil, which has a wider adaptability and grows in all parts of the chervil-growing world where the soil is fertile and with sufficient moisture.

8.2.2 Climate

Chervil is a hardy plant and may thrive in much cooler climates provided it finds a warm location, but as a cold weather crop, chervil is susceptible to frost and should be planted in a sheltered area. In temperate climates, it can also be grown as a summer season crop. Under such conditions, it prefers partial shade. It is helped by having the leaves cut off, so they can shoot up again. The plants are not robust and soon wither and die. In other parts of the world, it is mainly grown as a cold season crop.

8.2.3 Propagation

Chervil can be propagated only through seeds. For this purpose, the seeds must be bedded in damp sand for a few weeks before being sown, otherwise their germination is slow. In temperate region the seeds are usually sown in March–April, whereas in tropical or subtropical parts they are sown during October by drill or scattered in well-prepared land and mixed with well-decomposed farmyard manure. The recommended seed rate is roughly 3 kg/ha which is sown in rows. Seeds should be grown in the spring in shallow drills 30 cm apart. When the seedlings are about 7–8 cm high, the plants should be thinned to 8–10 cm apart. The seedlings are too fragile to be transplanted. In the South, the seeds are usually sown in the autumn, but they may not germinate until spring. In the North, the seeds may be sown in the autumn to germinate in the spring; or the plant may be started indoors in later winter and transplanted to open ground later on.

Seed vernalization induces rapid bolting and flowering under long days; without vernalization, bolting is very slow under all conditions. Vernalization also decreases yield. But higher yields were obtained when varnalized seeds were germinated at 20°C. Later adjustments in sowing dates in field resulted in higher yields.

8.2.4 Manure and fertilizers

Chervil prefers to be grown organically with the application of well-decomposed farmyard manure or leaf mould at about 8–10 tonnes per hectare. However, to obtain higher yields, its inorganic fertilizer requirement needs to be assessed.

8.2.5 Weeding and irrigation

Hand weeding in the initial stages is recommended. But if the labour is a problem and weed population is heavy, weedkillers like influtalin and ethafluralin (1.1 kg/ha), sethoxydim (4.5 kg/ha), linuron (1 kg/ha), chlorobromuron (4.5 kg/ha) and thiobencarb (6–8 kg/ha) can be used to control weeds. Since chervil is a herbaceous crop, it requires frequent irrigation. It grows poorly in hot, dry conditions. Regular watering is therefore essential. Chervil should be protected from summer sun, wherever it is grown as a summer season crop.

8.2.6 Intercultural operations

Soil should be earthed up to loosen it and to enhance aeration for better growth. Once the plants are established, they will self-seed. The flowers should be picked as soon as they appear as it helps to make stalks to shoot rapidly. It is better to follow the practice of cutting flower stems before they bloom in order to get denser foliage.

8.2.7 Intercropping

Chervil and radishes planted together produce hotter radishes, since chervil prefers light shade. Chervil can be intercropped with *Rauvolfia serpentina* or *Mentha arvensis* or *Salvia scalrea*.

8.2.8 Pests and diseases

Among pests, aphids occasionally cause damage and are generally controlled by spraying Melathion (0.5%) two or three times during the infestation.

There are a few botanical insecticides such as rotenone that are sold as 1 or 2% dust, which controls aphids, thrips and some soft-bodied sucking insects. Rotenone is available at 40% liquid concentrate, which is diluted in water and sprayed, and other botanicals such as Pyrethrum, Ryania sulfur, etc., also recommended. *Trichoderma* spp. are used to control such diseases as root rot.

Pyrethrum is a botanical obtained from the dried flower of *Chrysanthemum cinerarifolium* and is used as an insect control agent. It provides a rapid knockdown of a wide range of insects. Pyrethrum is very expensive and has a very short residual effect. Therefore, it is usually used in combination with other insecticides such as rotenone and with an activator or synergist such as piperonyl cyclonene or piperonyl butoxide.

Among the diseases, powdery mildew can be noticed at the flowering and early seedling stages. It can be controlled by spraying wettable sulphur (0.2%) two or three times at weekly intervals. *Fusarium* species cause root rot disease and which can be controlled by following phytosanitary measures, seed treatment with Agroson (@ 3 g/kg of seed) and by foliar spray of Bavistin (0.1%). Chervil may be infected with the virus for anthriscus yellows and also reported to exhibit mottling, leaf necrosis, dwarfing and malformation due to viral infections.

8.2.9 Harvesting

Harvesting of chervil should be properly timed and it mainly depends on the purpose of harvesting, whether it is for salads or vegetables or for obtaining the seeds. If the chervil is being harvested for salad or vegetable, the flowers shgould be removed well before harvesting to obtain maximum shoots. The leaves can generally be cut six to eight weeks after sowing. After the required leaves have been harvested, the plant should be cut down to the ground to allow more growth to occur.

After picking, the leaves and stems can be dried on wire racks in a cool, ventilated, shady place. Once the leaves are dried, they become brittle (either whole or crumbled) and can be stored in an airtight container. Fresh chervil may be chopped and frozen with water in ice cube trays.

If the plants are to be harvested for seed purposes, they should be allowed to mature until there are seeds in the field. Then the harvested material is dried in the field until the fruits are easily threshed. The threshed fruits should be spread in a thin layer and frequently turned over until they are thoroughly dry.

8.2.10 Yield

A herbage yield of about 2.5 to 3.0 tonnes per hectare can be obtained in case of leaf crops and 500–700 kg of seeds per hectare can be obtained.

8.3 Uses

Chervil has been used for several medicinal purposes throughout history by herbalists. The first-century Roman scholar Pliny and the seventeenth-century herbalist Nicholas Culpeper believed that chervil, as Culpeper put it, 'does much please and warm old and cold somachs'. Chervil drink has been used as an expectorant, a stimulant, a dissolver of congealed blood, a healer of eczema, a digestive, and a cure for high blood pressure, gout, kidney stones, pleurisy, dropsy and menstrual problems. Of these properties, the most persistently recognized to this day has been the ability to lower blood pressure, but no clinical studies support this or any of the claims.

The tender young leaves of chervil have been used in spring tonics for thousands of years, dating back to the ancient Greeks. A combination of chervil, dandelion and watercress rejuvenates the body from the deficiency brought on by winter and lack of fresh greens, because of all their vitamins and minerals. Even today European herbalists recommend this tonic. In Norway and France bowls of minced fresh chervil leaves often accompany meals. People liberally sprinkle the chopped leaves on salads, soups and stews. As with most herbs, chervil is an aid to sluggish digestion. When brewed as a tea it can be used as a soothing eye wash. The whole plant reportedly relieves hiccoughs, a practice still tried by some people.

Chervil is one of the staples of classic French cooking. Along with chives, tarragon and parsley, it is used as an aromatic seasoning blend called '*fines herbes*'. Most frequently it is used to flavour eggs, fish, chicken and light sauces and dressings. It also combines well with mild cheeses and is a tasty addition to herb butters. This blend is the basis for ravigote sauce, a warm herbed velouté served over fish or poultry. When a recipe calls for '*Pluches de cerfeuille*', it is leaves of chervil that are required. Chervil is what gives Bernaise sauce its distinctive taste. Chervil, being a spring herb, has a natural affinity for other spring foods: salmon trout, young asparagus, new potatoes, baby green beans, carrots and salads of spring greens.

Chervil's flavour is lost very easily, either by drying the herb, or from too much heat, so it should be added at the end of cooking or sprinkled on in its fresh, raw state. One way to keep chervil's flavour is to preserve it in white wine vinegar. Because its flavour is so potent, little else is needed as flavouring when added to foods. This makes it a low-calorie way to add interest to meals. Chervil's delicate leaves make it an attractive herb to use for garnishes. Despite its fragile appearance, it keeps well. Chervil will last up to a week in the refrigerator. Chervil has been over-looked in American cooking until recently, because most people have tasted only dried chervil, which is basically tasteless and musty and at best tastes sweet and grassy with a touch of liquorice.

Chervil is an effective seasoning to foods. Both the leaves and the stems can be used for cooking and whole sprigs make a delicate and decorative garnish. Blanched sprigs of chervil are occasionally used in soup.

8.4 Sources of further information

ANONYMOUS (1969), *The Wealth of India – Raw Materials*. Council of Scientific and Industrial Research (CSIR), New Delhi **8** 376–90.

ANONYMOUS (1985), *Report on Herbal Industry*. Industrial and Technical Consultancy Organization of Tamil Nadu Ltd, Chennai.

BUTTERFIELD, H.M. (1964), *Growing Herbs for Seasoning Food*. University of California, Berkeley.

FOX, H.M. (1970), *Gardening with Herbs – For Flavour and Fragrance*. Dover Publication Inc., New York.

HORE, A. (1979), 'Improvement of minor umbelliferous spices in India'. *Econ. Bot.* **33**(3) : 290–7.

LOW, S. (1978), *Herb Growing, a Visual Guide*. Diagram Group, Connecticut.

lowman, m.s. and birdseye, m. (1946), *Savory Herbs: Culture and Use*. USPA Farmers Bull No. 1977, US Department Agric. Washington, DC.

MACGILLIVARY, J.H. (1953), *Vegetable Production*. McGraw-Hill Book Co., New York.

PEPLOW, E. (1984), *The Herb Book*. W.H. Allen & Co., London.

SPLITTSTOESSER, W.E. (1984), *Vegetable Growing Hand Book*. AVI Publishing Co., Westport, Connecticut.

THOMPSON, H.C. and KELLY, W.C. (1957), *Vegetable Crops*. McGraw-Hill Book Co., New York.

WILLIAMS, L.O. (1960), *Drug and Condiment Plants*. USDA Agric. Handbook No. 172, US Department Agric., Washington, DC.

YAMAGUCHI, M. (1983), *World Vegetables*. AVI Publishing Co., Westport, Connecticut.

9
Coriander

M. M. Sharma and R.K. Sharma, Rajasthan Agricultural University, India

9.1 Introduction and description

Coriander *Coriandrum sativum* L. is an important spice crop and occupies a prime position in flavouring substances. It was one of the first spices to be used as a common flavouring substance. The stem, leaves and fruits all have a pleasant aromatic odour. The entire plant when young is used in preparing chutneys and sauces, and the leaves are used for flavouring continental curries and soups. The fruits are extensively employed as a condiment in the preparations of curry powder, pickling spices, sausages and seasonings. They are also used for flavouring pastry, biscuits, buns and cakes, and in flavouring liquors, particularly gin. Coriander seeds are also known for their medicinal properties and are considered carminative, diuretic tonic, stomachic antibilious, refrigerant and aphrodisiac. As such, coriander is a frequent ingredient in the preparation of *Ayurvedic* medicines and is a traditional home therapy for different ailments. The new value-added products obtained from seeds are also in large demand in international markets. The volatile oil is also used in flavouring liquors and for obscuring the bad smell of medicines.

9.1.1 Botanical description
Coriandrum sativum L. ($2n = 22$) belongs to the family Umbelliferae with botanical classification:

- Division   Angiospermae
- Class   Dicotyledonae
- Sub-class   Polypetalae
- Series   Calyciflorae
- Order   Umbellales
- Genus   Apiaceae
- Species   Umbelliferae

Purseglove *et al.* (1981) have given a detailed botanical description of the plant. There are two distinct morphological types: one erect and tall with a comparatively stronger main shoot and shorter branches, the other bushy with a relatively weaker main shoot and longer, spreading branches. The plants attain heights from 30 to 100 cm, depending upon the variety. The crop comes to bloom in 45–60 days after sowing and matures in 65–120 days, depending upon the variety and cropping situation. Each branch as well as the main shoot terminates in a compound umbel (determinate growth) bearing 3–10 umbels, each umbel containing 10–50 pentamerous flowers. The flowers are small, protoandrous and difficult to manipulate for controlled pollination. Like other umbelliferous plants, coriander is also a cross-pollinated crop. The degree of cross-pollination has been reported to range from 50% by Ramanujam *et al.* (1964) to 60% by Dimri *et al.* (1977). Anuradha Hore (1979) considered poor seed set as a major constraint to yield. Pillai and Nambiar (1982) considered coriander to be andromonoecious. Singh and Ramanujam (1973) reported significant varietal differences in distribution of male and perfect flowers in the umbels. Hermaphrodite flowers opened earlier than males. Selection of a higher proportion of hermaphrodite flowers was considered an effective criterion for higher seed set.

9.2 Origin and distribution

It is believed that coriander originated from around the Mediterranean. Two species are found: only *Coriandrum sativum* L. is cultivated widely, mainly in the tropics. India has the prime position in the cultivation and production of coriander: it is cultivated over an approximate area of 5.25×10^5 hectares with an annual production of 3.10×10^5 tonnes. The main coriander growing states in India are Andhra Pradesh, Rajasthan, Madhya Pradesh, Karnataka, Tamil Nadu and Uttar Pradesh. In addition to India, coriander is also cultivated in Morocco, Rumania, France, Spain, Italy, the Netherlands, Myanmar, Pakistan, Turkey, Mexico, Argentina and, to some extent, in the UK and the USA.

9.3 Chemical composition

Seed spices contain a variable amount of proteins, fats, carbohydrates, fibres, minerals and vitamins. However, owing to the very small quantity used in the foods, their contribution to nutrient requirements is not significant. Proteins, carbohydrates, minerals and vitamins are thus less important in delineating the quality of spices.

Coriander green leaves contain 87.9% moisture, 3.3% protein, 0.6% fat, 6.5% carbohydrates and 1.7% mineral matter. The mature dry seeds are tan to brownish-yellow and have 6.3–8.0% moisture, 1.3% protein, 0.3–1.7% volatile oil, 19.6% non-volatile oil, 31.5% ether extract, 24.0% carbohydrates, 5.3% mineral matter and vitamin A 175 IU per 100 g.

In unripe fruits/seeds and vegetative parts of the plant, aliphatic aldehydes predominate in the steam-volatile oil and are responsible for the peculiar, fetid-like aroma. On ripening, the seeds acquire a more pleasant and sweet odour, mainly because of an increase in linalool content. Dried ripe coriander seeds contain both steam-volatile oil and fixed oil. The aromatic odour and taste of coriander fruit is due to its volatile oil, which is a clear, colourless to light yellow liquid. The flavour of the oil is warm, spicy-aromatic, sweet and fruity. The oil contents of seeds vary widely with geographical origin. Higher volatile oil content is found in Norwegian coriander (1.4–1.7 %) followed by Bulgarian coriander (0.1–0.5%). Indian seeds are poor in volatile oil content (0.1–0.4%) (Agrawal and Sharma, 1990). Major components of essential oil are linalool (67.7%), followed by α-pinene (10.5%), γ-terpinene (9.0%), geranyl acetate (4.0%), camphor (3.0%) and geraniol (1.9%). Minor components in the oil are β-pinene, camphene, myrcene, limonene, *p*-cymol, dipentene, α-terpinene, *n*-decylaldehyde, borneol and acetic acid esters.

Coriander oil has the following physical properties:

- specific gravity at 25°C: 0.863–0.875.
- refractive index at 20°C: 1.463–1.472.
- optical rotation: +8 to +15°.

Small coriander seed is characterized by relatively high volatile oil content (exemplified by the Russian and North European types) whereas the bold seed types, mainly Moroccan and Indian types, are reported to possess relatively low oil content. However, exceptions to these are not uncommon as the bold-seeded Indian varieties CS-4 and CS-6 contain higher oil content compared with the small-seeded variety RCr.41. The small-seeded, high-oil-yielding types of coriander are generally late in flowering and maturity (Kumar *et al.,* 1977). High oil types, which are generally small seeded, are preferred for distillation purposes. Bold-seeded types have better appearance and are more suited for spice usage. Indian coriander oil differs from the European oil in possessing lower linalool content and comparatively higher ester (linalyl acetate) content (Rao *et al.,* 1925).

Hirvi *et al.* (1986) observed differences in the constitution of volatile oil extracted through steam distillation, CO_2 and other three commercial extractions. Boelens *et al.* (1989) reported higher linalool content (70.4%) and lesser monoterpene hydrocarbons (23.6%) in volatile oil extracted by hydrodistillation than in that extracted by hydrodiffusion where linalool content (66.2%) and monoterpene hydrocarbons (27.3%) were observed.

9.4 Cultivation and post-harvest practices

9.4.1 Climate
Coriander is a tropical crop; it requires a cool and comparatively dry frost-free climate, particularly at the time of flowering and seed formation stages, for good quality and high yields. Frost following the flowering stage reduces production drastically. High temperature and high wind velocity during anthesis and seed formation enhances sterility and reduces yield. Cloudy weather at the time of flowering increases the number of aphids and disease.

9.4.2 Soils
Coriander is grown as an irrigated crop on loamy to moderately heavy soils. It is also cultivated as an unirrigated crop with conserved moisture on black cotton or heavy soil types with high moisture retention capacity. Saline, alkaline and even sandy soils are not suitable for coriander cultivation.

9.4.3 Field preparation
The coriander field is brought to a fine tilth by two or three repeated ploughings, preferably by first ploughing with a soil-turning plough. If soil moisture level is low, a light irrigation may be given before ploughing. For an unirrigated crop, field preparation should be done when the moisture level in the soil falls to an optimum level following the preceding rains and thoroughly planked to check the moisture losses until the sowing time arrives.

9.4.4 Sowing time
The optimum temperature for germination and early growth of coriander is 20–25°C. In the

Indian subcontinent the main crop is sown from the last week of October to the first or second week of November in north India. Seed germination and early growth are adversely affected by high temperature if the crop is sown earlier. In south India a second crop is also taken during the summer, when sowing is done after Rabi/winter seasons (March–June) from 15 May to 15 June. Delay in sowing reduces the plant growth and increases the incidence of diseases and pests.

9.4.5 Seed rate and sowing
To achieve optimum plant density in irrigated conditions, a seed rate of 12–15 kg/ha, depending upon the seed size, with a slightly higher seed rate for bold-seeded type, is sufficient. For unirrigated conditions, a seed rate of 25 to 30 kg/ha is recommended. Seeds are divided into two halves and treated with 1.0 g Bavistin per kg of seed or with any of the Agroson GN, Thiram or any other mercurial fungicide at the rate of 2.0 g per kg of seeds.

9.4.6 Sowing method
Sowing is done by scattering or in 30 cm apart, shallow rows behind the plough. Line sowing facilitates intercultural operations in the standing crop. In heavy soils or under high soil fertility conditions 40 cm row spacing is recommended. An optimum plant-to-plant distance in rows is 10 cm. Care should be taken in both methods of sowing that seeds are uniformly covered with soil no deeper than 4 cm.

9.4.7 Manure and fertilizers
At the time of field preparation, about 10–20 tonnes/ha of farmyard manure (FYM) or compost should be applied. In addition to the FYM/compost, 20–30 kg nitrogen, 30 kg phosphate and 20 kg potash per hectare should be applied in the form of fertilizers at the time of sowing. In irrigated conditions, an additional dose of 40 kg nitrogen/ha should be applied with irrigation in two equal portions, first at 30 days and second at 75 days after sowing.

9.4.8 Weed control
Initial growth of coriander is slow. The first hoeing and weeding should be done 30 days after sowing. Thinning to remove excessive plants may be done at this stage. The second hoeing and weeding may be done between 50 to 60 days after sowing depending upon the regrowth of weeds. Chemical control of weeds with a pre-planting application of the herbicide Fluchloralin at the rate of 0.75 kg/ha or a pre-emergent application of Oxyfluorfen at the rate of 0.15 kg/ha or of Pendamithalin at the rate of 1.0 kg/ha dissolved in 400–500 litres of water is very effective.

9.4.9 Irrigation
Depending upon the climatic conditions, moisture-retaining capacity of soil and variety used, four or five irrigations are required after germination. The first irrigation should be given at 30–35 days after sowing, the second at 60–70, the third at 80–90, the fourth at 100–105 and the fifth at 110–150 days. Besides the schedule, one light irrigation may sometimes be needed between five and eight days after sowing to facilitate proper germination.

9.4.10 Harvesting

Harvesting should be done as soon as the colour of seeds starts turning from green to yellow. To obtain good lustre of seed with maximum yield, the harvesting should be done when 50% seeds are yellow. The harvested material should be dried in the shade to retain seed colour and quality; if it is not possible then the material should be kept in bundles upside down to avoid direct sun rays on the seeds, which adversely affects the colour of the produce. After drying the harvested material, the seeds are separated by light beating with sticks and winnowing. To obtain extra income, leaf plucking to the extent of 50% at 75 days after sowing without reducing in seed yield may be done.

9.4.11 Yield

Under good management practices and use of high-yielding varieties, an average yield of 1200–1500 kg/ha under irrigated conditions and 700–800 kg/ha under unirrigated conditions can be easily obtained.

9.4.12 Post-harvest management

Proper care should be exercised on the post-harvest operations to ensure proper quality of the produce. The threshing of the dried bundles should be done on a clean floor or on tarpaulin. The produce should be properly cleaned with vacuum gravity separators or a Distoner spiral gravity separator and graded. The graded material should be packed in lint-free bags and stored in a damp-free aerated storehouse to insure insect-free conditions.

9.5 Uses

Coriander was one of the first spices to be used as a common flavouring substance. The stem, leaves and fruits have a pleasant aromatic odour. The entire plant, when young, is used in preparing chutneys and the leaves are used for flavouring curries, sauces and soups. The dried fruits are extensively used in preparation of curry powder, pickling spices, sausage and seasoning. The seeds are used in medicine as a carminative, refrigerant, diuretic and aphrodisiac. It is used in the preparation of many household medicines to cure bed cold, seasonal fever, nausea, vomiting and stomach disorders. It is also used in *the Ayurvedic* system of medicines for curing their unpleasant odour and taste. Coriander oil and oleoresin are primarily used in seasonings for sausage and other meat products. They find application in baked goods, condiments, chewing gums and alcoholic/non-alcoholic beverages and also function as essential ingredients in curry mixes.

9.6 Diseases, pests and the use of pesticides

The following are the major important diseases that affect coriander:

No.	Disease	Causal organism
1.	Wilt	*Fusarium oxysporum* Schlecht f. sp. *corianderi*
2.	Powdery mildew	*Erysiphe polygoni* DC
3.	Blight	*Alternaria poonensis*
4.	Stem gall	*Protomyces macrosporus* Unger

9.6.1 Wilt

The coriander wilt caused *by Fusarium oxysporum* Schlecht f. sp. *conrianderi* may infect the plant at any stage of the crop, but more frequently affects younger plants. The infected plants wilt and dry up, resulting in a loss of up to 10% of the yield.

Adopting the following measures can reduce the incidence of disease:

- Seed treatment with Thiram and Bavistin with the proportion of 1 : 1 at the rate of 3.0 g/kg seed.
- . Use of Trichoderma 4.0 g per kg seed as seed treatment.
- Deep summer ploughings.
- Use of tolerant/resistant varieties.
- Adoption of at least three-year crop rotations.
- Use of disease-free and healthy seeds obtained from wilt-free seed crop.

9.6.2 Powdery mildew

Powdery mildew is caused by *Erysiphe polygoni* DC. A white powdery mass appears on the leaves and twigs of the plants in the initial stage; later on, the whole plant is covered with the whitish powder. It affects the number and size of seed and in severe conditions the infected plants do not even produce seed. The disease makes its appearance when the atmosphere gets damp, particularly during the flowering stage of the crop. The mildew can completely ruin the crop if adequate control measures are not taken in time.

Mildew can be controlled by the following:

- Dusting of sulphur powder at the rate of 20–25 kg/ha.
- Spraying 0.2% wettable sulphur or 0.1% Kerathane L.C. or 0.05% Calixin at the rate of 500–700 l/ha. Spraying or dusting should be repeated after 15 to 20 days.
- The harvesting of the mature crop should not be delayed; the seeds may be stored in gunny bags with paper lining and cloth bags for seed purposes.

9.6.3 Blight

Blight is caused by *Alternaria poonensis,* which appears in the form of dark brown spots on the stem and the leaves. It can be controlled by spraying a 0.2 % solution of Indofil M-45 or 0.1% Bavistin at the rate of 500–700 l/ha.

9.6.4 Stem gall

Stem gall is caused by *Protomyces macrosporus* Unger. In infected plants, blisters appear on the leaves and the stem and the infected plants produce deformed, hypertrophied and hard seeds. The yield as well as quality of the produce is reduced.

Stem gall can be controlled by the following:

- Seed treatment with Agrosan GN at the rate of 2 g, Thiram + Bavistin (1 : 1) at the rate of 3 g/kg seed is recommended.
- Use tolerant/resistant varieties such as Indian variety RCr.41.
- Spray 500–700 l solution of 0.1% Bavistin at the appearance of stem gall and repeat the spray twice or three times after 20 days interval till the disease is completely controlled.

In addition to these major diseases, some other minor diseases are also reported and appear sporadically in coriander, such as bacterial soft rot of leaves caused by *Erwinia aroida,* bacterial disease caused by *Xanthomonas tranlucens*, stem rot caused by *Rhizoctonia solani* Kuhn, root rot caused by *Curvularia pallescens* Boedjin and seed-borne mycoflora such as *Phoma multirostrata, Alternaria alternata, Fusarium moniliforme, F. semitectum* and *F. solani*, etc. The diseases may be controlled through seed treatment and other measures adopted for control of major diseases. These are of less economic importance because of their minor and sporadic appearance.

9.6.5 Aphids

Aphids (*Hyadaphis corianderi*) suck the plant sap from tender parts, leaves and flowers. The infected plants turn yellow, which results in shrivelled and poor seed formation, with reduced yield and quality of the produce.

When controlling aphids in coriander, remember that honeybees are the main pollinating agent; care must therefore be taken in choosing the insecticide such that it does not damage the honeybees. A spray of 500–700 l/ha Endosulfan (35 EC) at the rate of 0.07% or a spray 500–700 l emulsion of 0.03% Dimethoate or 0.03% Phosphamidon (85 WSC) or 0.1% Malathion (50%) or 0.03% Methyldemeton per hectare in evening hours does not cause much damage to honeybees because at this time, activity of honeybees in the field is slow. Spray can be repeated as the need arises but in any condition must be suspended one month before crop harvesting.

9.6.6 Mites

The mite *Petropbia lateens* frequently attacks the coriander crop at the stage of seed formation. The whole plant becomes whitish yellow and appears sickly: infestation is more severe on the young inflorescence. Yield is reduced and seeds become shrivelled if infestation is not checked quickly.

Mites can be controlled by the following:

- Spray the crop with 0.025% emulsion of Ethion (50 EC) or spray 0.07% solution of Dicofol (18.6 EC) or 0.03% of Phosphamidon (85 WSC).
- The systemic insecticides generally used against sucking insects may also be used for control of mites if the specific accaricides are not available.

9.6.7 Frost damage

The coriander crop is most vulnerable to frost damage at the flowering and early seed formation stages. The incidence can be minimized by adopting the following control measures:

- Spray 0.1% solution of sulphuric acid.
- Irrigate crop just before the incidence of frost.
- Set up wind breaks obstructing the cool waves.
- Create smoke cover in the early morning to dilute the effect of cold waves.

9.7 Quality issues

9.7.1 Quality of produce

Quality plays a vital role in all walks of life and its importance in coriander, like other seed spices, needs no separate emphasis. The quality of any product is assessed by means of its intrinsic as well as extrinsic qualities. The quality of coriander relates to size, shape, appearance, colour, odour and aroma characteristics. These characteristics vary widely, depending upon the variety, agro-climatic conditions existing in the area of production and harvest and post-harvest operations. Moisture, volatile oil, oleoresin content and major chemical constituents present in coriander determine the intrinsic quality. The customer need not, however, accept the high degree of intrinsic qualities alone as the final quality of the produce. The produce must be safe and free from any health hazard substances and contaminants. These are classified into three categories and known as defects.

Physical contaminants
Physical contaminants are termed macro-contaminants and decide the extrinsic quality (seed size, shape, appearance and colour) of the produce. The major defects coming under this class include immature or shrivelled seeds, insect-infested/defiled products, presence of live or dead insects, excreta of mammals (rodents, cattle, etc.), excreta of other animals such as insects and birds, extraneous foreign matter and filth. Extraneous matter can be of coriander itself or any other plant parts. Filth can be classified as heavy filth, including sand and mud particles, and light filth, including parts of insects, birds or animals, which are considered to be unacceptable in any food material.

Chemical contaminants
Among chemical contaminants, defects due to the presence of added colouring material, preservatives, antioxidants, fumigants (SO_2, ethyl oxide, methyl bromide), aflatoxin, trace metals (lead, arsenic, chromium, cadmium, copper, zinc, etc.) and pesticide residue are important.

Microbial contaminants
The prominent microbial contaminants are due to the presence of *Salmonella*, *Escherichia coli*, total variable plate count (TCP) or aerobic plate count (APC), yeast and mould. These contaminants cause severe health hazards.

9.7.2 Factors influencing seed quality

Effect of production practices
The major thrust of seed spices was on higher production or productivity; the quality considerations were generally poor in the developing countries mainly because of the following reasons/factors:

- Immediate benefit to the farmers attained through yield increase.
- Scientific grading based on intrinsic quality was not adequately developed mainly because the different levels of quality did not fetch differential prices.
- Lack of understanding of the elusive characters of quality.
- Lack of facilities to evaluate quality objectives.

As the quality picture is slowly clearing owing to the advent of modern chromatographic techniques, there is greater awareness and fierce competition for quality in the international market for export-oriented seed spices including coriander, and due attention is now being given to the development of high-quality varieties with appropriate post-harvest management techniques.

Effect of climatic conditions
The role of climatic conditions on the biosynthesis pattern of volatile oil constituents in coriander is well known. Lytkin (1953) has observed that the cooler and drier climate of northern Europe produces more linalool in coriander than the tropical climate of India and Morocco. Hotin (1957) observed that the fruit ripens at high humidity, volatile oil contents in the fruit may be high but its organoleptic quality is poorer owing to lower linalool contents and more aldehyde contents.

Effect of soil and soil fertility
Coriander is grown under a wide range of conditions; however, best yield and high volatile oil content are obtained on a medium to heavy soil in a sunny location with good drainage and well-distributed moisture. Coriander is usually grown as a rain-fed crop, and judicious use of fertilizers has also been shown to benefit the volatile oil contents as well as seed yield of coriander (Pillai and Bhoominathan, 1975; Prakash Rao *et al.*, 1983; Rahman *et al.*, 1990).

Effect of weed
Adulteration of weeds to the spices affects the quality. The presence of weed seeds in coriander seed will adversely affect the volatile oil contents and test weight of coriander.

Effect of diseases
Diseases such as wilt, powdery mildew and stem gall attack the crop causing heavy loss of yield and deteriorate the quality of the produce. Following good management practices and control measures, these diseases can be controlled (Sharma *et al.*, 1996).

9.7.3 Quality and marketing
Coriander grown in different places varies considerably in extrinsic quality. The seed may contain damaged seed, shrivelled seed and other foreign matter. This foreign matter may be a stalk, dirt, cereals, etc. Adulteration of superior grade with inferior grade is common. This unhealthy practice spoils the quality. The quality of the produce depends upon the quality of the raw material and the practices adopted in processing, packaging, storing and transportation. As quality is the most urgent challenge facing the industry, there is a need to ensure that the product for the market, either for domestic or export purposes, is completely free of pesticide residue, aflatoxin, other mycotoxin and unfavourable microbial contamination. After processing coriander should be graded according to the International Organization for Standardization (ISO) or according to the requirement of the importing country. Most of the importing countries have their own grades. Therefore, grading and standardization become the essential prerequisites ensuring quality.

Quality assurance through an effective and efficient quality control system is pivotal to augment the sale of spices and its products. Therefore, coriander to be exported should conform to the quality standards demanded by the importing countries. (Sharma and Agrawal, 1998). The relative importance of quality is dependent upon the end use of the spices. For whole seed entering the grocery trade, the appearance of the seeds is the primary quality determinant. The appearance is of less importance when coriander is intended for industrial extraction purpose, that is, for essential oil and oleoresin purpose. In these cases, the quantity and quality of volatile oil and its constituents are more important.

The major coriander-importing countries, viz. the UAE, Sri Lanka, Singapore, Malaysia, the UK, the USA and South Africa, are quality conscious and have strict quality standards. The USA, Japan, Canada, Australia and the European countries have their own stringent

Table 9.1 American Spice Trade Association (ASTA) Cleanliness Specifications for coriander

*Whole insects, dead (by count)	Excreta, mammalian (by mg/lb)	Excreta, mammalian (by mg/lb)	Mould (% by wt)	Insect defiled/ infested (% by wt)	Extraneous/ foreign matter† (% by wt)
4	3	10	1	1	0.5

*Whole insects, dead: Cannot exceed the limits shown.
†Extraneous matter: Includes other plant material, e.g. foreign leaves.

Table 9.2 Cleanliness Specification for spices in major importing countries

Country	Extraneous matter (%/wt)	Moisture (%/wt)	Total ash (%/wt)	Acid insoluble ash (%/wt)
Germany	–	–	8.0	2.5
Netherlands	1.5	10.0	8.0	2.5
UK	2.0	10.0	8.0	1.5

Source: Specifications in Germany, Netherlands and the UK (importers' specifications).
Methodology used in setting standards.
1. Moistures   ISO939
2. Total ash   ISO928
3. Acid insoluble ash ISO930
4. Volatile oil   ISO6571

Refer to the above methods when analysing the products.

food laws and regulations. The main objectives of the law are to protect the health and the safety of their citizens. The importers prescribe grade specifications for various spices depending upon the end use. The exporting countries must adhere to the practices for cultivation, post-harvest operations, packaging and storage, to maintain high-quality standards to compete in the international markets.

9.7.4 Limits of contaminants in importing countries

Spices exported to any country must conform to the cleanliness specifications stipulated by that country. These countries set limits for cleanliness specifications such as number of dead insects, amount of mammalian excreta and other excreta in the sample. If the exporting country does not fulfil these requirements, the consignment may be detained for reconditioning or be rejected.

The most popular specification for spices and herbs the world over is the 'ASTA Cleanliness Specifications for Spices, Seeds and Herbs'. The unified ASTA, US FDA Cleanliness Specifications for Spices, Seeds and Herbs was made effective from 1 January 1990. Major producing countries have built up their facilities to meet the requirements as per ASTA Cleanliness Specification (Table 9.1). Countries such as the UK, Germany and Netherlands have laid down cleanliness specification for spices (Table 9.2).

The European Spice Association (ESA), comprising the members of the European Union, has come out with the 'quality minima for herbs and spices' (Table 9.3). This serves as guideline specifications for member countries in the European Union. The European Union has yet to finalize the cleanliness specification for spices and spice products. The importing countries, where they do not have specifications for spices, used to request the exporting countries to supply spices as per the ASTA Specification.

Table 9.3 European Spice Association (ESA) Specifications of quality minima for coriander

Subject	Specifications
Extraneous matter	1%
Sampling	(For routine sampling) Square root of units/lots to a maximum of 10 samples. (For arbitration purposes) Square root of all containers e.g. 1 lot of coriander may = 400 bags, therefore square root = 20 samples.
Foreign matter	Maximum 2%
Ash	7 (ISO) Ash % w/w max
Acid insoluble ash (AIA)	1.5 (ISO) AIA % w/w max
H_2O	12 (ISO) H_2O % w/w max
Packaging	Should be agreed between buyer and seller. If made of jute and sisal, it should conform to the standards set by CAOBISCO Ref C502-51 -sj of 20-02-95. However, these materials are not favoured by the industry, as they are a source of product contamination, with loose fibres from the sacking entering the product.
Heavy metals	Shall comply with national/EU legislation
Pesticides	Shall be utilized in accordance with manufacturers' recommendations and good agricultural practices and comply with existing national and/or EU legislation.
Treatments	Use of any EC-approved fumigants in accordance with manufacturers' instructions, to be indicated on accompanying documents. (Irradiation should not be used unless agreed between buyer and seller.)
Microbiology	*Salmonella* absent in (at least) 25 g. Yeast and moulds 105/g target, 106/g absolute maximum. *E. coli* 102/g target, 103/g absolute maximum. Other requirements to be agreed between buyer and seller.
Off-odours	Shall be free from off-odour or taste
Infestation	Should be free in practical terms from live and/or dead insects, insect fragments and rodent contamination visible to the naked eye (corrected if necessary for abnormal vision).
Aflatoxins	Should be grown, harvested, handled and stored in such a manner as to prevent the occurrence of aflatoxins or minimize the risk of occurrence. If found, levels should comply with existing national and/or EU legislation.
Volatile oil	0.3 (ESA) V/O % v/w min
Adulteration	Shall be free from.
Bulk density	To be agreed between buyer and seller.
Species	To be agreed between buyer and seller.
Documents	Should provide details of any treatments the product has undergone; name of product; weight; country of origin; lot identification/batch number; year of harvest.

In addition to the cleanliness specification, the importing countries insist on the specification for parameters such as pesticide residues, aflatoxin, trace metal contamination and microbial contamination. Individual member countries in the European Union have fixed maximum residue levels (MRLs) for pesticide residues (Appendices I and II). The European Union has not prescribed the limits for pesticide residues in spices and spice products. The USA and Japan have prescribed the MRLs in spices. Under the Codex, MRLs for pesticide residues have not been prescribed. Some countries have prescribed pesticide residual limits for some specific spices. India has taken the initiative to fix the MRLs for spices at the Codex level. The European Union has prescribed limits for aflatoxin as 5 ppb, for Aflatoxin B1 and 10 ppb for aflatoxin total. Member countries in the European Union and others have fixed limits for aflatoxin varying from 1 ppb to 20 ppb (Table 9.4).

Importing countries are cautious about the microbial contamination in spices at the time

Table 9.4 Summary of legislation on aflatoxins in ESA member countries and other major importing countries

Country	Permitted levels	For which products	Comments
Austria	B1 < 1 ppb	All foodstuffs (except mechanically prepared cereals in the case of B1)	
Belgium	< 5 ppb for peanuts. EU legislation is expected		In Belgian law, aflatoxins (and toxins in general) may not present in foodstuffs, ie not detectable
Germany	B1+B2+G1+G2 < 4 ppb	All foodstuffs	
Denmark	B1 < 2 ppb		
Netherlands	B1 < 5 ppb	All foodstuffs	No control on B2
Switzerland	B1 < 1 ppb	All foodstuffs (except maize)	
	B2+G1+G2 < 5 ppb	All foodstuffs	
UK	< 50 ppb	Chilli	Only aflatoxin regulations on nuts/nut products
	< 10 ppb	Peanuts	Dried figs/dried fig products, which when sold to the consumer must contain < 4 ppb total aflatoxin
	< 4 ppb	Other nuts/dried figs, etc.	
Spain	B1 < 5 ppb B1+B2+G1+G2 < 10 ppb	All foodstuffs	
Sweden	B1+B2+G1+G2 < 5 ppb	All foodstuffs	
Finland	B1+B2+G1+G2 < 5 ppb	All foodstuffs	
Italy+France	< 10 ppb for B1		No regulations
USA	< 20 ppb	All foodstuffs	Guideline FDA

Source: EU Draft Legislation.

Table 9.5 General microbiological specification: Germany and Netherlands

Parameter	Standard value	Danger value
Germany		
Total aerobic bacteria	1×10^5/g	1×10^6/g
E. coli	Absent	Absent
Bacillus cereus	1×10^4/g	1×10^5/g
Staphylococcus aureus	1×10^2/g	1×10^3/g
Salmonella	Absent in 25 g	Absent in 25 g
Sulphite-reducing Clostridea	1×10^4/g	1×10^5/g
Netherlands		
Bacilus cereus	Absent in 20 g	Danger value similar to that of Germany
Escherichia coli	Absent in 20 g	
Clostridum perfringens	Absent in 20 g	
Staphylococcus aureus	Absent in 20 g	
Salmonella	Absent in 20 g	
Total aerobic bacteria	1×10^6/g	
Yeast and mould	1×10^3/g	
Coliform	1×10^2/g	

of import. Almost all the importing countries have fixed the limits for *Salmonella* as absent in 25 g. Specifications have been prescribed by major importing countries for the microbial parameters such as total plate count (TPC), *E. coli*, yeast, mould, coliforms, etc. The limits for the above parameters vary from country to country (Table 9.5).

It is obvious from the above that the utmost care in the production practices and post-harvest technology of coriander is essential for any country interested in exporting coriander.

9.8 Value addition

Spices are valued as ingredients of incense, embalming preservatives, perfumes, cosmetics and medicines. The use of coriander dates back to the day humans learnt to use fire for preparing food. But for a very long time the seed spices were used as freshly harvested/dried form. Much later, people realized the possibility of producing essential oil by pressing the plant parts. This was used for medicine and fragrance.

The beginning of industrialization at the end of the 19th century changed the habits of people considerably. People moved from agricultural living areas to urban areas where fresh food was not available so easily. This was the beginning of the food industry. The primary aim of the food industry was to give cheap nourishment. The question of good taste was of secondary importance. This has dramatically changed in the last four to five decades: with the change in life styles and urbanization, the popularity of fast food (convenience food) has increased. Food brands were created but the brands demanded consistent quality and long shelf-life of the food product. This could be achieved by using ingredients of the highest hygienic standards. This has led to the development of value-added products.

Value addition can be as simple as presenting a commodity in a cleaned graded form, which would instil confidence in the consumers for its quality image. On the other hand, it can be a completely different product such as oil, oleoresins, etc. Apparent value addition by image building is a marketing strategy successfully adopted in this area. The value-added form of spices has become the area with tremendous growth potential. The global market is increasingly shifting away from the commodity form towards the value-added form of consumer-packed branded spices, which overcome the disadvantages of raw spices.

Spices in raw forms have certain disadvantages. Whole or ground spices do not impart their total flavour readily. They are bulky for storage and often unhygienic owing to bacterial contamination. The price fluctuations for commodities are also very high. Some of these defects can be reduced by extracting oils through steam distillation and by preparing oleoresins using organic solvents.

Coriander can be used as value-added form like other seed spices as volatile oil, oleoresin, ground spices, curry powder, consumer packed spices and organic spices.

9.8.1 Volatile oil

The volatile oil is aromatic and is primarily recovered from the dried ripe seeds. To produce the oil, the dried seeds are placed in stainless steel distillation vessels equipped with steam inlet, vapour outlet, condenser and separator assembly. Live steam is introduced below the charge; the steam rising through the plant charge carries the volatile oil. The volatile oil is condensed and separated from water. The advantages of using essential oil are that it has uniform flavour quality, is free from enzymes and tannins and does not impart colour to the end product.

9.8.2 Oleoresin

Oleoresin represents the complete flavour and non-volatile resinous fraction present in the spices. The resinous fraction comprises heat components, fixative, natural antioxidant and pigments. Hence, oleoresin is designated as the true essence of the coriander.

Oleoresin in coriander seeds is obtained by solvent extraction of the ground seed and is a brownish-yellow liquid with a fruity, aromatic, slightly balsamic flavour. Oleoresin from roasted seeds has a more rounded and slightly caramellic flavour. Volatile oil in the oleoresin ranges from 2 to 12 ml per 100 g.

In coriander the volatile oil is found only in very small quantities, therefore the volatile oil content and oleoresin make less of a contribution as a value addition than the others.

9.8.3 Ground spices

These are the whole spices milled to a certain degree of fineness required by the food processor. The grinding technique should be studied in more detail in order to evolve efficient methods to prevent changes with respect to flavour and pungency. Ground spices can be incorporated into food dishes more uniformly than can whole spices. In spite of these attributes they have limited shelf-life and are subject to oxidation, flavour loss and degradation on long storage owing to microbial contamination.

9.8.4 Curry powder

Curry powder is an indigenous seasoning made from various spices. The number of spices varies from 5 to 20 depending on the powder's end use. Various spices, namely turmeric, garlic, chillies, coriander, cumin, fennel, fenugreek and black pepper, constitute the raw materials used in quality curry powder. The ingredients of curry change according to different needs. The colour form and taste of various curries are in accordance with the customs of various nations and regions. Consumers all over the world demand different curry powders. The international trade in curry powder is around 9000 Mt per annum. The export trade in curry powder at present is dominated by India.

9.8.5 Consumer-packed spices

The exported spices are consumed in three main segments, namely industrial, institutional and retail. Different packaging media are used according to the consumer's preference. Packaging has gained considerable importance as it increases the shelf-life of spices. The development of new and improved plastic films, aluminium foil, laminations, high-speed film-sealing machines, etc. has created new opportunities for packaging the spices as instant spices, spice pastes and spice powder, etc. Exporting consumer packed spices can earn higher unit value for the same quantity. The prices of such retail spice packs are higher – between 50 and 100% as compared with prices of bulk spices. The weights of retail packs generally range between 30 g and 500 g. However, institutional packs range between 500 g and 1 kg in weight. It is important to note that, with the stiff competition that India is facing in the spice market, building brand image is essential, particularly in the packed spices.

9.8.6 Organic spices

With the trend towards pre-processed foods (convenience foods), the demand for organic spices is increasing. Organic agriculture has gained importance in modern societies. This had led to the development of international trade for organic spices. Europe, the USA and Japan are by far the largest markets, though there are smaller but interesting markets in many other countries, including a few developing countries. The importance of organic agriculture can be inferred from the fact that some European countries are supporting organic agriculture by giving subsidies for conversion. As a matter of fact, organic products are more expensive than the conventional counterparts and fetch a premium in the international market. Prices may be higher by 20–50% but gaining certification from recognized international agencies is a costly affair.

9.9 Future research trends

- Enhancement of germplasm collections, their cataloguing and conservation.
- Developing varieties with high yield, quality and tolerance to pests diseases and drought.
- Strengthening the breeding programme to evolve high-yielding varieties with multiple resistances to biotic and abiotic stresses suitable for organic farming.
- Research on the integrated use of fertilizers and organics for sustainable yield and quality.
- Research on effective integrated pest and disease management strategies with emphasis on biological control that are environmentally friendly and ecologically sound.
- Developing appropriate farm processing technologies in spices.
- A well-organized extension programme supported with a sound seed production programme for speedy adoption of improved varieties.
- Development of a production technology of spices and spices based cropping systems.
- Disease and pest forecasting.
- Development of storage technology for reducing post-harvest losses.

9.10 References

AGARAWAL, S. and SHARMA, R.K. (1990), 'Variability in quality aspect of seed spices and future strategy'. *Indian Cocoa, Arecanut and Spices Journal* **13**: 127–9.
ANURADHA, HORE (1979), 'Improvement of minor (Umbelliferous) spices in India'. *Economic Botany* **33**(3): 290–7.
BOELENS, MAN H., VALVERDE, F., SEQUEIROS, L. and JIMENEZ, R. (1989), 'Ten years of hydro diffusion oils'. *Proc. of 11th International Congress of Essential Oils, Fragrances and Flavours,* New Delhi (India), 121–126.
DIMRI, B.P, KHAN, M.N.A. and NARAYAN, M.R. (1977), 'Some promising selection of Bulgarian coriander (*Coriandrum sativum* Linn.) for seed and essential oil with a note on cultivation and distillation of oil'. *Indian Perfumer* **20**(1A): 14–21.
HIRVI, T., SALOVAARA, I., OKSANEN, H. and HONKANEN, E. (1986), 'Volatile constituents of coriander fruits cultivated at different localities and isolated by different methods'. In Brunke, E.J., *Progress in Essential Oil Research,* Berlin, Walter de Gruyter, 111–16.
HOTIN, A.A. (1957), 'Biological basis of essential oil development'. Krasnodar, Dissertation.
KUMAR, C.R., SARWAR, M. and DIMRI, B.P. (1977), 'Bulgarian coriander in India and its future prospects in export trade'. *Indian Perfumer* **21**(3): 146–50.
LYTKIN, I.A. (1953), 'An experiment on the cultivation of coriander in Siberia'. *Agrobiologiya* **4**: 151–152. (*Hort. Abs.* **24**: 2887).
PILLAI, F.K.T. and NAMBIAR, M.C. (1982), *Cultivation and Utilization of Aromatic Plants.* Jammu-Tawi (India), CSIR, Regional Research Laboratory, 167–89.
PILLAI, O.R. and BHOOMINATHAN, H. (1975), 'Effect of N P K fertilizers on the yield of coriander'. *Arecanut and Spices Bulletin* **6**(4): 82–3.
PRAKASH RAO, E.V.S., CHANDRASHEKHARA, G. and PUTTANA, K. (1983), 'Biomass accumulation and nutrient uptake pattern in coriander var. Cimpo S-33'. *Indian Perfumer* **27**: 168–70.

PURSEGLOVE, J.W., BROWN, E.G., GREEN, C.L. and ROBBINS, S.R.J. (1981), *Spices*, Vol. II. Longman, London and New York, 736–88.

RAHMAN MD, O., HARIBABU, R.S. and SUBBA RAO, N. (1990), 'Effect of graded level of nitrogen on growth and yield of seed and essential oil of coriander'. *Indian Cocoa, Arecanut and Spices Journal* **13**: 130–3.

RAMANUJAM, S., JOSHI, B.S. and SAXENA, M.B.L. (1964), 'Extent and randomness of cross pollination in some umbelliferous spices in India'. *Indian J. Genet* **24**(1): 62–7.

RAO, B.S., SUDBOROUGH, J.J. and WATSON, H.E. (1925), 'Notes on some Indian essential oils'. *J. Indian Inst. Sci.* **8A**: 182.

SHARMA, R.K. and AGRAWAL, S. (1998), 'Export of seed spices – constraints and prospects'. *Proc. of National Seminar on Agricultural Development and Marketing,* Jobner (India).

SHARMA, R.K., DASHORA, S.L., CHOUDHARY, G.R., AGRAWAL, S., JAIN, M.P. and SINGH, D. (1996), *Seed Spices Research in Rajastha.* Bikaner (India), Directorate of Research, Rajasthan Agricultural University, 1–54.

SINGH, V.P and RAMANUJAM, S. (1973), 'Expression of andromonecy in coriander *Coriandrum sativum* L.'. *Euphytica* **22**: 181–8.

Appendix I Maximum pesticide residue limits in the Netherlands and the UK

Active substance	Limiting values in ppm	
	Netherlands	UK
HCH without lindane	0.02	0.02
Lindane	0.02	–
Hexachorobenzene	–	0.01
Aldrin and Dieldrin	0.03	0.01
Sum of DDT	0.15	0.05
Malathion	0.05	8.00
Dicofol	0.05	0.50
Chlorpyrifos	0.01	–
Ethion	0.01	–
Chlordan	0.01	0.02
Parathion	0.10	1.00
Parathion methyl	0.10	0.20
Mevinphos	0.05	–
Sum of Endosulfan	0.02	0.10
Phosalon	1.00	0.10
Vinclozolin	–	0.10
Dime;thoat	0.01	0.05
Quintozen	–	1.00
Metacriphos		
Heptachlor and epoxide	0.21	0.01
Methidathion		
Diazinon	0.05	0.05
Fenitrothion	0.05	0.05
Bromophos		
Mecarbam		
Methoxychlor	0.05	
Omethoat		0.20
Dichlorvos	0.05	
Phosmet	0.01	
Methylbromide		0.10
Tetradifon		

Appendix II Maximum residue levels fixed for spices as per the German legislation and pesticide residue limits prescribed by Spain

Germany*		Spain	
Active substance	Highest limit (mg/kg)	Name of pesticides	MRL (mg/kg)
Aldrin and Dieldrin	0.1	Acephate	0.10
Chlordane	0.05	Atrazine	0.10
Sum of DDT isomers	1.0	Bendiocarb	0.05
Endrin	0.1	Carbaryl	0.10
HCH without lindane	0.2	Carbosulfan	0.10
Heptachlor and epoxide	0.1	Chlorpyrifos	0.05
Hexachlor benzol	0.1	Chlorpyrifos – methyl	0.05
Lindane	0.01	Cipermethrin	0.05
HCN and cyanides	15.0	Diasinon	0.05
Bromides	400.0	Dicofol	0.02
Carbaryl	0.1	Dimethoate	0.05
Carbofuran	0.2	Etion	0.10
Chlorpyrifos	0.05	Fentoato	0.05
Methyl chlorpyrifos	0.05	Fenitrothion	0.05
Cypermethrin**	0.05	Fenthron	0.05
Deltamethrin	0.05	Melathron	0.50
Diazinol	0.02	Metalaxyl	0.05
Dichlorvos	0.1	Methamidophos	0.01
Diclofop methyl	0.1	Monocrotophos	0.02
Dicofol**	0.02	Omethoate	0.10
Dimethoate	0.5	Phosalone	0.10
Disulfoton	0.02	Pirimicarb	0.05
Dithiocarbamate	0.05	Pirimiphos – methyl	0.01
Endosulfan**	0.05	Profenofos	0.02
Ethion	0.05	Prothiofos	0.02
Fenitrothion	0.05	Pyrazphos	0.01
Fenvalarate**	0.05	Terbuconazole	0.05
Copper-based pesticides	40.0	Tolclophos – methyl	0.01
Malathion	0.05	Triazophos	0.01
Methyl bromide		Vinclozolin	0.05
Mevinphos	0.05		
Omethoate	0.05		
Parathion and para oxon	0.1		
Methyl parathion and methyl para oxon	0.1		
Phorate	0.05		
Phosalone	0.05		
Phosphamidon	0.05		
Pyrethrin	0.5		
Quinalphos	0.01		
Quintozen	0.01		

*Of the above, the limits mentioned against the first ten pesticides are specific for spices and the remaining are the general regulations for all plant foods.
**Sum of isomers.

10

Geranium

M. T. Lis-Balchin, South Bank University, UK

10.1 Introduction

Geranium oil is known as the 'poor-man's rose' and is extracted from the leaves of various cultivars of *Pelargonium* species, which originate in Southern Africa and not from the genus *Geranium* (Lis-Balchin, 2002a). The latter consists of many species, all hardy, found in European hedgerows, and rarely odorous, except for *G. robertianum* (Herb Robert) and *G. macrorhizum* (yielding Zdravetz oil, in Bulgaria). The confusion with the genus *Geranium* originated before Linnaeus (1753), as the two genera were originally under the one genus: *Geranium*. Acceptance of re-classification by Sweet (1820) has not improved the confusion, as garden centres still sell pelargoniums as geraniums. The primary sources of geranium oil are now Egypt, China and the Comores, with some recent production from plants grown in India and S. Africa; plants had previously been grown in southern France, Morocco and Tunisia. Geranium oil is primarily used by fragrance companies but some is employed in the food industry.

10.2 Chemical composition

This was reviewed in full by Williams and Harborne (2002). There is a wide variation in the flavonoid constituents among the ten taxa (Bate-Smith 1973) with a preponderance of myricetin, kaempferol and quercetin in species from sections *Hoarea* and *Pelargonium*. Later, quercetin was found to be universally present. Myricetin and kaempferol were detected in 71% of the taxa in section *Glaucophyllum* (van der Walt *et al.*, 1990) and myricetin was found in most species in section *Cortusina* (Dreyer *et al.*, 1992) but was absent from sections *Chorisma* (Albers *et al.*, 1995) and *Jenkinsonia* (van der Walt *et al.*, 1997). The major components of the section *Chorisma* were flavonols but two flavones, luteolin and apigenin, were additionally detected in two taxa. In a survey of 58 *Pelargonium* species from 19 sections, Williams *et al.* (2000) confirmed that flavonols are the major leaf vacuolar flavonoid constituents in the genus. Both quercetin 3-methyl ether and isorhamnetin were detected in 10% of the sample; however, apigenin was not detected in any taxon.

The floral flavonoids consist of pelargonin (pelargonidin 3,5-diglucoside), first found it in its pure form in the salmon pink petals of a zonal by Robinson and Robinson (1932). Pelargonidin and malvidin 3,5-diglucoside were accompanied by small amounts of the relatively rare peonidin 3,5-diglucoside (Harborne 1961). *P.* × *hortorum* petals contained all six common anthocyanidins (including delphinidin, petunidin and cyanidin) as the 3,5-diglucosides in different colour forms (Asen and Griesbach 1983). A range of colourless flavonol glycosides based on kaempferol and quercetin occur with the above anthocyanins in these petals.

Exudate lipophilic flavonoids as well as terpenoid constituents are produced by trichomes on the leaves. These were detected in 35% of the *Pelargonium* taxa surveyed by Williams *et al.* (1997), but mostly only in trace amounts. Exudate flavones were found in some taxa, also flavonol (quercetin and kaempferol) methyl ethers.

The genus *Pelargonium,* like *Geranium,* is unusual in synthesizing both hydrolysable (ellagitannins) and non-hydrolysable (proanthocyanidins) tannins in abundance in many of its species. Some *Pelargonium* species produce free ellagic acid in the absence of ellagitannins (Williams *et al.*, 2000). Gallic acid was also recorded from 62% of the species: their co-occurrence with flavonoids has some taxonomic and evolutionary significance (Williams and Harborne, 2002). Four coumarins – the common scopoletin, the rare 7-hydroxy-5,6-dimethoxycoumarin and its 7-methyl ether and its 7-glucoside – were identified in roots of *Pelargonium reniforme* and detected in roots of 11 other species (Wagner and Bladt, 1975). Salicylic acid derivatives have been found in *Pelargonium* × *hortorum* leaves, which are largely resistant to attack by the two-spotted spider mite, *Tetranychus urticae* Koch, owing to the production of a toxic, sticky exudate from glandular trichomes on both surfaces of the leaves (Craig *et al.*, 1986), identified as 6-[(Z)-10'-pentadecenyl]salicylic acid and 6-[(Z)-12'-heptadecenyl]salicylic acid (Walters *et al.*, 1988). Tartaric acid is a characteristic constituent of the genus *Pelargonium,* making up over 1.5% of the dry weight of the aerial parts of 23 species and hybrids (Stafford 1961). Oxalic acid has been isolated from *P. peltatum,* a plant potentially poisonous to livestock since oxalic acid is present in a water-soluble form instead of the more usual insoluble calcium salt. It occurs together with malic, tartaric and succinic acids in this species.

Screening studies on *Pelargonium* species revealed the presence of alkaloids, especially in 'zonal' cultivars (Lis-Balchin, 1997); alkaloids had previously only been found in the *Erodium* genus of the Geraniaceae (Lis-Balchin and Guittoneau, 1995). Both simple amines, tyrosine and tryptamine (Lis-Balchin *et al.*, 1996b) and then more complex indole alkaloids – elaeocarpidine and isoelaeocarpidine – were identified. They were found in all zonal *Pelargonium* cultivars, but not in the pure ivy leaf or regals (Lis-Balchin, 1996), and appear to have an insect repellent activity against whitefly (Woldermarian *et al.*, 1997; Simmonds, 2002) and be concentrated in the darker zonal area of the leaves. Essential oils consist mainly of geraniol, citronellol, citronellyl and geranyl esters, limonene, linalool and characteristic sesquiterpenes: γ-epi-eudesmol (Egyptian) or guaia-6,9-diene in the Bourbon and China oils (Table 10.1).

10.3 Production and cultivation

10.3.1 Production
The main producing areas are currently Egypt, China and the Comores, with some recent production from plants grown in India; plants were previously grown in southern France, Morocco, Algeria and Tunisia. Some fine-quality geranium oil is now emerging from new

Table 10.1 Physico-chemical characteristics of geranium oils from different sources, their components and their sensitisation potentials

	Bourbon	Morocco	Egypt
Relative density at 20°C	0.884–0.892	0.883–0.900	0.887–0.892
Refractive index at 20°C	1.462–1.468	1.464–1.472	1.466–1.470
Optical rotation at 20°C	–8 to –14	–8 to –13	–8 to –12
Acid value maximum	10	10	6
Ester value	52–78	35–80	42–58
Ester value after acetylation	205–230	192–230	210–235
Carbonyl value expressed as iso-menthone	58	58	Not given
Apparent citronellol (rhodinol) content	42–55	35–58	40–58

Normal range of main components for commercial geranium oils (%)

Citronellol	28–58
Geraniol	7–19
Linalool	3–10
Isomenthone	4–7
Citronellyl formate	5–12
Geranyl formate	1–4
10-epi-χ-eudesmol	3–7 (Egyptian)
guaia-6,9-diene	1–7 (Bourbon, China)

EC regulations 2002 (CHIP) governing its major components

Geranium oil, CAS No. 8000-46-2; EEC No. 290-140-0; Hazard symbol: Xn; Risk phase: R65; H/C 15%; Safety phase S62

D-Limonene, CAS No. 5989-27-5; EEC No. 227-813-5; Hazard symbol: Xn N; Risk phase: R10, 38, 43, 50/53; H/C 100%; Safety phase S24, 37, 60, 61

L-Limonene, CAS No.5989-54-8; EEC No. 228-813-5; Hazard symbol: Xn N; Risk phase: R10, 38, 43, 50/53; H/C 100%; Safety phase S24, 37, 60, 61

Linalool, CAS No.78-70-6; EEC No. 201-134-4; Hazard symbol: none; Risk phase: none; H/C: none; Safety phase: none

Citral, CAS No. 5392-40-5; EEC No. 226-394-6; Hazard symbol: Xi ; Risk phase: R38, 43; H/C: none; Safety phase: S24/25, 37

Citronellol, CAS No.106-22-9; EEC No. 203-375-0; Hazard symbol: Xi N; Risk phase: R38, 43, 51/53; H/C: none; Safety phase S24, 37, 61

Geraniol, CAS No.106-24-1; EEC No. 203-377-1; Hazard symbol: Xi ; Risk phase: R38, 43; H/C: none; Safety phase: S24, 37

Maximum levels of fragrance allergens in aromatic natural raw materials:

European Parliament and Council Directive 76/768/EEC on Cosmetic Products, 7th Amendment 2002: The presence of the substances must be indicated in the list of ingredients when its concentration exceeds 0.001% in leave-on products and 0.01% in rinse-off products.

Total sensitizers: citral, 1.5; geraniol, 18; linalool, 10; = 29.5 (EFFA)

plantations in South Africa. World demand is estimated at around 200 tonnes pa, but there are huge fluctuations probably because of harvest failures and thereby rise in price. The major markets for geranium oil are the USA, France, Germany, the UK, other European countries and Japan. France is a major re-exporter of geranium oil, which is often further distilled and re-blended to client specifications (Demarne, 2002). Total US imports for the years 1993 to 2001 fluctuated between 38 and 107 megatonnes per year; the highest being in1995 and lowest in 2000. The total US exports varied from 58 megatonnes in 1996 to 11 megatonnes in 2000 and 2001.

10.3.2 Cultivation
Cultivars obtained from *P. capitatum* × *P. radens* and *P. capitatum* × *P. graveolens* must be

tested for potential yield and oil quality in different parts of the world as well as their odour, as mint-scented cultivars are common (van der Walt and Demarne, 1988; Demarne, 2002). The plants are treated as perennials but last only about three to five years, owing mainly to fungal, bacterial and other infestations. The plants are heterozygous and highly polyploid and are often sterile so are generally propagated from cuttings; this ensures standardization as hybridization would affect the growth of the plants and the essential oil quality. Micropropagation and tissue culture have been attempted (Charlwood and Lis-Balchin, 2002), but although most species of *Pelargonium* could be micropropagated, the biotechnological production of the essential oil was not commercially successful.

Herbaceous terminal stem cuttings, 12–20 cm long are cut above a node, that end being ideally treated with indole-butyric acid, 0.1–0.2%, and captan, 10% in talcum powder, and rooted under shade in a nursery with mist irrigation; day and night temperatures are 21 and 12°C to ensure rapid profuse root development for transplantation within 40–60 days (Demarne, 2002). Unrooted cuttings are often planted directly in the field, but wastage may be high. Requirements for cultivation include: abundant sunshine, well-drained fertile soil containing organic matter, temperatures above 2°C, no frosts and preferably cheap labour for the intensive husbandry required: taking cuttings, fertilization, hoeing for weed control and manual harvesting. About 35 000 plants per hectare with spacing of plants between 80 × 30 and 100 × 60 cm^2, depending on the soil and climate, should ensure profitability. Growth of the cuttings in the first six months is slow, thus encouraging weed proliferation: intercropping is therefore advisable with legumes or maize (Narayana *et al.*, 1986). Plants can grow in various different locations, such as Andra Pradesh (S. India), Karnatka State in Bangalore, Pulney Hills, the Nilgiri Hills and Uttar Pradesh (N. India) (Rajeswara Rao, 2002). Altitudes varied from 120 to 2400 m and climatic conditions from semi-arid, subtropical to cool, with actual temperatures from 5 to 40°C.

10.3.3 Fertilization, watering and weeding

A high-yield crop can produce 7 tonnes of dry matter/ha/year; this is about 18–20% of the total biomass and requires fertilization with 100 kg N, 32 kg P_2O_5, 165 kg K_2O, 250 kg CaO, 28 kg MgO, 15 kg Na and 10 kg S. Yields on Reunion Island were confirmed in India (Prakasa Rao *et al.*, 1986, 1988). Geranium oil crops respond linearly to nitrogen and phosphate application, when in a balanced fertiliser. Addition of trace elements (B, Cu, Zn, Mo) gives better oil production (Prakasa Rao *et al.*, 1984). Lime and organic manure are traditionally used where there is no intercrop. Rose geranium is fairly drought resistant and dislikes an excess of water, so irrigation is usually left to nature. When there is a risk of waterlogging, the cuttings are planted on ridges to ensure good drainage, as in the Nile valley in Egypt. Smothering by fast-growing weeds severely damages the crop, giving poor yields. Manual weeding and hoeing are practised where labour is cheap, even though herbicides are used (Demarne, 2002).

10.3.4 Harvesting

Geranium oil is contained in glandular trichomes (Demarne and Van der Walt, 1989) located on both surfaces of the young leaves, on the young stems, on the buds and on different parts of the inflorescences. Oil is thus obtained from the top young parts of the herb (Demarne, 2002) and the crop is best harvested by hand, six to eight months after planting. Subsequent harvests are made at intervals of three to five months, depending on plant development, weather conditions, crop management and labour availability. Harvesting requires clear, sunny days; the plant material is usually left for the day on the inter-row to wilt and is then transported to the distillery. Heaping up or chopping the plant material is inadvisable to avoid fermentation and distillation must be done quickly. In practice, there is only a market for water-distilled essential oil (Denny, 2002). Only Egypt produces and markets small quantities of geranium concrete and/or absolute (Demarne, 2002).

Details of the cultivation, varieties and sales of *Pelargonium* plants grown for ornamental use in the UK and worldwide are given by James (2002) and Lis-Balchin (2002b).

10.3.5 Organic geranium oil
Organic geranium oils are produced in various parts of the world, including South Africa and Egypt. A comparative study of the essential oil quantity and quality of organic versus normal geranium oil has not been made, but there is a small study on lavender and lavandin (Charles *et al.*, 2002), which shows no great difference in the essential oil composition; however, the absence of pesticides would be welcomed. The price charged for organic essential oils is often treble that of normal produce and is inexcusable. The essential oil industry for the food and cosmetics industry are therefore not very interested in organic produce, so the market is very small and reserved for aromatherapists.

10.3.6 Pests
Little information is available on pests in Egypt and China, but in Reunion Island geranium is attacked by at least 14 different species belonging to the Hemiptera (six species), Coleoptera (three species) and Lepidoptera (five species) (Quilici *et al.*, 1992). Among the most important pests are white grubs of *Hoplochelus marginalis* Fairmaire (Coleoptera, Fam. Scarabaeidae), cockchafers *Cratopus humeralis* Boh. and *C. angustatus* Boh. (Coleoptera, Fam. Curculionidae), the whitefly *Trialeurodes vaporariorum* Westwood (Hemiptera, Fam. Aleyrodidae), scale insect *Pseudaulacaspis pentagona* Targioni-Tozzetti (Hemiptera, Fam. Diaspididae) and the defoliating caterpillar of *Lobesia vanillana* (Lepidoptera, Fam. Tortricidae). An efficient integrated control of these insects includes a combination of light trapping, agronomic controls (minimum tillage or cover-crop), chemical controls (insecticide spraying or chemical trapping) and biological controls (Quilici *et al.*, 1992). Nematodes have also been reported as important pests, especially in India (Rajeswara Rao, 2002) and damage the crop, inflicting yield losses of up to 75.8%. Wilt, dieback, leaf blight, leaf spot, root and stem rot and anthracnose are common (*Colletotrichum, Botrytis, Septoria, Cercospora, Armillaria, Rosellinia, Phomopsis, Pythium, Fusarium* and *Pseudomonas solanacaerum*). Certain of those pathogens can lead to total destruction of the crop and the impossibility of growing geranium again on the same plot.

10.4 Main uses in food processing and perfumery
Virtually all essential oils produced are used in the food and perfumery/cosmetics industries.

10.4.1 Present uses in food
Reported uses (Fenaroli, 1998) are, in ppm: baked goods, 13.0; frozen diary, 7.11; soft candy, 11.39; gelatin and pudding, 5.41; non-alcoholic beverages, 3.47; alcoholic beverages, 2.08; hard candy, 295.2; and chewing gum, 308.4. The leaves of many of the scented

plants are used in domestic baking and the mint-flavoured *P. tomentosum* is used to make tea.

10.4.2 Novel uses of geranium oil and extracts in food processing and their possible uses as food preservatives

Pelargonium, essential oils (EOs), obtained from different species, with a wide spectrum of chemical compositions, have shown considerable potential as antimicrobial agents (Lis-Balchin *et al.*, 1995). Studies have been of 18 different *Pelargonium* petroleum spirit extracts (Lis-Balchin *et al.*, 1998), as well as the more hydrophylic extracts in methanol, against four bacteria: *Staphylococcus aureus*, *S. epidermidis*, *Proteus vulgaris* and *Bacillus cereus*. They showed that 'Attar of Roses', similar to commercial geranium oil with the main components citronellol and geraniol, was a very potent antibactericide, as was 'Lemon Fancy', owing to its high neral and geranial content. The petroleum spirit extracts resembled the activity of steam-distilled samples. Hydrophilic extracts proved to have more potent antibacterial activity, suggesting that flavonoids, tannins and other phenolics in the herb are the effective antimicrobial agents (Lis-Balchin and Deans, 1996; Lis-Balchin *et al.*, 1996c).

Using a quiche filling as a model food system, the antimicrobial activity of different scented *Pelargonium* EOs was investigated against *Salmonella enteriditis*, *Listeria innocua*, *Saccharomyces ludwigii* and *Zygosaccharomyces bailii*. The EOs, in concentrations ranging from 250 to 500 ppm (Lis-Balchin *et al.*, 2000; Lis-Balchin, 2002e) showed similar inhibition to that of thyme oil, a strong antimicrobial agent.

Activity against *Staphylococcus aureus* in a porridge system indicated that *Pelargonium* oil (at 1000 ppm) was effective against both *S. aureus* and *E. coli*, but the hydrosols were ineffective, similarly to clove and cinnamon. The complex interaction of the essential oils and extracts with different model food systems is discussed by Lis-Balchin (2002e). The results, however, suggest that the *Pelargonium* essential oils, including commercial geranium oil, could be used not just as a flavouring, but also as a novel food antimicrobial agent.

10.4.3 Perfumery usage

Geranium oil and concoctions using geranium oil components have long been used in making artificial rose oil or 'rose extenders'. *Rhodinol* ex *Geranium* is used with hydroxycitronellol, linalool, geraniol, dimethyl benzyl carbide, cinnamic alcohol, phenyl ethyl alcohol, geranyl and linalyl esters in modern perfumery and cosmetic products. Geranium oil is frequently used in masculine fragrances often in conjunction with lavender in, for example, Moustache (Rochas), also classical *fougère* blends. Geranium also appears in women's fragrances, such as Ivoire and Balmain, as well as featuring in classical *chypres* such as Cabochard, Gres, and the original *chypre*, Coty. Giorgio, Armani, is a combination of mandarin and geranium (Wells and Lis-Balchin, 2002).

10.5 Functional properties

Functional properties include: antimicrobial, insecticidal, pharmacological, physiological and miscellaneous.

10.5.1 Pharmacological effects

Many *Pelargonium* species have been used in the past as traditional medicines in Southern

Africa with mainly anti-dysenteric properties (Watt and Breyer-Brandwijk, 1962), e.g. root of *P. transvaalense* and *P. triste* and the leaves *of P. bowkeri* and *P. sidaefolium*. Some *Pelargonium* species were also used to treat specific maladies, e.g. *P. cucullatum* for nephritis; *P. tragacanthoides* for neuralgia, *P. luridum* and *P. transvaalense* root for fever; *P. minimum, P. reniform* and *P. grossularioides* for menstrual flow (Pappe 1868; Watt and Breyer-Brandwijk, 1962). The latter was also used as an emmenagogue and abortifacient by both Zulus and Boers and has recently been studied further (Lis-Balchin and Hart, 1994) and shown to have spasmogenic properties on the uterus and smooth muscle preparations *in vitro*. *Pelargonium reniforme* and *P. sidoides* extracts are currently used in the herbal remedy Umckaloabo® (produced in Germany) for respiratory ailments, owing to its strong antimicrobial properties; it also has immunomodulatory properties, leishmanicidal activity and interferon-like properties (Kolodziej, 2002).

Imaseki and Kitabatake (1962) found an antispasmodic action of citronellol, geraniol and linalool on mouse small intestine, but *Pelargonium* EOs were not studied until recently. Results from experiments on isolated guinea pig ileum demonstrate that the majority of *Pelargonium* oils, and their components, produce a relaxation of smooth muscle through a mechanism involving adenylate cyclase and a rise in the concentration of the second messenger, cAMP (Lis-Balchin and Hart, 1997, 1998; Hart and Lis-Balchin, 2002); there is some evidence of calcium channel blockade, but only at concentrations higher than those required to produce a significant spasmolytic effect, in contrast to other essential oils (Hills and Aaronson, 1991). Preliminary results using more hydrophilic (methanolic) extracts of *Pelargonium* species and cultivars, and their teas, indicate that most have a contractile effect initially, which is followed by a relaxation (Hart and Lis-Balchin, 2002). There is also some evidence that a few methanolic extracts use calcium channels at normal concentrations.

The essential oils of *P. grossularioides*, as well as its water-soluble and methanolic extracts, were all spasmogenic on guinea pig ileum and on the rat uterus, in contrast to all other geranium oils and their components, such as geraniol and linalool and all other commercial oils studied, which had a spasmolytic action on the uterus. Action on skeletal muscle (chick biventer and rat phrenic nerve diaphragm) showed an increase in tone and reduction of contraction. Alkaloid extracts obtained from the zonals (Lis-Balchin, 1996,1997; Lis-Balchin *et al.*, 1996b) were also all spasmolytic on guinea pig ileum; as were methanolic, water-soluble extracts (teas), and alkaloid fractions of *P. luridum* (root) and the leaves of *P. inquinans* and *Pelargonium* cultivars.

10.5.2 Antimicrobial action

The antimicrobial action of geranium oil was reviewed recently by Deans (2002). Deans and Ritchie (1987) studied 50 commercial volatile oils at four concentrations against a range of 25 bacterial genera: 'geranium oil' was most effective against the dairy products organism *Brevibacterium linens* and the toxin-producing *Yersinia enterocolitica* but, in contrast with *Klebsiella pneumoniae* and *Escherichia coli*, its presence resulted in enhancement of growth. Pattnaik *et al.* (1995) tested geranium oil for antibacterial activity against 22 bacteria (Gram-positive cocci and rods, Gram-negative rods) and 12 fungi (3 yeast-like, 9 filamentous) by disc diffusion. Only 12 bacterial strains were inhibited by the geranium oil, but all the fungi were inhibited. Lis-Balchin *et al.* (1996c) found that antibacterial activity against 25 different bacteria varied among samples of commercial oil, ranging from 8 to 19 inhibited, which could not be correlated with the chemical composition of the samples. The action of the geranium oils against 20 strains of *Listeria monocytogenes* was again very variable, the number of strains affected ranging from 3 to 16 out of 20 (Lis-Balchin and

Deans, 1997a); in antifungal studies, the geranium samples showed from 0 to 94% inhibition of *Aspergillus niger*, 12 to 95% against *A. ochraceus* and 40 to 86% against *Fusarium culmorum*, in agreement with the results of an earlier study (Lis-Balchin *et al.*, 1995) wherein 24 cultivars were tested for antimicrobial activity against 25 test bacteria and *A. niger*.

In a study into the potential usage of mixtures of plant volatile oils as synergistic antibacterial agents in foods, Lis-Balchin and Deans (1997b) included 'geranium oil' in a mixture with nutmeg and bergamot oils, but no synergistic effect was found (similarly with other combinations). The antimicrobial activity of *Pelargonium* oil components (Dorman and Deans, 2000) showed the following ranking order of activity: linalool > geranyl acetate > nerol > geraniol > menthone > β-pinene > limonene > α-pinene. Compared with more phenolic compounds, these activities are relatively modest. The bacteria showing the greatest level of inhibition were *Clostridium sporogenes* > *Lactobacillus plantarum* > *Citrobacter freundii* > *Escherichia coli* > *Flavobacterium suaveolens*.

Pelargonium × *hortorum* leaves (unscented) were reported as being most active against *Candida albicans*, *Trichophyton rubrum* and *Streptococcus mutans*, organisms causing common dermal, mucosal or oral infections in humans (Heisey and Gorham, 1992). *Pelargonium* species, including the commercial 'Geranium oil' have been shown to have antioxidative properties (Dorman *et al.*, 2000; Fukaya *et al.*, 1988; Youdim *et al.*, 1999), though these properties had very variable activities in different commercial samples of 'Geranium oil' (Lis-Balchin *et al.*, 1996a).

10.5.3 Physiological action

There is little direct scientific evidence for the physiological effectiveness of geranium oil apart from the pharmacological studies and studies in the brain (Torii *et al.*, 1988; Manley, 1993). There are miscellaneous physiological reactions attributed to a geranium component, linalool: a hypoglycaemic effect in normal and streptozotocin-diabetic rats (Afifi *et al.*, 1998); a hepatic peroxysomal and microsomal enzyme induction in rats (Roffey *et al.*, 1990), as does geraniol (Chadba and Madyastha, 1984) and choleretic and cholagogic activity of a mixture of linalool and α-terpinol (Peana *et al.*, 1994; Gruncharov, 1973). Linalool's dose-dependent, sedative effect on the central nervous system of rats could be caused by its inhibitory activity on glutamate binding in the cortex (Elisabetsky *et al.*, 1995a,b).

10.5.4 Paramedical usage

Geranium oil is commonly used in aromatherapy, owing to the misinterpretation of aromatherapists of old English herbals (Culpeper 1653), which referred to the real *Geranium* genus (e.g. *G. robertianum*) and not *Pelargonium*. The actual usages of the *Geranium* extracts mentioned in old herbals are mainly (antidiarrhoeal) associated with their tannin content and other water-soluble chemicals, e.g. flavonoids, in the leaves. Essential oils, on the other hand, are steam-distilled volatiles and do not contain these components. Valnet (1982) gave geranium oil's major attributes as 'its vulnerary powers and its power to mend fractures and eliminate cancers' taken straight out of the old herbals; his directions for oral use are given as for 'Herb Robert', a real *Geranium* (Lis-Balchin, 2002d)! The mistake was then perpetuated. No evidence has been provided by clinical studies, e.g. in childbirth (Burns and Blaney, 1994), for the current usage of geranium oil, while there is pharmacological evidence for the decrease and even cessation of uterine contractions in animal experiments, which could prove harmful (Hart and Lis-Balchin, 2002). Other clinical studies (using mainly lavender oil) have not shown any extra benefit of using essential oils with massage, as massage in itself provides a beneficial effect, e.g. Dunn *et al.* (1995). Essential oil inclusion may have a detrimental effect due to sensitization (Schaller and Korting, 1995; Anderson *et al.*, 2000). Furthermore, although 'geranium' oils are very active on many different animal tissues *in vitro* (Lis-Balchin *et al.*, 1997), there is no proof as yet whether minute amounts (as used in aromatherapy massage) can have direct action on target organs or tissues rather than through the odour pathway (Vickers, 1996), despite some evidence that certain essential oil components can be absorbed either through the skin or lungs (Jager *et al.*, 1992; Buchbauer *et al.*, 1993).

10.5.5 Psychological and physiological effects of geranium oil

The main action of essential oils is probably on the primitive, unconscious, limbic system of the brain, which is not under the control of the cerebrum or higher centres (Kirk-Smith, 2002). Many fragrances have been shown to have an effect on mood and in general, pleasant odours generate happy memories, more positive feelings and a general sense of well-being (Warren and Warrenburg, 1993). Some essential oils have also been used in hospitals and hospices to create a more happy and positive atmosphere and also in offices and factories to enhance productivity. Many essential oil vapours have been shown to depress contingent negative variation (CNV) brain waves in human volunteers (i.e. sedative); others increase CNV (i.e. stimulant): these parameters were often in agreement with the effect on mouse motility and the direct effect of the essential oil on smooth muscle *in vitro*. However, geranium oil has both a sedative and stimulant effect on the CNV (Lis-Balchin, 2002d).

10.5.6 Toxicology of the essential oil of geranium

Geranium oil Bourbon, Algerian, Moroccan were granted GRAS (generally recognized as safe) status by FEMA (1965) and approved by the US Food and Drug Administration (FDA) for food use. The Council of Europe included geranium oil in the list of spices, seasonings, etc. deemed admissible for use with a possible limitation of the active principle in the final product.

10.5.7 Biological toxicity studies

Acute toxicity: oral LD50 in rats > 5 g/kg); dermal in rabbits, 2.5 g/kg (Moreno, 1973).

Irritation: applied undiluted to abraded or intact rabbit skin for 24 h under occlusion was found to be moderately irritant (Moreno, 1973), but applied to backs of hair-less mice, it was not irritating (Urbach and Forbes, 1972). Human patch test (closed) to 10% geranium oil in petrolatum produced no irritation after 48 h (RIFM, 1974).

Sensitization: a maximization test on 25 volunteers, using 10% in petrolatum produced no sensitization (RIFM, 1974).

Phototoxicity has not been found for geranium oil.

10.5.8 Toxicity of *Pelargonium* species

There are very few, scattered, references to any toxicity, and all references are to contact dermatitis and sensitization. Most of the references are to the geranium oil and the main components geraniol (Lovell, 1993). *Pelargonium* plants themselves have caused hand dermatitis (Anderson, 1923) and sensitization (Rook, 1961; Hjorth, 1969).

10.5.9 Toxicity of components

Patch tests to geraniol proved negative but dermatitis to perfumes containing geranium oil has been shown in a few cases (Klarmann, 1958). Ointments containing geraniol, e.g. 'Blastoestimulina', were reported to cause sensitization when used in the treatment of chronic leg ulcers (Romaguera *et al.*, 1986; Guerra *et al.*, 1987), although the patients were also sensitive to other ointments that contained no essential oils.

Sensitization to geraniol using a maximization test proved negative (Opdyke1975), but the allergen may be geraniol as cross-reactions often occurred with citronella (Keil, 1947); however, the main sensitizer in citronella is citronellal, with citronallol less reactive; geraniol was even weaker, as was citral. In two cases, strong reactions were obtained with 1% solutions of citronellal and weaker ones with citronellol, geraniol, geranyl acetate. In 23 out of 23 cases no response was found using lemon oil, suggesting specificity of the response. In a lemon oil sensitization case, α-pinene gave a greater response than β-pinene: this is because of the close similarity between limonene and α-pinene (due to an exposed methylene radical).

Recent Japanese studies, on patients with ordinary cosmetic dermatitis and pigmented cosmetic dermatitis, who showed a positive allergic responses to a wide range of fragrances (Nakayama, 1998), gave rise to a list of common cosmetic sensitizers and primary sensitizers, which included geranium oil, geraniol, sandalwood oil, artificial sandalwood, musk ambrette, jasmine absolute, hydroxycitronellal, ylang ylang oil, cinnamic alcohol, cinnamaldehyde, eugenol, balsam of Peru and lavender oil. Geraniol was found to give a positive patch test in over 1.2% cases when used at 1% in white petrolatum with 5% sorbitan sesquioleate (Frosch, 1998). D-Limonene, although present in small quantities in geranium oil has shown many sensitization reactions.

The European Council and the European Commission have now issued the 7th Amendment to their Cosmetic Directive, 2002, and have included geraniol, limonene and citronellol in its list of sensitizers (see Table 10.1).

10.6 Quality issues and adulteration

10.6.1 Quality specification of the essential oil (CAS: 8000-46-2)

The International Organization for Standardization or ISO, defines geranium oil as 'the oil obtained by steam distillation of the fresh or slightly withered herbaceous parts of *Pelargonium graveolens* L'Heritier ex Aiton, *Pelargonium roseum* Willdenow and other undefined hybrids which have given rise to differing ecotypes in the various geographical areas' (International Standard 4731: 1972). The colour is various shades of amber-yellow to greenish-yellow. The odour is given as characteristic of the origin, rose-like with a varying minty note. The specification does not include the Bulgarian geranium oil distilled from *Geranium macrorrhizum*, known as Zdravetz oil, containing mainly sesquiterpenes of which half is apparently germacrone (Ognyanov, 1985). ISO 4731 has set the concentration for citronellol content at a minimum 42%/maximum 55% for Bourbon geranium oil; 35/58 for Moroccan; 40/58 for Egyptian and 40/58 per cent for Chinese oils. Other physicochemical values are given in Table 10.1, but these may now be academic as the greatest production is from China (Quinhua, 1993), the oil resembling Bourbon, and considerable variation is found in the chemical composition (Lawrence, 1976–1978; 1979–1980; 1981–1987; 1988–1991; 1994–1995) with notable incidence of apparent adulteration (Lis-Balchin, 2002c).

10.6.2 Adulteration

Geranium oil contains mainly citronellol and geraniol and their esters, and therefore can be easily concocted from cheaper essential oils and adjusted to the recommended ISO standards. The antimicrobial activity of such essential oils is much greater than that of some authentic oils but has a similar pharmacological effect on smooth muscle (spasmolytic) and the actual odour can be even more appreciated by perfumers than the real essential oil (Lis-Balchin, 2002c). The essential oil composition of this geranium oil differs completely from that of a true *Geranium robertianum* oil (Pedro *et al*., 1992) or that of *G. maccrorhizum* (Ognyanov, 1985). The most expensive geranium oil was always Bourbon (Guenther, 1950), and it increased in tonnage as well as value over some years, surprisingly, on the small volcanic island of Reunion; this was partly due to the increase in geranium oil production in China, which being very similar to that of Bourbon would often get accepted as such (Verlet, 1992). Recent geranium oil production in China is restricted to the region of Binchuan, 450 km from Kunming (Cu, 1996) and there are two harvests, a summer one, which yields an oil that is relatively similar to Bourbon, and the winter harvest, which gives a low-grade oil with only 4% geraniol, compared with the summer 7% and the Bourbon with 14%. The citronellol content is, however, much greater than that of Bourbon geranium and is virtually doubled. The characteristic sesquiterpene is guaia-6,9-diene as in Bourbon oil.

Adulteration of geranium oil is perhaps encouraged by the ISO requirements themselves and the comparatively low price of synthetics compared with the low yield of geranium oil (less than 0.3%). Adulteration, as with all essential oils, occurs to a considerable extent, with diluents such as propylene glycol, triacetin, triethyl citrate or benzyl alcohol, ethyl alcohol and, in the case of aromatherapy oils, with fixed oils such as almond oil. Adulteration also includes giving the wrong source on the labelling, e.g. Bourbon, if it came from another country or was synthetic, or even when a geranium oil leaflet (Body Shop) stated that it originated from *Geranium maculatum* (which is not only the wrong species but has no odour).

10.6.3 Detection of adulteration

Carrier or fixed oil or solvents can be detected by simple gas chromatography. Chiral columns must be employed for other adulterations (when fractions of other oils or synthetic components are used), as ordinary gas chromatography, with or without mass spectrometry (MS) and other identification facilities, such as infra-red (IR), are not sophisticated enough. Detection of such adulteration was perfected by the use of special enantiomeric or chiral columns, mainly composed of an α-cyclodextrin phase (Ravid *et al*., 1992; Lis-Balchin *et al.*, 1999). One of the major components, citronellol occurs in the (–)-form in geranium and rose oils and has a finer rose odour than the (+) enantiomer, and a sweet, peach-like flavour. The (+)-citronellol enantiomer has been found in citronella oils from Ceylon and Java, *Cymbopogon winterianus, Boronia citridora, Eucalyptus citriodora*, Spanish verbena and other essential oils. These two enantiomers are starting materials for numerous chiral pheromones and flavours (Ravid *et al*., 1992). Analyses of, for example, commercial Egyptian geranium oil yielded almost a racemic mixture of citronellol enantiomers, while a true Bourbon oil gave a highly concentrated S-(–)-citronellol.

Recent studies on Australian geranium oils grown from specific *Pelargonium* clones showed that by using ten key chiral components, and calculating a so-called 'chiral excess', it was possible to distinguish geographically different and seasonally different essential oils as well as adulteration (Doimo *et al*., 1999). Chiral columns can, however, be used by synthetic chemists and by those involved in adulterating essential oils, as the same type of column can be used to separate out the enantiomers, which could then be added in the correct proportion for a given essential oil, making detection impossible.

A comparison of chemical composition with bioactivity also yielded a useful indication of adulteration. The apparent geographical source had, on the whole, no correlation with the chemical composition of commercial geranium oils (Lis-Balchin et al., 1996a) except for the presence or absence of the relevant sesquiterpene, i.e. 10-epi-χ-eudesmol in Egyptian oils (3–7%) and guaia-6,9-diene (1–7%) in the Bourbon and China oils; a Moroccan oil contained both these sesquiterpenes. The proportion of the main components, i.e. citronellol, geraniol, linalool, iso-menthone, citronellyl formate and geranyl formate, was not consistent for any geographical source. The bioactivity, as determined by the action of the oils against 25 different bacterial species, 20 different *Listeria monocytogenes* cultivars, 3 different fungi and also their anti-oxidant action, was not correlated with the geographical source of the geranium oil specimens or their chemical composition. The increased activity of the synthetic components was compared to that of the pure geranium oil, suggesting possible adulteration of many commercial oils with synthetic components (Lis-Balchin et al., 1996a).

10.7 References

AFIFI, E.U., SAKET, M. and JAGHABIR, M. (1998), Hypoglycaemic effect of linalool in normal and streptozotocin diabetetic rats. *Acta Tecnol Legisa Medicam* **9**: 101–6.

ALBERS, F., VAN DER WALT, J.J.A., GIBBY, M., MARSCHEWSKI, D.E., PRICE, R.A. and DU PREEZ, G. (1995), A biosystematic study of *Pelargonium* section *Chorisma*. *S. Afr. J. Bot.* **61**: 339–46.

ANDERSON, C., LIS-BALCHIN, M. and KIRK-SMITH, M. (2000), Evaluation of essential oil therapy on Childhood atopic eczema. *Phytother. Res.* **14**: 452–6.

ANDERSON, J.W. (1923), Geranium dermatitis. *Arch. Dermatol. Syphilology* **7**: 510–1.

ASEN, S. and GRIESBACH, R. (1983), High pressure liquid chromatographic analysis of flavonols in *Geranium* florets as an adjunct for cultivar identification. *J. Am Soc. Hortic. Sci* **108**: 845–50.

BATE-SMITH, E.C. (1973), Chemotaxonomy of *Geranium*. *Bot. J. Linn. Soc.* **67**: 347–59.

BUCHBAUER, G., JAGER, W., JIROVETZ, L., ILMBERGER, J. and DIETRICH, H. (1993), Therapeutic properties of essential oils and fragrances. In R. Teramishu, R.G. Buttery and H. Sugisawa (eds), *Bioactive Volatile compounds from Plants*, ACS Symposium Series 525. American Chemical Society, Washington DC, pp. 159–65.

BURNS, E. and BLANEY, C. (1994), Using aromatherapy in childbirth. *Nurs. Times* **90**: 54–8.

CHADBA, A. and MADYASTHA, K.M. (1984), Metabolism of geraniol and linalool in the rat and effects on liver and lung microsomal enzymes. *Xenobiotica* **14**: 365–74.

CHARLES, D.J., RENAUD, E.N.C. and SIMON, J.E. (2002), Comparative study of the essential oil quantity and quality of ten cultivars of organically grown lavender and lavandin. In M. Lis-Balchin (ed.) *Lavender; The Genus Lavandula: Medicinal and Aromatic Plants – Industrial Profiles*. Taylor and Francis, London, pp. 232–42.

CHARLWOOD, B. and LIS-BALCHIN, M. (2002), Micropropagation and biotechnological approaches to tissue culture of *Pelargonium* species and essential oils of scenteds. In: M. Lis-Balchin (ed.) *Geranium and Pelargonium; The Genera Geranium and Pelargonium: Medicinal and Aromatic Plants – Industrial Profiles*. Taylor and Francis, London, pp. 218–233.

CHARLWOOD, B.V. and CHARLWOOD, K.A. (1991), *Pelargonium* spp. (*Geranium*): *In vitro* culture and the production of aromatic compounds. In Y.P.S. Bajaj (ed.) *Biotechnology in Agriculture and Forestry, Vol. 15, Medicinal and Aromatic Plants III*. Springer-Verlag, Berlin and Heidelberg, pp. 339–52.

CHARLWOOD, B.V., and MOUSTOU, C. (1988), Essential oil accumulation in shoot-proliferation cultures of *Pelargonium* species: In R.J. Robins and M.J.C. Rhodes (eds) *Manipulating Secondary Metabolism in Culture*. Cambridge University Press, Cambridge, pp. 187–94.

CRAIG, R., MUMMA, R.O., GERHOLD, D.L., WINNER, B.L. and SNETSINGER, R. (1986), Genetic control of a biochemical mechanism for mite resistance in geraniums. In M.B. Green, and P.A. Hedin (eds) *Natural Resistance of Plants to Pests*. American Chemical Society, Washington, DC, pp. 168–76.

CU, J-Q. (1996), Geranium oil from Yunnan, China. *Perf. Flav.* **21**: 23–4.
CULPEPER, N. (1653), *The English Physitian Enlarged*. George Sawbridge, London.
DEANS, S.G. (2002), Antimicrobial properties of *Pelargonium* extracts contrasted with that of *Geranium*. In M. Lis-Balchin (ed.) *Geranium and Pelargonium The Genera* Geranium *and* Pelargonium*: Medicinal and Aromatic Plants – Industrial Profiles*. Taylor and Francis, London, pp. 132–46.
DEANS, S.G. and RITCHIE, G. (1987), The antibacterial properties of plant essential oils. *Int. J. Food Microbiol.* **5**: 165–80.
DEMARNE, F.E. (2002), 'Rose-scented geranium' a *Pelargonium* grown for the perfume industry. In M. Lis-Balchin (ed.) *Geranium and Pelargonium; The genera* Geranium *and* Pelargonium*: Medicinal and Aromatic Plants – Industrial Profiles*. Taylor and Francis, London, pp. 193–211.
DEMARNE, F.E. and VAN DER WALT, J.J.A. (1989), Origin of rose-scented pelargonium cultivar grown on Reunion Island. *S. Afr. J. Bot.* **55**: 184–91.
DENNY, E.F.K. (2002), Distillation of the lavender type oils: theory and practice. In M. Lis-Balchin (ed.) *Lavender; The Genus* Lavandula*: Medicinal and Aromatic Plants – Industrial Profiles*. Taylor and Francis, London, pp. 100–16.
DOIMO, L., FLETCHER, R.J. and D'ARCY, B.R. (1999), Chiral excess: measuring the chirality of geographically and seasonally different geranium oils. *J. Essent. Oil Res.* **11**: 291–9.
DORMAN, H.J.D. and DEANS, S.G. (2000), Antimicrobial agents from plants: antibacterial activity of plant volatile oils. *J. of Appl. Microbiol.* **88**: 308–16.
DORMAN, H.J.D., SURAI, P. and DEANS, S.G. (2000), *In vitro* antioxidant activity of a number of plant essential oils and phytoconstituents. *J. Essential Oil Res.* **12**: 241–8.
DREYER, L.L., ALBERS, F., VAN DER WALT, J.J.A. and MARSCHEWSKI, D.E. (1992), Subdivision of *Pelargonium* sect. *Cortusina* (Geraniaceae). *Pl. Syst. Evol.* **181**: 83–97.
DUNN, C., SLEEP, J. and COLLETT, D. (1995), Sensing an improvement: an experimental study to evaluate the use of aromatherapy, massage and periods of rest in an intensive care unit. *J. Adv. Nursing* **21**: 34–40.
ELISABETSKY, E., COELHO DE SOUZA, G.P., DOS SANTOS, M.A.C., SIQUIEIRA, I.R., AMADOR, T.A. and NUNES, D.S. (1995a), Sedative properties of linalool. *Fitoterapia* **66**: 407–14.
ELISABETSKY, E., MARSCHNER, J. and SOUZA, D.O. (1995b), Effects of linalool on glutamatergic system in the rat cerebral cortex. *Neurochem. Res.* **20**: 461–5.
ELISABETSKY, E, BRUM, L.F.S. and SOUZA, D.O. (1999), Anticonvulsant properties of linalool in glutamate-related seizure models. *Phytomedicine* **6**: 107–13.
FEMA (FLAVORING EXTRACT MANUFACTURER'S ASSOCIATION) (1965), Survey of flavoring ingredient usage levels. No. 2508. *Food Technol.* **19**: part 2, 155.
FENAROLI, G. (1998), *Fenaroli's Handbook of Flavor Ingredients,* vol. 1, 3rd ed. CRC Press, Boca Raton.
FROSCH, P.J. (1998), Are major components of fragrances a problem? In P.J. Frosch, J.D. Johansen and I.R. White (eds) *Fragrances: Beneficial and Adverse Effects*. Springer Verlag, Berlin, pp. 92–9.
FUKAYA, Y., NAKAZAWA, K., OKUDA, T. and IWATA, S. (1988), Effect of tannin on oxidative damage of ocular lens. *Jap. J. Ophthalmol.* **32**: 166–75.
GRUNCHAROV, V. (1973), Clinico-experimental study on the choleretic and cholagogic action of Bulgarian lavender oil. *Vutr. Boles.* **12**: 90–6.
GUENTHER, E. (1950), *The Essential Oils*, Vol. 4. van Nostrand Co., New York.
GUERRA, P., AGUILAR, A., URBINA, F., CRISTOBAL, M.C. and GARCIA-PEREZ, A. (1987), Contact dermatitis to geraniol in a leg ulcer. *Contact Dermat.* **16**: 298–9.
HARBORNE, J.B. (1961), The anthocyanins of roses. Occurrence of peonin. *Experientia* **17**: 72–73.
HARBORNE, J.B. and WILLIAMS, C. (2002), Phytochemistry of the genus *Geranium*. In M. Lis-Balchin (ed.) *Geranium and Pelargonium; The Genera* Geranium *and* Pelargonium*: Medicinal and Aromatic Plants – Industrial Profiles*. Taylor and Francis, London, pp. 20–9.
HART, S and LIS-BALCHIN, M. (2002), Pharmacology of *Pelargonium* essential oils and extracts *in vitro* and *in vivo*. In M. Lis-Balchin (ed.) *Geranium and Pelargonium; The Genera* Geranium *and* Pelargonium*: Medicinal and Aromatic Plants – Industrial Profiles* Taylor and Francis, London, pp. 116–31.
HEISEY, R.M. and GORHAM, B.K. (1992), Antimicrobial effects of plant extracts on *Streptococcus mutans, Candida albicans, Trichophyton rubrum* and other microorganisms. *Lett. Appl. Microbiol.* **14**: 136–9.
HILLS, J.M. and AARONSON, P.I. (1991), The mechanism of action of peppermint oil on gastrointestinal smooth muscle. An analysis using patch clamp electrophysiology and isolated tissue pharmacology in rabbit and guinea-pig. *Gastroenterology* **101**: 55–65.

HJORTH, N. (1969), Plant dermatitis. *Contact Dermat., Newsletter* **6**: 126.
HOUGHTON, P. and LIS-BALCHIN, M. (2002), Chemotaxonomy of *Pelargonium* based on alkaloids and essential oils. In M. Lis-Balchin (ed.) *Geranium and Pelargonium; The Genera* Geranium *and* Pelargonium: *Medicinal and Aromatic Plants – Industrial Profiles*. Taylor and Francis, London, pp.166–73.
IMASEKI, I. and KITABATAKE, Y. (1962), Studies on effect of essential oils and their components on the isolated intestines of mice. *J. Pharmaceut. Soc. Japan.* **82**:1326–8.
JAGER, W., BUCHBAUER, G., JIROVETZ, L. and FRITZER, M. (1992), Percutaneous absorption of lavender oil from a massage oil. *J. Soc. Cosmet. Chem.* **43**: 49–54.
JAMES, J. (2002), Cultivation and sales of *Pelargonium* plants for ornamental use in the UK and worldwide. In M. Lis-Balchin (ed.) *Geranium and Pelargonium; The Genera* Geranium *and* Pelargonium: *Medicinal and Aromatic Plants – Industrial Profiles*. Taylor and Francis, London, pp. 80–91.
KEIL, H. (1947), Contact dermatitis due to oil of citronella. *J. Investig. Dermatol.* **8**: 327–34.
KIRK-SMITH, M. (2002), The psychological effects of lavender. In M. Lis-Balchin (ed.) *Lavender; The Genus* Lavandula: *Medicinal and Aromatic Plants – Industrial Profiles*. Taylor and Francis, London, pp. 155–70.
KLARMANN, E.G. (1958), Perfume dermatitis. *Ann. Allergy* **16**: 425–34.
KLIGMAN, A.M. (1966), Report to RIFM, 31 October.
KOLODZIEJ, H. (2002), *Pelargonium reniforme* and *Pelargonium sidoides*: their botany, chemistry and medicinal use. In M. Lis-Balchin (ed.) *Geranium and Pelargonium; The Genera* Geranium *and* Pelargonium: *Medicinal and Aromatic Plants – Industrial Profiles*. Taylor and Francis, London, pp. 262–90.
LAWRENCE, B.M. (Lawrence 1976–1978; 1979–1980; 1981–1987; 1988–1991; 1992–1994), *Essential Oils*, Allured Pub. Corp., IL.
LIS-BALCHIN, M. (1996), A chemotaxonomic reappraisal of the Section Ciconium *Pelargonium* (Geraniaceae). *S. Afr. J. Bot.* **62**: 277–9.
LIS-BALCHIN, M. (1997), A chemotaxonomic study of the *Pelargonium* (Geraniaceae) species and their modern cultivars. *J. Hort. Sci.* **72**: 791–5.
LIS-BALCHIN, M. (2002a), History of nomenclature, usage and cultivation of *Geranium* and *Pelargonium* species. In M. Lis-Balchin (ed.) *Geranium and Pelargonium; The Genera* Geranium *and* Pelargonium: *Medicinal and Aromatic Plants – Industrial Profiles*. Taylor and Francis, London, pp. 5–10.
LIS-BALCHIN, M. (2002b), Growing pelargoniums in the garden, conservatory and for shows. In M. Lis-Balchin (ed.) *Geranium and Pelargonium; The Genera* Geranium *and* Pelargonium: *Medicinal and Aromatic Plants – Industrial Profiles*. Taylor and Francis, London, pp. 92–8.
LIS-BALCHIN, M. (2002c), Geranium essential oil: standardisation, ISO, adulteration and its detection using GC, enantiomeric columns and bioactivity. In M. Lis-Balchin (ed.) *Geranium and Pelargonium; The Genera* Geranium *and* Pelargonium: *Medicinal and Aromatic Plants – Industrial Profiles*. Taylor and Francis, London, pp. 184–92.
LIS-BALCHIN, M. (2002d), Geranium oil and its use in aromatherapy. In M. Lis-Balchin (ed.) *Geranium and Pelargonium; The Genera* Geranium *and* Pelargonium: *Medicinal and Aromatic Plants – Industrial Profiles*. Taylor and Francis, London, pp. 234–46.
LIS-BALCHIN, M. (2002e), New research: 1. Possible uses of various *Pelargonium* leaf oils and extracts as food preservatives. In M. Lis-Balchin (ed.) *Geranium and Pelargonium; The Genera* Geranium *and* Pelargonium: *Medicinal and Aromatic Plants – Industrial Profiles*. Taylor and Francis, London, pp. 251–61.
LIS-BALCHIN, M. and DEANS, S.G. (1996), Antimicrobial effects of hydrophylic extracts of *Pelargonium* species (Geraniaceae). *Lett. Appl. Microbiol.* **23**: 205–7.
LIS-BALCHIN, M. and DEANS, S.G. (1997a), Bioactivity of selected plant essential oils against *Listeria monocytogenes. J. Appl. Microbiol.* **82**: 759–62.
LIS-BALCHIN, M. and DEANS, S.G. (1997b), Studies on the potential usage of mixtures of plant essential oils as synergistic antibacterial agents in foods. *Phytother. Res.* **12**: 1–4.
LIS-BALCHIN, M. and GUITTONNEAU, G.-G. (1995), Preliminary investigations on the presence of alkaloids in the genus *Erodium* L'Herit. (Geraniaceae). *Acta Bot. Gallica* **141**: 31–5.
LIS-BALCHIN, M. and HART, S. (1994), A pharmacological appraisal of the folk medicinal usage of *Pelar-gonium grossularioides* and *Erodium cicutarium* (Geraniaceae). *Herbs, Spices Med. Plants* **2**: 41–8.
LIS-BALCHIN, M. and HART, S. (1997), Correlation of the chemical profiles of essential oil mixes with

their relaxant or stimulant properties in man and smooth muscle preparations *in vitro*. Proc. 27th Int. Symp. Ess. Oils, Vienna, Austria, 8–11 Sept. 1996. Ch. Franz, A. Mathé and G. Buchbauer (eds). Allured Pub. Corp., Carol Stream, IL., pp. 24–8.

LIS-BALCHIN, M. and HART, S. (1998), Studies on the mode of action of scented-leaf Pelargonium (Geraniaceae). *Phytother. Res.* **12**: 215–7.

LIS-BALCHIN, M., HART, S., DEANS, S.G. and EAGLESHAM, E. (1995), Potential agrochemical and medicinal usage of essential oils of *Pelargonium* species. *Herbs, Spices Med. Plants* **3**: 11–22.

LIS-BALCHIN, M., DEANS, S.G. and HART, S. (1996a), Bioactivity of commercial geranium oil from different sources. *J. Essent. Oil Res.* **8**: 281–90.

LIS-BALCHIN, M., HOUGHTON, P. and WOLDERMARIAM, T. (1996b), Elaeocarpidine alkaloids from *Pelargonium* (Geraniaceae). *Nat. Products Letters* **8**: 105–12.

LIS-BALCHIN, M., HART, S., DEANS, S.G. and EAGLESHAM, E. (1996c), Comparison of the pharmacological and antimicrobial action of commercial plant essential oils. *J. Herbs, Spices, Med. Plants* **4**: 69–86.

LIS-BALCHIN, M., HART, S., DEANS, S.D. and EAGLESHAM, E. (1997), Comparison of the pharmacological and antimicrobial action of commercial plant essential oils. *J. Herbs, Spices, Med. Plants*, **4**, 69–86.

LIS-BALCHIN, M., DEANS, S.G. and EAGLESHAM, E. (1998), Relationship between the bioactivity and chemical composition of commercial plant essential oils. *Flav. Fragr. J.* **13**: 98–104.

LIS-BALCHIN, M., OCHOCKA, R. J., DEANS, S. G. ASZTEMBORSKA, M. and HART, S. (1999), Differences in bioactivity between the enantiomers of α-pinene. *J. Essent. Oil Res.* **11**: 393–7.

LIS-BALCHIN, M., BUCHBAUER, G., ASTRID BRANDSTETTER, A. and ANDREA BAUER, A. (2000), Comparative antimicrobial activity of *Pelargonium* and other selected plant essential oils and their respective hydrosols in a model food system. Paper presented at: *Int. Ess. Oil. Symp.*, Hamburg, 10–13 Sept.

LOVELL, C.R. (1993), *Plants and the Skin*. Blackwell Scientific Publ., Oxford.

MANLEY, C.H. (1993), Psychophysiological effect of odor. *Crit. Rev. Food Sci. Nutr.* **33**: 57–62.

MITCHELL, K.A., MARKHAM, K.R. and BOASE, M.R. (1998), Pigment chemistry and colour of *Pelargonium* flowers. *Phytochemistry* **47**: 355–61.

MORENO, O.M. (1973), Report to RIFM, 25 July.

NAKAYAMA, H. (1998), Fragrance hypersensitivity and its control. In P.J. Frosch, J.D. Johansen and I.R. White (eds) *Fragrances: Beneficial and Adverse Effects*. Springer Verlag, Berlin, pp. 83–91.

NAKAYAMA, H., HARADA, R. and TODA, M. (1976), Pigmented cosmetic dermatitis. *Int. J. Derm.* **15**: 673–5.

NARAYANA, M.R., PRAKASA RAO, E.V.S., RAJESWARA RAO, B.R. and SASTRY, K.P. (1986), Geranium cultivation in India: potentials and prospects. *Pafai J.* **8**: 25–9.

NAVES, J.-R., LAMPARSKY, D., and OCHSNER, P. (1961), Études sur les matières végétales volatiles CLXXIV (1). Présence de tétrahydropyrannes dans l'huile essentielle de géranium. *Bull. Soc. Chim. France*, 645–647.

NAVES J.-R., OCHSER, P., THOMAS, A.F. and LAMPARSKY, D. (1963), Études sur les matières végétales volatiles CLXXXVI (1). Présence d'acétonyl-2-méthyl-4-tétrahydropyranne dans l'huile essentielle de géranium. *Bull. Soc. Chim. France*, 1608–11.

OGNYANOV, I. (1985), Bulgarian Zdravetz oil. *Perf. Flav.* **10**: 38–44.

OPDYKE, D.L.T. (1975), Monographs on fragrance raw materials. *Food Cosmet. Toxicol.* **13**: 451.

PAPPE, L. (1868), *Florae Capensis Medicae, Prodromus;*, 3rd ed., Cape Town.

PATTNAIK, S., SUBRAMANYAM, V.R., KOLE, C.R. and SAHOO, S. (1995), Antibacterial activity of essential oils from *Cymbopogon*: Inter- and intra-specific differences. *Microbios* **84**: 239–45.

PEANA, A., SATTA, M., MORETTI, M.D.L. and ORECCHIONI, M. (1994), A study on choleretic activity of *Salvia desoleana* essential oil. *Planta. Med.* **60**: 478–9.

PEDRO, L. G., PAIS, M.S.S. and SCHEFFER, J.J.C. (1992), Composition of the essential oil of *Geranium robertianum* L. *Flav. Fragr. J.* **7**: 223–6.

PRAKASA RAO, E.V.S., SINGH, M, NARAYANA, M.R. and CHANDRASEKHARA, G.C. (1984), Micronutrient studies in geranium (*Pelargonium graveolens* l'Hérit.) and davana (*Artemisia pallens* Wall.). Indian Perfume **28**: 88–90.

PRAKASA RAO, E.V.S., SINGH, M. and GANESH RAO, R.S. (1986), Effect of nitrogen fertilizer on geranium (*Pelargonium graveolens* L'Hérit ex. Ait.), cowpea and blackgram grown in sole cropping and intercropping systems. *Intern. J. Trop. Agric.* **4**: 341–345.

PRAKASA RAO, E.V S., SINGH, M. and GANESH RAO, R.S. (1988), Effect of plant spacings and nitrogen levels on herb and essential oil yields and nutrient uptake in geranium (*Pelargonium graveolens* L'Hérit. ex Ait.). *Intern. J. Trop. Agric.* **6**: 95–101.

QUILICI, S., VERCAMBRE, B. and BONNEMORR, C. (1992), Les insects ravageurs in le géranium rosat à la Réunion. C. A. Hauts. Saint-Denis (Reunion Island), *Graphica*, 79–90.
QUINHUA, Z. (1993), China's Perfumery Industry Picks Up. *Perf. Flav.* **18**: 47–8.
RAJESWARA RAO, B.R. (2002), Cultivation and distillation of geranium oil from *Pelargonium* species in India. In M. Lis-Balchin (ed.) *Geranium and Pelargonium; The Genera* Geranium *and* Pelargonium: *Medicinal and Aromatic Plants – Industrial Profiles*. Taylor and Francis, London, pp. 212–17.
RAVID, U., PUTIEVSKY, E., KATZIR, I., IKAN, R. and WEINSTEIN, V. (1992), Determination of the enantiomeric composition of citronellol in essential oils by chiral GC analysis on a modified χ-cyclodextrin phase. *Flav. Fragr. J.* **7**: 235–8.
RIFM (1974), Monograph on geranium oil (Bourbon), *Food Cosmet. Toxicol.* **12**: Suppl., Special issue 1, 883–4.
ROBINSON, M. and ROBINSON, R. (1932), A survey of anthocyanins. Part II. *Biochem. J.* **26**: 1647–64.
ROFFEY, S.J., WALKER, R. and GIBSON, G.G. (1990), Hepatic peroxisomal and microsomal enzyme induction by citral and linalool in rats. *Food Chem. Toxicol.* **28**: 403–8.
ROMAGUERA, C., GRIMALT, F. and VILAPLANA, J. (1986), Geraniol dermatitis. *Contact Dermat.* **14**: 185–6.
ROOK, A. (1961), Plant dermatitis – botanical aspects. *Trans. St. John's Dermatol. Soc.* **46**: 41–7.
SCHALLER, M. and KORTING, H.C. (1995), Allergic airborne contact dermatitis from essential oils used in aromatherapy. *Clin. Exp. Dermatol.* **20**: 143–5.
SIMMONDS, M.S.J. (2002), Interactions between arthropod pests and pelargoniums. In M. Lis-Balchin (ed.) *Geranium and Pelargonium; The Genera* Geranium *and* Pelargonium: *Medicinal and Aromatic Plants – Industrial Profiles*. Taylor and Francis, London, pp. 291–8.
STAFFORD, H.A. (1961), Distribution of tartaric acid in the Geraniaceae. *Am. J. Bot.* **48**: 699–701.
TORII, S., FUKUDA, H., KANEMOTO, H., MIYANCHIO, R., HAMAUZU, Y. and KAWASAKI, M. (1988), Contingent negative variation and the psychological effects of odor. In S. Toller and G.H. Dodds (eds) *Perfumery: The Psychology and Biology of Fragrance*. Chapman & Hall, New York.
URBACH, F. and FORBES, P.D. (1972), Report to RIFM, 22 September.
VALNET, J. (1982) *The Practice of Aromatherapy*. C.W. Daniels Co. Ltd., Saffron Walden.
VERLET, N. (1992), Geranium Bourbon: quel avenir? *Parf Cosmet. Aromes* **108**: 49–51.
VICKERS, A. (1996), *Massage and Aromatherapy. A Guide for Health Professionals*. Chapman & Hall, London.
WAGNER, H. and BLADT, S. (1975) Coumarine aus südafrikanischen *Pelargonium*-arten. *Phytochemistry* **14**: 2061–4.
VAN DER WALT, J.J.A. and DEMARNE, F.E. (1988), *Pelargonium graveolens* and P. *radens*: a comparison of their morphology and essential oils. *S. Afr. J. Bot.* **54**: 617–22.
VAN DER WALT, J.J.A., ALBERS, F. and GIBBY, M. (1990), Delimitation of *Pelargonium* sect. *Glaucophyllum* (Geraniaceae). *Pl. Syst. Evol.* **171**: 15–26.
VAN DER WALT, J.J.A., ALBERS, F., GIBBY, M., MARSCHEWSKI, D.E., HELLBRUGGE, D., PRICE, R.A. and DER MERWE, A.M. (1997), A biosystematic study of *Pelargonium* section *Ligularia*: 3. Reappraisal of section *Jenkinsoniaa*. *S. Afr. J. Bot.* **63**: 4–21.
WALTERS, D.S., MINARD, R., CRAIG, R. and MUMMA, R.O. (1988), Geranium defensive agents. III. Structural determination and biosynthetic considerations of anacardic acids of geranium. *J. Chem. Ecol.* **14**: 743–51.
WARREN, C. and WARRENBURG, S. (1993), Mood benefits of Fragrance. *Perf. Flavorist* **18**: 9–16.
WATT, J.M. and BREYER-BRANDWIJK, A. (1962), *The Medicinal Plants of Southern Africa*. Livingstone Ltd, Edinburgh.
WELLS, R. and LIS-BALCHIN, M. (2002), Perfumery and cosmetic products utilizing geranium oil. In M. Lis-Balchin (ed.) *Geranium and Pelargonium; The Genera* Geranium *and* Pelargonium: *Medicinal and Aromatic Plants – Industrial Profiles*. Taylor and Francis, London, pp. 247–50.
WILLIAMS, C. and HARBORNE, J.B. (2002), Phytochemistry of the genus *Pelargonium*. In M. Lis-Balchin (ed.) *Geranium and Pelargonium; The Genera* Geranium *and* Pelargonium: *Medicinal and Aromatic Plants – Industrial Profiles*. Taylor and Francis, London, pp. 99–115.
WILLIAMS, C.A., HARBORNE, J.B., NEWMAN, M., GREENHAM, J. and EAGLES, J. (1997), Chrysin and other leaf exudate flavonoids in the genus *Pelargonium*. *Phytochemistry* **46**: 1349–53.
WILLIAMS, C.A., NEWMAN, M. and GIBBY, M. (2000), The application of leaf phenolic evidence for systematic studies within the genus *Pelargonium* (Geraniaceae). *Biochem. Syst. Ecol.* **28**: 119–32.
WOLDEMARIAN, T.Z., HOUGHTON, P.J., LIS-BALCHIN, M. and SIMMONDS, S.J. (1997), Whitefly toxins. *Pharmaceut. J.* **259**: 481.

YOUDIM, K.A., DORMAN, H.J.D. and DEANS, S.G. (1999), The antioxidant effectiveness of thyme oil, α-tocopherol and ascorbyl palmitate on evening primrose oil oxidation. *J. Essential Oil Res.* **11**: 643–8.

11

Lavender

M. T. Lis-Balchin, South Bank University, UK

11.1 Introduction

A total of 32 species of *Lavandula* have been described in the literature, plus a number of infraspecific taxa and hybrids (Upson, 2002). They are distributed from the Canary Islands, Madeira, Mediterranean Basin, North Africa, South West Asia, Arabian Peninsula, and tropical NE Africa and India. Chaytor (1937) had classified the genus into five sections: all the common commercial plants belong to two main sections: *Stoechas (Lavandula stoechas, L. dentata, L. viridis* and *L. pedunculata)* and *Spica (L. officinalis,* syn. *L. angustifolia, L. latifolia* and *L. lanata);* most are probably hybrids between *L. angustifolia* and spike, *L. latifolia;* there is confusion with the naming of lavenders round the world, owing to differences in their appearance under different climatic and/or husbandry conditions (Lis-Balchin, 2002a).

11.2 Chemical composition

11.2.1 Phytochemistry of the genus *Lavandula*
This genus *Lavandula* is relatively rich in phenolic constituents, with 19 flavones and 8 anthocyanins (Harbourne and Williams, 2002). Characteristic of the family are various glycosides of hypolaetin and scutellarein. Triterpenoids include ursolic acid (Le Men and Pourrat, 1953). Leaf flavonoids are mainly flavone glycosides and their individual distribution among the taxa shows some taxonomic significance.

In a survey of anthocyanin pigments in flowers, Saito and Harborne (1992) showed that *L. dentata* and *L. stoechas* were characterized by eight floral pigments, e.g. delphinidin and malvidin (purplish). Two hydroxycinnamic acid esters, rosmarinic acid and chlorogenic acid, are regularly present in the leaves. Coumarins and 7-methoxycoumarin (herniarin) have been detected in the volatile oil fractions.

11.2.2 Chemistry of the essential oils of different lavenders
The commercial hybrids, lavandins, have variable concentrations of 1,8-cineole and camphor,

absent from *Lavandula angustifolia* P. Miller, which provide the harsher notes. The 'rhodinol content', consisting of citronellol, geraniol, nerol, neryl acetate and geranyl acetate, which amounts to a very small percentage of the total composition, gives a sweet, rose-like odour to the lavandin oils, with small differences between the cultivars (Lis-Balchin, 2002b). The chemical composition of *L.* 'Grosso' varies with the method of extraction: steam-distilled and CO_2-extracted samples showed differences in linalool and linalyl acetate compared to an absolute.

Lavandula latifolia Medicus, the spike oil of commerce, with a high yield, 0.8–1.2% has variable compositions (Lawrence, 1976–1978; 1979–1980; 1981–1987; 1988–1991; 1994–1995). More than 300 components have been identified and the main ones are linalool (19–48%), 1,8-cineole (21–42%) and camphor (5–17%).

Lavandula angustifolia of commerce (Naef and Morris, 1992), whose main components are linalool (25–38%) and linalyl acetate (25–45%) shows some considerable differences between the subspecies *L. angustifolia* ssp. *pyrenaica* (DC), growing wild in NE Spain (Garcia-Vallejo *et al.*, 1989), whose three main components were: linalool (20–66%), borneol (6–32%) and camphor (2–14%), making it unacceptable as normal lavender oil.

Lavandula lanata Boisse is morphologically similar to *L. latifolia* but has a very high concentration of camphor (43–59%) and variable amounts of lavandulol (3–27%); *L. dentata* L. grows wild along the Mediterranean coast of Spain (Garcia-Vallejo *et al.*, 1989) and has two chemotypes: 1,8-cineole/β-pinene and β-pinene/α-pinene; *L. multifida* has carvacrol and β-bisabolene. *Lavandula stoechas* L. ssp. *pedunculata* (Miller) Samp. ex Roziera (*L. pedunculata* Cavanilles) and ssp. *sampaioana* (*L. stoechas* L. ssp. *sampaioana* Roziera) had two chemotypes: camphor/fenchone and β-pinene/camphor/fenchone; *L. stoechas* L. ssp. *stoechas* has camphor and fenchone (with 1,8-cineole). Four wild populations of *Lavandula stoechas* L. ssp. *stoechas* in Crete had different percentages of α-pinene, 1,8-cineole, fenchone, camphor and myrtenyl acetate (Skoula *et al.*, 1996). *Lavandula lusieri* (Rozeira) Rivas-Martinez (*L. stoechas* ssp. *luisieri* (Rozeira) Rozeira) has two chemotypes with an unidentified ester as their main component; *L. viridis* has a high concentration of 1,8-cineole, camphor and α-pinene (Garcia-Vallejo *et al.*, 1989). *Lavandula pinnata* L. il. var. *pinnata* grown on Madeira (Figuereido *et al.*, 1995) has a high percentage of monoterpenes (37–80%) and a relatively small proportion of sesquiterpenes (13–22%). A further comparison of the composition of essential oils from leaves of different species is given by Wiesenfeld (1999) and Lis-Balchin (2002g).

11.3 Production

11.3.1 Lavender grown for oil production

Lavandula angustifolia is mainly propagated by seed, sown in spring or autumn, depending on the severity of the winters in the region (Weiss, 1997). Sowing can be directly into fields but more often is in nursery beds, where the plants remain for about a year. Clonal plants are made via cuttings. Healthy mother plants are cut down near ground level and the branches can be stored for months before preparing the cuttings of 10–15 cm with one or two branchlets. These are also planted in a nursery, usually in the spring, for a year. Green cuttings can be used but these require tender care, growth hormones and misting. The plants are planted out in rows 1.5 m apart with 0.4–0.4 m between rows; giving 10 000 plants per ha for *L. angustifolia* and about half for the hybrids (Weiss, 1997). Husbandry has now improved the lavender crops (Lis-Balchin, 2002c) and include fertilizers, often as ash

(Chaisse and Blanc, 1990). The soil is loosened superficially two or three times a year to remove weeds, or else weedkillers are used.

There are many lavender pests and diseases and this reduces a possible 15–20 year life span to 3 years. Root rot due to *Armillaria mellea* is a very serious fungal disease; *Thomasiniana lavandulae* (Diptera) is the most serious insect as its larvae feed under the bark, causing damage to the tops of branches. Other diseases are due to the fungus *Rosellinia necatrix*, the Homoptera *Hyalesthes obsoletus, Cechenotettix martini, Eucarazza elegans*; Coleoptera include *Arima marginata, Chrysolina americana* and *Meligethes subfumatus*; Lepidoptera include *Sophronia humerella; Argyrotaenia pulchellana, Pyterophorus spicidactyla* and many others (Chaisse and Blanc, 1990).

Harvesting was done by hand, especially in the mountains, using a sickle, but mechanical harvesters are now fully developed, cutting 7500 kg per day compared with hand harvesters cutting 500 kg. The yield of lavender oil is 40 kg/ha and lavandin is up to 120 kg. Spike lavender yields 50 kg/ha. The harvested lavender is left in the fields for a few days then steam distilled (Denny, 2002) or extracted with CO_2 or other solvent.

11.3.2 Production of lavender oils

The recent primary sources of lavender oil include: France, Bulgaria, China and Spain. Most of the lavender plants were originally grown and distilled in the higher areas of Mediterranean France (600–1500 m). Exact figures for the production of the oil are difficult to obtain owing to the immense amount of adulteration, mixing, cutting and addition of synthetics or simply synthetic lavender oil itself. In 1984, world production of lavender oil was 200 tonnes; Bulgaria produced 100–129 tonnes; France 55, USSR 35, Australia 5. Recently US imports varied from 303 to 555 tonnes per year, peaking in 2000. US exports varied from 52 tonnes from 1996 to 2001, peaking at 121 in 2000. More than 30 different types of lavender oils and blends are traded on world markets, but there are only a few that are sold in bulk, mainly *L. angustifolia* oil and a few lavandins.

11.3.3 Organic lavender oil

Organic essential oils, especially lavender oil, are produced in various parts of the world, including the UK, Australia and the USA; a comparative study of the essential oil quantity and quality of ten cultivars of organically grown lavender and lavandin is provided by Charles *et al.* (2002). There does not seem to be any great difference in the essential oil composition of organic compared with conventionally grown lavender, except for some percentages of enantiomers; however, the absence of pesticides would be welcomed. Farmers in the UK must comply with European Council Regulation (EEC) No. 2092/91, enforced 22 July 1991, regarding organic production and the rules governing the processing and sale of organic products. Land must be put into conversion prior to full-scale organic production and then applications must be made for status with the Soil Association, which inspects the sites. This takes around three years, and this, together with a lower yield, due to loss by natural predation, increases the cost. The premium charged, however, is often treble that of normal produce and reflects the gross over-commercialization of the produce. Organic essential oils have not been widely accepted by the main dealers for the food and cosmetics industry and the market is small, reserved mainly for aromatherapists. In France the certification is ECO-Cert, Qualité France, SOCOTEC, brought in recently to control the expanding organic market.

11.3.4 Lavenders grown for gardens, pot-pourri and drying

There are hundreds of different lavenders grown for the garden, perfumes, aromatherapy, pot-pourri and for drying on the stalk, etc. Different types require different conditions, especially regarding the temperature they are grown at throughout the year (Charlesworth, 2002). Lavenders can be grown in Australia, Europe and the USA, but require sunshine and dryness for maximum growth and perennial habit.

Very hardy lavenders are traditional lavenders: 'true lavender' (*L. angustifolia*) and lavandin (*L. × intermedia*). *Lavandula angustifolia* is the most popular species grown in England for oil extraction, yielding high-quality oils used for perfumes, aromatherapy, pot-pourri and drying on the stalk. Cultivars include: 'Ashdown Forest', 'Compacta', 'Folgate', 'Loddon Blue', 'Munstead', 'Nana Alba', 'Royal Purple'.

Lavendula × intermedia (Lavandin) is a sterile hybrid of *L. angustifolia* and *L. latifolia* (spike Lavender). Its camphoraceous oils are used in soaps, cosmetics and detergents and also for drying off the stalk and for pot-pourri, e.g. 'Vera', 'Grappenhall', 'Grosso', 'Hidcote Giant', 'Old English'. They can withstand −10°C.

Frost-hardy lavenders all have 'ears' on top, which are sterile bracts (coma). They will survive to −5°C and often lower. Most have a camphoraceous foliage, but no appreciable scent to the flowers. They include some of the subspecies of *L. stoechas* and the species *L. viridis* and their hybrids, e.g. *L. stoechas* subsp. *pedunculata* (Spanish lavender), also known as 'Papillon'; *L. stoechas* subsp. *stoechas* (French lavender); 'Kew Red', 'Fathead', 'Helmsdale', 'Marshwood'. Half-hardy lavenders will thrive above 0°C and include: *L. dentata* (fringed lavender) and *L. lanata* (woolly lavender), and one hybrid *L. lanata* with *L. dentata*, 'Goodwin Creek Gray'.

Tender lavenders need to be brought in before the first frosts and kept warm at around 5°C. All have spiralling triple flower spikes in a trident formation but no scent, e.g. *L. buchii* varietas *buchii* , *L. × christiana* (a sterile hybrid of *L. canariensis* and *L. pinnata*), *L. minutolii* and *L. pinnata*.

11.4 Uses in food processing, perfumery and paramedical spheres

11.4.1 Natural food flavours

Lavandin oil, lavender oil, spike lavender oil and lavender absolute and even concrete are used as natural food flavours. Reported uses in the food industry (Fenaroli, 1998) include: baked goods, frozen dairy, soft candy, gelatin, pudding, non-alcoholic and alcoholic beverages from 4 to 44 ppm. Lavenders are also included in tissanes or teas; booklets of Norfolk Lavender, UK, suggest many recipes for cooking with lavender at home, e.g. herring or trout stuffed with lavender sprigs, and Vickers (1991) offers further recipes for cooking and garnishing foods and use as crystallized flowers.

11.4.2 Perfumery and cosmetic uses

Lavender and lavandin oils are used in colognes, lavender-waters, *fougères, chypres, abres*, floral and non-floral perfumes. They blend well with bergamot and other citrus oils, clove, patchouli, rosemary, etc. (Wells and Lis-Balchin, 2002). Lavandin is used in cheaper products, such as soaps.

11.4.3 Paramedical uses

Lavender drops were used for fainting, and red lavender (lavender mixed with rosemary and cinnamon bark, nutmeg and sandalwood and macerated in spirit of wine for several days) was used for indigestion (Grieve, 1937). The *British Pharmacopoeia* (BPC) officially recognized red lavender 200 years ago. In the 18th century it was known as palsy drops and red hartshorn. BPC products included: Compound Lavender Tincture BPC 1949 (dose: 2–4 ml) and Lavender Spirit BPC 1934 (dose: 0.3–1.2 ml).

Paramedical uses appear in many modern books, e.g. Potter (1988), where *Lavandula angustifolia* is stated to be a carminative, spasmolytic, tonic and antidepressant. Bertram (1995) suggests numerous uses for *L. angustifolia*, which are identical to those suggested both by Culpeper (1653) and Gerard (1633), both of whom were referring to a different species! These include: nervous headache, neuralgia, rheumatism, depression, insomnia, windy colic, fainting, toothache, sprains, sinusitis, stress and migraine. The use of *Lavandula latifolia*, with its high camphoric content was recently suggested as an expectorant by Charron (1997). Aromatherapy should be defined as 'treatment with odours' (Buchbauer, 1992) but different definitions abound. Many of the attributes of lavender oil were mistakenly taken from herbals, e.g. Culpeper (1653), who used alcoholic extracts or teas, not distilled essential oils; there was also an interest in astrology, hence every plant had an assigned planet: *Lavandula angustifolia* has Mercury and now also a 'yang' quality (Tisserand, 1985). The species referred to was also misinterpreted (see above). René-Maurice Gattefossé (1937), the so-called pioneer of modern aromatherapy, actually used perfumes or at most deterpenated essential oils and not pure natural plant essential oils. Aromatherapy involves massage using a very diluted essential oil or mixture of essential oils (1–2%) in a carrier oil such as almond oil or addition of essential oil to the bath or a basin of hot water, or using burners (Lis-Balchin, 2002d).

11.5 Functional properties and toxicity

Lavender has antimicrobial, pharmacological, physiological and miscellaneous functions.

11.5.1 Pharmacological effects

Plant (1920) applied 'waters of lavender' to the intestine of dogs *in vivo* and reported increased activity, which was sometimes followed by relaxation and decreased peristaltic activity. Linalool was reported to relax the small intestine of the mouse (Imaseki and Kitabatake, 1962) while Shipochliev (1968) observed a spasmolytic action on rabbit and guinea pig gut by the essential oil of lavender (*L. spica* L.). Reiter and Brandt (1985) report that linalool relaxes the longitudinal muscle of guinea pig ileum. A spasmolytic activity of *L. dentata* L. oil and its components 1,8-cineole and α- and β-pinene, has been observed on rat duodenum. Izzo *et al.* (1996) showed that the essential oil of *L. angustifolia* Mill. relaxed both longitudinal and circular muscle of the guinea pig ileum. There appears therefore to be good agreement that the oils of lavender are spasmolytic on intestinal muscle but Lis-Balchin *et al.* (1996a, 1996b) and Lis-Balchin and Hart (1999) reported that, with some commercial samples, the spasmolytic action is preceded by a contraction on guinea pig ileum.

Recent experiments using three different extracts of several *Lavandula* species, including a cold methanolic extract, a tea (made with boiling water) and a hydrosol (the water remaining after steam/water distillation) showed that the methanolic extracts of *L. angustifolia* dried flowers, *L. angustifolia* fresh flowers and fresh leaves, assessed separately, *L. stoechas* leaves and *L. viridis* leaves have a spasmolytic action on the guinea pig ileum. All the teas and hydrosols, except for *L. angustifolia* dried flowers and *L. angustifolia* fresh leaves, were also spasmolytic, while the water-soluble tea extract of *L. angustifolia* dried flowers and the leaves of *L. angustifolia* showed an initial spasmogenic action (Hart and Lis-Balchin, 2002). Brandt (1988) reported the spasmolytic actions of linalool on tracheal muscle.

Action on skeletal muscle of the essential oil of *L. angustifolia* Miller and also linalool and linalyl acetate produced a reduction in the size of the contraction in response to stimulation of the phrenic nerve and also when the muscle was stimulated directly (Lis-Balchin and Hart, 1997a). Thus the action would appear to be myogenic; however, Ghelardini *et al.* (1999) interpret their similar results as showing a local anaesthetic action; similarly, Re *et al.* (2000) conclude from experiments on mouse neuromuscular junction that linalool has a local anaesthetic action. Linalyl acetate also caused an increase in baseline or resting tone (Lis-Balchin and Hart, 1997a), while limonene caused a rise in tone, with a decrease in the size of the contractions.

Lavender oil, linalool, linalyl acetate, α and β-pinene and 1,8-cineole reduce uterine activity at concentrations that are spasmolytic on intestinal muscle (Lis-Balchin and Hart, 1997b).

Mode of action
All essential oils of different lavenders showed a post-synaptic effect on the guinea pig ileum and none possesses atropine-like activity (Lis-Balchin and Hart, 1999) or appears to stimulate adrenoceptors. Lavender oil and linalool, appear to mediate a spasmolytic action on intestinal smooth muscle via a rise in cAMP (Lis-Balchin and Hart, 1999). There is no evidence of the use of calcium channels except at very high concentrations. This is in contrast to other essential oils (Vuorela *et al.*, 1997). There is no evidence for potassium channel opening. The essential oil from *L. dentata* L., and its component 1,8-cineole, has been shown to inhibit calcium-induced contraction of rat duodenum. There is recent evidence to show that the methanolic extracts of *L. angustifolia* (dry flowers, fresh flowers and fresh leaves) are calcium channel blockers, as are the leaves of *L. viridis* and *L. stoechas* (Hart and Lis-Balchin, 2002).

The fact that some extracts of *L. angustifolia* have a strong spasmogenic action (dried flowers and fresh leaves) is somewhat disturbing as so many modern herbal and aromatherapy books state that the teas are sedative and are often prescribed for upset stomachs. The results support the findings (Castle and Lis-Balchin, 2002; Lis-Balchin, 2002a,d) that the information on lavender has been mistakenly transcribed from early herbals, such as those of Culpeper (1653), where *L. spica*, a more camphoric lavender, was used medicinally and not the very floral *L. angustifolia*. The spasmolytic results shown for the water-soluble extracts of the more camphoraceous *L. stoechas* again supports the well-quoted action of the camphoraceous spike lavender over the centuries and emphasizes the confusion.

11.5.2 Physiological effect
Evidence for the sedative properties of the EO of lavender after inhalation in animals is provided by Buchbauer *et al.* (1991, 1993) as it significantly decreased the motility of 'normal' test mice as well as that of animals rendered hyperactive or 'stressed' by an intraperitoneal caffeine. The main constituents of this oil, linalool and linalyl acetate, elicited a similar effect, which was dose related. The absorption of linalool from percutaneous application of lavender oil (Jager *et al.*, 1992) provided some evidence for the aromatherapeutical use of lavender. Stress and travel sickness of pigs was reduced by lavender straw, measured by concentrations of cortisol in the pigs' saliva (Bradshaw *et al.*,

1998). Linalool, which has a dose-dependent, sedative effect on the central nervous system of rats (Elisabetsky *et al.*, 1995a), may be caused by its inhibitory activity on glutamate binding in the cortex (Elisabetsky *et al.*, 1995b). Potentiation of GABAA receptors expressed in *Xenopus* oocytes by perfumes and phytocides, including lavender oils and lavender perfumes, (shown by benzodiazepine, barbiturates, steroids and anaesthetics, which induce an anxiolytic, anticonvulsant and sedative effect) was investigated by Aoshima and Hamamoto (1999).

Swiss mice showed sedation after lavender oil (1/60 in olive oil) was orally administered (Guillemain *et al.*, 1989). Lavender inhalation showed a similar effect (Komori *et al.*, 1997). The positive effects of lavender oil as treatment for insomnia was indicated in a limited study of four elderly people (Hardy *et al.*, 1995). A Japanese patent application for the usage of several monoterpenes (which can be incorporated into food such as chewing gums) as brain stimulants and/or enhancers of brain activity was filed by Nakamatsu (1995). Certain central neurotropic effects of lavender essential oil were shown by Atanassova-Shopova and Roussinov (1970). A more detailed account of physiological and other effects is given by Buchbauer (2002).

11.5.3 Psychological effects

Scientific research into the psychological (often referred to as psychopharmacological) effects of lavender is limited; however, there is a long history of it being regarded, and used, as a sedative or calming agent (Kirk-Smith, 2002). The effects on cells and brain tissues also suggest both reduction in electrical activity and an anti-convulsant effect. Both laboratory and clinically based studies reveal that responses to lavender may be determined not only by these pharmacological sedative effects, but by individual, situational and expectational factors independent of the lavender odour itself.

Many fragrances have been shown to have an effect on mood and, in general, pleasant odours generate happy memories, more positive feelings and a general sense of well-being (Warren and Warrenburg, 1993). Inhalation of lavender was found to have a sedative effect on people (judging by CNV studies) (Kubota *et al.*, 1992; Torii *et al.*, 1988; Manley, 1993). This was in agreement with the reduced motility in mice (Buchbauer, 1992; Jager *et al.*, 1992; Kovar *et al.*, 1987; Ammon, 1989).

Inhalation studies in people, of rosemary oil versus lavender oil using EEG and simple maths computations, showed that lavender increased α-power, suggesting drowsiness, while rosemary instigated decreased frontal alpha and beta power, suggesting increased alertness with faster and more accurate results in the maths (Diego *et al.*, 1998). These results seem to show that odour has an effect on performance *per se*, but Knasko *et al.* (1990), who lied to their subjects that odour would be given, also showed an improvement in carrying out tasks, i.e. mind over matter! Karamat *et al.* (1992), however, found that lavender had a stimulant effect on decision times in human experiments. Subjects in a group given an ambient odour of dimethyl sulphide were less happy than those in the lavender group on both odour and non-odour days (Knasko, 1992). Ambient odours of lavender and cloves given to 72 volunteers (Ludvigson and Rottman, 1989) showed that lavender adversely influenced arithmetic reasoning. Lavender (at imperceptible levels) reduced the number of errors made in the arithmetical and concentration tasks compared to jasmine (Degel and Koster, 1999) and reduced stress in flight controllers (Leshchinskaia *et al.*, 1983).

Most clinical studies initiated by aromatherapists used lavender oil, and showed little, if any, benefit (Vickers, 1996; Cooke and Ernst, 2000; Lis-Balchin, 2002d). There was no significant difference shown between the use of aromatherapy (with lavender), massage and periods of rest in an intensive care unit (Dunn *et al.*, 1995). Aromatherapy massage on four patients with severe dementia and disturbed behaviour proved detrimental for most (Brooker *et al.*, 1997).

The main action of essential oils is probably on the primitive, unconscious, limbic system of the brain (Lis-Balchin, 1997), which is not under the control of the cerebrum or higher centres and has a great subconscious effect on the person. Mood and behaviour could be influenced by odours, and memories of past odour associations could also be dominant, an area that needs to be fully explored before aromatherapy is used by psychologically unqualified persons in the treatment of Alzheimer's or other ageing diseases. Aromatherapy can, however, be effective in reducing stress and improving moods of terminally ill patients, but only in association with touch and the time to listen to the patient, as aromatherapy, like other alternative medicines, has a placebo effect owing to the greater time spent by the therapist with the patient, the belief imparted by the therapist and the willingness of the patient to believe in the therapy (Benson and Stark, 1996).

11.5.4 Antimicrobial effects

The antimicrobial activity of lavender oil against different bacterial species of lavender is moderate, in contrast to the considerable antimicrobial status awarded to lavender by aromatherapists (Deans, 2002). Lavender was found to be most effective against *Enterococcus faecalis* out of 25 bacteria, but *Klebsiella pneumoniae* enhanced growth! The genus *Bacillus* has been shown to be susceptible to lavender volatile oil by Jeanfils *et al.* (1991) and Lis-Balchin *et al.* (1998), the latter also showing differences in activity of different lavenders against 25 bacteria. Similarly, using 20 strains of *Listeria monocytogenes*, Lis-Balchin and Deans (1997) showed a wide variation in activity of different commercial lavenders. Vokou *et al.* (1993) suppressed potato sprout growth using crude herb material. Lavender also possesses antifungal properties, e.g. against *Aspergillus niger*, *A. ochraceus* and *Fusarium culmorum,* which all reacted differently to the oils (Lis-Balchin *et al.*, 1998).

11.5.5 Other properties of lavender oil or its components

A study on mast cell-mediated immediate-type allergic reactions induced by an irritant in test animals showed a dose-dependent beneficial effect of lavender oil administered either topically or intradermally (Kim and Cho, 1999). Lavender flowers had a protective effect against enzyme-dependent lipid peroxidation (Hohmann *et al.*, 1999). Lipid peroxidation and lipid metabolism studies in patients with chronic bronchitis showed normalization of the level of total lipids by lavender oil (Siurin, 1997). Inhalation of lavender oil had no effect on the content of cholesterol in the blood, but reduced its content in the aorta and atherosclerotic plaques (Nikolaevskii *et al.*, 1990). Linalool showed only marginal effects on lipid peroxidation of polyunsaturated fatty acids (PUFAs) (Reddy and Lokesch, 1992). Yamada *et al.* (1994) showed anticonvulsive effects of inhaling lavender oil vapour and Elizabetsky *et al.* (1999) showed similar effects for linalool in glutamate-related seizure models.

A hypoglycaemic effect of various species of lavender was shown by Gamez *et al.* (1987a,b). Linalool leads to a hepatic peroxysomal and microsomal enzyme induction in rats (Roffey *et al.*, 1990; Chadba and Madyasthe, 1984) and choleretic and cholagogic activity of Bulgarian lavender oil and a mixture of linalool and α-terpinol was found by Peana *et al.* (1994) and Gruncharov (1973). Some periodontal diseases can be treated with a mixture of

EOs, including lavender (Sysoev and Lanina, 1990). Lavender oil was said to be suitable for prevention and treatment of decubitus ulcers, insect bites, athletes' foot and skin rash and can also be used for the topical treatment of acne, prevention of facial scarring and blemishes of the face and body (Hartwig 1996). The EO of lavender was used in a mixture as a hair growth stimulant and for the treatment of *Alopecia areata* (Hay *et al.*, 1998) and in a pilot study to determine possible novel, safe pediculicides in children. Skin penetration enhancers, especially for the transdermal absorption of various drugs and medicaments have included lavender oil with Nifedipine (Thacharodi and Rao, 1994). Research into cell cultures of *L. vera* for rosmarinic acid production was discussed by Ilieva-Stoilova *et al.* (2002).

There are numerous miscellaneous uses for lavender flowers, both fresh and dried (Lis-Balchin, 2002d) e.g. herbal pillows, lavender bags, household cleaning products and scented candles. Spike lavender is included in some veterinary shampoos and other products as an insect, especially flea repellent (Potter, 1988). Lavender oil is used as a component in topical formulations to relieve the pain associated with rheumatic and musculo-skeletal disorders, acting as a potent radical scavenger (Billany *et al.*, 1995).

Perillyl alcohol, a minor component of lavender and the most important metabolite of D-limonene, is a chemo-preventative and chemo-therapeutic agent (Reddy *et al.*, 1997; Bellanger, 1998), e.g. against rat liver cancer and rodent mammary and pancreatic tumours). Pancreatic tumours were inhibited completely by geraniol at 20 g/kg diet and 50% by perillyl alcohol at 40 g/kg diet in hamsters. Patents have been taken out for various uses of perillyl alcohol including: antibiotic and anti-fungal action (US Patent 5,110,832) and carcinoma regression (US Patent 5,414,019).

Contemporary patents for lavender include: wound treatment (US Patent 4,318,906); treating skin and scalp conditions (US Patent 4,855,131); minor skin irritations, promoting healing, resisting insects (US Patent 5,620,695); fly and mosquito attractant (US Patent 5,635,174) and control of dermatomycoses and dermatophytoses of skin ailments with Tinea pedis (US Patent 5,641,481).

11.5.6 Toxicity of lavender essential oils

Culpeper (1653) said that lavender (*L. vera*) 'provokes menses of women, and expels both a stillborn child and afterbirth' (the only reference to lavender as an abortifacient).

The BIBRA Working Group (1994) showed little or no irritation to human and animal skin, but it has caused sensitization, photosensitization and pigmentation. Patch tests have shown a few allergies due to photosensitization and also pigmentation (Brandao, 1986; Nakayama *et al.*, 1976). Its principal effect following administration by oral, injection or inhalation routes to rodents was sedation. Linalool was irritant to the skin of various species of laboratory animals. There was the danger of causing dermatitis in sensitive people (Rudzki *et al.*, 1976), e.g. an occupational allergy to a lavender shampoo used by a hairdresser (Brandao, 1986; Menard, 1961). Facial 'pillow' dermatitis due to lavender oil allergy was described by Coulson and Khan (1999). Facial psoriasis caused by contact allergy to linalool and hydroxycitronellal in an after-shave was described by De Groot and Liem (1983). Patch testing using lavender oil at 20% in petrolatum on patients suspected of suffering from cosmetic contact dermatitis over a nine-year period in Japan showed a dramatic increase in 1997, which coincided with the importation of the aromatherapy trend for using lavender oil and dried flowers. There is also the danger of airborne contact allergic dermatitis through overuse of essential oils and their storage (Schaller and Korting, 1995), which produced a severe response in a man who had been active with essential oils.

11.5.7 D-Limonene toxicity

Although present in small quantities in most lavenders, except *L. stoechas*, the dangers of D-limonene sensitization have become more prominent as it is used in so many industrial processes, e.g. degreasing metal before industrial painting, cleaning assemblies and as a hand cleanser. It oxidizes to *R*-(–) carvone, *cis* and *trans*-isomers of limonene oxide and hydroperoxides, all potential contact allergens (Karlberg *et al.*, 1994). Two per cent of dermatitis patients gave a positive patch test to D-limonene (Karlberg and Dooms-Goossens, 1997), especially when aged (Chang *et al.*, 1997). Pulmonary exposure of human volunteers to D-limonene caused a decrease in the lung vital capacity (Falk-Filipsson *et al.*, 1993). The major volatile component of lactating mothers' milk in the USA contained D-limonene (von Burg, 1995), thus making it possible that the baby could develop an allergic response soon after birth. Cats and dogs, too, are very susceptible to insecticides and baths containing D-limonene.

In contrast to all the toxicity, anticarcinogenic properties of D-limonene were shown *in vitro*, when applied subcutaneously to mice that were then injected with benzopentaphene, but although the lung tumours took longer to develop and therefore the animals lived longer, it did not prevent them forming (Homburger *et al.*, 1971).

11.6 Quality issues and adulteration

11.6.1 Quality specifications of essential oils of lavender and solvent extracts

Boelens (1995) reviewed the chemical and sensory evaluation of *Lavandula* oils. The true oil is almost colourless and has a sweet, floral, herbaceous, refreshing odour with a pleasant, balsamic-wood undertone and a fruity-sweet top-note.

Definition of lavender and lavandula oils
The International Organization for Standardization (ISO) defines Oil of French Lavender, ISO 3515 as 'The oil obtained by steam distillation of recently picked lavender flowers (*Lavandula angustifolia* P. Miller) either growing wild or cultivated in France'. The established chromatographic profile includes the main identifying components (Table 11.1).

Spike lavender (*Lavandula latifolia* (L.) Medikus) has a separate ISO (4719: 1992), as does Oil of Lavandin abrialis (*Lavandula angustifolia* P. Miller × *Lavandula latifolia* (L.) Medikus), France. The latter has a requirement for a minimum linalyl acetate content of 27%/37% maximum and linalool 28%/38% with camphor at 7%/11% maximum. Oil of Lavandin *grosso* (*Lavandula angustifolia* P. Miller × *Lavandula latifolia* (L.) Medikus), France also has an ISO.

Terpeneless lavender oil is produced by careful vacuum distillation; a 'topping off' of about 10% of the oil is sufficient to make it mellower, softer and more soluble in dilute alcohol. Of course, it has increased stability and is more useful in foods.

11.6.2 Lavandin oil

This was first produced in the late 1920s, but has since escalated well above that of true lavender. Many different hybrids, growing all over the world, give a higher yield than the shorter lavender. The oil is pale yellow to almost colourless and has a strongly herbaceous odour with a distinctive top-note which is fresh camphene cineole-like (Arctander, 1960). Lavandin oil is used in large quantities for a fresh note in perfumes and in detergents.

Table 11.1 ISO composition of *Lavandula angustifolia* P. Miller, ISO 3515 (a), its components (b), EC Regulations (c), and sensitization values (d)

(a)	Optical rotation	−11 to −7	
	Ester min.	38%; max. 58% as linalyl acetate	
(b)	**Components**	**Min**	**Max**
	trans-β-ocimene	2	6
	cis-β-ocimene	4	10
	Octanone-3	–	2
	1,8-cineole	–	1.5
	Limonene	–	0.5
	Camphor	–	0.5
	Linalool	25	38
	Linalyl acetate	25	45
	Terpinen-4-ol	2	6
	Lavandulol	0.3	
	Lavandulyl acetate	2	
	α-terpineol	–	1

(c) **EC regulations 2002 (CHIP)**

Lavender oil, CAS No. 8000-28-0; EEC No. 289-995-2; Hazard symbol: Xn; Risk phase: R65; H/C 15%; Safety phase S62

Lavandin oil, CAS No. 8022-15-9; EEC No. 294-470-6; Hazard symbol: none; Risk phase: none; H/C: none; Safety phase: none

Lavender spike oil, CAS No. 8016-78-2; EEC No. 284-290-6; Hazard symbol: none; Risk phase: R10; H/C: none; Safety phase: none

D-Limonene, CAS No. 5989-27-5; EEC No. 227-813-5; Hazard symbol: Xn N; Risk phase: R10, 38, 43, 50/53; H/C 100%; Safety phase S24, 37, 60, 61

L-Limonene, CAS No. 5989-54-8; EEC No. 228-813-5; Hazard symbol: Xn N; Risk phase: R10, 38, 43, 50/53; H/C 100%; Safety phase S24, 37, 60, 61

Linalyl acetate, CAS No.115-95-7; EEC No. 204-727-6; Hazard symbol: none; Risk phase: none; H/C: none; Safety phase: none

Linalool, CAS No.78-70-6; EEC No.201-134-4; Hazard symbol: none; Risk phase: none; H/C: none; Safety phase: none

Maximum levels of fragrance allergens in aromatic natural raw materials:

European Parliament and Council Directive 76/768/EEC on Cosmetic Products, 7th Amendment 2002: The presence of the substances must be indicated in the list of ingredients when its concentration exceeds 0.001% in leave-on products and 0.01% in rinse-off products.

(d) **Sensitisers present in lavender oils (EFFA)**

Lavender: coumarin: below 0.1%; geraniol, 1.1; limonene, 0.6; linalool, 38;   Total: 39.7

Lavender and lavandin absolute: coumarin: 6; geraniol, 0.3; limonene, 0.7; linalool, 28; Total: 35

Lavandin oil: coumarin: below 0.1%; geraniol, 0.4; limonene, 1; linalool, 37;Total: 38.4

Spike lavender: coumarin: below 0.1%; geraniol, below 0.1%; limonene, 1; linalool, 46; Total: 47

11.6.3 Lavender and lavandin absolute and concrete

Lavandula angustifolia P. Miller (or *L. officinalis*) absolute is produced from direct extraction of the herb with solvents and thence extraction with absolute alcohol after chilling and this is then evaporated continuously under reduced vacuum; it can also be produced from the distillation water by extraction with benzene or petroleum ether and thence re-extracted with alcohol.

Lavandin absolute, like the lavender absolute, is a viscous, dark green liquid of herbaceous odour, resembling the flowering plant. Both are sweeter than the essential oil and are used in similar fragrances (Wells and Lis-Balchin, 2002).

11.6.4. Adulteration of lavender oil
Adulteration of lavender oils is primarily with lavandin oils and its fractions (as it is so much cheaper, being produced in at least a ten-fold excess), but other synthetic and natural fractions occur. Adulterants include: acetylated lavandin, synthetic linalool and linalyl acetate, fractions of ho leaf oil and rosewood oil, terpinyl propionate, isobornyl acetate, terpineol, fractions of rosemary, aspic oil, lavandin, etc. (Arctander, 1960; Lis-Balchin, 2002e).

Ordinary gas chromatography can be used to detect diluting solvents; however, GC, with or without mass spectrometry (MS) or other identification facilities, such as infra-red (IR), are not sophisticated enough to find most adulterations when fractions of other oils or synthetic components are used. Synthetic adulteration with linalool and/or linalyl acetate could often be detected by the presence of dehydrolinalool, dihydrolinalool, dehydrolinalyl acetate and dihydrolinalyl acetate, but detection was perfected by the use of enantiomeric (chiral) columns), mainly composed of an α-cyclodextrin phase (Ravid et al., 1992; Lis-Balchin, 2002e). Pure lavender oil had either (3R)-(–)-linalyl acetate or R-(–)-linalyl acetate. Chiral columns can also be used by those involved in adulteration of essential oils to separate out the enantiomers, then add them in the correct proportion for a given essential oil!

11.7 References

AMMON, H.P.T. (1989), Phytotherapeutika in der Kneipp-therapie, *Therapiewoche*, **39**, 117–27.
AOSHIMA, H. and HAMAMOTO, K. (1999), Potentiation of GABAA receptors expressed in *Xenopus* oocytes by perfume and phytoncid, *Biosci. Biotechnol. Biochem.*, **63**, 743–8.
ARCTANDER, S. (1960), *Perfume and Flavor Materials of Natural Origin*, Elizabeth, NJ.
ATANASSOVA-SHOPOVA, S. and ROUSSINOV, K.S. (1970), On certain central neurotropic effects of lavender essential oil, *Izvest. Inst. Fiziol. Sofiia*, **13**, 69–77.
BELLANGER, J.T. (1998), Perillyl alcohol: application in oncology, *Altern. Med. Rev.*, **3**, 448–57.
BENSON, H. and STARK, M. (1996), *Timeless Healing. The Power and Biology of Belief*, Simon & Schuster, London.
BERTRAM, T. (1995), *Encyclopaedia of Herbal Medicine*, 1st ed., Grace Publishers, Dorset.
BIBRA WORKING GROUP (1994), Lavender oil: BIBRA toxicity profile of lavender oil, *Govt. Reports Announcements & Index* (GRA & I), Issue **19**, 1996.
BILLANY, M.R., DENMAN, S., JAMEEL, S. and SUGDEN, J.K. (1995), Topical antirheumatic agents as hydroxyl radical scavengers, *Int. J. Pharm.*, **124**, 279–83.
BOELENS, M.H. (1995), Chemical and sensory evaluation of *Lavandula* oils, *Perf. Flav.*, **20**, 23–51.
BRADSHAW, R.M., MARCHANT, J.N., MEREDITH, M.J. and BROOM, D.M. (1998), Effects of lavender straw on stress and travel sickness in pigs, *J. Altern. Complement. Med.*, **4**, 271–5.
BRANDAO, F.M. (1986), Occupational allergy to lavender, *Contact Dermatitis*, **15**, 249–50.
BRANDT, W. (1988), Spasmolytische wirkung atherischer Ole, *Zeitschrift fur Phytotherapie*, **9**, 33–9.
BROOKER, D.J., SNAPE, M., JOHNSON, E., WARD, D. and PAYNE, M. (1997), Single case evaluation of the effects of aromatherapy and massage on disturbed behaviour in severe dementia, *Br. J. Clin. Psychol.*, **36**, 287–96.
BUCHBAUER, G. (1992), Biological effects of fragrances and essential oils, *Perfumer Flavorist*, **18**, 19–24.
BUCHBAUER, G. (2002), Lavender oil and its therapeutic properties, in M. Lis-Balchin (ed.) *Lavender; The Genus* Lavandula: *Medicinal and Aromatic Plants – Industrial Profiles*, Taylor and Francis, London, pp. 124–39.

BUCHBAUER, G., JIROVETZ, L., JAGER, W., DIETRICH, H. and PLANK, C. (1991), Aromatherapy: evidence for sedative effects of the essential oil of lavender after inhalation, *Z. Naturforsch*, **46**, 1067–72.
BUCHBAUER, G., JAGER, W., JIROVETZ, I., ILMBERGER, J. and DIETRICH, H. (1993), Therapeutic properties of essential oils and fragrances, in R. Teramishu, R.G. Buttery and H. Sugisawa (eds) *Bioactive Volatile Compounds from Plants*, ACS Symposium Series 525, American Chemical Society, Washington DC, pp. 159–65.
VON BURG, R. (1995), Toxicology update, *J. Appl. Toxicol.*, **15**, 495–9.
CASTLE, J. and LIS-BALCHIN, M. (2002), History of usage of *Lavandula* species, in M. Lis-Balchin (ed.) *Lavender; The Genus Lavandula: Medicinal and Aromatic Plants – Industrial Profiles*, Taylor and Francis, London, pp. 35–50.
CHADBA, A. and MADYASTHA, K.M. (1984), Metabolism of geraniol and linalool in the rat and effects on liver and lung microsomal enzymes, *Xenobiotica*, **14**, 365–74.
CHAISSE, E. and BLANC, M. (1990), Les ravageurs de la lavande et du lavandin, *Phytoma*, **419**, 45–6.
CHANG, Y., CHING, A., KARLBERG, A-T. and MAIBACH, H.I. (1997), Allergic contact dermatitis from oxidised D-limonene, *Contact Dermatitis*, **37**, 308–9.
CHARLES, D.J., RENAUD, E.N.C. and SIMON, J.E. (2002), Comparative study of the essential oil quantity and quality of ten cultivars of organically grown lavender and lavandin, in M. Lis-Balchin (ed.) *Lavender; The Genus Lavandula: Medicinal and Aromatic Plants – Industrial Profiles*, Taylor and Francis, London, pp. 232–42.
CHARLSWORTH, S. (2002), The retail lavender industry, in M. Lis-Balchin (ed.) *Lavender; The Genus Lavandula: Medicinal and Aromatic Plants – Industrial Profiles*, Taylor and Francis, London, pp. 60–75.
CHARRON, J.M. (1997). Use of *Lavandula latifolia* as an expectorant, *Chem. Senses*, **22**, 237–48.
CHAYTOR, D.A. (1937), A taxonomic study of the genus *Lavandula*. *Linn. Soc.*, **51**, 153–204.
COOKE, B. and ERNST, E. (2000), Aromatherapy: a systematic review, *Br. J. Gen. Pract.*, **50**, 493–6.
COULSON, I.H. and KHAN, A.S.A. (1999), Facial 'pillow' dermatitis due to lavender oil allergy, *Contact Dermatitis*, **41**, 111.
CULPEPER, N. (1653) *The English Physitian Enlarged,* George Sawbridge, London.
DE GROOT, A.C. and LIEM, D.H. (1983), Facial psoriasis caused by contact allergy to linalool and hydroxycitronellal in an after-shave, *Contact Dermatitis*, **9**, 230–2.
DEANS, S.G. (2002), Antimicrobial properties of lavender volatile oil, in M. Lis-Balchin (ed.) *Lavender; The Genus Lavandula: Medicinal and Aromatic Plants – Industrial Profiles*, Taylor and Francis, London, pp. 171–9.
DEGEL, J. and KOSTER, E.P. (1999), Odors: implicit memory and performance effects, *Chem. Senses*, **24**, 317–25.
DENNY, E.F.K. (2002), Distillation of the lavender type oils: theory and practice, in M. Lis-Balchin (ed.) *Lavender; The Genus Lavandula: Medicinal and Aromatic Plants – Industrial Profiles*, Taylor and Francis, London, pp. 100–16.
DIEGO, M.A., JONES, N.A., FIELD, T., HERNANDEZ-REIF, M., SCHANBERG, S., HUHN, C., MCADAM, V., GALAMAGA, R. and GALAMAGA, M. (1998) Aromatherapy positively affects mood, EEG patterns of alertness and math computations, *Intern. J. Neuroscience*, **96**, 217–24.
DUNN, C., SLEEP, J. and COLLETT, D. (1995), Sensing an improvement: an experimental study to evaluate the use of aromatherapy, massage and periods of rest in an intensive care unit, *J. Adv. Nurs.*, **21**, 34–40.
ELISABETSKY, E., BRUM, L.F. and SOUZA, D.O. (1999), Anticonvulsant properties of linalool in glutamate-related seizure models, *Phytomedicine*, **6**, 107–13.
ELISABETSKY, E., COELHO DE SOUZA, G.P., DOS SANTOS, M.A.C., SIQUIEIRA, I.R., AMADOR, T.A. and NUNES, D.S. (1995a), Sedative properties of linalool, *Fitoterapia*, **66**, 407–14.
ELISABETSKY, E., MARSCHNER, J. and SOUZA, D.O. (1995b), Effects of linalool on glutamatergic system in the rat cerebral cortex, *Neurochem. Res.*, **20**, 461–5.
FALK-FILIPSSON, A., LOF, A., HAGBERT, M. *ET AL.* (1993), D-Limonene exposure to humans by inhalation: uptake, distribution, elimination, and effects on the pulmonary function, *J. Toxic Environ. Health*, **38**, 77–88.
FENAROLI, G. (1998), *Fenaroli's Handbook of Flavor Ingredients,* vol. 1, 3rd ed, CRC Press, Boca Raton, FL.
FIGUEREIDO, A.C., BARROSO, J.G., PEDRO, L.G., SEVINATE-PINTO, I., ANTUNES, T., FONTINHA, S.S., LOOMAN, A. and SCHEFFER, J.J.C. (1995), Composition of the essential oil of *Lavandula pinnata* L. fi, var. *pinnata* grown on Madeira, *Flav. Fragr. J.*, **10**, 93–6.
GAMEZ, M.J., JIMENEZ, J., RISCO, S. and ZARZUELO, A. (1987a), Hypoglycaemic activity in various

species of the genus *Lavandula*. Part 1. *Lavandula stoechas* L. and *Lavandula multifida* L., *Pharmazie*, **42**, 706–7.

GAMEZ, M.J., ZARZUELO, A., RISCO, S., UTRILLA, P. and JIMENEZ, J. (1987b), Hypoglycaemic activity in various species of the genus *Lavandula*. Part 2. *Lavandula dentata* and *Lavandula latifolia, Pharmazie*, **43**, 441–2.

GARCIA-VALLEJO, M.C., GARCIA-VALLEJO, I. and VELASCO-NEGUERUELA, A. (1989), Essential oils of the Genus *Lavandula* L. in Spain, *Proc. ICEOFF*, New Delhi, **4**, pp. 15–26.

GATTEFOSSE, R.M. (1937), *Aromatherapy* (translated 1993), C.W. Daniel and Co. Ltd., Saffron Walden.

GERARD, J. (1633), *The Herbal or General History of Plants; the Compleete 1633 edition as revised and enlarged by Thomas Johnson* (Facsimile edition: Dover Publications Inc., New York, 1975).

GHELARDINI, C., GALEOTTI, N., SALVATORE, G. and MAZZANTI, G. (1999), Local anaesthetic activity of the essential oil of *Lavandula angustifolia*, *Planta Med.*, **65**, 700–3.

GRIEVE, M. (1937), *A Modern Herbal*, reprinted 1992, Tiger Books International, London.

GRUNCHAROV, V. (1973), Clinico-experimental study on the choleretic and cholagogic action of Bulgarian lavender oil, *Vutr. Boles.*, **12**, 90–6.

GUILLEMAIN, J., ROUSSEAU, A. and DELAVEAU, P. (1989), Effets neurodepresseurs de l'huile essentielle de *Lavandula Angustifolia* Mill, *Ann. Pharm. Fr.*, **47**, 337–43.

HARBORNE, J.B. and WILLIAMS, C. (2002), Phytochemistry of the genus *Lavandula*, in M. Lis-Balchin (ed.) *Lavender; The Genus* Lavandula*: Medicinal and Aromatic Plants – Industrial Profiles*, Taylor and Francis, London, pp. 86–99.

HARDY, M., KIRK-SMITH, M.D. and STRETCH, D.D. (1995), Replacement of chronic drug treatment for insomnia by ambient odour, *Lancet*, **346**, 701.

HART, S. and LIS-BALCHIN, M. (2002), Pharmacology of *Lavandula* essential oils and extracts *in vitro* and *in vivo*, in M. Lis-Balchin (ed.) *Lavender; The Genus* Lavandula*: Medicinal and Aromatic Plants – Industrial Profiles*, Taylor and Francis, London, pp. 140–54.

HARTWIG, G. (1996), Essential oils for prevention and treatment of decubitus ulcers. *Germany Offen.* 19518836; *Chem. Abstr.*, **126**, 5, 65430 e.

HAY, I.C., JAMIESON, M. and ORMEROD, A.D. (1998), Randomized trial of aromatherapy: Successful treatment for *Alopecia areata*, *Arch. Dermatol.*, **134**, 1349–52.

HOHMANN, J., ZUPKO, I., REDEI, D., CSANYI, M., FALKAY, G., MATHE, I. and JANICSAK, G. (1999), Protective effects of the aerial parts of *Salvia officinalis*, *Melissa officinalis* and *Lavandula angustifolia* and their constituents against enzyme dependent and enzyme-independent lipid peroxidation, *Planta Med.*, **65**, 576–8.

HOMBURGER, F., TREGER, A. and BOGER, E. (1971), Inhibition of murine subcutaneous and intravenous benzo(rst)pentaphene carcinogenesis by sweet orange oils and D-limonene, *Oncology*, **25**, 1–10.

ILIEVA-STOILOVA, M.P., PAVLOV, A.I. and KOVATCHEVA-APOSTOLOVA, E.G. (2002), Further research into *Lavandula* species: cell cultures of *L. vera* and rosmarinic acid production, *in* M. Lis-Balchin (ed.) *Lavender; The Genus* Lavandula*: Medicinal and Aromatic Plants – Industrial Profiles*, Taylor and Francis, London, pp. 214–26.

IMASEKI, I. and KITABATAKE, Y. (1962), Studies on effect of essential oils and their components on the isolated intestines of mice, *J. Pharmaceut. Soc. Japan*, **82**, 1326–8.

IZZO, A.A., CAPASSO, R., SENATORE, F., SECCIA, S. and MORRICA, P. (1996), Spasmolytic activity of medicinal plants used for the treatment of disorders involving smooth muscle, *Phytother. Res.*, **10**, S107–8.

JAGER, W., BUCHBAUER, G., JIROVETZ, L. and FRITZER, M. (1992), Percutaneous absorption of lavender oil from a massage oil, *J. Soc. Cosmet. Chem.*, **43**, 49.

JEANFILS, J., BURLION, N. and ANDRIEN, F. (1991), Antimicrobial activities of essential oils from different plant species. *Landbouwtijdschrift-Revue de l'Agriculture* **44**, 1013–1019.

KARAMAT, E., ILMBERGER, J., BUCHBAUER, G. ET AL. (1992), Excitatory and sedative effects of essential oils on human reaction time performance, *Chem. Senses*, **17**, 847.

KARLBERG, A.-T. and DOOMS-GOOSSENS, A. (1997), Contact allergy to oxidised D-limonene among dermatitis patients, *Contact Dermatitis*, **36**, 201–6.

KARLBERG, A.-T., SHAO, L. P., NILSSON, U. ET AL. (1994), Hydroperoxides in oxidised D-limonene identified as potent contact allergens, *Arch. Dermatol. Res.*, **286**, 97–103.

KIM, H.M. and CHO, S.H. (1999), Lavender oil inhibits immediate-type allergic reaction in mice and rats, *J. Pharm. Pharmacol.*, **51**, 221–6.

KIRK-SMITH, M. (2002), The psychological effects of lavender, in M. Lis-Balchin (ed.) *Lavender; The Genus* Lavandula*: Medicinal and Aromatic Plants – Industrial Profiles*, Taylor and Francis, London, pp. 155–70.

KNASKO, S.C. (1992), Ambient odours effect on creativity, mood and perceived health, *Chem. Senses*, **17**, 27–35.
KNASKO, S,C., GILBERT, A.N. and SABINI, J. (1990), Emotional state, physical well-being and performance in the presence of feigned ambient odour, *J. Appl. Soc. Psychol.*, **20**, 1345–7.
KOMORI, T., TAMIDA, M., KIKUCHI, A., SHOJI, K., NAKAMURA, S. and NOMURA, J. (1997), Effects of odorant inhalation on pentobarbital-induced sleep time in rats, *Human Psychopharmacol.*, **12**, 601.
KOVAR, K.A., GROPPER, B., FRIESS, D. and AMMON, H.P.T. (1987), Blood levels of 1,8-cineole and locomotor activity of mice after inhalation and oral administration of rosemary oil, *Planta Med.*, **53**, 315–8.
KUBOTA, M., IKEMOTO, T., KOMAKI, R. and INUI, M. (1992), Odor and emotion-effects of essential oils on contingent negative variation, *Proc. 12th Int. Congress on Flavours, Fragrances and Essential oils*, Vienna, Austria, 4–8 Oct., pp. 456–61.
LAWRENCE, B. (1976–1978; 1979–1980; 1981–1987; 1988–1991; 1992–1994), *Essential Oils*, Allured Pub. Corp., IL.
LE MEN, J. and POURRAT, H. (1953), Distribution of ursolic acid among the Labiatae, *Ann. Pharm. Franç.*, **11**, 190–2.
LESHCHINSKAIA, I.S., MAKARCHUK, N.M., LEBEDA, A.F., KRIVENKO, V.V. and SGIBNEV, A.K. (1983), Effect of phytoncides on the dynamics of the cerebral circulation in flight controllers during their occupational activity, *Kosm. Biol. Aviakosm. Med.*, **17**, 80–3.
LIS-BALCHIN, M. (1997), Essential oils and 'Aromatherapy': their modern role in healing, *J. R. Soc. Health*, **117**, 324–9.
LIS-BALCHIN, M. (2002a), History of nomenclature and location of *Lavandula* species, hybrids and cultivars, in M. Lis-Balchin (ed.) *Lavender; The Genus* Lavandula*: Medicinal and Aromatic Plants – Industrial Profiles*, Taylor and Francis, London, pp. 51–6, 80–5.
LIS-BALCHIN, M. (2002b), Essential oils from different *Pelargonium* species and cultivars and their chemical composition (by GC and GC/MS), in M. Lis-Balchin (ed.) *Lavender; The Genus* Lavandula*: Medicinal and Aromatic Plants – Industrial Profiles*, Taylor and Francis, London, pp. 147–65.
LIS-BALCHIN, M. (2002c), Miscellaneous uses of lavender and lavender oil: use in hair products, food flavouring, tissanes, herbal pillows, medicinal products, in M. Lis-Balchin (ed.) *Lavender; The Genus* Lavandula*: Medicinal and Aromatic Plants – Industrial Profiles*, Taylor and Francis, London, pp. 200–5.
LIS-BALCHIN, M. (2002d), Lavender oil and its use in aromatherapy, in M. Lis-Balchin (ed.) *Lavender; The Genus* Lavandula*: Medicinal and Aromatic Plants – Industrial Profiles*, Taylor and Francis, London, pp. 180–93.
LIS-BALCHIN, M. (2002e), Lavender essential oil: standardisation, ISO; adulteration and its detection using GC, enantiomeric columns and bioactivity, in M. Lis-Balchin (ed.) *Lavender; The Genus* Lavandula*: Medicinal and Aromatic Plants – Industrial Profiles*, Taylor and Francis, London, pp. 117–23.
LIS-BALCHIN, M. (2002f), New research into *Lavandula* species, hybrids and cultivars, in M. Lis-Balchin (ed.) *Lavender; The Genus* Lavandula*: Medicinal and Aromatic Plants – Industrial Profiles*, Taylor and Francis, London, pp. 206–13.
LIS-BALCHIN, M. and DEANS, S.G. (1997), Bioactivity of selected plant essential oils against *Listeria monocytogenes*, *J. Appl. Microbiol.*, **82**, 759–62.
LIS-BALCHIN, M. and HART, S. (1997a), A preliminary study of the effect of essential oils on skeletal and smooth muscle *in vitro*, *J. Ethnopharmacology*, **58**, 183–7.
LIS-BALCHIN, M. and HART, S. (1997b), Pharmacological effect of essential oils on the uterus compared to that on other different tissue types, *Proc. 27th Int. Symp. Ess. Oils*, Vienna, Austria, 8–11 Sept. 1996, Ch. Franz, A Mathé & G. Buchbauer (eds), Allured Pub. Corp., Carol Stream, Il., pp. 29–32.
LIS-BALCHIN, M. and HART, S. (1999), Studies on the mode of action of the essential oil of lavender (*Lavandula angustifolia* Miller), *Phytother. Res.*, **13**, 540–2.
LIS-BALCHIN, M., HART, S., DEANS, S. D. and EAGLESHAM, E. (1996a), Comparison of the pharmacological and antimicrobial action of commercial plant essential oils, *J. Herbs, Spices Med. Plants*, **4**, 69–86.
LIS-BALCHIN, M., DEANS, S. and HART, S. (1996b), Bioactivity of New Zealand medicinal plant essential oils, *Proc. Int. Symp. Medicinal and Aromatic Plants*, L. E. Craker, L. Nolan and K. Shetty (eds), 13–27.
LIS-BALCHIN, M.., DEANS, S.G. and HART, S. (1997), A study of the changes in the bioactivity of essential oils used singly and as mixtures in aromatherapy, *J. Alt. Complement. Med.*, **3**, 249–55.

LIS-BALCHIN, M., DEANS, S.G. and EAGLESHAM, E. (1998), Relationship between the bioactivity and chemical composition of commercial plant essential oils, *Flav. Fragr. J.,* **13**, 98–104.

LUDVIGSON, H.W. and ROTTMAN, T.R. (1989), Effects of ambient odours of lavender and cloves on cognition, memory, affect and mood, *Chem. Senses,* **14**, 525–36.

MANLEY, C.H. (1993), Psychophysiological effect of odor, *Crit. Rev. Food Sci. Nutr.,* **33**, 57–62.

MENARD, E. (1961), Les dermatoses profesionelles, *Concours Medicale,* **83**, 4308–11.

NAEF, R. and MORRIS, A.F. (1992), Lavender–Lavandin-A comparison, *Riv. Ital. EPPOS,* **3**, special issue, Feb., 364–77.

NAKAMATSU, Y. (1995), Brain stimulants containing terpenes, *Japan Kokai Tokkyo Koho,* **95**, 258113; *Chem. Abstr.,* **124**, 2, 15522 z.

NAKAYAMA, H., HARADA, R. and TODA, M. (1976), Pigmented cosmetic dermatitis, *Int. J. Derm.,* **15**, 673–5.

NIKOLAEVSKII, U.V., KONONOVA, N.S., PERTSOVSKII, A.I. and SHINKARCHUK, I.F. (1990), Effect of essential oils on the course of experimental atherosclerosis, *Patol. Fiziol. Eksp. Ter.,* **Sept/Oct.**, 52–3.

PEANA, A., SATTA, M., MORETTI, M.D.L. and ORECCHIONI, M. (1994), A study on choleretic activity of *Salvia desoleana* essential oil, *Planta Med.,* **60**, 478–9.

PLANT, O.H. (1920), The effect of carminative volatile oils on the muscular movements of the intestine, *J. Pharmacol. Exper. Therap.,* **16**, 311–25.

POTTER, N. (1988), *Potter's New Cyclopaedia of Botanical Drugs and Preparations,* Revised ed., E.M. Williamson and F.J. Evans (eds), C.W. Daniel Co. Ltd, Saffron Walden

RAVID, U., PUTIEVSKY, E., KATZIR, I., IKAN, R. and WEINSTEIN, V. (1992), Determination of the enantiomeric composition of citronellol in essential oils by chiral GC analysis on a modified γ-cyclodextrin phase, *Flav. Fragr. J.,* **7**, 235–8.

RE, L., BAROCCI, S., MENCARELLI, A., VIVANI, C., PAOLUCCI, G., SCARPANTONIO, A., RINALDI, L. and MOSCA, E. (2000), Linalool modifies the nicotinic receptor-ion channel kinetics at the mouse neuromuscular junction, *Pharmacol. Res.,* **42**, 177–82.

REDDY, A.C. and LOKESCH, B. (1992), Studies on spice principles as antioxidants in the inhibition of lipid peroxidation of rat liver microsomes, *Molec. Cell. Biochem.,* **111**, 117–24.

REDDY, B.S., WANG, C.X., SAMAHA, H., LUBET, R., STEELE, V.E., KELLOFF, G.J. and RAO, C.V. (1997), Chemoprevention of colon carcinogenesis by dietary perillyl alcohol, *Cancer Res.,* **57**, 420–25.

REITER, M. and BRANDT, W. (1985), Relaxant effects on tracheal and ileal smooth muscles of the guinea-pig, *Arzneim.-Forsch/Drug Res.,* **35**, 408–14.

ROFFEY, S.J., WALKER, R. and GIBSON, G.G. (1990), Hepatic peroxisomal and microsomal enzyme induction by citral and linalool in rats, *Food Chem. Toxicol.,* **28**, 403–8.

RUDZKI, E., GRZYWA, Z. and BRUO, W. S. (1976), Sensitivity to 35 essential oils, *Contact Derm.,* **2**, 196–200.

SAITO, N. and HARBORNE, J.B. (1992), Correlations between anthocyanin type, pollinator and flower colour in the Labiatae, *Phytochemistry,* **31**, 3009–15.

SCHALLER, M. and KORTING, H.C. (1995), Allergic airborne contact dermatitis from essential oils used in aromatherapy, *Clin. Exp. Dermatol.,* **20**, 143–5.

SHIPOCHLIEV, T. (1968), Pharmacological investigation into several essential oils. First communication. Effect on the smooth musculature, *Vet. Med. Nauki,* **5**, 63–9.

SIURIN, S.A. (1997), Effects of essential oil on lipid peroxidation and lipid metabolism in patients with chronic bronchitis, *Klin. Med. Moskau,* **75**, 43–5.

SKOULA, M., ABIDI, C. and KOKKALOU, E. (1996), Essential oil variation of *Lavandula stoechas* L. ssp. *stoechas* growing wild in Crete (Greece), *Biochem. System. Ecol.* **24**, 255–60.

SYSOEV, N.P. and LANINA, S.I. (1990), The results of sanitary chemical research into denture base materials coated with components from essential oil plants, *Stomatologia,* **July/Aug.**, 59–61.

THACHARODI, D. and RAO, D.P. (1994), Transdermal absorption of nifedipine from microemulsions of lipophilic skin penetration enhancers, *Int. J. Pharm.,* **111**, 235–40.

TISSERAND, R. (1985), *The Art of Aromatherapy,* revised ed., C.W. Daniel Co. Ltd, Saffron Walden.

TORII, S., FUKUDA, H., KANEMOTO, H., MIYANCHIO, R., HAMAUZU, Y. and KAWASAKI, M. (1988), Contingent negative variation and the psychological effects of odor, in S. Toller and G.H. Dodds (eds) *Perfumery: The Psychology and Biology of Fragrance,* Chapman & Hall, New York.

UPSON, T. (2002), The taxonomy of the genus *Lavandula* L, in M. Lis-Balchin (ed.) *Lavender: The Genus Lavandula: Medicinal and Aromatic Plants – Industrial Profiles,* Taylor and Francis, London, pp. 2–34.

VICKERS, L. (1991), *The Scented Lavender Book*, Edbury Press, London.
VICKERS, A. (1996), *Massage and Aromatherapy. A Guide for Health Professionals*, Chapman & Hall, London.
VOKOU, D., VARELTZIDOU, S. and KATINAKIS, P. (1993), Effects of aromatic plants on potato storage: sprout suppression and antimicrobial activity, *Agric. Ecosystems Environ.*, **47**, 223–5.
WARREN, C. AND WARRENBURG, S. (1993), Mood benefits of Fragrance, *Perf. Flavorist*, **18**, 9–15.
WEISS, E.A. (1997), *Essential Oil Crops*, CAB International, Abington.
WELLS, R. and LIS-BALCHIN, M. (2002), Perfumery uses of lavender and lavandin oils, in M. Lis-Balchin (ed.) *Lavender; The Genus* Lavandula*: Medicinal and Aromatic Plants – Industrial Profiles*, Taylor and Francis, London, pp. 194–9.
WIESENFELD, E. (1999), Aroma profiles of various *Lavandula* species, Noville, South Hackensack, NJ, USA; http://www.sisweb.com.referenc/applnote/noville.htm
YAMADA, K., MIMAKI, Y. and SASHIDA, Y. (1994), Anticonvulsive effects of inhaling lavender oil vapour, *Biol. Pharm. Bull.*, **17**, 359–60.

12
Mustard

J. Thomas, K. M. Kuruvilla and T. K. Hrideek, ICRI Spices Board, India

12.1 Introduction and description

Mustard is among the oldest recorded spices as seen in Sanskrit records dating back to about 3000 BC (Mehra, 1968) and was one of the first domesticated crops. Originally it was the condiment that was known as mustard and the word was derived from the Latin *mustum*. *Must*, the expressed juice of grapes or other fruits mixed with ground mustard seeds to form *mustum ardens* ('hot or burning *must*') was a Roman speciality condiment. Romans' love for mustard carried the same throughout Europe where it became popular for seasoning meat and fish. Apart from its use as a condiment, its medicinal value also was recognized early, as it was mentioned by Pythagoras in 530 BC as a remedy for scorpion bites. Mustard seeds were used for entombing their kings in Egypt. Some say that mustard was used for flavouring food to disguise the taste of degraded perishables.

Most mustard was prepared in the early days by pounding the seeds in a mortar and moistening them with vinegar. Dijon in France produced the famous mustard by using '*verjus*', a unique grape juice of the Bourgogne region. The modern era of mustard, however, began in 1720 when Mrs Clements of Durham, England, found a way of milling the heart of the seed to fine flour. Other entrepreneurs experimented with combining various types of mustard seeds to create superb mustard powder. Today there are countless mustard varieties available throughout the world, each reflecting local, regional and national cuisine.

Three types of mustard seeds are popularly used as condiments: pale yellow or white mustard (*Sinapsis alba* syn. *Brassica hirta* Moench or *Brassica alba*); brown or oriental mustard (*Brassica juncea*); and black or dark brown mustard (*Brassica nigra*). Apart from their use as a spice, mustards are widely used as green vegetables, as a salad crop, as an important oil seed crop (particularly in India where rape seed-mustard is the largest vegetable oil next to groundnut), green manure or as fodder crop and for industrial oil purposes.

12.1.1 Botany
Mustards are members of the Cruciferaceae or Brassicaceae family. The genus *Brassica* consists of 150 species of annuals or biennial herbs, several of which are cultivated as

oilseed crops or as vegetables or as fodder. Seeds of *Sinapsis alba* (syn. *Brassica hirta*), *Brassica juncea* and *Brassica nigra* have only condiment value.

Sinapis alba (L.) Syn. *Brassica alba* (L) *B. hirta* Moench
This is commonly called yellow mustard. Yellow/white mustard seeds are also known as *sufed rai* (Hindi), *moutarde blache* (French), *senape biancha* (Italian), *biji sawi* (Malaysian, Indonesian) or *mostaza blanca/mostaza silvstre* (Spanish). Yellow or white mustard is indigenous to southern Europe. Presently it is widely cultivated in Australia, China, Chile, Denmark, Italy, Japan, the UK, Netherlands, North Africa, Canada and the USA. (Farrell, 1985). *Brassica alba* is an annual herbaceous plant. Leaves are alternate, long, bristly branched, irregularly toothed, petiolate, hairy on both sides. Flowers are small, yellow with four petals, cruciform; stamens tetradynamous; pistil bicarpellate. The fruit is a bristly siliqua, round, ribbed, swollen at the seeds, and with a long ensiform beak, pods spreading in the raceme. Seeds are globular and yellowish. They are about 1.5–3 mm, minutely pitted, seed coat is thin, endosperm meagre and invisible to the naked eye; embryo large, yellowish, with curved hypocotyls, radicle partially surrounded by two folded cotyledons. Yellow mustard seed does not have any odour when crushed in water (Parry, 1969).

Brassica juncea (L.) Czern and Coss
Brown mustard was originally introduced from China into northern India from where it has extended to Afghanistan via Punjab (Sambamurthy and Subramanyam, 2000). It is popularly known as *rai* or Indian mustard, *moutarde de Chine* (French), *Indischer senf* (Greman), *senape Indiana* (Italian) and *mostaza India* (Spanish). This species originated from the hybridization of *Brassica nigra* with *Brassica campestris* and this probably happened in southwestern Asia and India where the natural distribution of the two species overlaps (Saucer, 1993). Brown mustard comprises two varieties, viz. 'Oriental', which is mostly used by Chinese, and the other darker and stronger 'brown' variety that is used by Indians. It is an annual herbaceous erect and much branched plant and is the main source for pungency among the cultivated mustards (Fig.12.1). Flowers are small and bright yellow in colour. Seeds are small and contain 35% oil. In the USA, prior to the 1940s, *B. juncea* was considered inferior to *B. nigra*. However, with the introduction of a new yellow-seeded variety of *B. juncea* from China, it became highly popular since the crop is amenable to combine harvesting.

Brassica nigra (L.) Koch
Black mustard seeds are called true mustard. They are also known as *senafich* (Amharic), *zwarte mosterd* (Dutch), *moutarde noir* (French), *rai* (Hindi), *senape near* (Italian), *biji sawi hitam* (Malaysian, Indonesian), *mostarda preta* (Portuguese), *abba* (Singalese) and *mostaza negra* (Spanish).

Black mustard is probably endemic in the southern Mediterranean region. *Brassica nigra* is of importance not only as a crop plant but it also contributed to the evolution of several species in the genus *Brassica*. It is an annual herbaceous plant which grows to a height of about 1 m. Leaves are petiolate, alternate and dark green hairy. Lower leaves are large, rough, irregularly sinuate – dentate, pinnate with terminal lobe large and small lower lobes. Upper leaves are smooth and moderately lobed. The flowers are small bright yellow, cruciform with four petals, stamens tetradynamus, pistil bicarpellate. The fruit is siliqua, quadrangular, smooth with a short slender beak. Seeds are small, red-brown to black in colour and minutely pitted. The seeds differ in outward appearance from those of brown mustard seeds. They are about 2 mm or less in size but tend to be a little more oblong than spherical, varying in colour from dark-reddish brown to black, more or less covered with

Fig. 12.1 *Brassica juncea*.

white pellicle, smaller and much more pungent than the white. Black mustard is not as popular in the USA or Europe because of difficulties in harvesting (Uhl, 2000).

12.2 Chemical composition

White or yellow mustard *(S. alba)* contains the glucosinolase **sinalbin** which on hydrolysis by enzyme (myrosin or glucosinolases) yields *p*-hydroxy benzyl- isothiocynate, *p*-hydroxy benzylamine known as the 'white principles' and other similar compounds (protein, fixed oils, mucilage, etc.) as brown mustard.

The most important constituent in brown mustard is a glucosinolate, **sinigrin** (potassium myronate), and the enzyme myrosin (myrosinase), sinapic acid; sinapine (sinapic acid choline ester); fixed oils (25–37%), consisting mainly of glycerides of erucic, eicosenoic, arachidic, nonadecanoic, behenic, oleic and palmitic acids (Leung and Foster, 1996). Sinigrin on hydrolysis by myrosin (myrosinase) yields allyl isothiocyanate, glucose and potassium bisulphate. Allyl isothiocyanate is volatile, its yield from *B. juncea* is 0.25–1.4%. Minor volatile components that are also set free by enzymatic hydrolysis include methyl, isopropyl, *sec*-butyl, butyl, 3-butenyl, 4-pentenyl, phenyl, 3-methylthopropyl, benzyl and β-phenylethyl isothiocyanates.

Black mustard (*B. nigra*) contains similar constituents as *B. juncea*, predominantly 2-propenyl (allyl) glucosinate (**sinigrin**), which on hydrolysis yields allyl isothiocyanate

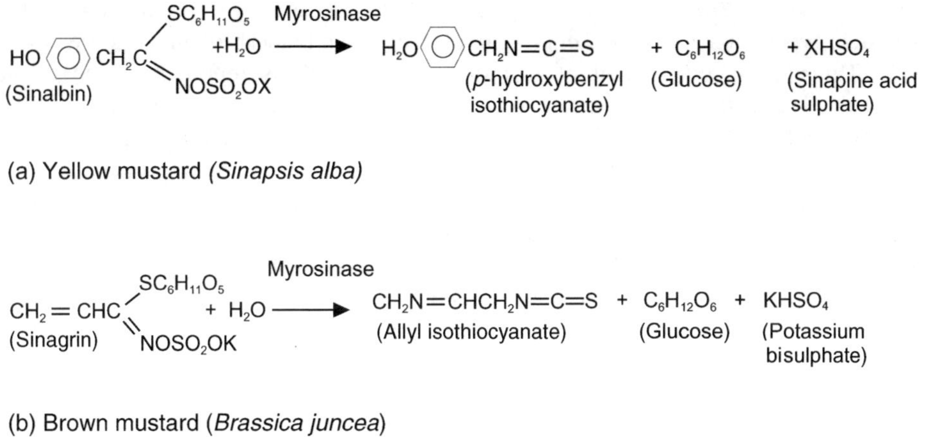

(a) Yellow mustard *(Sinapsis alba)*

$$\text{HO-C}_6\text{H}_4\text{-CH}_2\text{-C}(\text{SC}_6\text{H}_{11}\text{O}_5)=\text{NOSO}_2\text{OX} + H_2O \xrightarrow{\text{Myrosinase}} H_2O\cdot\text{C}_6\text{H}_4\text{-CH}_2\text{N}=\text{C}=\text{S} + C_6H_{12}O_6 + XHSO_4$$

(p-hydroxybenzyl isothiocyanate)   (Glucose)   (Sinapine acid sulphate)

(b) Brown mustard (*Brassica juncea*)

$$CH_2=CHC(SC_6H_{11}O_5)=NOSO_2OK + H_2O \xrightarrow{\text{Myrosinase}} CH_2N=CHCH_2N=C=S + C_6H_{12}O_6 + KHSO_4$$

(Sinagrin)   (Allyl isothiocyanate)   (Glucose)   (Potassium bisulphate)

Fig. 12.2 Chemical reaction of mustard. (a) Sinalbin in the presence of water reacts with myrosinase to form *p*-hydroxybenzyl isothiocyanate (sharp taste without pungent aroma). (b) Sinigrin in the presence of water reacts with myrosinase to form allyl isothiocyanate (pungent irritating odour). (*Source*: Tainter and Grenis 2001.)

known colloquially as volatile oil. The difference between the mustard types is the components responsible for the reaction and the end products produced. The reactions are well illustrated by Tainter and Grenis (2001) and are shown in Fig. 12.2.

12.2.1 Nutritional value

The nutritional data of the mustard seed are presented in Table 12.1. The moisture levels recommended by ASTA are a maximum of 11%.

Brown mustard seeds, have a caloric value of 541, a little less than that of groundnut, which is 561. Mustard oil contains many fatty acids of which eruvic and lenoleic acid are

Table 12.1 Nutritional composition of mustard seed per 100 g

Composition	USDA Handbook 8.2" (Yellow)*	ASTA (powder)**
Water (g)	6.86	3.0
Food energy (kcal)	469	580
Protein (g)	24.94	32
Fat (g)	28.76	42.6
Carbohydrates (g)	34.94	18.5
Ash (g)	4.51	4.0
Calcium (g)	0.521	0.3
Phosphorus (mg)	841	790
Potassium (mg)	682	700
Sodium (mg)	5	10
Iron (mg)	9.98	8.3
Thiamine (mg)	0.543	0.65
Riboflavin (mg)	0.381	0.45
Niacin (mg)	7.890	8.5
Ascorbic acid (mg)		22
Vitamin A activity (RF)	6	6

*Composition of foods spices & herbs, USDA Agricultural Handbook 8–2, January 1977.
**The nutritional composition of spices, ASTA Research Committee, February 1977.

very important. In *Brassica juncea* (L.) Czern and coss, oil content is usually 30–38% but in certain types, viz. *Lahi* and *Lahta* cultivated in Uttar Pradesh of India, possess 42–43%. The volatile oil content of *Brassica juncea* seeds is reported to 2.9%. The characteristics of the Indian mustard volatile oil are as follows: specific gravity 0.995; refractive index 1.5185; optical rotation 0°12'; but these characteristics differs in black mustard (*B. nigra*) volatile oil; specific gravity 1.015–1.025; refractive index 1.5267–1.5291. The volatile oil is optically inactive and consists almost entirely of allyl isothiocyanate (93–99%). The specification for pharmaceutical oil are (BPC), specific gravity 1.014–1.025; n_{20} 1.525–1.530 and allyl isothiocyanate content is not less than 92%. Mustard oil is a harmful because of its high of allyl isothiocyanate content. Fresh seeds or mustard powder do not possess essential oil and hence preparations made from these do not contain allyl isothiocyanate.

12.3 Production and cultivation

Around the globe, during 2002, mustard was grown over an area of 663 697 ha with a total seed production of 468 725 Mt (FAO database). Of the total world mustard seed sale, about 60% accounts for seeds of *S. alba* and the rest by *B. juncea*. In the USA the present consumption of mustard is more than any other spice except pepper. The important mustard-growing countries in the world are Canada, Nepal, the USA, Russian Federation, Myanmar, Czech Republic, Romania, Slovakia, Germany and France. But the condiment manufacturing is concentrated mainly in the USA, France, Germany, Japan, Canada and the UK. In the recent past, the area under brown mustard is increasing to the cost of other *Brassica*s, owing to its enhanced production and tolerance to biotic and abiotic stresses.

Mustard is a cool season crop, well suited to a short growing season. Mustard is grown in drier regions because of the better seed quality obtained under these conditions (Rosengarten, 1969). It prefers well-aerated soils that do not become waterlogged and are drought tolerant. Poor aeration in the root zone permanently stunts their growth. Mustard performs best in soil with a near-neutral pH, but will tolerate alkaline and slightly saline soils. Yellow mustard varieties mature in 80–85 days while brown and oriental types require about 90–95 days.

Mustard seeds are small and must be planted in moist, firm and shallow seed bed to ensure rapid germination and emergence. A seed rate of 6 kg/ha for brown and 10–12 kg/ha for yellow mustard is usually followed. Five weeks after emergence, the plants will begin to bud and after 10 days the plant will flower. Good moisture supply favours a long blossoming period and a longer flowering period ensures better yields.

Harvesting is invariably by direct combining in Europe, but it is a common practice to swathe the crop in Canada to promote the drying process. Yellow mustard can be straight combined if the crop has matured uniformly. Many growers prefer to straight combine while the crop is still tough (12–13% moisture) and artificially dry. Brown and oriental mustards are generally more susceptible to shattering and are usually swathed. Swathing should begin when 75% of the seeds have reached their mature colour. Yield realization under commercial farming is around 1000 kg/ha for yellow and around 1500 kg/ha for brown mustard. Seeds can be stored for long periods if the moisture content is less than 10%. While drying the mustard it is important to ensure that the seed temperature never exceeds 52°C or damage to endogenous enzymes may result, which on processing will impair hydrolysis of the glucosinolate to the isothiocyanate, the principle imparting the 'hot' characteristic.

12.4 Uses

Mustards are a versatile group of plants, which were used historically in a variety of ways. The leaves, seeds and oil are useful.

12.4.1 As vegetables

Mustard greens
Tender green plant as well as the green pods can be eaten as vegetables or salads. Mustard leaves, called as mustard greens, have a radishy taste. Dried/dehydrated mustard greens are available in the market as vegetables (Pruthi, 2001).

12.4.2 As flavouring

As whole seed
The English enjoy brown mustard with roast beef and ham. The Japanese use the oriental brown variety as a dip for raw fish. The Barbadians and other populations in the Caribbean use yellow or brown mustard with fruits and chilli peppers for great tasting sauces, marinades and stews. In Indian cooking, especially in vegetarian meals in the south, whole brown or black mustards seeds are 'popped' in heated *ghee* or oil to bring out their nuttiness, and are then added to sauces, chutneys, pickles, curries, *sambars* and *dals*. Black mustard is sometimes used to flavour *ghee* in south India.

Ground mustard or flour
Ground mustard seeds provide flavour and consistency in Bengali fish curries. Ground mustard and mustard flour are used in seafood cocktail sauces, barbecue sauces, cheese dishes, spice cakes and cookies, devilled eggs, baked beans, ham dishes, roast pork, meat loaf, ham salad, salad dressings, chowders and bisques, Chinese dish accompaniments, and on beets, cabbage and cucumbers. Sprouts from the mustard seeds are used in salads in many Asian recipes.

Compounded mustard or mustard compound
Compounded mustard generally consists of the flavour of mixture of seeds of *B. nigra*, *B. juncea* and *S. alba* and in addition it contains turmeric powder, starch or wheat flour (not exceeding 10%) and spices, etc.

Spice blends
A variety of blends are available in the market, which find their way into kitchens of every nationality:

- *American ballpark-style* mustard is made from the white seeds and blended with sugar and vinegar and coloured with turmeric.
- *Bordeaux mustard* is made from black seeds blended with unfermented wine. The seeds are not husked, producing a strong, aromatic, dark brown mustard often flavoured with tarragon.
- *Dijon mustard* is made from the husked black seeds blended with wine, salt and spices. This is the mustard generally used in classic French mustard sauces, salad dressings and mayonnaise.

- *English mustard* is hot, made from white seeds and is sometimes mixed with wheat flour for bulk and turmeric for colour.
- *German mustard* is usually a smooth blend of vinegar and black mustard, varying in strength.
- *Meaux mustard* is the partly crushed, partly ground black seed mixed with vinegar, producing crunchy, hot mustard that perks up bland foods.

12.4.3 Therapeutics and folklore

The ancient Greeks knew that mustard could be used as an antidote to scorpion and snake bites. Mustard plasters were used to stimulate blood circulation and to warm cold feet, to relax stiff muscles and to treat arthritis and rheumatism. Mustard treats skin diseases because of its high sulphur content. Mustard also stimulates the flow of salivary and gastric juices and promotes appetite. It has been used as a laxative, as a treatment for asthma and to induce vomiting or relieve coughs.

In *Ayurveda*, the Indian medicine, as well as *Yunani*, mustard and its oil are extensively used (Krishnamurthy, 1993). Seeds are useful in itching skin diseases and the diseases of viscera and worm infection. Mustard oil is simulative, pungent and enhances digestion. Excess use of oil causes impotency in males. It also forms an ingredient in many *Ayurvedic* medicated oils used as liniment or massage in many paralytic diseases of the nervous system.

Mustard is considered as diuretic, emetic, rubefacient and stimulant. Mustard relieves congestion by drawing the blood to the surface as in head afflictions, neuralgia and spasms. Mustard plaster is used externally for many afflictions, such as arthritis and rheumatism.

12.4.4 Industrial

Mustard is used for the manufacture of blown oil, which is an oxidized and viscous oil. Animal skins contain a certain amount of fat in their cells, which is removed during tanning, thus making the leather slightly hard. In order to make the leather soft or pliable, mustard oil is incorporated in the hides. Oil cake is used as cattle feed and manure. Mustard soil is used in soap making and as a lubricant and illuminant (Leung and Foster, 1996).

12.4.5 Other uses

Glucosinolates found in *Brassica* sp. are of interest due to the potential for using their degradation products as fumigants; isothiocyanates and nitriles have been demostrated to control fungi, bacteria and nematodes (Mojtahedi *et al.*, 1991). Biofumigation with mustard could be integrated to provide environmentally friendly and affordable control of soil-borne pests and diseases under integrated pest management systems.

The plants are used as green fodder for cattle. Mustards are very important honey crops in the Lompoc valley of California where the mustard is grown commercially. Honeybees forage on mustard plants during the peak flowering season and produce substantial quality of mild-flavoured light-coloured honey. Mustard is agriculturally used as a cover crop. Its oil is used as cat and dog repellents.

12.5 Properties

Brown mustard seed is spherical, medium in size and has a nutty, sweeter and mellow burning flavour. The whole mustard seed has no flavour, but can provide a pungent taste after chewing (Hirasa and Takemasa, 1998). The heat experienced in yellow mustard is on the tongue, whereas in brown and black mustard the heat is also felt in the nose and eyes. The severity of pungent aroma varies with different mustards. The white or yellow type has a less pungent aroma than brown mustard seeds, which have a very pungent aroma. Black mustard seeds have the highest pungency.

In ground mustard, aroma does not persist. However, flavour and pungency are experienced when enzymatic action is triggered in the presence of water, which releases mustard's flavour or pungency. It is due to a variety of isothiocyanate compounds that exist in mustard tissue as glycosides. The major pungent compound of black and brown mustard is allyl isothiocyanate. The release of sensation, especially in brown and black mustard, is delayed and begins at the back of the mouth, with a shooting sensation to the sinuses, owing to the activation of an enzyme, myrosinase. The enzyme myrosinase, in the presence of water, breaks down the glycoside (sinalbin) in yellow mustard or sinigrin in black or brown mustard to *para*-hydroxybenzyl isothiocyanate, which is responsible for the characteristic pungent aroma. The odours last until the enzyme activities ceases.

Yellow mustard flavour and pungency, like brown or black mustard, can be fully experienced only by triggering the enzyme myrosinase action, which releases them. The most effective enzymatic trigger is in the presence of water at room temperature, although other low-acid liquids such as milk and beer also work. Acidic liquids such as wine, vinegar and lemon juice are poor triggers of mustard's overall flavours, but are good subsequent preservatives of the flavour and they extend the penetrating odour. When water, vinegar, milk, wine or beer is added to mustard, mixed and left to stand for a few minutes, different degrees of flavour sensations are produced. With water a very sharp and hot taste is produced, while with vinegar milder flavour is induced. With milk a milder spicier and pungent flavour is created. With beer a very hot flavour is brought out (Uhl, 2000).

Mustard flour has preservative and antioxidant properties in addition to providing flavour and colour. In salad dressings, the most important property of the spice is its emulsifying function, binding water and oil phases as well providing viscosity. Mustard's fixed oil, which amounts to 30–35% by weight, is extracted by the cold pressed method. The oil is used for cooking in India and other Asian countries, including China and Japan

Mustard oil is extracted from the black mustard seeds, which have been macerated in warm water by steam or water distillation. Crude oil is dark brown in colour and contains a large proportion of free fatty acids. The refined oil is bland and light brown in colour. The characteristic odour of mustard oil is due to sulphur-containing essential oils produced by the hydrolysis of glucosides contained in the seeds. The quality of the mustard oil depends on the contents of the fatty acids and their percentage therein. Mustard oil is hazardous because of its high content of allyl isothiocyanate.

Essential oil is obtained by steam distillation of pressed cakes obtained after extraction of mustard oil (brown seed), after it has been hydrolysed by the enzyme myrosinase to release the allyl isothiocyanate from the glucoside. The essential oil of yellow mustard is obtained by solvent extraction of the press cake because it contains little or no volatile oil. Again, the press cake must first be hydrolysed to release the end products caused by the action of enzymes. Hydrolysis in either case is brought about by maceration of the press cake with warm water.

Oleoresin of mustard seed is usually obtained from a blend of the three different types of mustard to provide a balanced flavour. It is usually a yellow to light brown oily type of liquid with a volatile oil content of 5 ml per 100 g. Two kilograms is equivalent to 45.45 kg of the mustard spice.

Of the spices, condiments and herbs studied with respect to their effect on yeast fermentation in wines, etc., mustard flour was easily the most effective. It was found stronger than the two chemical preservative tried, viz. benzoic acid and sulphur dioxide (Pruthi, 1992). Mustard and its constituent allyl isocyanate have bacteriostatic and bactericidal properties (Charalambous, 1994).

12.6 Quality specifications

The quality of whole mustard seed is determined by the quantity of mature, undamaged seeds. Ground mustard quality is dependent on its final use. As per federal specifications, whole mustards are the seed of *Brassica hirta* (white mustard), *Brassica juncea* or varieties, or closely related varieties of the types of *Brassica nigra* and *Brassica juncea*.

Brassica hirta, or yellow-seeded varieties of *Brassica hirta*, shall be small, globular, yellow, clean-looking hard seeds and shall not possess any volatile oil or 'nose heat' when crushed and mixed with water. *Brassica juncea* shall be small, globular, and yellow to brown or brown hard seeds. Seeds, when ground and mixed with water on a one to three basis, must be capable of liberating a sharp, piercing, irritating odour and a very pungent taste.

If whole or ground mustard is purchased, it is generally designated as no. 1, no. 2 or no. 3 grade. This generally denotes the number of dark seeds present and other parameters. The International Standard states that seeds, when ground, must produce an odour free of mustiness and rancidity. The seed must have no more than 0.7% extraneous material and no more than 2% damaged or shrivelled seed. *Brassica nigra* and *B. juncea* must yield a minimum of 1.0% and 0.7% allyl isothiocyanate, respectively, and *S . alba* a minimum of 2.3% 4-hydroxybenzyl isothiocyanate.

According to the federal specification the condiment mustard is classified into three types: whole, ground and flavour. Whole mustard shell contains 5% total ash, 1% acid insoluble ash and 10% moisture. Ground mustard is the powder prepared by grounding whole yellow mustard seed. The finished product of this represents the seed ground in its whole form and without the outer hulls of husk removed and without removal of fixed oils. It shall be uniformly ground to allow for a minimum of 5% by weight to pass through a US standard no. 15 sieve. It shall contain not more than 5% total ash, 1% acidic insoluble ash and 6% moisture. The mustard flavour is a white yellow powder prepared from blend of powder derived from milling the endosperm or interior portion of seed of whole mustard and ground mustard. It shall contain not more than 5% total ash, 5% acidic insoluble ash, 9% crude fibre, 10% moisture and not less than 2.5% non-volatile ether esters. The US Department of Agriculture limits the use of ground mustard in sausages, owing to its high protein content.

12.7 References

CHARALAMBOUS, G. (ed.) (1994), *Development in Food Science* – 34 series. Elsevier Science B.V., Amsterdam, The Netherlands, pp. 265–70.

FAO DATABASE: www.fao.org/FAOSTAT .

FARRELL, K.T. (1985), *Spices, Condiments and Seasonings*. AVI Publishing Co., Westport, Connecticut. pp. 150–5.

HIRASA, K. and TAKEMASA, M. (1998), *Spice Science and Technology*. Marcel Dekker Inc. New York, pp. 9–16.

KRISHNAMURTHY, K.H. (1993), *Seasoning Herbs*: Health series: Traditional Family Medicine. Books for All, New Delhi, pp. 5–29

LEUNG, A.Y. and FOSTER, S. (1996), *Encyclopedia of Common Natural Ingredients used in Food, Drugs and Cosmetics*, 2nd edition. John Wiley and Sons, Inc., New York, pp. 379–81.

MEHRA, K.L. (1968), History and ethiobotany of mustard in India. *Adv. Front. Pl. Sci.* **19**: 57.

MOJTAHEDI, H.G. SANTO, HANG, A. and WILSON, J. (1991), *J. Nematology* **23**: 2174.

PARRY, J.W. (1969), – Vol. I & II. Chemical Publishing Company, INC, New York. pp. 199–200 (Vol. I) and pp. 82–6 (Vol. 2).

PRUTHI, J.S. (1992), *Spices and Condiments*. National Book Trust of India, New Delhi, pp. 160–4.

PRUTHI, J.S. (2001), *Minor Spices and Condiments – Crop Management and Post Harvest Technology*. ICAR, New Delhi, pp. 242–54.

ROSENGARTEN, F. (1969), *The Book of Spices*. Livingston Publishing Co., Wynnewood, Pennsylvania, pp. 299–305.

SAMBAMURTHY, A.V.S.S and SUBRAMANYAM, N.S. (2000), In *Economic Botany of Crop Plants*. Asiatec Publishers Ltd, New Delhi, pp. 103–5.

SAUCER, J.D. (1993), *Historical Geography of Crop Plants – a Select Roster*. CRC Press, Boca Raton, Florida.

TAINTER. D.R. and GRENIS, A.T. (2001), In *Spices and Seasoning. A Food Technology Handbook*, II Edition, John Wiley & Sons Inc., New York, pp. 111–16.

UHL, S.R. (2000), *Handbook of Spices, Seasonings and Flavourings*. Technomic Publishing Co., Inc, Lancaster, pp. 132–6.

13
Nigella

S. K. Malhotra, National Research Centre on Seed Spices, India

13.1 Introduction and description

The genus *Nigella* contains about 20 species of annual herbs, the most popular of which is *Nigella sativa* L. It is native to the Mediterranean region through West Asia to northern India and has long been domesticated. It can be frequently found growing wild as a weed in cultivated crops. *Nigella* as black cumin is mentioned in ancient Greek, Roman and Hebrew texts as a condiment and component of herbal medicines and was reportedly introduced to Britain in 1548. It is a minor seed spice cultivated from Morocco to Northern India; in sub-Saharan Africa, particularly Niger and eastern Africa, especially Ethiopia, where it is also reportedly used as a fish poison (Jansen, 1981) and in Russia, Europe and North America. In South-East Asia, *Nigella* seeds are mainly used for medicinal purpose. *Nigella* has been used since antiquity by Asian herbalists and pharmacists and was used for culinary purposes by the Romans. The seeds of *nigella* were found in the tomb of Tutankhamun in ancient Egypt. Dioscorides, a Greek physician of the first century AD, recorded that black cumin seeds were taken to treat headaches, nasal catarrh, toothache and intestinal worms, as a diuretic and to increase breast milk. The name *Nigella* derives from the Latin *nigellus* or *niger*, meaning black. It is commonly called as black cumin and is popular by different names in different countries. It is called as black cumin or small fennel in English; *cheveux de venus, nigelle, cumin noir* or *poivrette* in French; *nigella* in Italian; *schwarzkummel* in Germany; *neguilla* or *pasinara* in Spanish; *kolongi* in Turkish; *jinten hitan* in Indonesia and Malaya; *kala zira, kalongi, krishanjirka, mangrail* and many other vernacular names in India.

Nigella occurs wild in India and has been used as a condiment from ancient times. *Nigella* is quoted as black cumin in many texts and, because of similarity in common names, may be confused with other spices of family Apiaceae, viz. *Siah Zira* (literally black cumin – *Carum carvi* L.), *Kala Zira* (literally black cumin – *Bunicum persicum* Bioss. Fedtsch syn. *Carum bulbocastum* Koch.). Botanically and structurally, *Nigella* seed is altogether different from the above seed spices and belongs to a different family. To avoid such confusion, it is most appropriate to call the spice *Nigella*. The seeds of *Nigella* have been used as spices from ancient times in India when preparing pickles, as one of the ingredients, has the properties of a preservative. India is known to be the largest producer of *Nigella* in the world. The other producing countries are Sri Lanka, Bangladesh, Nepal, Egypt, Iraq and Pakistan. In India, it is commercially cultivated in Punjab, Himachal Pradesh, Madhya Pradesh, Bihar, Jharkhand, Assam, West Bengal and Andhra Pradesh (Vijay and Malhotra, 2002). Exact information on its area, production and productivity is not available, but it is estimated to be produced in an area of about 9000 ha area, with production of about 7000–8000 tonnes in India. During the year 2000–2001, about 1960 tonnes of *Nigella* seed, valuing Rs. 1053 × 10^5, was exported from India (Selven, 2002).

Nigella sativa L. belongs to the buttercup family (Ranunculaceae) and the order Ranales. As per the conventional classification of spices, out of five types, viz. hot spices, mild spices, aromatic spices, herbs and aromatic vegetables, *Nigella* is classified as a mild spice and, on the basis of plant organs used, *Nigella* is illustrated as seed because the dried seeds are mostly used as spices.

The *Nigella* plant is an erect, herbaceous annual plant, with height ranging from 30 to 60 cm. The leaves are compound 2–3 pinnatisect cut into linear or linear-lanceolate segments, the leaves are greyish green, fine and feathery. The flowers are pale green when young and light blue when mature, becoming pale blue or white later. The flowers are bluish-white, solitary and terminal long peduncle without an involucre, beautiful because of the development of carina, five sepals, petaloid, corolla absent, stamens numerous, five carpels, partially united. The fruit is a capsule having many nectaries, generally 10, pocket-like, epicalyx present. The seeds are trigonous, black, rugulose-tubercular. The seeds are small, matt-black grains with a rough surface and an oily white interior. They are roughly triangulate, 1.5–3 mm long (Chopra, 1998; Malhotra, 2004a). *Nigella* seeds possess an aroma resembling strawberries when crushed. The seeds are similar to onion seeds. The seeds are slightly bitter and peppery with a crunchy texture. The other closely related species, *Nigella damascena* L. and *Nigella arvensis* L., are mostly used as ornamental plants and in medicines. India is the largest producer and exporter of *Nigella* seeds in the world.

13.2 Chemical structure

Nigella seeds are aromatic and contain a disagreeable odour. The composition varies with the variety, region and the age of the product. The nutritional constituents of *Nigella* seed from Europe and Ethiopia are given in Table 13.1.

The different chemical constituents present in *Nigella* seeds are :

- 0.5% volatile oil in seeds and seven main constituents (approx.) are *p*-cymene 31%, thymoquinone 25%, ethyl linoleate 9%, α-pinene 9%, ethyl hexadecanoate 3%, ethyl oleate 3% and β-pinene 2% (Weiss, 2002)
- The other chemical constituents found in *Nigella* seed are glucosides, melanthin and melanthingenin, bitter substances and a crystalline active principle nigellone, essential oils, fixed oil, resins and tenins.
- The amino acids present in dormant seeds are crystine, lysine, aspartic acid, glutamic acid, alanine and tryptophan.
- The fatty acids of the oil present are myristic, pimitic, stearic, oleic and linoleic. The component glycerides of the oil are trillinolein, oleodilinolein, dioleolinolein, palmito-oleo-linolein and stearo-oleolinolein. Glycerides of some volatile acids are present in the oil in small quantities, tannins, resins, proteins, reducing sugars, cystine, lysine, aspartic acid, leucine but asparagines are not present (Prajapati *et al.*, 2003).

The chemical structure of active principle nigellone is shown in Fig. 13.1 below (Harborne *et al.*, 1999).

Table 13.1 Nutritional constituents of *Nigella* seed (per 100 grams)

Constituents	European seed	Ethiopian seed
Moisture (g)	4	6.6
Protein (g)	22	13.8
Fat (g)	41	32.2
Carbohydrate (g)	17	–
Fibre (g)	8	16.4
Ash (g)	4.5	7.5
N (g)	–	2.2
Na (g)	0.5	–
K (g)	0.5	–
Ca (g)	0.2	0.5
P (g)	0.5	0.6
Fe (mg)	10	17
Thiamine (mg)	1.5	0.62
Niacin (mg)	6	9.5
Pyridoxine (mg)	0.7	–
Tocopherol (mg)	34	–

Source: Takruri and Damah (1993), Nergiz and Ottles (1993).

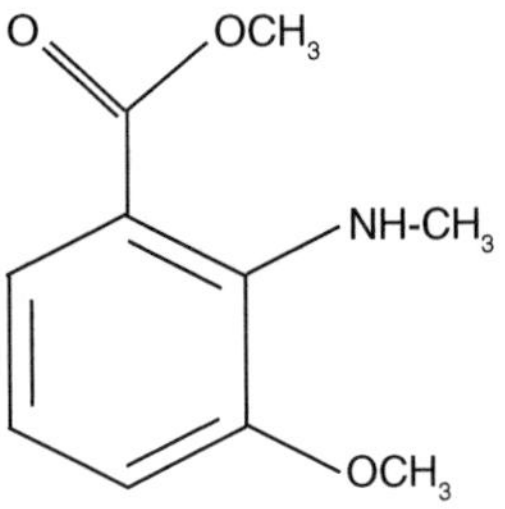

Fig. 13.1 Nigellone (chemical formula: $C_{10}H_{13}NO_3$ and molecular weight: 195.22).

13.3 Cultivation

Nigella is known to grow wild and cultivated in India, Egypt and the Middle East. It is primarily exported from India and Egypt. Plants are frost sensitive at any growth stage and this limits its range in Europe and in highland areas of the tropics. In the Northern Hemisphere, *Nigella* is sown in late spring–early summer, but in regions with wet and dry seasons, just after the first rains. Regional cultivars can be grown from sea level to 2500 m. Cultivars able to withstand considerable moisture stress have developed in North Africa and West Asia. *Nigella* is often intercropped with barley, wheat in Ethiopia and strip cropped in North Africa and elsewhere (Weiss, 2002). *Nigella* is a cool season crop, requiring a frost-free growing season and is cultivated in the northern plains, central and peninsular region of India during the winter season. Fairly warm weather during sowing with a temperature of 20–25°C is desirable. Cold weather is congenial for the early growth period and the crop requires warm sunny weather during seed formation (Pruthi, 2001; Malhotra, 2002).

Nigella can thrive on a wide range of soils, which are rich in organic matter and free from waterlogging. However, loamy, medium to heavy soils with a better fertility level are most suitable. *Nigella* is propagated by seed. Seeds are sown at row spacing of 30 cm and plant spacing of 15–20 cm and a seed rate of 8 kg/ha is required. Under Indian conditions, sowing during the month of October has been found appropriate. The ripe seeds germinate relatively quick and germination time is normally 12 days (Malhotra, 2001, 2002).

Light irrigation should be given immediately after sowing if initial moisture is low. Irrigation should be given at 5–6 day intervals initially and thereafter at 10–15 days, depending upon the weather conditions and soil type. Flowering and seed formation are two important stages requiring irrigation. The general manure and fertilizer recommendations under normal field conditions may be followed. At the time of land preparation 10–15 tonnes/ha well-rotted manure, 30 kg N, 60 kg P_2O_5 and 20 kg K_2O/ha should be applied. Another dose of 30 kg nitrogen in two portions after 40 and 60 days of sowing should be top dressed (Malhotra, 2002). No serious disease except root rot has been observed. This is caused by a *Rhizoctonia* and *Fusarium* complex. The symptoms start with yellowing and drying of leaves, resulting in premature drying of plants, which drastically reduces the yield. No perfect control measures are available but incidence can be minimized by treating seeds before sowing, deep summer ploughing and adopting proper crop rotation. Aphids have also been observed. In Ethiopia, the larvae of armyworm *Spodoptera litura* and *Cercospora* leaf spot (*Cercospora nigellae*) have been reported to cause damage to the crop. Suitable chemical controls measures are available. The *Nigella* crop takes 140–160 days to reach maturity. The crop should be harvested when the seed have attained full maturity in capsule and have turned to full black colour. Delay in harvesting may cause shattering of seeds. An average yield of 600–800 kg can be obtained from one hectare of land (Malhotra, 2003b).

13.3.1 Cultivars

The flowers of *Nigella sativa* L. are protandrous and self-fertile. Being entomophilous, cross-pollination occurs through insects. The variability for yield and quality characters are seen frequently among the cultivars and within the same cultivar. The farmers cultivate local cultivars and the available cultivars have been developed through selection from land races. As well as India and Egypt, *Nigella* is cultivated in Sri Lanka, Bangladesh, Nepal, Egypt, Iraq and Pakistan on a smaller scale. Regional cultivars are more popular in these areas. India is the largest producers of *Nigella* and farmers are reported to grow local cultivars. Recently, in India two high-yielding varieties, Azad Kalongi from Kanpur (Srivastava and Tripathi, 2000) and AN-1 from the National Research Centre on Seed Spices, have been developed (Malhotra, 2004a).

13.4 Main uses in food processing

The dried seeds are the only commercially important product and the essential oil is of minor importance. *Nigella* seeds are used in India and the Middle East as a spice and condiment and occasionally in Europe as both a pepper substitute and a spice. They are widely used in Indian cuisines, particularly in mildly braised lamb dishes such as korma. *Nigella* is also added to vegetable and *dhal* (lentil) dishes as well as in chutneys. The seeds are sprinkled on to naan bread before baking. *Nigella* is an ingredient of some garam masalas and is one of the five spices in *panch phoran*. This famous Bengali origin spice mix consists of equal parts of five spices such as cumin, fennel, mustard, fenugreek and *Nigella* seeds mixed together without roasting or grinding. In the Middle East *Nigella* is added to bread dough and is an essential constituent of the Middle East *choereg* rolls. The dried seeds of *Nigella* are the major commercial product being used in foods, pickles, baked goods, confectionery, pharmaceutical and perfumery industry. Owing to preservative qualities, the seeds of *Nigella* have been used as a spice from ancient times in the preparation of pickles, and seeds are scattered between folds of linen and wool to stop insect attack. The major processed products from *Nigella* seed are *Nigella* oil and fixed oils.

Nigella essential oil can be extracted from the crushed seeds by the steam-distillation method. The two kinds of oils, essential oil (volatile oil) and non-volatile fatty oils, are extracted. *Nigella* seed contains 0.5% volatile oil and about 31% of the fatty oil also called fixed oil. The essential oil is yellowish-brown in colour and has an unpleasant odour. The fatty oil obtained by the expression of seeds is reported to be used for edible purposes. Extraction with benzene and subsequent steam distillation of the extract to remove the volatile yields the fixed oil. *Nigella* oleoresin can also be prepared but is not popular due to its low commercial value. *Nigella* has its use in folk medicines in India and Greece, as explained below. *Nigella* seed, powder and oil are used as adjuncts for flavouring foods, as preservative in confectioneries and in the pharmaceutical industry. *Nigella* oils are used as a stabilizing agents for edible fats (Pruthi, 2001). Every Indian house uses *Nigella* seeds as a preservative in different types of homemade pickles.

13.5 Functional properties and toxicity

Nigella seed and oil are known to possess several pharmacological properties such as detergent, sedative, anti-inflammatory and expectorant. From ancient times, *Nigella*, because of its insect repellent properties, has been used as a seed spread in woollens and silk clothes to protect them from insects and used like moth balls. The presence of the carboxyl compound nigellone and non-carboxyl fractions are reported to protect guinea pigs against histamine-induced broncho-spasm and phenolic fractions obtained from seeds have been reported to be antibacterial. In Vitilago, *Nigella* powder is used as vinegar and applied on spots followed by exposure to sunlight. A decoction of seeds mixed with sesame oil is used externally in various skin eruptions. They are also used against scorpion sting. Preliminary clinical trials indicate *Nigella*'s possible therapeutic use in some conditions of cough and bronchial-asthma. Alcoholic extracts of the seeds show antibacterial activity against *Micrococcus pyogenes* var. *aureus* and *Escherichia coli* (Pruthi, 2001).

Nigella sativa L. has not shown the specific inhibitory activity against tyrosinase (Mukherjee *et al.*, 2001). The oil has microbial activity and has been investigated as antimicrobial (Minakshi and Banerjee, 1999), antiococeptive (Abdel-Fattah *et al.*, 2000) and carminative (El-Dakhakhny, 2000). *Nigella* oils have played a significant role for altering the liver damage induced by *Schistosoma mansoni* infection in mice and helped in improving the immunological host system and to some extent with its antioxidant effect (Mahmoud *et al.*, 2002). Recent studies had revealed that extract of *Nigella sativa* L. has a strong immunomodulatory and interferon-like activity (Medenica *et al.*, 2000). It inhibits cancer and endothelial cell progression, and decreases the production of the angiogenic protein fibroblastic growth factor made by tumour cells.

Nigella is used in Indian medicine as a carminative and stimulant and is used against indigestion and bowel complaints. It is also used to induce post-natal uterine contraction and to promote lactation (Barbara, 2000). *Nigella* seeds are known from ancient Greece as a remedy for headaches, toothaches and intestinal parasites. Prajapati *et al.* (2003) have reviewed *Nigella* with many medicinal properties such as thermogenic, aromatic, carminative, diuretic, emmenagogue, anodyne, antibacterial, anti-inflammatory, deodorant, appetizing, digestive, anthelmintic, constipating, sudorific, febrifuge, stimulant, galactagogue and expectorant. It is also useful in skin diseases, haemorrhoids, cephalalgia, jaundice, inflammation, fever, paralysis, ophthalmia, halitosis, anorexia, dyspepsia, flatulence, diarrhoea, dysentery, cough, amenorrhoea, dysmenorrhoea, helminthiasis especially tapeworm, strangury, intermittent fevers, agalaetia and vitiated conditions of *vāta* and *kapha* in the Indian *Ayurvedic* system of medicines.

The seeds are found to contain melanthin, a substance allied to helleborin and, like saponin, possessing emulsifying powers. The seeds are employed as a corrective of purgatives and other medicines and are believed to possess diuretic, anthelmintic and emmenagogue properties, useful in indigestion, loss of appetite, fever, diarrhoea, dropsy, puerperal diseases, etc. They have a definite action as a galactagogue and are therefore given to recently delivered women in combination with other medicines. The use of the seeds to protect clothing from insect damage is common all over India. For this purpose the seeds are mixed with powdered camphor as a preservative. The seeds have also antibilious property and are administered internally to arrest vomiting. The seeds are fried, bruised, tied in a muslin bag and smelt to give relief from cold and catarrh of the nose by constant inhalation.

Some of the native Indian medical preparations as reviewed by Nadkarni (2001) are given below:

- In intermittent fever *Nigella* seeds slightly roasted are recommended to be given in two-drachm doses with the addition of an equal quantity of treacle.
- In doses of 10–20 grains, *Nigella* seeds have a well-marked emmenagogue effect, useful in dysmenorrhoea and in large doses may induce abortion.
- In loss of appetite and distaste for food, a confection made of *Nigella* seeds, cumin seeds, black pepper, raisins, tamarind pulp, pomegranate juice and sonchal salt with treacle and honey is said to be very useful.
- In the after-pains of puerperal women, the administration of *Nigella* seeds with the addition of long-pepper, sonchal salt and wine have proved useful.
- In puerperal diseases such as fever, loss of appetite and disordered secretions after delivery, the following preparation called *pancha jiraka paka* is used. It consists of seeds of *Nigella*, cumin, anise, ajowain, carum, *Anethum sowa*, fenugreek, coriander, ginger, long pepper, long pepper root, plumbago root, habusha (an aromatic substance), dried pulp of *Ziziphus jujuba*, root of *Aplotaxis auriculate* and Kamala powder. To each 10 g, add treacle 1000 g, milk one seer (about 1 litre), butter 40 g. Boil them together and prepare a confection. Dose is about a drachm every morning.

Other uses as reviewed by Weiss (2002) are as follows:

- Crushed seeds in vinegar are applied to skin disorders such as ringworm, eczema and baldness.
- In Egypt, a tea made from powdered *Nigella* seeds fenugreek, garden cress, *Commiphora* spp. and dried leaves of *Cleome* spp., *Abrosia maritina* L. and *Centaurium pulchellum* (SW) Druce is used to treat diabetes.

The *Nigella* seed yields a volatile oil containing melanthin, nigelline, damascene and tannin. Melanthin is toxic in large dosages and nigelline is paralytic, so this spice must be used in moderation. The traditional use of *Nigella* seeds have been supported by Zaoui *et al.* (2002a) for treatment of dyslipidaemia; the hyperglycaemia and related abnormalities, however, indicate a relative toxicity of this plant. In another report, Zaoui *et al.* (2002b) have reported acute and chronic toxicity of *Nigella sativa* fixed oil. The methanol extract from related species *Nigella damascene* seeds showed a high oestrogenic activity. Among the purified phenolic compounds tested, the phenolic ester 1-o-(2,4-dihydroxy) benzolglycerol showed the strongest oestrogenic activity owing to the presence of flavonoid compounds (Agardi *et al.*, 2000). It has been said that love-in-the-mist (*Nigella damascene*) should never be used as a substitute for *Nigella sativa* (Chevallier, 2001).

13.6 Quality specifications and adulteration

13.6.1 Specification for whole seed

The quality of *Nigella* seed mainly depends on appearance: matt-black seeds, oily white interior and roughly triangulate, 1.5–3 mm long, uniformity in size, shape and texture. The odour in *Nigella* seeds when crushed resembles strawberry. Some authors have mentioned its smell is similar to oregano or carrots.

The Indian Agmark grade specifications for *Nigella* seeds with minimum specific quality indices as laid down under the Prevention of Food Adulteration Indian Act (PFA standards) for *Nigella* seeds are given below:

- Seed moisture = not more than 11% by weight.
- Total ash = not more than 6% by weight.
- Ash insoluble in acid = not more than 1% by weight.
- Organic extraneous matters = not more than 3% by weight.
- Inorganic extraneous matters = not more than 2% by weight.
- Volatile oil = not less than 1% (v/w).
- Ether extract (crude oil) = not less than 35% (v/w).
- Alcoholic acidity as oleic acid = not more than 7% (v/w).

Nigella seed has not received a place in ASTA, ESA and other ISO specification lists. It has lesser demand in European and American countries. India has been exporting to neighbouring countries, the Middle East and Gulf states and satisfies a demand from the many expatriate Asian workers.

Nigella powder is produced by grinding dried, cleaned and sterilized seed. After sieving through the required mesh size, the powder should be packed in airtight containers. The freeze grinding technique can be used to avoid the loss of flavour during heat grinding. The powder is whitish creamy in colour with an aroma like strawberry. The whole *Nigella* seed specification should be strictly followed in addition to seed powder quality specifications.

13.6.2 Volatile oil and fixed oils

The volatile oil content of *Nigella* seed averages 0.5% to 1.4% and it contains, primarily, glucosides, melanthin and melathingenin, a bitter substance and a crystalline active principle nigellone. The aroma of *Nigella* oil is warm, spicy and fatty and its flavour is strawberry-like with a burning sensation. The volatile oil of *Nigella* is yellowish-brown with an unpleasant flavour. The physiological properties of *Nigella* oil are given below (Pruthi, 2001):

- Specific gravity at 15°C = 0.875 to 0.886.
- Refractive index at 20°C = 1.4836 to 1.4844.
- Optical rotation at 20°C = +1.43 to +2.86.
- Acid value = up to 1.9.

- Ester value = 1.0 to 21.6.
- Ester value (after acetylation) = 15 to 73.
- Solubility = 2 to 4.5 or more volumes of alcohol.

The fixed oils are also extracted from *Nigella* seeds. The fatty oil obtained from seeds is used for edible purposes. Extraction with benzene and subsequent steam distillation of extract to remove volatile oil gave about 31% of reddish-brown, semi-drying oil with following characteristics:

- specific gravity at 25°C = 0.9152.
- refractive index at 21°C = 1.4662.
- acid value = 42.83.
- saponification value = 199.6.
- iodine value = 17.6.
- Reichert–Meissl value = 3.9%.
- unsaponifiable matter = 0.03%.

Nigella oleoresins can be extracted from seed but have little commercial value. Such oleoresins do not find a place in ASTA, ESA and other ISO specification lists.

13.6.3 Adulteration

Nigella seed is available both as whole or in ground form. The whole seed is subject to adulteration by onion seeds, because of their similarity with *Nigella* seeds. Onion seeds lose viability after one year and such unused seeds are adulterated with *Nigella* seeds. The exhausted seed or spent seed after oil extraction is also adulterated in whole seed or ground form. The chaff oil is also added to the essential oil extracted from seeds. The range of essential oil is 0.5–1.4% and should contain melangin as the major content, which should not go below 30%. A high ratio of eicosadienoic acid to eicosamonoenoic acid, combined with a high level of CO_2 fatty acids, is a characteristic of *Nigella* seed oils and could be used to identify genuine oil (Weiss, 2002). The adulterants can be detected through chromatographic techniques using the specifications explained here.

The quality standards as laid down under the Prevention of Food Adulteration (PFA) Act and Rules summed up to 1997 by Ministry of Health, Government of India for whole *Nigella* seeds and powder are given below (Anon., 1998).

Whole seed:
- Whole means the dried seeds of *Nigella sativa* L.
- Extraneous matter should not exceed 7% by weight.
- The edible seeds other than cumin black shall not exceed 5% by weight.
- Seed should be free from added colouring matter.

Powder:
- Powder means grinding the dried seeds of *Nigella sativa* L.
- Moisture: not more than 12% by weight.
- Total ash: not more than 7% by weight.
- Ash soluble in dilute HCL: not more than 1.5% by weight.
- Volatile oil: not less than 0.5% (v/w).
- It should be free from added colouring matter.

13.7 References

ABDEL-FATTAH, A.F. (2000), Antiococeptive effects of *Nigella sativa* oil and its major component thymoquinone in mice. *European J. Pharmacology* **400**(1): 89–97.
AGARDI, E, CILLO, F, FICO, G. AND TOME, F. (2000), Estrogenic activity of flavanoid compounds of *Nigella damascena* seeds. *Phytomedicine* **7**(Suppl II): 69.
ANON. (1998),*PFA Specifications*, upto 1997–98, DGHS, Ministry of Health, Government of India, A.05.10.01–10.
CHEVALLIER, A.(2001), *Encyclopaedia of Medicinal Plants.* Dorling Kindersley, London, p. 215.
CHOPRA , G.L. (1998), *Angiosperms.* Pradeep Publications, Jalandhar, India, pp. 55–6.
EL-DAKHAKHNY, M. (2000), Effect of *N. sativa* oil on gastric secretion and ethanol induced ulcer in rats. *J. Ethnopharmacology* **72**(1/2): 299–304.
HARBORNE, J.B., BAXTER, H. AND MOSS, G.P. (1999), *Phytochemical Dictionary – A Handbook of Bioactive Compounds.* Taylor and Francis, London, p. 351.
JANSEN, P.C.M. (1981), *Spices, Condiments and Medicinal Plants in Ethiopia, Their Taxonomy and Agricultural Significance.* Centre for Agricultural Publishing and Documentation, Wageningen, The Netherlands.
MAHMOUD, M.R., EL-ABHAN, H.S. AND SALEH, S. (2002), The effect of *N. sativa* oil against the liver damage induced by *Schistosoma mansoni* infection in mice. *J. Ethnopharmacology* **79**(1): 1–11.
MALHOTRA, S.K. (2001), Research activities. *Seed Spices Newsletter* **I**(1):1–6.
MALHOTRA, S.K. (2002),*Nigella cultivation practices* (in Hindi). NRCSS, Ajmer. Extension Folder No. 7, pp. 1–4.
MALHOTRA, S.K. (2004a), Underexploited seed spices. In *Spices, Medicinal and Aromatic Crops.* J. Singh (ed.) University Press, Hyderabad, India (in press).
MALHOTRA, S.K. (2004b), Minor seed spices 2 – Parsley, caraway, black,caraway and nigella. In *Fifty Years of Spices Research in India.* P.N. Ravindran (ed.) IISR, Calicut, India (in press).
MEDENICA, R, JANSSENS, J, TARASENKO, A., LAZOVIC, G., CORBITT, W., POWELL, D., JOCIC, D. AND MUJOVIC, V. (2000), Anti-angiogenic activity of nigella sativa plant extract in cancer therapy. *Proc. Annual Meeting Am. Assoc. Cancer Res.* **38**: A1377, 1997.
MINAKSHI, D.C. AND BANERJEE, A. (1999), Antimicrobial screening of some Indian spices. *Phytotherapy Research* **13**(7): 616–18.
MUKHERJEE, P.K., BADANI, S., WAHILE, A.M., RAJANI, S. AND SURESH, B. (2001), Evaluation of tyrosinase inhibitary activity of some Indian spices. *J. Natural Remedies* **1**(2): 125–9.
NADKARNI, K.M. (2001), *Indian Plants and Drugs with their Medicinal Properties and Uses.* Asiatic Pub. House, Delhi, India, pp. 259–60.
NERGIZ, C. AND OTTLES, S. (1993), Chemical composition of *N. sativa* seeds. *Food Chemistry* **48**: 257–61.
PRAJAPATI, N.D., PUROHIT, S.S., SHARMA, A. AND KUMAR, T. (2003), *A Handbook of Medicinal Plants.* Agribios India, Jodhpur, India, pp. 362–3.
PRUTHI, J.S. (2001), *Minor Spices and Condiments.* ICAR, New Delhi, pp. 1–782.
SRIVASTAVA, J. P. AND TRIPATHI, S.M. (2000), Breeding of seed spices – *Nigella* (*Nigella saliva* L.) In *Proc. Centennial Conference on Spices and Aromatic plants*, 20–23 September, 2000, held at IISR Calicut, p. 84.
SELVEN, T.M. (2002),*Arecanut and Spices Database.* Directorate of Arecanut and Spices Development, Calicut, Kerala, India, pp. 1–105.
TAKRURI, H.R. AND DAMAH, M.A. (1993), Study of nutritional value of black cumin seeds (*N. sativa*). *J. Science of Food and Agri.* **76**: 404–10.
VIJAY, O.P. AND MALHOTRA, S.K. (2002), Seed spices in India and world. *Seed Spices Newsletter* **2**(1): 1–4.
WEISS, E.A. (2002), *Spices Crops.* CABI Publishing, Wallingford, pp. 356–60.
ZAOUI, A., CHERRAH, Y., ALAOUI, K., MAHSSINA, N., MAROUCH, H. AND HASSAN, M. (2002a), Effect of *N. sativa* fixed oil on blood homoeostasis in rat. *J. Ethnopharmacy* **79**(1): 23–4.
ZAOUI, A., CHERRAH, Y., MAHSSINA, N., ALAOUI, K., MAROUCH, H. AND HASSAN, M. (2002b), Acute and Chronic toxicity of *N. sativa* fixed oil. *Phytomedicine* **9**(1): 69–74.

14

Oregano

S. E. Kintzios, Agricultural University of Athens, Greece

14.1. Introduction and description

Oregano is the common name for a general aroma and flavour primarily derived from a plethora of plant genera and species used all over the world as a spice, but usually refers to the genus *Origanum*, the European oregano, the name of which is derived from the Greek words *oros*, mountain and hill, and *ganos*, ornament. At least 61 species of 17 genera belonging to six families are mentioned under the name oregano. The family Lamiaceae (Labiatae) is considered to be the most important group containing the genus *Origanum* that provides the source of well-known oregano spices – Turkish and Greek types. Two genera of the Verbenaceae family (*Lanata* and *Lippia*) are used for production of oregano herbs. The other families (Rubiaceae, Scrophulariaceae, Apiaceae and Asreraceae) have a restricted importance. However, we frequently encounter the herbs of the above-mentioned families under the name of oregano in the market (Bernath, 1996).

14.1.1 Botanical characteristics

Oregano is generally considered as a perennial herb, with creeping roots, branched woody stems and opposite, petiolate and hairy leaves (Grieve, 1994). The flowers are in corymbs with reddish bracts, a two-lipped pale purple corolla and a five-toothed calyx. In moderate climates, the flowering period extends from late June to August. Each flower produces, when mature, four small seed-like structures. The foliage is dotted with small glands containing the volatile or essential oil that gives the plant its aroma and flavour (Simon *et al.*, 1984).

14.1.2 Taxonomy and geographical distribution

During the past 150 years, more than 300 scientific names have been given to fewer than 70 presently recognized *Origanum* species, subspecies, varieties and hybrids. Within the genus *Origanum*, and based on a diverse palette of morphological characters, such as length of stems, indumentum of stems and leaves, number of sessile glands on leaves, arrangement of verticillasters, arrangements, number and length of branches, Ietswaart (1980) recognized 3 groups, 10 sections, 38 species, 6 subspecies and 17 hybrids. Since then, 5 more species (Duman *et al.*, 1995; Danin and Künne, 1996; Skoula and Harborne, 2002) and one more hybrid (Duman *et al.*, 1998) have been recognized, raising the number of species to 43 and the number of hybrids to 18.

Ietwaart's three groups are classified as follows.

- Group A has two or one-lipped, rather large, calyces 4–12 mm long. Bracts are rather large 4–25 mm long, membranous, usually purple, sometimes yellowish-green, more or less glabrous.
- Group B has two or one-lipped, rather small, calyces 1.3–3.5 mm long. Bracts are rather small 1–5 mm, leaf-like in texture and colour, more or less hairy.
- Group C has calyces with five (sub)equal teeth.

The members of the genus are mainly distributed around the Mediterranean region: 35 out of 43 occur in the East Mediterranean, exclusively (Greuter *et al.*, 1986); four species are found restricted in the West Mediterranean, while three are endemic to Libya. In addition, hybrids that have been found when *Origanum* species co-occur, either in natural or in artificial conditions. Often hybrids have been considered initially as species, as in the case of *Majorana leptoclados* (*Origanum* × *minoanum*), *Origanum paniculatum* Koch. (*Origanum* × *aplii* Boros), *Amaracus lirius* Hayek (*Origanum* × *lirium* Heldreich ex Halacsy) and others (Skoula and Harborne, 2002).

14.2 Chemical structure

14.2.1 Chemical composition of *Origanum* species and their volatile oils

Although abundant chemical compounds have been isolated from oregano, the most important group, from a commercial and application point of view, refers to its volatile oils, basically composed of terpenoids. A comprehensive review of the composition of a 'standard' essential oil is given in Table 14.1. However, composition may vary significantly among different genotypes. Oregano species are rich in phenolic monoterpenoids such as carvacrol (Fig. 14.1) (and secondarily thymol, Fig. 14.2), while species rich in bicyclic monoterpenoids *cis*- and *trans*-sabinene hydrate (Fig. 14.3) are commercially designated as marjoram. It is quite easy to distinguish the difference between the pungent smell of oregano and the sweet smell of marjoram. In the first group are a number of chemically related compounds such as γ-terpinene (Fig. 14.4), *p*-cymene, thymol and carvacrol methyl ethers, thymol and carvacrol acetates; also compounds such as *p*-cymenene, *p*-cymen-8-ol, *p*-cymen-7-ol, thymoquinone and thymohydroquinone are also present. In the second group, α-thujene, sabinene, *cis*- and *trans*-sabinene hydrate acetates, *cis*- and *trans*-sabinol and sabina ketone can also be found (Skoula and Harborne, 2002).

Other chemical groups that are commonly detected in *Origanum* species are acyclic monoterpenoids such as geraniol, geranyl acetate, linalool, linalyl acetate and β-myrcene; bornane-type compounds such as camphene, camphor, borneol, and bornyl and isobornyl acetate; and sesquiterpenoids, such as β-caryophyllene, β-bisabolene, β-bourbonene, germacrene-D, bicyclogermacrene, α-humulene, α-muurolene, γ-muurolene, γ-cadinene, allo-aromadendrene, α-cubebene, α-copaene, α-cadinol, coryophyllene oxide and germacrene-D-4-ol.

Table 14.1 Comprehensive composition of oregano essential oil

Cymyl- compounds	**Sesquiterpenoids**
p-cymene	*allo*-aromadendrene
p-cymenene	β-bisabolene
p-cymen-8-ol	β-bourbonene
Carvacrol	γ-cadinene
Carvacrol acetate	α-cadinol
Carvacrol methylether	β-caryophyllene
γ-terpinene	Caryophyllene oxide
Thymol	α-copaene
Thymol acetate	β-cubonene
Thymohydroquinone	Germacrene-D
Thymoquinone	Germacrene-D-ol
	Bicyclogermacrene
Sabinyl- compounds	α-humulene
Sabinene	α-muurolene
Sabinene hydrate	γ-muurolene
cis-sabinene hydrate	
trans-sabinene hydrate	**Diterpenoids**
cis-sabinene hydrate acetate	Akhdarenol
trans-sabinene hydrate acetate	Akhdardiol
cis-sabinol	Akhdartriol
trans-sabinol	Isoakhdartriol
Sabina ketone	
Sabinyl acetate	**Triterpenoids**
Thujene	β-amyrin
	Betulic acid
Acyclic compounds	Betulin
Geraniol	Methyl-3β-21α-dihydroxyurs-12-en-28-olic acid
Geranyl acetate	
Linalool	Oleanolic acid
Linalyl acetate	Ursolic acid
β-myrcene	Uvaol
Bornyl- compounds	
Borneol	
Bornylacetate	
Camphene	
Camphor	
Isoborneol	
Isobornyl aceate	

14.2.2 Chemotaxonomy

From a chemotaxonomical point of view, the qualitative variation of the volatile compounds at the infrageneric level is quite considerable. At the infraspecific level, it has been reported that *O. vulgare* ssp. *hirtum* plants produce fewer essential oils during the cool and wet vegetative period and more during the warm and dry flowering period, and essential oil yield decreases thereafter, as leaves get older and drier. In addition, the concentration of *p*-cymene and γ-terpinene fluctuate enormously according to season (Poulose and Croteau, 1978; Skoula *et al.* unpublished data; Skoula and Harborne, 2002). The decline in total essential oil and of thymol or carvacrol, which occurs in the autumn naturally, can be mimicked by growing *O. syriacum* in short days (Putievsky *et al.*, 1996); Similarly, *O. majorana* grown

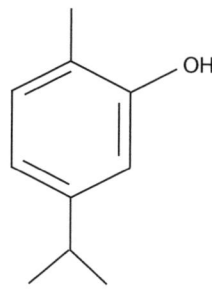

Fig. 14.1 Carvacrol.

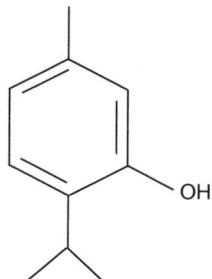

Fig. 14.2 Thymol.

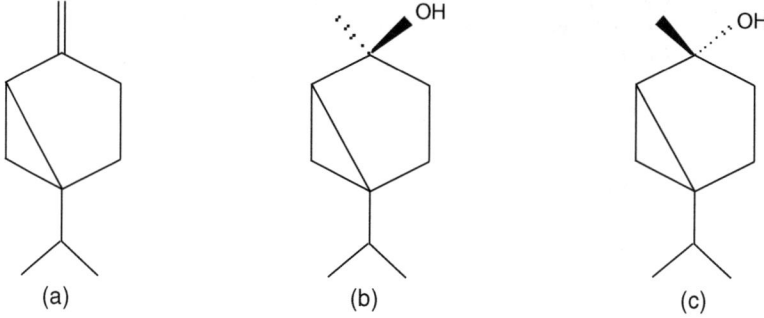

Fig. 14.3 (a) Sabinene, (b) *cis*- and (c) *trans*-sabinene hydrates.

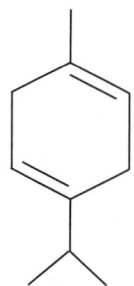

Fig. 14.4 γ-terpinene.

in controlled conditions under short days yield fewer essential oils (Circella *et al.*, 1995). Kokkini *et al.* (1997) reported high content of *p*-cymene in the essential oils of wild *O. vulgare* ssp. *hirtum* collected in autumn.

Besides the qualitative variation of the volatile compounds at the infrageneric level, there is considerable quantitative variation at the infraspecific level. Remarkable chemical variations have been observed not only between but also within populations and accessions. For example, single plant investigations of a grouping of *O. vulgare* ssp. and their offspring resulted in an unexpected differentiation into chemotypes, including one with a marjoram-like profile, but growth characteristics and winter hardiness of a *O. vulgare* ssp. (Marn *et al.*, 1999), whereas carvacrol contents ranging from traces up to 95% of the essential oil are described (Kokkini *et al.*, 1991).

14.3 Production and cultivation

14.3.1 Growth habit of wild oregano populations

As a perennial species, oregano grows spontaneously in areas across the Mediterranean region, particularly in high locations. In these areas oregano is harvested mainly from wild populations, once or twice a year, at flowering stage. The reported life zone of marjoram (*Origanum majorana* L.), is 6–28°C with an annual precipitation of 0.5–2.7 m and a soil pH of 4.9–8.7. The plant is adapted to well-drained, fertile loam soils. The cold-sensitive plant cannot survive northern climates.

Origanum vulgare L., and the subspecies *O. vulgare* subsp. *vulgare*, *O. vulgare* subsp. *viride* and *O. onites*, originate from the Mediterranean and are closely related to marjoram. They grow to a height of about 20 cm, with woody stems and dark green leaves around 2 cm long. The plants protect the inclined soils, and are quite tolerant to cold and dryness. During the winter the aerial parts are destroyed, but the roots maintain their vitality for revegetation in spring.

Oregano grows in medium soils, and in areas with high elevation and cool summer (Makri, 2002). Plants seed in warm soil in late summer and can be moved outdoors after three to four months. Oregano is best treated as an annual in cold climates where it will not over-winter well. When grown as a perennial, roots should be divided every three years for best growth and flavour. Older plants will do well as a potted plant as long as they receive sufficient sunlight. As with most herbs, desiccated plant parts should be removed as frequently as necessary (Sarlis, 1994). Commercial material of oregano (*O. vulgare*) is partially collected from wild plants even today. To avoid the disadvantages of exploiting oregano directly from the wild, efforts have been made in its domestication and cultivation. Growing wild oregano is rather easy. It grows well in shade; the cultivated subspecies *O. v. hirtum* does not.

14.3.2 Cultivation

For cultivation, marjoram is both seeded directly and transplanted into fields. Oregano has a spreading root system and is usually propagated by seed or cuttings, the latter being removed in late spring once the leaves are firm enough to prevent wilting when placed in sand (average shoot length: 30 cm). Well-rooted cuttings are placed in the ground about 30 cm apart or planted outside in pots. If seeds are used, they should be sown in a seedbox in spring and planted outside when seedlings are 7.5 cm tall. Old wood that becomes leggy should be cut out at the end of winter and plants should be replaced every four years or so to prevent legginess. Pungency declines in rich soils, and after flowering.

Ploughing of the soil and fertilization with ammonium phosphate during November to December is sufficient for oregano cultivation; under normal conditions, pest control can be reduced to a simple weeding out (manually or by using pesticides) (Kintzios, 2002a; Makri, 2002) although aphids, thrips and red spider mites may occasionally present a problem (Csizinszky, 1992). In addition, *O. majorana* can be severely affected by *Alternaria* and *Fusarium*. There is scarce documentation on biological pest control in oregano, which needs frequent (e.g. at least four times a year) mechanical weed control (Chiapparo, 1997; Hammer and Junghanns, 1997). The lifespan of oregano is about five or six years and usually one harvest is done in the first year and two in the following years. On average, the yield ranges from 2.5 to 3.5 t/ha and the essential oil yield ranges from 0.5 to 1.5% of dry weight (Bernath, 1996).

Cultivation practices may differ from one country to another: for example, Hungarian and German farmers prefer to establish oregano plantations by means of seed propagation (Hammer and Junghanns, 1997; Bernath, 1996), whereas their colleagues in the Mediterranean Basin, Slovenia and the Federal Republic of Yugoslavia prefer to use stem cuttings (Macko and Cok, 1989; Baricevic 1996; Putievsky *et al.*, 1996). The percentage of seed germination does not exceed 75% and declines rapidly with time. Germination occurs over a relatively low temperature range, with an optimum temperature around 15–20°C (Kozlowski and Szczyglewska, 1994; Thanos *et al.*, 1995). Seedlings are usually planted with a spacing of 50–60 cm between rows and 20–25 cm within rows (spacing within rows may reach 40–45 cm in dry areas), therefore allowing for a plant density of approximately 3000 plants/ha. In humid areas, however, or under conditions of frequent irrigation, plant densities up to 63 000 plants/ha have been reported. Irrigation is required only at the time of planting and a few other times in the first year. In the following years, plants have developed an efficient root system and thus no further irrigation is usually needed.

14.3.3 Harvest

Depending on irrigation frequency and, subsequently, yield, two or three harvests of the crop are allowed annually. Harvesting the leaves and stem tips should start when plants are at the flowering stage, beginning 10 cm from the ground. In dry climates, the best harvest time to collect the highest amount of essential oil is when 50% of the plants in the field have started flowering. In relatively small fields, harvest is usually done manually, mechanical harvesting being recommended only for large fields.

Harvesting is generally accomplished at full bloom. Plant material is often dried in drying sheets to avoid direct sunlight and thus preserve the green colour and aroma (Sarlis, 1994; Makri, 2002).

After harvesting, plants are dried in the shade. Although drying under natural conditions is a common procedure, drying ovens operating at 30–35°C can also been used in commercial-scale production. Moisture content of 7% (min.) to 12% (max.) is required (Kitiki, 1996). Leaves should be dried in a warm, dry, shaded place, and stored in an airtight container.

Pääkkönen *et al.* (1990) studied the effect of different drying and storage methods on oregano and marjoram. Marjoram was harvested at the time of bud formation and oregano when in bloom. Herbs were either dried immediately after harvesting or were frozen and stored at –20°C and freeze-dried within two weeks. For convection drying temperature was 35–37°C and for freeze-drying 30°C. The corresponding drying times were 24 and 12 hours. The moisture content of fresh herbs was 85% for marjoram and 75% for oregano. After drying with heated air, the moisture content for marjoram was 9%, and 7% for oregano, but only 5% after freeze-drying. The drying method did not affect the water-holding capacity of the dried product, neither did it have any effect on either odour or the taste of dried oregano. A detrimental effect of elevated storage temperature was obvious. In another study (Hälvä, 1987), the concentration of volatile oils in oregano decreased from 2.55% to 1.94% in the drying process.

14.3.4 Breeding

As already mentioned in Section 14.2.2, there is yet limited knowledge in the biosynthesis of the essential oil compounds and their inheritance, which would be useful for a more effective selection and establishing a targeted breeding programme, and only some key enzymes have been identified so far for carvacrol, thymol and linalool synthesis in *Origanum* (Croteau and Karp, 1991; Franz and Novak, 2002). For all these reasons, wild collection accompanied with quality and species maintaining assurance systems (sustainability, good horticultural practice) and/or field production of reliable genotypes are the future methods of choice for quality products. The enormous inter- and infraspecific chemical polymorphism of *Oregano* sp. offers a wide range for selection towards the production of specific monoterpenes as fine chemicals, new odour and flavour profiles, etc.

Crop improvement is highly recommended considering oregano's widespread use and the great difficulties that non-uniform material may cause to the commercial sector. Taking into consideration both producers' and users' needs, efforts of any oregano breeding programme should be directed to the improvement of the following targets: yield-related parameters, e.g. growth habit, leaf/stem ratio, stress (salt, cold) tolerance, resistance to diseases and quality-related parameters, e.g. better aromatic characteristics, colour (green is preferred to grey), essential oil content (usually more than 2%) and composition, antioxidant and antimicrobial properties. In particular, and as far as the composition of essential oils is concerned, a high carvacrol or *cis*-sabinene-hydrate content is desired in oregano or marjoram, respectively. Among agronomical traits, yield is one of the most important parameters securing the necessary productivity for competing on the market. The variation between single plants can range between approx. 10 g dried leaf/flower-fraction per plant up to 250 g (Marn *et al.*, 1999). This progress in yield can also be obtained within a relatively short time of breeding.

To achieve these goals, selection and hybridization methods, combined with analytical controls on the variability encountered in the material, are the most appropriate tools for crop improvement. Local strains of *Origanum vulgare* subspecies and *O. majorana* (*Majorana hortensis*), as well as spontaneous hybrids (*Origanum* × *majoricum*, *Origanum* × *intercedens*), are traditionally cultivated in many countries. In addition, several ornamental varieties are also present on the market. Breeding of oregano started in relatively recent times. Breeding work has focused mainly on *O. majorana*, *O. syriacum*, *O. virens*, *O. vulgare* subsp. *hirtum* and some hybrids, by using chemotaxonomy results and male sterility as tools for controlled crossings (Kheyr-Pour, 1981). In this context it is worth mentioning that oregano belongs to the species with the smallest fruits, weighing only approx. 60 µg per seed (thousand seed mass = 0.06 g) (Thanos *et al.*, 1995). Because of this, direct sowing of oregano is difficult and up to now planting has been preferred (Franz and Novak, 2002). Artificial pollination is also difficult because of the small flower size and the high number of flowers within an inflorescence. Selecting for higher seed weight will be the first step to enhancing the production technique of direct sowing, since seed quality, germinability and vigour depend on it.

Systematic breeding programmes in several, mostly Mediterranean, countries are using indigenous wild species as starting material and have already produced promising results (Franz, 1990; Franz and Novak, 2002). Although few studies have yet been reported on the tissue culture of oregano (Matsubara *et al.*, 1996; Baricevic *et al.*, 1997), such studies enable biotechnological methods to be used for the enhancement of breeding activities (e.g. *in vitro* selection for disease resistance, exploitation of somaclonal variation) (for a review see Kintzios, 2002b).

14.4 Main uses in food processing and medicine

14.4.1 General

Oregano is used in meat, sausages, salads, stewings, dressings and soups. The food industry uses oregano oil and oregano resin both in foods and in beverages and also in cosmetics. Oregano oil is used in alcoholic beverages, baked goods, meats and meat products, condiments and relishes, milk products, processed vegetables, snack foods, and fats and oils. It is the most common spice for pizza. Along with black pepper, it is a common ingredient of dressings and a good substitute for table salt. Marjoram, too, is used in many foods and beverages in food industry; meat sauces, canned foods, vinegar, vermouths and bitters are often seasoned with marjoram. It increases aroma in such vegetable dishes as pea soup and other pea dishes, squash and stews made from mixed vegetables, mushrooms and asparagus.

14.4.2 Dietary value

The dietary value of oregano is quite high: it contains significant amounts of vitamins E, B_6, riboflavin, niacin, folate, pantothenate and biotin (Holland *et al.*, 1991). Relatively high values (expressed as mg/100 g fresh leaves) have also been reported for vitamin C (45), thiamin (0.07) and carotene (0.81). Lagouri and Boskou (1996) detected α-, β-, γ- and δ-tocopherol in a non-polar fraction of oregano extracts, with the γ-tocopherol content being significantly higher than other tocopherol homologues. Oregano is also rich in mineral elements such as potassium, calcium, magnesium, phosphorus, zinc, manganese, iron, copper, sulphur, chlorine, iodine and selenium, whereas its sodium content is low. However, Brune *et al.* (1989) reported that oregano inhibits iron absorption and the effect is caused by its galloyl substances and the inhibition is in proportion to its content of galloyl groups. Oregano also has a relatively modest energy and fat content (66 kcal/100 g and 2 g fat/100 g, respectively). According to Gray *et al.* (1997), the concentration of oregano in food can increase or reduce its palatability and intake compared with an unseasoned control food.

14.4.3 Food-preserving properties

Apart from its dietary value, oregano is an effective antioxidant additive in different types of foods, such as mayonnaise and French dressing (Chipault *et al.*, 1956; Nakatani and Kikuzaki, 1987; Baratta *et al.*, 1998). This property is usually attributed to the high carvacrol content of the spice (Tsimidou and Boskou, 1994), although additional compounds, such as flavonoids may also be responsible (Vekiari *et al.*, 1993).

14.4.4 Medicinal uses

There are various reports on the traditional medicinal uses European oregano has as a carminative, diaphoretic, expectorant, emmenagogue, stimulant, stomachic and tonic. In addition, it has been used as a folk remedy against colic, coughs, headaches, nervousness, toothaches and irregular menstrual cycles. Turkish villagers have traditionally used *kekik* water, the aromatic water obtained after removing essential oil from the distillate of oregano herbs, which has in recent years become a commercial commodity (Baser, 2002; Kintzios, 2002a). Although the monograph documentation of *O. vulgare* was submitted to the German Ministry of Health, the staff responsible for phytotherapeutic medicinal domain – Commission E – evaluated *Origani vulgaris herba* negatively (Banz. No. 122 from 6th July 1988), because of lack of scientific proof for a number of indication areas (Blumenthal, 1998). Nevertheless, many of the studies confirmed benefits of oregano for human health and its use for the treatment of a vast list of ailments, including respiratory tract disorders such as cough or bronchial catarrh (as expectorant and spasmolitic agent), in gastrointestinal disorders (as choleretic, digestive, eupeptic and spasmolitic agent), as an oral antiseptic, in urinary tract disorders (as diuretic and antiseptic) and in dermatological affections (alleviation of itching, healing crusts, insect stings), viral infections and even cancer (for a detailed review, see Baricevic and Bartol, 2002).

14.4.5 Microbiological quality and safety considerations

Although oregano can cause aversion symptoms during pregnancy (Hook, 1980), its consumption is considered safe from the chemical point of view. However, considerations have been frequently raised on the microbiological quality of preserved oregano. For example, Mäkinen *et al.* (1986) and Malmsten *et al.* (1991) tested the microbiological quality of marjoram and of oregano and detected moulds and aerobic spore-formers, especially *Bacillus cereus*, in most samples (although at concentrations not high enough to cause food poisoning). Coliforms and faecal streptococci were found in both freeze-dried and air-dried samples, but only sporadically and at very low counts. Moulds and yeasts were found in almost all samples, while increasing the storage time from one year to two increased tenfold the number of aerobic spore-formers in freeze-dried and in air-dried oregano. However, as demonstrated below, microbial contamination of oregano is not a common source of concern, owing to the antimicrobiological properties of the herb.

14.5 Functional properties

14.5.1 Antioxidant properties

Oregano extracts have documented antioxidant and antimicrobial properties (Dorofeev *et al.*, 1989; Mirovich *et al.*, 1989; Deighton *et al.*, 1993), which have been presumably attributed to phenylcarboxylic acids, such as cinnamic, caffeic, *p*-hydroxybenzoic, syringic, protocatecholic and vanillic acids. Dietary supplies of antioxidants from *Origanum* species have been considered as effective scavengers of the free radicals that are generated by metabolic pathways in the body; however, limited industrial applications are often ascribed to the characteristic oregano aroma and flavour that influence the sensorial characteristics of processed food, so deodorization steps would be required (Nguyen *et al.*, 1991; Moure *et al.*, 2001).

Taking these limitations into consideration, practical considerations on the use of oregano as stabilizers of edible oils or of finished meat products have been made by several research groups (Baricevic and Bartol, 2002). Dry leaves of *Origanum vulgare* ssp. *hirtum* showed a high antioxidant activity in olive oil and, besides their stabilizing effect, the organoleptic quality of the olive oil was significantly improved by addition of oregano, as assessed by Mediterranean consumer acceptability studies (Antoun and Tsimidou, 1997; Charai *et al.*, 1999). A significant increase in the oxidative stability of fried chips, measured as the rate of peroxide formation during storage at 63°C, was achieved both by addition of ground oregano or its petroleum ether extracts (Lolos *et al.*, 1999). In contrast with the significant antioxidative and stabilizing effects of oregano extracts in lard and various oils, no effect on the quality or shelf-life of the fat obtained from animals fed with oregano additives or of meat and fat-containing food was observed (Vichi *et al.*, 2001).

14.5.2 Antimicrobial properties
In conjunction with the antioxidant properties of the herb, there are abundant reports on the microbial inhibitory effects of oregano essential oil or its components. These effects are generally classified either as antifungal or antibacterial. According to general consensus, there is a relationship between the chemical structure of the most abundant essential oil components and their antifungal and anti-aflatoxigenic potency, which is, in addition, strongly correlated with the concentration of the essential oil or active ingredient and pH of the testing medium *in vitro* (Deans and Svoboda, 1990; Thompson, 1990; Biondi *et al.*, 1993; Baricevic and Bartol, 2002). Phenols are believed to be the most potent antimicrobials, followed by alcohols, ketones, ethers and hydrocarbons (Bullerman *et al.*, 1977; Hitokoto *et al.*, 1980; Hussein, 1990; Daw *et al.*, 1994; Charai *et al.*, 1996). In more practical terms, ground oregano (at 2% concentration) was found to possess a strong antifungal potential against several food-contaminating moulds, such as *Alternaria alternata* Keissler, *Fusarium oxysporum* Schlecht, *Penicillium citrinum*, *P. roqueforti*, *P. patulum*, *Aspergillus flavus* and *A. parasiticus* (Azzouz and Bullerman, 1982; Schmitz *et al.*, 1993).

Phenolic compounds are probably responsible for the high inhibitory activity of carvacrol/thymol chemotypes of oregano against fungal growth, conidial germination and production of *Penicillium* species, such as *P. digitatum* (Daferera *et al.*, 2000). In particular, monoterpene components seem to have more than an additive effect in fungal inhibition. Phenolic derivatives, present in essential oils, may also be involved in inhibition of yeast sporulation through depletion of cellular energy by reduction of respiration (Baricevic and Bartol, 2002). Curtis *et al.* (1996) reported that carvacrol or thymol, when applied in concentrations of more than 100 ppm led to a complete inhibition of fungal growth *in vitro*.

Although the antibacterial properties of oregano extracts are far less documented, Hammer *et al.* (1999) found that *O. vulgare* (Australian origin) yielded one of the most potent antibacterial agents among 52 investigated essential oils, which considerably inhibited the growth of all tested microorganisms. Other reports (Biondi *et al.*, 1993; Izzo *et al.*, 1995) demonstrated the inhibitory effects of oregano extracts against a number of Gram-positive (such as *Staphylococcus aureus* and *Bacillus subtilis*) and Gram-negative bacteria (such as *Proteus vulgaris* and *Escherichia coli*). These activities have been mainly attributed to thymol and carvacrol. However, as also shown for the antifungal properties of the species, it seems more appropriate to combine the antimicrobial efficacy of different food-preservative compounds, creating synergistic effects, such as those reported by Pol and Smid (1999) for carvacrol and nisin (a bactericidal peptide, used as a biopreservative in certain foods) against *Bacillus cereus* and *Listeria monocytogenes in vitro*.

The antimicrobial value of oregano may exceed its scope of applications beyond the food industry: a therapeutic potency of essential oil of *Origanum vulgare* L. subsp. *hirtum* against experimentally induced dermatophytosis in rats (infection with *Trichophyton rubrum*) was found by Adam *et al.* (1998). Other studies demonstrated the promising applications of oregano, its essential oil or isolated compounds, in plant protection, in post-harvest crop/fruit protection or in apiculture, where species-specific fungi endanger the production systems (for an extensive and detailed review, see Baricevic and Bartol, 2002).

14.6 Quality specifications and commercial issues

Although oregano has been known and used for centuries, it has only lately gained mass popularity, largely because of its relationship with marjoram (*O. marjorana*), the popular and botanical terms for both species having long been confused. While sweet marjoram was one of the most popular herbs during the Middle Ages, oregano was scarcely cultivated, probably because of the plant's tendency to compete against other plants growing nearby. On the other hand, wild oregano has been traditionally collected in Mediterranean countries and in Mexico for use in many of the favourite dishes (e.g. for tomato-based sauces, lamb, seafood, chilli peppers and almost any garlic flavoured dish). The rest of the world discovered oregano after World War II, with the expansion of pizza consumption (and to a lesser-degree, Mexican-style foods). Oregano consumption boomed from almost nil to a consumption volume of over 500 000 tonnes, demonstrating a per capita increase of importation into the USA of 3800% from 1940 to 1985 (Kintzios, 2002a). The European Union imported more than 1000 tonnes of oregano in 1999 (Tsagadopoulos, 2002).

Product prices depend heavily on quality. The overall market of oregano is expanding, and oregano is by far the biggest-selling herb today. Latest estimates put worldwide production at about 10 000 tonnes. Turkey has a dominant position in the worldwide trade of oregano (over two-thirds of the total production, with 3392 tonnes exported to the USA in 1995), followed by Mexico, Greece and other Mediterranean countries. Greece has long been a leading source and its product has traditionally commanded the highest prices; nevertheless it has not always met demand. Though Italy harvests large amounts of oregano, most of it is consumed domestically. The Mediterranean-type of product, as compared with the Mexican, is a smaller leaf of somewhat lighter green colour and milder, sweeter flavour. Compared with sweet marjoram, however, it is much stronger flavoured. The harvesting and processing of oregano are similar in Mediterranean and Mexican areas. It is generally accepted that the Greek oregano has the best essential oil quality, the main constituents of which are carvacrol (the compound responsible for characterizing a plant as of the oregano type) and/or thymol, accompanied by *p*-cymene and γ-terpinene. Mexican oregano oil contains approximately equal amounts of carvacrol and thymol and smaller amounts of 1,8-cineole and other compounds.

The herb is often sold by mesh size, indicating average particle size. In the USA, oregano imports are roughly equal from both Mediterranean and Mexican species. Mexican oregano is a much stronger, more robustly, 'wild' flavoured oregano. After cleaning, the leaves of Mediterranean oregano come into a size of 30 or 60 mesh, with larger leaf particles giving the choicest, more refined appearance. In Mexico, shippers often refer to their most refined product as 'Greek cut'. In the USA the herb is offered as ground or whole leaf oregano (although not always in the original whole form). Beyond that, various mesh sizes may also be available, each being the most appropriate choice for a particular use. Other important species collected and marketed as European oregano include *Thymus capitatus* (Spanish oregano), *Origanum syriacum* (*Origanum maru* Syrian marjoram or *zatar*) and *Origanum virens*. Additional species used in Mexico oregano include *Lippia palmeri* and *Lippia origanoides*.

The original fresh material is the essential factor determining the quality of the dried herbs. Nevertheless, the drying method, type of packaging and storage conditions also have clear effects on the microbiological quality of the herbs. Blending of oregano with substitute spices is very popular, in particular when high essential oil concentrations (> 3%) are desired. Quality evaluation is usually based on colour of the traded spice, while in several instances sensory (organoleptic) tests are carried out. Quality criteria also include the relative contribution of leaves to the dried product, since they contain a large number of glandular hairs. Leaves should be uniform and have a relative moisture content of less than 15%. The essential oil concentration should not be lower than 0.5% (w/w). Products should be free of impurities, in particular biologically unsafe components such as insects, animal hair and excretions. Recent, optical and electron microscopy has been applied for the quality assessment of commercial oregano in Greece (Tsagadopoulos, 2002).

Growers currently enjoy increased market prices owing to the limited product availability, as a result of the exhaustion of wild oregano populations due to intensive collection. A recent survey in Greece (Papanagiotou *et al.*, 2001) indicated that, for a given average yield of 1850 kg per hectare and an average product price of 4.1 euro per kg, the net profit for the grower is 2500 euro per hectare, a value considerably higher then for most crop and horticultural species. Labour (1260 man–hours/hectare) was estimated to reach 64% of the total production cost.

14.7 References

ADAM K., SIVROPOULOU A., KOKKINI S., LANARAS T. and ARSENAKIS M. (1998), 'Antifungal activities of *Origanum vulgare* subsp. *hirtum*, *Mentha spicata*, *Lavandula angustifolia*, and *Salvia fruticosa* essential oils against human pathogenic fungi,' *J. Agric. Food Chem.*, **46**(5), 1739–45.

ANTOUN N. and TSIMIDOU M. (1997), 'Gourmet olive oils: stability and consumer acceptability studies', *Food Res. Int.*, **30**(2), 131–6.

AZZOUZ M.A. and BULLERMAN L.B. (1982), 'Comparative antimycotic effects of selected herbs, spices, plant components and commercial antifungal agents,' *J. Food Protection*, **45**(14), 1298–1301.

BARATTA M.T., DORMAN H.J.D., DEANS S.G., BIONDI D.M. and RUBERTO G. (1998), 'Chemical composition, antimicrobial and antioxidative activity of laurel, sage, rosemary, oregano and coriander essential oils,' *J. Ess. Oil Res.*, **10**(6), 618–27.

BARICEVIC D. (1996), 'Experiences with oregano (*Origanum* spp) in Slovenia,' *Proceedings of the IPGRI International Workshop on Oregano*, CIHEAM, Italy.

BARICEVIC D. and BARTOL T. (2002), 'The biological/pharmacological activity of the oregano genus' in Kintzios S., *Medicinal and Aromatic Plants – Industrial profiles – Oregano: The Genera Origanum and Lippia*, London, Taylor & Francis, 177–214.

BARICEVIC D., ZUPANCIC A., ERZEN-VODENIK M. and SELISKAR A. (1997), '*In situ* and *ex situ* conservation of natural resources of medicinal and aromatic plants in Slovenia,' *Sjemenarstvo*, **14**, 23–9.

BASER K.H.C. (2002), 'The Turkish *Origanum* species' in Kintzios S., *Medicinal and Aromatic Plants – Industrial profiles – Oregano: The Genera Origanum and Lippia*, London, Taylor & Francis, 109–26.

BERNATH J. (1996), 'Some scientific and practical aspects of production and utilisation of oregano in central Europe,' *Proceedings of the IPGRI International Workshop on Oregano* CIHEAM, Italy.

BIONDI D., CIANCI P., GERACI C., RUBERTO G. and PIATTELLI M. (1993), 'Antimicrobial activity and chemical composition of essential oils from Sicilian aromatic plants,' *Flavour Fragrance J.*, **8**(6), 331–7.

BLUMENTHAL M. (1998), 'The Complete German Commission E Monographs' in Blumenthal M., Busse W.R., Goldberg A. and Gruenwald J., *Therapeutic Guide to Herbal Medicines*, Austin, American Botanical Council, 358–9.

BRUNE M., ROSSANDER L. and HALLBERG L. (1989), 'Iron absorption and phenolic compounds: importance of different phenolic structures,' *Eur. J. Clin. Nutr.*, **43**(8), 547–57.
CHARAI M., MOSADDAK M. and FAID M. (1996), 'Chemical composition and antimicrobial activities of two aromatic plants: *Origanum majorana* L and *O compactum* Benth,' *J. Essential Oil Res.*, **8**(6), 657–64.
CHARAI M., FAID M. and CHAOUCH A. (1999), 'Essential oils from aromatic plants (*Thymus broussonetti* Boiss, *Origanum compactum* Benth, and *Citrus limon* (L) NL Burm) as natural antioxidants for olive oil,' *J. Essential Oil Res.*, **11**(4), 517–21.
CHIAPPARO D. (1997), 'Cultivating oregano in Italy: The case of "Bioagricola A. Bosco" a Sicilian firm,' *Proceedings of the IPGRI International Workshop on Oregano*, CIHEAM, Italy.
CHIPAULT J.R., MIZUNO G.R. and LUNDBERG W.O. (1956), 'Antioxidant and antimicrobial constituents of herbs and spices,' *Food Technol.*, **10**, 209–11.
CIRCELLA G., FRANZ C., NOVAK J. and RESCH H. (1995), 'Influence of day length and leaf insertion on the composition of marjoram essential oil,' *Flavour Fragrance J.*, **10**, 371–4.
CROTEAU R. and KARP F. (1991), 'Origin of natural odorants' in Müller P.M. and Lamparsky D., *Perfumes – Art, Science and Technology*, London, Elsevier, 32–56.
CSIZINSZKY A.A. (1992), 'The potential for aromatic plant production with plastic mulch culture in Florida,' *First World Congress on Medicinal and Aromatic Plants for Human Welfare (WOCMAP)*, Netherlands.
CURTIS O.F., SHETTY K., CASSAGNOL G. and PELEG M. (1996), 'Comparison of the inhibitory and lethal effects of synthetic versions of plant metabolites (anethole, carvacrol, eugenol and thymol) on food spoilage yeast (*Debaromyces hansenii*),' *Food Biotechnol.*, **10**(1), 55–73.
DAFERERA D.J., ZIOGAS B.N. and POLISSIOU M.G. (2000), 'GC-MS analysis of essential oils from some Greek aromatic plants and their fungitoxicity on *Penicillium digitatum*,' *J. Agric. Food Chem.*, **48**(6), 2576–81.
DANIN A. and KÜNNE I. (1996), '*Origanum jordanicum (Labiatae)*, a new species from Jordan, and notes on the other species of sect Campanulaticalyx,' *Willdenowia*, **25**, 601–11.
DAW Z.Y., EL-BAROTY G.E. and EBTESAM A.M. (1994), 'Inhibition of *Aspergillus parasiticus* growth and aflatoxin production by some essential oils,' *Chem. Mikrobiol. Technol. Lebensm.* **16** (5/6), 129–35.
DEANS S.G. and SVOBODA K.P. (1990), 'The antimicrobial properties of marjoram (*Origanum majorana* L) volatile oil,' *Flavour Fragrance J.*, **5**(3), 187–90.
DEIGHTON N., GLIDEWELL S.M., DEANS S.G. and GOODMAN B.A. (1993), 'Identification by EPR spectroscopy of carvacrol and thymol as the major sources of free radicals in the oxidation of plant essential oils,' *J. Sci. Food Agric.*, **63**, 221–5.
DOROFEEV A.N., KHORT T.P., RUSINA I.F. and KHMEL'NITSKII Y.V. (1989), 'Search for antioxidants of plant origin and prospects of their use,' *Sbornik Nauchnykh Trudov Gosudarstvennyi Nikitskii Botanicheskii Sad*, **109**, 42–53.
DUMAN H., AYTEC Z., EKICI M., KARAVELIOGULLARI E.A., DONMEZ A. and DURAN A. (1995), 'Three new species (Labiatae) from Turkey,' *Flora Mediterranea*, **5**, 221–8.
DUMAN H., BASER K.H.C. and AYTEC Z. (1998), 'Two new species and a new hybrid from Anatolia,' *Tr. J. Botany*, **22**, 51–5.
FRANZ C. (1990), 'Sensorial versus analytical quality of marjoram,' *Herba-Hungarica*, **29**(3), 79–86.
FRANZ C. and NOVAK J. (2002), 'Breeding of oregano' in Kintzios S., *Medicinal and Aromatic Plants – Industrial profiles – Oregano: The Genera* Origanum *and* Lippia, London, Taylor & Francis, 163–74.
GRAY R.W., MITCHELL C.J., TRUE S. and YEOMANS M.R. (1997), 'Independent effects of palatability and within-meal pauses on intake and appetite ratings in human volunteers,' *Appetite*, **29**, 61–76.
GREUTER W., BURDET H.M. and LONG G. (1986), *Med-Checklist*, Vol 3, Editions de Conservatoire de Jardin Botaniques de la Ville de Geneve, Switzerland.
GRIEVE M. (1994), *A Modern Herbal*, London, Tiger, 519–21.
HÄLVÄ S. (1987), 'Studies on production techniques of some herb plants. I. Effect on Agryl P 17 mulching on herb yield and volatile oils of basil (*Ocimum basilicum L*) and marjoram (*Origanum majorana L*),' *J. Agric. Sci. Finland*, **59**(1), 31–6.
HAMMER K. and JUNGHANNS W. (1997), '*Origanum majorana* L – some experiences from Eastern Germany,' *Proceedings of the IPGRI International Workshop on Oregano*, CIHEAM, Italy.
HAMMER K.A., CARSON C.F. and RILEY T.V. (1999), 'Antimicrobial activity of essential oils and other plant extracts,' *J. Appl. Microbiol.*, **86**(6), 985–90.
HITOKOTO H., MOROZUMI S., WAUKE T., SAKAI S. and KURATA H. (1980), 'Inhibitory effects of spices on growth and toxin production of toxigenic fungi,' *Appl. Environ. Microbiol.*, **39**, 818–22.

HOLLAND B., UNWIN I. and BUSS D. (1991) *The Composition of Foods: Vegetables, Herbs and Spices,* Bath, McCance & Widdowson's.

HOOK E.B. (1980), 'Influence of pregnancy on dietary selection' *Intern. J. Obesity,* **4**(4), 338–40.

HUSSEIN A.S.M. (1990), 'Antibacterial and antifungal activities of some Libyan aromatic plants,' *Planta Medica,* **56**, 644–5.

IETSWAART J.H. (1980), *A Taxonomic Revision of the Genus* Origanum *(Labiatae),* The Hague, Leiden University Press.

IZZO A.A., CARLO G., BISCARDI D., FUSCO R., MASCOLO N., BORRELLI F., CAPASSO F., FASULO M.P. and AUTORE G. (1995), 'Biological screening of Italian medicinal plants for antibacterial activity,' *Phytotherapy Res.,* **9**(4), 281–6.

KHEYR-POUR A. (1981), 'Wide nucleo-cytoplasmic polymorphism for male sterility in *Origanum vulgare* L,' *J. Hered.,* **72**, 45–51.

KINTZIOS S. (2002a), 'Profile of the multifaceted prince of the herbs' in Kintzios S., *Medicinal and Aromatic Plants – Industrial profiles – Oregano: The Genera* Origanum *and* Lippia, London, Taylor & Francis, 3–8.

KINTZIOS S. (2002b), 'The biotechnology of oregano' in Kintzios S., *Medicinal and Aromatic Plants – Industrial profiles – Oregano: The Genera* Origanum *and* Lippia, London, Taylor & Francis, 237–42.

KITIKI A. (1996), 'Status of cultivation and use of oregano in Turkey,' *Proceedings of the IPGRI International Workshop on Oregano,* CIHEAM, Italy.

KOKKINI S., VOKOU D. and KAROUSOU R. (1991), 'Morphological and chemical variation of *Origanum vulgare* L in Greece,' *Botanica Chronica,* **10**, 337–46.

KOKKINI S., KAROUSOU R., DARDIOTI A., KRIGAS N. and LANARAS T. (1997), 'Autumn essential oils of Greek oregano,' *Phytochemistry,* **44**(5), 883–6.

KOZLOWSKI D. and SZCZYGLEWSKA D. (1994), 'Biology of germination of medicinal plant seeds Vd *Origanum vulgare* L seeds,' *Herba Polonica,* **40**(3), 79–82.

LAGOURI V. and BOSKOU D. (1996), 'Nutrient antioxidants in oregano,' *Int. J. Food Sci. Nutr.,* **47**(6), 493–7.

LOLOS M., OREOPOULOU V. and TZIA C. (1999), 'Oxidative stability of potato chips: effect of frying oil type, temperature and antioxidants,' *J. Food Agr.,* **79**(11), 1524–8.

MACKO V.H. and COK H. (1989), 'Efficiency and selection of herbicides in oregano (*Origanum heracleoticum* L),' *Proceedings of Yugoslav Conference on the Application of Pesticides,* Yugoslavia.

MÄKINEN S., HÄLVÄ S., PÄÄKKÖNEN K., HUOPALAHTI R., HIRVI T., OLLILA P., NYKÄNEN I. and NYKÄNEN L. (1986), *Maustekasvitutkimus,* The Academy of Finland, Report SA 01/813, Helsinki, Finland.

MAKRI O. (2002), 'Cultivation of oregano' in Kintzios S., *Medicinal and Aromatic Plants – Industrial profiles – Oregano: The Genera* Origanum *and* Lippia, London, Taylor & Francis, 153–62.

MALMSTEN T., PÄÄKKÖNEN K. and HYVÖNEN L. (1991), 'Packaging and storage effects on microbiological quality of dried herbs,' *J. Food Sci.,* **56**(3), 873–5.

MARN M., NOVAK J. and FRANZ CH. (1999), 'Evaluierung von Nachkommenschaften von *Origanum vulgare*,' *Z. Arzn. Gew. pfl.,* **4**, 171–6.

MATSUBARA S., INO M., MURAKAMI K., KAMADA M. and ISHIHARA I. (1996), 'Callus formation and plant regeneration of herbs in Perilla family,' *Scientific Rep. Faculty Agriculture, Okayama University,* **85**, 23–30.

MIROVICH V.M., PESHKOVA V.A., SHATOKHINA R.K. and FEDOSEEV A.P. (1989), 'Phenolcarboxylic acids of *Origanum vulgare*,' *Khimiya Prirodnykh Soedinenii,* **25**(6), 850–1.

MOURE A., CRUZ J.M., FRANCO D., DOMÍNGUEZ J.M., SINEIRO J., DOMÍNGUEZ H., NÚÑEZ M.J. and PARAJÓ J.C. (2001), 'Natural antioxidants from residual sources,' *Food Chem.,* **72**(2), 145–71.

NGUYEN U., FRAKMAN G. and EVANS D.A. (1991), 'Process for extracting antioxidants from Labiatae herbs,' *United States Patent,* US5017397.

PÄÄKKÖNEN K., MALMSTEN T. and HYVÖNEN L. (1990), 'Drying, packaging, and storage effects on quality of basil, marjoram and wild marjoram,' *J. Food Sci.,* **55**(5), 1373–7.

PAPANAGIOTOU E., PAPANIKOLAOU K. and ZAMANIDIS S. (2001), 'The cultivation of aromatic and medicinal plants in Greece: 1 The economic dimension,' *Agriculture,* **1**, 36–42 (in Greek).

POL I.E. and SMID E.J. (1999), 'Combined action of nisin and carvacrol on *Bacillus cereus* and *Listeria monocytogenes*,' *Lett. Appl. Microbiol.,* **29**(3), 166–70.

POULOSE A.J. and CROTEAU R. (1978), 'Biosynthesis of aromatic monoterpenes: conversion of γ-terpinene to *p*-cymene and thymol in *Thymus vulgaris* L,' *Arch. Biochem. Biophys.,* **187**(2), 307–14.

SARLIS G. (1994), *Aromatic and Pharmaceutical Plants,* Athens, Agricultural University of Athens.

SCHMITZ S., WEIDENBÖRNER M. and KUNZ B. (1993), 'Herbs and spices as selective inhibitors of mould growth,' *Chem. Mikrobiol. Technol. Lebensm.*, **15**(5/6), 175–7.

SIMON J.E., CHADWICK A.F. and CRAKER L.E. (1984), *Herbs: An Indexed Bibliography 1971–1980. The Scientific Literature on Selected Herbs and Aromatic and Medicinal Plants of the Temperate Zone*, Hamden, CT, Archon Books.

SKOULA M. and HARBORNE J.B. (2002), 'The taxonomy and chemistry of *Origanum*' in Kintzios S., *Medicinal and Aromatic Plants – Industrial profiles – Oregano: The Genera* Origanum *and* Lippia, London, Taylor & Francis, 67–108.

THANOS C.A., KADIS C.C. and SKAROU F. (1995), 'Ecophysiology of germination in the aromatic plants thyme, savory and oregano (Labiatae),' *Seed Science Res.*, **5**(3), 161–70.

THOMPSON D.P. (1990), 'Influence of pH on the fungitoxic activity of naturally occurring compounds,' *J. Food Protection*, **53**, 428–9.

TSAGADOPOULOS A. (2002), *Development of a method for the quality evaluation of oregano (*Origanum *spp.) based on morphological characteristics*. Degree Thesis, Athens, Agricultural University of Athens.

TSIMIDOU M. and BOSKOU D. (1994), 'Antioxidant activity of essential oils from the plants of the *Lamiaceae* family' in Charalambous G., *Spices, Herbs and Edible Fungi,* The Netherlands, Elsevier, 273–84.

VEKIARI S.A., OREOPOULOU V., TSIA C. and THOMOPOULOS C.D. (1993), 'Oregano flavonoids as lipid oxidants,' *J. Am. Oil Chem. Soc.*, **70**(5), 483–7.

VICHI S., ZITTERL-EGLSEER K., JUGL M. and FRANZ C. (2001), 'Determination of the presence of antioxidants deriving from sage and oregano extracts added to animal fat by means of assessment of the radical scavenging capacity by photochemiluminescence analysis,' *Nahrung*, **45**(2), 101–4.

15

Parsley

D. J. Charles, Frontier Natural Products, USA

Parsley (Fig. 15.1) has the following names:

- English – Parsley.
- Chinese – *Heung choi.*
- Danish – *Persille.*
- Dutch – *Peterselie.*
- French – *Persil.*
- German – *Petersilie, Petersil; Peterwurz* (root).
- Greek – *Maintanos, Makedonisi, Petroselino.*
- Italian – *Prezzemolo.*
- Polish – *Pietruszka zwyczajna.*
- Portuguese – *Salsa.*
- Russian – *Petrushka.*
- Spanish – *Perejil.*
- Swedish – *Persilja.*

15.1 Introduction and description

Parsley (*Petroselinum crispum* (Mill.) Nymen ex A. W. Hill) belongs to the family (Umbelliferae) Apiaceae. Other scientific names are: *P. hortense* Hoffm., *P. sativum* Hoff., *Carum petroselinum* (L.) Benth. and Hook. F. Also included in this family are parsnip, celery, dill, carrot, lovage, and a number of other well-known herbs and vegetables. Parsley is an upright, much branched plant, reaching heights of 1 to ½ ft or 0.8 m with green leaves and yellow greenish flowers growing in clusters extending from the main stem. Parsley has thin, spindle-shaped roots, with erect, grooved, glabrous, angular stems. The upper leaves are dark green and divided pinnately into featherlike-sections.

The lower leaves are bi- or triternately divided. The small greenish yellow flowers have five petals on compound umbels. Parsley is mostly cultivated as an annual culinary herb and is widely grown in Europe and Western Asia. Parsley is to the Western world what *cilantro* (sometimes called Chinese parsley) is to the Eastern world. Most parts of the plant are used – the leaves, the above-ground herb and the seeds. The essential or volatile oils can be obtained through distillation. The volatile oil obtained from the above-ground herb has the

(a) (b)

Fig. 15.1 Fresh parsley.

aroma of the fresh herb and is marketed as herb oil. The volatile oil obtained from the seed has a unique distinctive flavour and is marketed as seed oil.

It is interesting to note that the name 'parsley' originates from the name *Petroselinum* ascribed by Dioscorides. Later on in the Middle Ages it became *Petrocilium* and later expressed in the English language as Petersylinge, Persele, Persley and finally Parsley as it is known today. The name *Petroselinum* is derived from the Greek word *petros* which means stone, referring to the plant's habit of growing in rocky places. *Selinon* was the Greek word for parsley in ancient history.

It is surprising to know that the common parsley, which we toss into our soups and sauces without a thought, has an interesting saga behind it and is also wrapped up in Greek mythology. According to Linnaeus, parsley originated in Sardinia and belongs to the family Apiaceae (Umbelliferae). Parsley was known in England around 1548, though some, such as Bentham and De Candolle, believe that the plant was a native of Eastern Mediterranean regions and also of Turkey, Algeria and Lebanon. During the 16th century parsley was naturalized in England, growing in old walls and rocks, hence the name 'rock selinon'. Greek mythology and culture have much to say about parsley as it was planted near graves and anyone near death was said 'to be in need of parsley'. In Greece, parsley had long been associated with death. According to legend, the fertility king, Archemorus, was the one from whose blood the plant sprouted. For centuries the association of parsley with death continued and every generation with its folklore and myriads of legends connected parsley with the 'gods' of the day. The Greeks have used parsley for funerals and placed wreaths of parsley on tombs. Parsley is also said to have been used as a decorative garland for the Head of Hercules, signifying his victory as a pillar of strength. In Roman culture parsley was often used as a deodorant to mask the smell after consuming garlic.

Parsley, being associated with death, later on became associated with evil, Satan. It was powerful with all its devilish powers and, for those who undermine its powers, there were negative consequences. Virgins could not plant parsley without losing their virginity to the devil. The safest day to plant parsley was Good Friday, and it was usually done by a male head of the household to ward off any evil effects the Devil might generate. Since parsley seeds have a slow rate of germination the popular belief was that the seeds had to travel to hell and back two, three, seven or even nine times, before they could germinate.

Parsley is also used in the Hebrew celebration of Passover as a reminder of the grief and sorrow of Hebrews, which came to an end that day. Parsley is also associated with Catherine de Medici and Charlemagne as one of the plants in their garden. It is said that Medici was responsible for the popularity of parsley when she brought it to France from Italy. In ancient times parsley was used in medicinal concoctions for cure-alls, general tonics, poison antidotes, antirheumatics and formulations to relieve kidney and bladder stones.

15.2 Chemical composition

The essential oils of parsley leaves and seeds have been studied extensively (Garnero and Chretien-Bessiere, 1968; Kasting *et al*., 1972; Franz and Glasl, 1974; Freeman *et al*., 1975; Clark and Menary, 1983; Berger *et al*., 1985; MacLeod *et al*., 1985; Heide *et al*., 1986; Shaath *et al*., 1988; Simon and Quinn, 1988; Srinivas, 1986; Gbolade and Lockwood, 1989; Porter, 1989; Nitz *et al*., 1989; Kim *et al*., 1990; Spraul, 1991; Baritaux *et al*., 1992; Jung *et al*., 1992; Perineau *et al*., 1992; Zheng *et al*., 1992a,b).

Parsley seed contains 2–8% essential oil with α-pinene, apiol, myristicin and tetramethoxyallybenzene as the major constituents. It also contains 13–22% fixed oil consisting mainly of petroselinic acid, and smaller amounts of linoleic, myristic, myristolic, oleic, palmitic, stearic and 7-octadecenoic acids (Leung, 1980).

Parsley leaf oil contains 0.04–0.4% volatile oil with α-pinene, β-pinene, myrcene, limonene, α-phellandrene, *p*-cymene, α-terpinolene, menthatriene, α-terpineol, apiol and myristicin as the major constituents. Clark and Menary (1983) identified the following compounds in the oil of a plain leaf variety of *Petroselinum crispum*: α-pinene (22.71%), β-pinene (17.15%), myrcene (5.08%), β-phellandrene (11.57%), terpinolene (3.03%), 1,3,8-*p*-menthatriene (21.05%), *p*-cymene (2.24%), myristicin (4.86%) and apiol (7.48%).

Zheng *et al*. (1992a) showed myristicin as an effective cancer or tumour preventive agent. 1,3,8-*p*-Menthatriene is considered to be one of the major compounds to contribute to the parsley aroma (Kasting *et al*., 1972). Garnero *et al*. (1967) described it as having the odour of parsley leaves. MacLeod *et al*. (1985) believe that β-phellandrene, 1,3,8-*p*-menthatriene, 4-isopropenyl-1-methylbenzene and apiole are characteristic aroma constituents of parsley leaf. Parsley leaf (herb) also contains furocoumarins, xanthotoxin, isopimpinellin, flavonoids, proteins 2–22%, fats 4%, sugars and others (Leung, 1980).

Simon and Quinn (1988) studied the genetic variability of the major essential oil constituents of parsley leaf using the US Department of Agriculture (USDA) parsley collection. They found myristicin as a major compound (20%) and this was in disagreement with other findings (Lawrence, 1981/1982) that a high myristicin content (17%) indicated

Table 15.1 Variation in constituents based on origin

Compounds	Iran	Iraq	Syria	Turkey	USA	Yugoslavia
Myrcene	2.9–13.6	15.7	2.0–12.2	0.9–16.4	3.3	2.8–13.1
β-phellandrene	9.7–33.5	8.2	12.8–14.7	5.8–29.8	18.3	3.6–15.9
Terpinolene+ 1-methyl-4-isopropylbenzene	2.7–5.4	4.3	3.8–6.6	2.5–10.7	3.5	1.9–5.8
1,3,8-*p*-menthatriene	36.1–68.0	51.7	50.8–57.1	29.3–67.4	47.5	20.1–68.8
Myristicin	0.2–19.4	1.2	3.1–11.0	0.1–36.8	19.5	0.2–60.5
Apiol	0.1–0.3	0	4–5.6	0–22.1	0.1	0.1–1.4
MW 268 isomers	2.3–10.9	7.8	3.5–4.7	1.7–10.5	2.9	1.7–7.6

Source: Simon and Quinn (1988).

Table 15.2 Composition of commercial sample of parsley essential oil

Compounds	Parsley leaf oil (% of oil)	Parsley herb oil (% of oil)	Parsley seed oil (% of oil)
α-pinene	26.42	27.41	15.73
Sabinene	1.10	1.01	0.64
β-pinene	18.04	17.83	10.01
Myrcene	4.24	6.51	0.22
α-phellandrene	0.51	0.66	0.12
β-phellandrene	6.48	7.12	2.14
Terpinolene	2.52	2.65	0.01
p-mentha-1,3,8-triene	16.41	15.96	0.12
Myristicin	11.92	9.78	39.65
Elemicin	2.71	1.45	4.84
2,3,4,5-tetramethoxy-allylbenzene	0.72	2.62	7.82
Apiol	0.27	0.11	18.32

a partial seed origin. The changes in the oil composition based on parsley origin are presented in Table 15.1. MacLeod *et al.* (1985) also showed a high myristicin content in parsley leaf oil. Unique to the Apiaceae is 1,3,8-p-menthatriene (MacLeod *et al.*, 1985), and this suggests a strong phylogenetic relationship among the parsley lines (Simon and Quinn, 1988). The author analysed samples of commercial oils of parsley leaf, herb and seed and the results are presented in Table 15.2. The major constituents were α-pinene, β-pinene, β-phellandrene, p-mentha-1,3,8-triene and myristicin.

15.3 Production and cultivation

As with most herbs, parsley grows best in a sunny area that receives direct light for six to eight hours a day, although it can tolerate some shade. Plants will be more productive if grown in well-drained soil that is fairly rich in organic matter, with a pH range of 6 to 7.

15.3.1 Cultivars

In the USA three distinctive types of parsley are commonly grown, although there are several types of parsley cultivated in different parts of the world. The **curly-leaf** or **common parsley** var. *crispum* is widely cultivated in the USA, Germany, France, Hungary and Belgium. Common parsley types are Moss curled, Dark moss, Banquet, Colored, Market gardener, Decorator, Deep green, Improved, Sherwood. These types are used primarily as dried or dehydrated in food products, but mostly used fresh as garnish. These curly types are quite versatile, typically 8 to 14 inches (20–35 cm) tall, forming dense clumps, which are great for borders or inter-planting in the garden beds, not to mention growing in containers. The **Italian parsley** var. *neopolitanum* Danert is flat with crisped leaves and is also known as **flat-leaf** or **plain**, its types are known as Plain, Plain Italian Dark green. Italian parsley is used to flavour sauces, soups and stews. This can grow quite tall (2–3 ft; 1 m) and is more gangly in habit. The flat serrated leaves have a much stronger and sweeter flavour than the other varieties, making it desirable for cooking. The **turnip rooted** or **Hamburg parsley** var. *tuberosum* Bernh is used in specialized markets for its edible roots. Tall, fern-like leaves make up the foliage. **Japanese parsley**, *Cryptotaenia japonica*, is not commonly grown, it resembles the Italian parsley. It is mostly used in Oriental cooking and has a bitter taste.

15.3.2 Germination
Parsley seed numbers approximately 296 500/lb (134 000/kg). Since the germination rate of the smooth ribbed and ovate parsley seeds is very low and erratic, which is common to the Apiaceae family (Simon and Overly, 1986), a pretreatment soaking is recommended to increase the usual slow rate of germination which is three to six weeks. The seeds have a very slow rate of germination in wet soils characteristic of early spring. Propagation by seed grown from early spring to early summer and in autumn or by seed grown in spring (*P. crispum* var. *tuberosum*). The commercially produced parsley seeds are mericarps. The Italian variety is very slow to germinate.

15.3.3 Field preparation
Since parsley plants need a rich, moist soil and a good drainage, a pH of 5.3 to 7.3 is preferred (Simon *et al.*, 1984). Like other small-seeded crops, parsley needs a fine seedbed and the soil should be finished after ploughing and disc harrowing with rototillers and bed shapers (Simon *et al.*, 1984). For proper germination, the seeds should be covered a quarter inch (0.5 cm) but in case of a heavy soil rich in minerals, the seeds should be covered with leaf mould or sand so that no crust is formed (Simon *et al.*, 1984).

15.3.4 Sowing
Sowing rates may differ based on environmental and soil conditions from 12 to 20 lb (5–9 kg) to as high as 40 to 60 lb (18–27 kg) in some areas (Simon *et al.*, 1984). Usually to avoid freezing conditions in the north, plants can be grown indoors and then transplanted to open prepared fields. In spring, as soon as the soil is ready, the parsley seeds are sown into 60 inch (150 cm) raised beds with three or four rows 18 to 22 inches (45–55 cm) apart. Transplants could be spaced 4 to 8 inches (10–20 cm) apart on 36 inch (90 cm) rows. The highest yields can be obtained with very high plant populations (Simon *et al.*, 1984).

15.3.5 Fertilization
The soil type and prior cropping history determine the suggested fertilization NPK ratio of 1–1–1 or 3–1–2. One application of $N–P_2O_5–K_2O$ at a rate of 120–120–120 lb per acre (22–22–22 kg/ha) for heavy textured soils should be sufficient. Usually one-third of the fertilizer is applied, which is followed by two side-dressings of NPK fertilizer and supplemented with N according to crop needs and cropping (Simon *et al.*, 1984).

15.3.6 Irrigation
Parsley, like other leafy green vegetables, needs overhead sprinklers or drip irrigation (Simon *et al.*, 1984).

15.3.7 Weed control
For weed control, most states in the USA use Stoddard solvent on parsley. When the seedlings are two inches (5 cm) tall and have three true leaves, the herbicide is applied at the rate of 60 gallons per acre (90 l/ha) (Simon *et al.*, 1984).

15.3.8 Insect and disease control

The parsley crop has many pests, which cause viral diseases such as Aster yellows. For pests such as aphids and cabbage looper, phosdium and methomyl are recommended but the local extension service should be consulted before using the insecticides. Other pests such as carrot weevil, corn earworm, flea beetles and leafhoppers can be treated with different pesticides. '*Septoria apiicola*' is one of the most important foliage diseases, it is seed-borne or splash disseminated (Simon *et al.*, 1984).

15.3.9 Harvesting

Depending on the crop quality, multiple harvests of parsley are possible by machine or hand. For multiple harvests parsley should be cut at least 1.25 inches (3 cm) above the crown. The machine-harvested fields are mechanically chipped 1–3 inches (2.5–7.5 cm) above the crown and transported to dehydrators. Hand labour is the preferred method for harvesting parsley and is intensive because there must be minimal crop damage for parsley best suited for fresh market use. Usually the workers bunch up a group of plants manually, slice the stalks with a knife and tie them up in bunches by slipping a rubber band around the stalks. Sometimes the bunching is done later. In Southern areas late summer and autumn sowings are harvested in winter. In the north, harvesting can be continuous from April to December. Summer sowing can be harvested in autumn; autumn sowing can be harvested during winter and spring; or early spring sowing can be harvested late spring or throughout summer. The highest price and quality are obtained from the earliest harvested spring-seeded crop. Hamburg parsley can grow in moderate freezing conditions. Prior to marketing, roots should be washed. Parsley should be shipped and packed in ice or coolers to maintain crispness and fresh appearance. Relative humidity of 95% and temperatures from 0 to 2°C is recommended for storage and handling (Simon *et al.*, 1984).

15.4 Organic farming

Farmers, consumers and policy makers have shown a renewed interest in organic farming as the objective of today's common agricultural policy – the sustainability of both agriculture and the environment without compromising food production and conservation of finite resources and protecting the environment so that the needs of people are met today and for generations to come.

Approximately 2% of the US food supply is grown using organic methods. Since 1990, sales of organic products have shown an annual increase of at least 20%, the fastest-growing sector of agriculture. In 2001 retail sales of organic food were projected to be $9.3 billion (*Organic Consumer Trends*, 2001). Organic foods can be found at natural food stores and major supermarkets, as well as through growers' direct marketing such as CSAs (community supported agriculture) and farmers' markets. Many restaurant chefs across the country are using organic produce because of its superior quality and taste.

Organic food is also gaining international acceptance, with nations such as Japan and Germany becoming important international organic food markets. Although in 2000 it represented only around 3% of the total European Union agricultural area, organic farming has in fact developed into one of the most dynamic agricultural sectors in the European Union. The organic farm sector grew by about 25% a year between 1993 and 1998 and, since 1998, it is estimated to have grown by around 30% a year. In some member states, however, it now seems to have reached a plateau.

Some of the essential characteristics of organic farming include: design and implementation of an 'organic system plan' – a detailed record-keeping system that tracks all products from the field to point of sale, and also the maintenance of buffer zones to prevent inadvertent contamination from adjacent conventional fields. Organic farming is an ecological production management system that promotes and enhances biodiversity, biological cycles and soil biological activity. The primary goal of organic agriculture is to optimize the health and productivity of interdependent communities of soil life, plants, animals and people. The holistic vision includes the maintenance of valuable relationships between soil, water, air, plants, animals and people.

Organic farmers build healthy soils by nourishing the living component of the soil, the microbial inhabitants that release, transform and transfer nutrients. Soil organic matter contributes to good soil structure and water-holding capacity. Organic farmers feed biota and build soil organic matter with cover crops, compost and biologically based soil amendments. These produce healthy plants that are better able to resist disease and insect predation.

The key principles used by organic farming systems are not to use chemo-synthetic mineral fertilizers and to minimize the use of permitted external fertility inputs, crop protection products and energy use – 'external' meaning that they are not produced on the farm as a group of collaborating farm businesses. Chemo-synthetic mineral fertilizers such as nitrogen and phosphorus fertilizers are not permitted and this prohibition has resulted from a range of considerations. Most importantly, such fertilizers are thought to substitute for natural mechanisms of nutrient acquisition by plants. Clover and other legumes have traditionally been used to enrich agricultural soils with nitrogen.

Legume plants have the unique ability to form symbiotic relationships with a specific group of soil bacteria called rhizobium. The symbiotic relationship between the legume plant and the rhizobium bacterium is extremely close in that the bacterium is taken up by the plant and 'housed' in a separate plant organ in the plant root (called the root nodule). Rhizobium growth and nitrogen fixation activity in the nodule is fuelled by transfer of carbohydrate from the plant to the root nodule. The bacterium focuses all the energy on transforming or fixing atmospheric nitrogen into ammonia. Application of mineral fertilizers to soil reduces soil microbial activity and, in particular, the activity of bacterial nitrogen fixation.

In soil, most phosphorus is usually present in non-water-soluble forms and therefore is not readily available for plants. In nature, most plants (this includes most crop plants) have developed symbiotic relationships with a specific group of soil fungi (called mycorrhizal fungi), which can access and make available non-water-soluble forms of phosphorus (phosphorus the plant could not access on its own) to the plant. The fungus colonizes the plant's roots and expands a web of 'hyphae' (microscopic fungal tubes) from the plant root into the soil. This web of hyphae greatly extends the area of soil the plant can reach in 'collaboration' with the fungus. Fungal enzymes and acids then allow the fungus to take up soil phosphorous in a water-soluble form. The phosphorous is then transported via the fungal hyphae to the root and taken up by the plant tissue, which has been colonized by the fungus. As with the rhizobium bacteria the plant supplies energy to the fungus in return for the phosphorus the fungus supplies to the plant.

In many respects the symbiotic relationship between plants and mycorrhizal fungi is even closer than that between legumes and the nitrogen-fixing rhizobium. For example, the mycorrhizal fungus cannot grow and multiply in soil without plant roots being present. The addition of chemo-synthetic, water-soluble phosphorus fertilizers (e.g. superphosphate) inhibits the development of the symbiotic relationship between mycorrhizal fungi and plant roots.

The use of chemo-synthetic nitrogen and phosphorus fertilizers increases the risk of environmental pollution, since they are highly water soluble. They can therefore cause pollution and environmental problems such as algal blooms when they are: (a) leached into ground water (especially nitrate) or (b) transported by run-off into the terrestrial and marine aquatic ecosystems. Organic farmers mostly use compost or manure to replenish the soil with minerals, as it is rich in beneficial soil microorganisms, which in turn slowly and steadily make minerals available to plants. Using cover crops or practising low till farming strictly observes soil conservation or also leaving unwanted portions, thus preventing soil erosion.

Organic farming uses a variety of methods to control fungus and harmful insects. Farmers often use the method of intercropping. This is done by planting different crops in alternating rows, thereby interrupting the movement of disease-causing organisms through a field. Sometimes crops are sprayed with bacteria, which in turn destroy the larvae of harmful insects.

Organic farmers also use pesticides derived from chemically unaltered plant, animal or mineral substance in which the active ingredient becomes non-toxic after being applied to the crops. For example, pyrethrum extracted from chrysanthemums and oil extracted from neem trees is widely used.

In organic farming, weed control is mainly done through mulching to smother weeds and by planting cover crops such as cereal rye and oat, which either inhibit weed seed germination or deprive them of the nutrients they need to grow. Sometimes tractor-drawn equipment is also used to uproot weeds.

Organic farming represents long-term savings and also maintains ecological balance and harmony. The crops are free from synthetic toxic chemicals and pesticides, being pure and natural and often believed to be more nutritious by organic farmers. Organic farming also preserves top soil so more crops can be grown in future without polluting the environment, and saving on fuels, pesticides and fertilizers make it more attractive than conventional farming.

'Certified organic' refers to agricultural products that have been grown and processed according to strict uniform standards, verified annually by independent state or private organizations accredited by the USDA. Certification includes inspection of farm fields and processing facilities. Farm practices inspected include long-term soil management, buffering between organic farms and any neighbouring conventional farms, product labelling and record keeping. Processing inspections include review of the facility's cleaning and pest control methods, ingredient transportation and storages, and record keeping and audit control.

The phrase 'certified organic' needs to be understood from the point of origin; these are products using organic agriculture. In organics, the focus is not so much on pushing the resources to produce tremendous yields and profits. Instead, organics foster the ecological processes that produce resources and add value to the resulting products: the added value of the designation, and the assurance that the products that consumers are buying are not just chemical-free products but that they are good for consumers, and are good for the environment as well. Organic products help minimize the adverse effects that agriculture can have on soil, water and air.

Organic food is as safe to consume as any other kind of food. Just as with any kind of produce, consumers should wash food before consuming it to ensure maximum cleanliness. Organic produce contains significantly lower levels of pesticide residues than conventional produce. It is a common misconception that organic food could be at greater risk of *Escherichia coli* contamination because of raw manure application, although conventional farmers commonly apply tonnes of raw manure as well with no regulation whatsoever. Organic standards set strict guidelines on manure use in organic farming: either it must be first composted or it must be applied at least 90 days before harvest, which allows ample time for microbial breakdown of any pathogens.

The cost of organic food is higher than that of conventional food because the organic price tag more closely reflects the true cost of growing the food: substituting labour and intensive management for chemicals, the health and environment costs of which are borne by society. These costs include clean-up of polluted water and remediation of pesticide contamination. Prices for organic food include costs of growing, harvesting, transportation and storage. In the case of processed foods, processing and packaging costs are also included. Organically produced foods must meet stricter regulations governing all these steps than conventional foods. The intensive management and labour used in organic production are frequently (though not always) more expensive than the chemicals routinely used on conventional farms. There is mounting evidence that, if all the indirect costs of conventional food production were factored into the price of food, organic foods would cost the same or, more likely, be cheaper than conventional food.

Organic growing of crops is a challenging, more detailed type of agriculture that requires the farmer to take responsibility for, and make a commitment to, the land even while pursuing the traditional business goal of making a profit. This commitment is embodied by the organic farming practices of weed, pest and disease control.

15.5 General uses

Parsley is mostly used in the culinary area but it has many other uses, such as chopped fresh leaves being used in soups, stuffings, minces, rissoles and also used as garnish over vegetable and salads. The leaves are cultivated extensively for the purpose of sending to markets fresh and also being dried and powdered to be used as a culinary flavouring especially in the winter months when the fresh supply is very low. The seeds, roots and even stems are used. The stem can be dried and powdered and used for culinary colouring and dyeing; the roots of the turnip-rooted variety are used as vegetable and flavoring; there is also a market for seeds to supply to nurserymen (Fig. 15.2).

Fig. 15.2 Dried parsley.

Parsley, with its mystic aura being wrapped in folk tradition, is said to increase female libido, also help in promoting menstruation and ease the difficulties of childbirth (Review of Natural Products, 1991; Tyler, 1994). Parsley juice can be used in treating hives and other allergy symptoms; it also inhibits the secretion of histamine. Parsley has also been used as a liver tonic and helped in the breaking up of kidney stones. The German Commission E has approved parsley as a preventive measure and also for treatment of kidney stones. The parsley root can be used as a laxative and also helps to eliminate bloating. It can reduce weight by reducing excess water gain. The root can be used to relieve flatulence and colic, due to its carminative action. Parsley is rich in such minerals as calcium, thiamin, riboflavin, potassium, iron and vitamins such as A, C and niacin (Review of Natural Products, 1991; Gruenwald, 1998; Blumenthal, 1998; Tyler, 1994, 1998; Marczal et al., 1977). Parsley can be used as a tasty breath freshener owing to its high chlorophyll content. It also speeds the healing of bruises and soothes tired and lustre-lacking eyes. The juice soaked in a pad can relieve earache and toothache. Parsley can be used as a face wash to lighten freckles. Parsley juice relieves itch and stings from insect bites; it works amazingly well as a mosquito repellent. Lactating women have used the leaves of parsley as poultice to relieve breast tenderness. The powdered seeds of parsley are a folk remedy for hair growth and scalp stimulation if massaged into the scalp for three days. It also has strong antioxidant properties (Pizzorna and Murray, 1985). Parsley has other uses: the essential oil is used in commercial food flavourings and perfumes for men. Head lice can be eradicated if parsley is used as a hair rinse.

15.5.1 Precautions
As a widely eaten food, parsley is generally regarded as safe. Though no interactions have been reported between parsley and standard allopathic medications, it may cause allergy in sensitive persons. Parsley contains furocoumarins – compounds that may cause photosensitivity in fair-skinned persons exposed to sunlight after coming in skin contact with the freshly harvested herb. An overdose of parsley's essential oil can lead to poisoning because of the toxicity in high doses. Persons with kidney diseases should not take parsley internally without consulting a physician because parsley is said to irritate the epithelial tissues of the kidney, hence enhancing the flow of blood and filtration rate. Pregnant or lactating women should not use parsley, as the oil-rich seeds contain a chemical, which is said to have abortifacient properties (Review of Natural Products, 1991; Tyler, 1994; Lagey et al., 1995; Stransky and Tsankov, 1980).

15.6 Essential oils and their physicochemical properties

15.6.1 Extraction
Parsley essential oil is obtained by steam distillation of the seeds or the above-ground parts of the plant. Commercially, there exist two types of parsley oil, viz., parsley herb oil and parsley seed oil.

15.6.2 Storage
Parsley seed oil and herb oil should be stored in full, preferably glass, tin-lined, or other suitably lined containers in a cool place protected from light.

15.6.3 Aroma profile

Parsley herb oil is a pale yellow or greenish yellow, rarely water-white liquid of a peculiar, warm-spicy, heavy-leafy, yet fresh-herb-like odour. It is very similar to the odour of the freshly cut herb (Arctander, 1982). Parsley seed oil is a yellowish to amber-colored or brownish liquid, more or less viscous. The odour is warm-woody, spicy, somewhat sweet herbaceous (Arctander, 1982).

15.6.4 Physiochemical properties

FCC (2000)

Parsley herb oil

- Optical rotation: +1 to –9.
- Refractive index: 1.503 to 1.530.
- Specific gravity: 0.908 to 0.940.

Parsley seed oil

- Optical rotation: –4 to –10.
- Refractive index: 1.513 to 1.522.
- Specific gravity: 1.040 to 1.080.

According to Guenther (1976) vol IV pp 656 to 663

French parsley seed oil

- Optical rotation: –4 to –10.
- Refractive index: 1.512 to 1.528.
- Specific gravity: 1.043 to 1.110.

French parsley herb oil

- Optical rotation: +6 to –6.10.
- Refractive index: 1.5029 to 1.526.
- Specific gravity: 0.9023 to 1.0157.

American parsley herb oil

- Optical rotation: –2.13 to –7.40.
- Refractive index: 1.5080 to 1.5179.
- Specific gravity: 0.945 to 1.046.

Hungarian parsley herb oil

- Optical rotation: –1.46 to –6.18.
- Refractive index: 1.5053 to 1.5250.
- Specific gravity: 0.948 to 0.987.

According to NF T75-230 and ISO 3527

Parsley fruit oil

- Optical rotation: –4 to –11.
- Refractive index: 1.510 to 1.522.
- Specific gravity: 1.043 to 1.083.

According to Fenaroli (1975) p. 428

Parsley seed oil

- Optical rotation: −4 to −11.
- Refractive index: 1.5100 to 1.5290.
- Specific gravity: 1.043 to 1.083.

Parsley herb oil

- Optical rotation: +6.
- Refractive index: 1.5029.
- Specific gravity: 0.911.

Parsley leaf oil

- Optical rotation: −2.55 to −6.10.
- Refractive index: 1.5087 to 1.5159.
- Specific gravity: 0.948 to 0.967.

15.7 References

ARCTANDER S. (1982), *Perfume and Flavor Materials of Natural Origin*. Elizabeth, NJ.
BARITAUX O., RICHARD H., TOUCHE J. and DERBESY M. (1992), Sechage et conservation des plantes aromatiques. *Rivista Ital. EPPOS (Numero Speciale)*, 416–26.
BERGER R.G., DRAWERT F., KOLLMANNSBERGER H. and NITZ S. (1985), Natural occurrence of undecaenes in some fruits and vegetables, *J. Food Sci.*, **50**, 1655–6.
BLUMENTHAL M. (1998), *The Complete German Commission E Monographs. Therapeutic Guide to Herbal Medicines*. Boston, MA: Integrative Medicine Communications, 179.
CLARK R.J. and MENARY R.C. (1983), The Tasmanian Essential Oil Industry-Production of Steam-Distilled Essential Oil Crops. Paper presented at the 9th International Essential Oil Congress, Singapore.
FCC (2000), Foods Chemical Codex. Washington, DC. National Academy Press.
FENAROLI G. (1975), *Handbook of Flavor Ingredients*. Vol. I, Cleveland, OH: CRC Press, 427–9.
FRANZ C. and GLASL H. (1974), *Ind. Obst. Gemueseverwert.*, **59**, 176; *Chem. Abstr.*, **81**, 62325r.
FREEMAN G.G., WHENHAM R.I., SELF R. and EAGLES J. (1975), Volatile favour components of parsley leaves (*Petroselinum crispum* (Mill.) Nyman). *J. Sci. Food Chem.*, **26**, 465.
GARNERO J. and CHRETIEN-BESSIERE Y. (1968), Contribution à l'étude de la composition chimique de l'huile essentielle de feuilles de persil de Yugoslavie, *Fr. Ses Parfums.*, **11**, 332.
GARNERO J., BENEZET L., PEYRON L. and CHRETIEN-BESSIERE Y. (1967), Sur la presence du *p*-menthatriene-1,3,8 dans l'huile essentielle de feuilles de persil de Yugoslavie, *Bull. Soc. Chim. Fr.*, **12**, 4679.
GBOLADE A.A. and LOCKWOOD G.B. (1989), Volatile constituents from parsley cultures, *Flavor Frag. J.*, **4**, 69–71.
GRUENWALD J. (1998), *PDR for Herbal Medicine*. 1st Ed. Montvale, NJ: Medical Economics, 1023–4.
GUENTHER E. (1976), *The Essential Oils*. Vol. IV:, New York: Robert E. Krieger, 656.
HEIDE R., DE VALOIS P.J., DE RIJKE D and BEDNARCZYK A.A. (1986), Acids and Phenols in Seven Spice Essential Oils. Paper presented at the ACS meeting, New York, 13–18 April.
INTERNATIONAL ORGANIZATION FOR STANDARDIZATION (ISO) (1975), ISO 3527, Oil of Parsley Fruit.
JUNG H.P., SEN A. and GROSCH W. (1992), Evaluation of potent odorants in parsley leaves (*Petroselinum crispum* (Mill.) Nym. Ssp. Crispum) by Aroma Extract Dilution Analysis, *Lebensm. Wiss. Technol.*, **25**, 55–60.
KASTING E., ANDERSON J. and SYDOW E. (1972), Volatile constituents in leaves of parsley, *Phytochemistry*, **11**, 2277.
KIM Y.H., KIM K.S. and HONG C.K. (1990), Volatile components of parsley leaf and seed (*Petroselinum crispum*), *J. Korean Agric. Chem. Soc.*, **33**, 62–7.

LAGEY K., DUINSLAEGER L. and VANDERKELEN A. (1995), Burns induced by plants. *Burns*, **21**, 542–3.
LAWRENCE B.M. (1981/1982), Parsley oils: leaf, seed and herb, *Perfumer and Flavorist*, **6**, 43.
LEUNG A.F. (1980), *Encyclopedia of Common Natural Ingredients. Uses in Food, Drugs, and Cosmetics*, New York: John Wiley, 257–9.
MACLEOD A.J., SNYDER C.H. and SUBRAMANIAN G. (1985), Volatile aroma constituents of parsley leaves, *Phytochemistry*, **24**(11), 2623.
MARCZAL G., BALOGH M. and VERZR-PETRI G. (1977), Phenol-ether components of diuretic effect in parsley. I. *Acta Agron. Acad. Sci. Hung.*, **26**, 7–13.
NF T75-230 (1996), *Huiles essentielles*, Vol 2, specifications, AFNOR.
NITZ S., KOLLMANNSBERGER H., SPRAUL M.H. and DRAWERT F. (1989), Oxygenated derivatives of menthatriene in parsley leaves, *Phytochemistry*, **28**(11), 3051.
ORGANIC CONSUMER TRENDS (2001), Natural Marketing Institute/Organic Trade Association.
PERINEAU F., GANOU L. and GASET A. (1992), Selective chimique lors de chydrodistillation du fruit de Persil commun (*Petroselinum sativum*). *Rivista Ital. EPPOS (Numero Speciale)*, 449–56.
PIZZORNO J.E. and MURRAY M.T. (1985), *A Textbook of Natural Medicine*. Seattle, Washington: John Bastyr College Publications.
PORTER N.G. (1989), Composition and yield of commercial essential oils from parsley 1: Herb oil and crop development, *Flav. Frag. J.*, **4**, 207–19.
REVIEW OF NATURAL PRODUCTS (1991), Facts and Comparisons; Parsley monograph. St. Louis, MO.
SHAATH N.A., GRIFFIN P., DEDEIAN S. and PALOYMPIS L. (1988), The chemical composition of Egyptian parsley seed, absolute and herb oil in *Flavors and Fragrances: A World Perspective*, Eds, B.M. Lawrence, B.D. Mookherjee and B.J. Willis, Amsterdam: Elsevier Science Publishers BV, pp 715–29.
SIMON J.E. and OVERLY M.L. (1986), A comparative evaluation of parsley cultivars. *The Herb, Spice and Medicinal Plant Digest*, **4**(1), 3–7.
SIMON J.E., CHADWICK A.F. and CRAKER L.E. (1984), *HERBS: An Indexed Bibliography, 1971–1980. The Scientific Literature on Selected Herbs, and Aromatic and Medicinal Plants of the Temperate Zone*. Archon Books.
SIMON J.E. and QUINN J. (1988), Characterization of essential oil of parsley, *J. Agric. Food Chem.*, **36**, 467–72.
SPRAUL M. (1991), Strukturaufklarung wertgebender inhaltsstoffe aus Petersilienblattern, -wurzelnund- samen sowie aus Dillbluten. Ph.D Thesis, Techn. Univ. Munchen.
SRINIVAS S.R. (1986). *Atlas of Essential Oils*. Published by the author, Bronx, NY.
STRANSKY L. and TSANKOV N. (1980), Contact dermatitis from parsley (*Petroselinum*). *Contact Dermatitis*, **6**, 233–4.
TYLER V. (1994), *Herbs of Choice*. Binghampton, NY: Pharmaceutical Product Press, 75–6.
TYLER V.E. (1998), *Herbs of Choice*. New York: Pharmaceutical Products Press, The Haworth Press Inc.
ZHENG G-Q., KENNEY P.M. and LAM L.K-T. (1992a), Myristicin: a potential cancer chemopreventive agent from parsley leaf oil, *J. Agric. Food Chem.*, **40**, 107–10.
ZHENG G-Q., KENNEY P.M. and LAM L.K-T. (1992b), Inhibition of benzo(a)pyrene-induced tumorigen- esis by myristicin, a volatile aroma constituent of parsley leaf oil, *Carcinogenesis*, **13**, 1921–3.

16
Rosemary

B. Sasikumar, Indian Institute of Spices Research, India

16.1 Introduction and description

Rosemary (*Rosmarinus officinalis* L.), family Lamiaceae, is a dense, evergreen, hardy, perennial aromatic herb of 90–200 cm height with small (2–4 cm) pointed, sticky and hairy leaves (Fig. 16.1). The upper surface of the leaf is dark green whereas it is white below; leaves are resinous. Branches are rigid with fissured bark and stem square, woody and brown. Pale blue small flowers appear in cymose inflorescence. The leaves, flowering tops and twigs yield an essential oil and oleoresin valued in traditional medicine, modern

Fig. 16.1 Rosemary.

medicine and aromatherapy as well as in the perfumes and flavour industries. Rosemary has culinary uses too. The leaves, twigs, value added products and whole plant extract are also valued as functional food (antioxidant) and botanical neutraceutical. Rosemary is also credited with insect repellent properties and is used in wardrobes to protect clothing. It is also used as an insect repellent herb (functional insecticide) in orchards, as a botanical pesticide, etc. Rosemary is tolerant to pruning and shaping, making it suitable for topiary, and is a valued potted indoor plant.

Rosemary can be grown either as a field crop or as an indoor plant. The plant thrives well in well-drained soils of pH 6.5–7.0 under warm, sunny weather (Doulgas, 1971). It will grow in a semi-arid tropical climate as well. The plant is, however, susceptible to severe cold and frost, though frost-resistant varieties are now available. (Domokos *et al.*, 1997).

Rosemary is a native of Mediterranean region and numerous cultivars and wild forms (chemotypes) are available in Mediterranean countries (Giugnolinini, 1985). Rosemary ecotypes with distinct morphological characters and oil quality occur in Italy (Mulas *et al.*, 1998). Male sterility is known to occur in natural populations of rosemary. Mitochondrial genome polymorphism linked to male sterility has been reported from Spain (Hidalgo-Fernandez *et al.*, 1999). This male sterility may be responsible for spontaneous population evolution and occurrence of chemotypes in rosemary. Though generally *Rosmarinus officinalis* is used for oil extraction, in Morocco *Rosmarinus eriocalyx* is also used for extracting essential oil (Elamrani *et al.*, 2000).

The word rosemary is derived from the Latin word '*rosmarinus*', meaning 'sea dew'. It was also called '*antos*' by the ancient Greeks, meaning the flower of excellence or 'libanotis' for its smell of incense (Giugnolinini, 1985).

There are many myths and folklores associated with rosemary. It is believed that placing rosemary sprigs under the pillow would ward off evil spirits and nightmares from the sleeper and that the aroma of rosemary would keep old age at bay (Rose, 1974). During the Middle Ages it was believed that burning rosemary leaves and twigs would scare away evil spirits and disinfect the surroundings. Many of these myths and beliefs had an underlying scientific logic behind it, as present-day studies reveal. Now it is clear that the essential oil and tannins present in rosemary produce an aromatic smoke of cleansing and purifying properties! However, the scientific logic of certain other customs and myths surrounding rosemary have yet to be unravelled. For example, in Hungary, ornaments made of rosemary were once used as a symbol of love, intimacy and fidelity of a couple. Rosemary was also used in bridal wreaths along with other herbs and flowers. Another belief associated with rosemary is that if rosemary thrives in home gardens, the woman rules the house! The presence of rosemary in one's body is believed to enhance clarity of mind and memory, akin to the belief surrounding sweet flag (*Acorus calamus*) in India. In certain beliefs, rosemary represents the sun and fire signs.

16.2 Chemical composition

The composition of rosemary oil is 1,8-cineol (30–40%), camphor (15–25%), borneol (16–20%), bornyl acetate (up to 7%), α-pinene (25%) as well as β-pinene, linalool, camphene, subinene, myrcene, α-phellandrene, α-terpinene, limonene, *p*-cymene, terpinolene, thujene, copalene, terpinen-4-ol, α-terpineol, caryophyllene, methyl chavicol, thymol, etc. The initial distillation fraction contains mostly α-thujene, α-pinene, camphene, β-pinene and 1,8-cineol, while camphor and bornyl acetate constitute the bulk of the later distillation (Prakasa Rao *et al.*, 1999).

Rosemary oil exhibits variation in composition, both profile and percentage, with respect to location and/or other factors such as source population and phenology (Guazzi et al., 2001; Porte et al., 2000; Ouahada, 2000; Boutekedjiret et al., 1999; Arnold et al., 1997). Pintore et al. (2002) reported a total of 58 compounds from rosemary oil from Sardinia and Corsica (Italy) based on gas chromatography retention index (GC–RI), GC mass spectroscopy (GC–MS) and carbon nuclear magnetic resonance (C–NMR) studies. A study of Moroccan rosemary oil not only demonstrated the existence of three different rosemary chemotypes but also identified a total of 91 compounds, based on GC and GC–MS studies (Elamrani et al., 2000).

16.3 Production and cultivation

Rosemary is grown in Algeria, China, France, Hungary, Italy, the Middle East, Morocco, Portugal, Russia, Romania, Serbia and Montenegro, Spain, Tunisia, Turkey, the USA, and to a limited extent in India. France, Spain and Tunisia are the important countries producing rosemary oil. The annual production of rosemary is oil now about 200–300 Mt.

16.3.1 Agrotechniques

Rosemary is a perennial herb propagated either by cuttings or seeds. Cuttings of 10–15 cm length from selected mother plants are ideal for vegetative propagation of rosemary. Treatments of cuttings with growth hormones such as indole butyric acid (IBA), indole acetic acid (IAA) or saponin are reported to enhance rooting of cuttings (Shah et al., 1996; Silva and Pedras, 1999). Among the different seasons, the end of winter is found the best season for rooting cuttings (Silva and Pedras, 1999).

Cuttings with the lower leaves removed are first planted in raised sand beds of convenient size, in a protected nursery. Regular watering is needed for good sprouting of the cuttings. It takes about 45–50 days for the cuttings to be transplanted to the main field.

Rosemary seeds are very small and black in colour. In India, the seed nursery is raised usually during September to November. Seed rate is about 0.2 to 2.5 g seed per 1 m^2 area (Farooqi and Sreeramulu, 2001). Raised seed beds with adequate shade, sufficient watering, good drainage and weeding ensure healthy seedlings. Seedlings are transplantable at 8–10 weeks. The usual spacing adopted for rosemary is 45 × 45 cm as a monocrop.

16.3.2 Soil

Soil properties are known to influence yield and composition of rosemary oil. Moretti et al. (1998a) reported that granite silt soils are better for herb yield and oil quality of rosemary than calcareous soils.

16.3.3 Fertilizers and growth regulators

A fertilizer dose of 40 kg P_2O_5, 40 kg K_2O and 20 kg N with 20 Mt farmyard manure ha^{-1} is recommended for rosemary in India (Farooqi and Sreeramulu, 2001). Further, N level can be raised to 300 kg ha^{-1} in different splits to maximize oil yield (Prakasa Rao et al., 1999). Studies conducted at Italy (Sardinia) on fertilizer dose and weed management revealed that applying 80 kg N + 60 kg P_2O_5 ha^{-1} coupled with hand weeding of the major weeds such as *Genista corsica* and *Cytisus* Spp. increased herbage yield and oil (Milia et al., 1996).

The beneficial effect of biofertilizers such as *Azospirillum*, *Azotobacter* and VAM applied in combination with inorganic fertilizers in increasing herb and oil yield of rosemary has also been reported (Anuradha *et al.*, 2002). Weed control in rosemary is achieved through occasional hand weeding and intercultivation.

Application of Fe as foliar spray in irrigated rosemary increased the verbenone concentration in the oil (Moretti *et al.*, 1998b).

Studies on the effect of growth regulators (brassinosteroids and uniconazole) on growth, yield and chemical composition of rosemary in Egypt revealed the beneficial effect of growth hormones (Tarraf and Ibrahim, 1999).

16.3.4 Irrigation

Irrigation is reported to be beneficial for herbage yield of rosemary. To establish the crop well in the field, two irrigations per week are recommended in India. Subsequently, irrigation once a week will be sufficient (Farooqui and Sreeramulu, 2001). Drip fertigation with water-soluble fertilizers (80% WSF) coupled with micronutrients are reported to increase growth, yield and quality traits of rosemary (Vasundhara *et al.*, 2002). Field experiments on water requirements of rosemary in Egypt revealed that irrigation once every 14 days resulted in high herbage and oil yield (Kandeel, 2001).

A study on the effect of soil moisture regime, irrigation water : cumulative pan evaporation (IW : CPE) ratios and nitrogen level on herbage and oil yield of rosemary on alfisol, Banglore, India indicated that soil moisture regime maintained at 0.50 IW : CPE ratio with 150 kg N ha^{-1} significantly increased herbage and oil yield of rosemary (Singh and Ramesh, 2000*)*. Unstressed rosemary plants are reported to yield higher levels of essential oil and phenolic compounds than those subjected to water/nutrient stress (Solinas *et al.*, 1996).

16.3.5 Pruning

Pruning of rosemary is advisable, though not essential, after two or three years to enhance shoot and leaf production (Farooqi and Sreeramulu, 2001). As rosemary is a perennial herb, aged plantations (10–12 years old) need to be rejuvenated by cutting back the plant to a height of 4–5 cm above ground coupled with fertigation for better herbage yield.

16.4 Post-harvest technology

Leaves, flowering tops, flowers and twigs are of economic importance. First harvesting is done about eight months after planting, with the onset of flowering or just before flowering. In the first year, two crops can be taken, whereas in the subsequent years two to four harvests are possible at an interval of 100–120 days. Generally, harvesting of the plants can be done up to 50% flowering. At above 90% flowering, harvesting is not desirable (Farooqi and Sreeramulu, 2001). Tender, non-hardy shoots are also harvested for distillation upon attaining full size.

Usually the harvested leaves, flowering tops and shoots are used for downstream processing without drying. However, the leaves and twigs can also be used after drying for oil extraction. Drying studies in normal condition at 50°C and in dehydrated air (30°C) revealed that, in terms of percentage values of the characteristic volatile oil compounds, the dehydrated air-drying product is on par with raw leaves (di Cesare *et al.*, 2001). However,

Ibanez *et al.* (1999) reported that, among different drying methods, the traditional method (drying in a ventilated room) is the best in terms of the yield and quality of the antioxidant principles of rosemary. Influence of genotype of the plant, age of leaf and other growing conditions are known to affect the oil quality especially the antioxidant principle, the carnosic acid, in oil (Hidalgo *et al.*, 1998). Boutekedjiret *et al.* (1999) reported a phenology-dependent variation in yield and quality. For best oil yield, the flowering stage of rosemary is the best, though the oil quality from such plant is slightly inferior. To get quality oil of good yield, one has to select an appropriate growth stage. In short, the flowering stage of the plant as well as pre-flowering plants are suitable for the production of oil, though there will be difference in the oil quality.

16.4.1 Extraction of oil

Rosemary oil is usually recovered by steam or water distillation, though super-critical fluid extraction using CO_2 as solvent is also in practice (Coelho *et al.*, 1997; Bicchi *et al.*, 2000). Most of the oil (90%) comes out within the first 60 minutes of distillation, though distillation can be continued for 120 minutes for full recovery of rosemary oil under field distillation conditions (Prakasa Rao *et al.*, 1999). Oil obtained from leaves and flowering tops is of better quality than the oil from whole plant distillation. Comparing steam and water distillation the former is found better in terms of yield and quality profile of rosemary oil (Boutekedjiret *et al.*, 1997). Blanching (microwave blanching for 1 min) is observed to have a positive effect on retention of the antioxidant principles, green colour and texture of rosemary though blanching leads to total loss of volatile oils (Singh *et al.*, 1996).

Though essential oil is traditionally extracted by steam or water distillation, experiments are in progress with new extraction processes such as controlled instantaneous decompression (DIC) (Rezzoug *et al.*, 2000). This processes involves exposing the rosemary leaves for a brief period of time to steam pressure varying from 0.5 to 3 bar followed by an instantaneous decompression to vacuum (about 15 mbar). This method is reported to be faster than the conventional method.

The oil content of fresh rosemary leaves is 1% and in shade-dried leaves it increases to 3% (Farooqi and Sreeramulu, 2001). From an hectare one can harvest approximately 10–12 t herbs year^{-1}; yielding 25–100 kg oil (Farooqi and Sreeramulu, 2001). However, in field distillation conditions the oil yield varies from 0.5 to 0.9% (Prakasa Rao *et al.*, 1999). Close spacing coupled with increased nitrogen dose result in higher herbage and oil yield (Prakasa Rao *et al.*, 1999).

16.4.2 Extraction of other active compounds

For the extraction of the other active compounds of rosemary, conventional solvent extraction techniques using solvents such as hexane, benzene, ethylene, chloroform, dioxane and methanol (Chang *et al.*, 1977), distillation and super-critical fluid extraction (SFE) are routinely employed. SFE, using CO_2 is found superior to liquid solvent sonication for maximum recovery of carnosic acid in pure form (Tena *et al.*, 1997). Recent research on improving the yield and quality of rosemary extract with new techniques showed encouraging results. Superheated water under pressure between 125 and 175°C has been shown not only to rapidly extract high-quality oxygenated fragrance and flavour compounds from rosemary but also to produce in higher yields than steam distillation (Basile *et al.*, 1998). Ibanez *et al.* (1999) proposed a two-step super-critical fluid extraction and fractionation of essential oil-rich oleoresin and antioxidant compounds by varying the pressure and temperature requirements. Enzyme assisted ensiling (ENLAC) prior to polyphenol extraction is reported to double the yield of polyphenol from rosemary (Weinberg *et al.*, 1999).

16.4.3 Biotechnology
In vitro studies of rosemary, such as tissue culture, *in vitro* selection and suspended cell culture, to produce active principles are in progress in several laboratories. Callus induction potential of various explants from rosemary, essential oil profile of *in vitro* cultures under salt stress, *in vitro* production of the pigment shisonin, *in vitro* production of carnosic acid in callus cultures and regenerated shoots are some of the biotechnological studies reported (Zhu-Ru Xing *et al.*, 1996; Tawfik, 1997; Hashimoto *et al.*, 1997; Yang-Rong Hui *et al.*, 1997; Caruso *et al.*, 2000), though large-scale downstream processing of value-added products of rosemary from *in vitro* techniques or scale-up of the process for commercial exploitation is in its infancy.

16.5 Uses

16.5.1 Food processing
Rosemary is a most effective herb with a wide range of uses in food processing. In Europe and the USA, rosemary is commercially available for use as an antioxidant, though not technically listed as natural preservative or antioxidant (Yanishlieva-Maslarova and Heinonen, 2001). Rosemary has potential application in the suppression of warmed over flavour (WOF) (Valenzuela and Nieto, 1996). The main antioxidant principles in rosemary are carnosic acid, 12-methoxy carnosic acid and carnosol as well as the antioxidative diterpenes such as epirosmarinol, isorosmanol, rosmaridiphenol, rosmariquinone and rosmarinic acid (Richheimer *et al.*, 1996).

The antioxidant properties of rosemary are attributed to its ability to scavenge superoxide radicals, lipid antioxidation, metal chelating, etc. Extracts and essential oil of rosemary can be used to stabilize fats, oils and fat containing foods, butter, etc. against oxidation and rancidity (Pokorny *et al.*, 1998; Zegarska *et al.*, 1996) and fermented meat product, etc. (Korimova *et al.*, 1998). Yanishlieva-Maslarova and Heinonen (2001) reviewed the literature on rosemary and sage antioxidant principles, extraction, properties, application and the chemistry involved. Deodorized liquid, commercial antioxidant formulations of rosemary either as monoherbal or as polyherbal (rosemary, thyme, sage, oreganum) formulations, viz. 'Herbor 025', 'Spice Cocktail', etc. are available (Aruoma *et al.*, 1996).

16.5.2 Medicine
Rosemary is credited as a carminative (flavanoids); antidepressant, antispasmodic (volatile oil); rubefacient (phenolics); antimicrobial (diterpenes); emmenagogue (oleanolic acid); anti-inflammatory (carnosol); carcinogen blocker and liver detoxifier (carnosol and whole plant extract); antirheumatic (ointment of rosemary oil); and abortifacient (aqueous extract). It has an emerging potential as a source of anticancer molecules and bioavailability enhancer of cancer drugs (Jones, 2002; Plouzek *et al.*, 1999).

Traditional medicine
In traditional medicine, herbalists recommend rosemary oil against pulmonary diseases, as

stomachic, antidiarrhoeic, wound healing (poultice), choleretic and colagogenic, antidiabetic, diuretic, antidepressant and antispasmodic (Giugnolinini, 1985; Oury, 1984; Erenmemisoglu *et al.*, 1997). Commercial herbal preparations such as 'Tinctura rosmarin', 'Extractum rosmarini 150'and 'Oleum rosmarini' are available (Wolski *et al.*, 2000). The whole plant, in the form of decoction, infusion, extract in ethanol (for external application) and essential oil, is administered against digestive disorders, vaginitis, leucorrhoea, respiratory diseases, varicose vein, heart pain, inflammation and dizziness by the native people of Mexico and Central America (Santos García-Alvarado *et al.*, 2001). In Russia and Central Asian Countries of the former Soviet Union, leaves of rosemary preparation (gallenical and powder made into cigarettes) are used to treat asthma (Mamedov and Craker, 2001). The abortifacient (anti-implantation) effect of rosemary extract is also known (Lemonica *et al.*, 1996).

HIV treatment
Liquid deodorized extract of rosemary ('Herbor 025') and oily extract of mixture of herbs such as rosemary, thyme, sage and oreganum ('Spice Cocktail') inhibited human immunodeficiency virus (HIV) infection at very low concentrations though they were cytotoxic. Carnosol and carnosic acid were found top be the main active constituents of the extracts (Aruoma *et al.*, 1996). Purified carnosol @ 8 µM exhibited anti-HIV activity besides being non-cytotoxic.

Cardiovascular effects
The cardiovascular effects of rosemary extract on the isolated intact rabbit heart demonstrated significant positive inotropic effect and coronary vasodilatation (Khatib *et al.*, 1998). Administering an infusion of dried rosemary leaves resulted in decrease of blood glucose levels in normoglycaemic and diabetic nice and had no toxic effects (Erenmemisoglu *et al.*, 1997).

Cancer treatment
Rosemary is now gaining importance in cancer treatment. The major anticancer compounds identified from rosemary are carnosol, carnosic acid, ursolic acid, befulinic acid, rosmaridiphenol and rosemanol (Jones, 2002). Carnosol is found to reduce cellular nitric oxide in mice, a free radical that can damage DNA (Chan *et al.*, 1995). It is also reported that administering a whole plant extract of rosemary is more effective in preventing the carcinogen 7,12-dimethyl benz(a)anthracene (DMBA) from binding to breast cell DNA in rats than administering carnosol or ursolic acid alone (Singletary *et al.*, 1996). Rosemary extract prevents binding of aflatoxins to human liver cells and benzo(a)pyrenen to bronchial tissue (Offord *et al.*, 1995). Extract of rosemary is found to increase the intracellular accumulation of the common chemotherapy drugs such as dexorubicin (DOX) and vinblastine (VIN) in drug-resistant MCF-7 human breast cancer cells leading to increased availability of the drugs for bioactivity (Plouzek *et al.*, 1999).

Diuretic effect
Haloui *et al.* (2000) reported a diuretic effect of aqueous extract of rosemary (8% concentration) on Wister rats.

Administration
Rosemary can be administered in the form of infusion, decoction, ethanol extract (external use), tinctures, rosemary wine, drug containing volatile oil, powdered drug, liquid extract,

dry extract, etc. The infusion is prepared by adding 2 g of herb to 1 l of boiling water (Santos Garcìa-Alvarado *et al.*, 2001).

16.5.3 Aromatherapy and cosmetics

Massaging a cocktail of essential oils of rosemary, thyme, cedar wood and lavender in carrier oils of jojoba and grapeseed over a period of seven months into the scalp of patients with *Alopecia areata* is found to be a safe and effective treatment (Hay *et al.*, 1998). Rosemary oil stimulates hair follicles and circulation in scalp and thus may be helpful in premature baldness too (http://www.holistic-online.com/Herbal-Med/_Herbs/h228.htm). Rosemary has cosmetic uses as well. The oil is useful in controlling dandruff, promoting hair growth and controlling greasy hair. Rosemary oil is also used in shampoo, as it is known to impart a black colour to hair. The flower distillate in water provides a soothing eye lotion. Rosemary is used in curing acne. Rosemary oil is a component in soaps, room fresheners, deodorants, perfumes, skin lotions, etc., either in formulations with other herb oils or singly. 'Hungary water' is a perfume based on rosemary oils (http://www.kevala.co.uk/aromatherapy/rosemary. cfm). Flower, calyx and leaves of rosemary are used in potpouri, tussie-mussies, herb pillows, etc. (Bonar, 1994).

16.5.4 Recipes

Rosemary leaves and flowering tops are used in lamb roast, mutton preparations, marinades, bouquet garni, with baked fish, rice, salads, occasionally with egg preparations, dumplings, apples, summer wine cups and fruit cordials, in vinegar and oil (Bonar, 1994). Though rosemary leaves are conventionally used with roasted meat or fish dishes, they can also be used in vegetable preparations as well. Rosemary goes with potatoes and is suited to vegetables fried in olive oil (http://www-ang.kfunigraz.ac.at/katzer/ engl/Rosm_off.html). As rosemary extract is known to have antioxidant properties, it is of use in bakery, beverages, savoury foods, for retarding rancidity in fats and oils, preventing flavour degradation, etc.

Some of the popular rosemary recipes are chicken salad with rosemary, sautéed scallops with rosemary and lemon (http://www. gardensablaze.com/HerbRosemary.htm), lamb roast with rosemary, rosemary cleansing cream for oily skin (Bonar, 1994), etc. Refreshing soft drinks can be prepared by adding fresh rosemary leaves or sprigs, e.g. rosemary lemonade.

16.5.5 Herbal pesticide

Rosemary is known to possess insect repellent and antimicrobial properties. Comparative laboratory studies on the effect of dusting different herbal powder, including rosemary powder, on stored grains of wheat and French bean against *Sitophilus granarius* and *Acanthoscelides obtectus* revealed that grain wheat can be very effectively protected against *S. granarius* with the dust of rosemary (Kalinovic *et al.*, 1997). Aphids are found to be repelled by the odour of rosemary and M*entha pulegium*. Based on GC–MS analysis, a few components of rosemary oil, viz. 1,8-cineole, D-1-camphor, α,1-camphor and α-pinene, etc. have been found to be the major principles repelling the aphids (Hori and Komastu, 1997). The anti-repellent property of rosemary oil to tobacco leaf aphids, *Myzus persiacea*, is attributed to 13 compounds; the important being linalool, D-1-camphor and α-terpineol (Hori, 1998). Rosemary oil (1%) is found to reduce the fecundity rate of the predacious mites *Amplyseius zaheri* and *A. barkeri* (Momen and Amer, 1999). Repellent and oviposition deterring activities of rosemary oil on the spider mites *Tetranychus usticae* and *Eutetranychus orientalis* have also been reported (Amer *et al.*, 2001). Essential oil of rosemary in vapour form is shown to reduce fecundity, decrease egg hatchability, increase neonatal larval mortality and adversely affect offspring emergence of *Acanthoscelides obtectus* (Papachristos and Stomopoulos, 2002).

Antimicrobial activity of the essential oil of rosemary against an array of bacterial and fungal species including *Listeria monocytogenes* and *Aspergillus niger* have been reported by Faliero *et al.* (1999) and Baratta *et al.* (1998). Gram-negative bacteria such as *Staphylococcus aureus* and *S. epidermidis* have been found to be more susceptible to rosemary oil than other Gram-negative bacteria such as *Escherichia coli* and *Pseudomonas aeruginosa* (Pintore *et al.*, 2002).

Rosemary leaves are found to be a source of antibacterial molecules. An active compound effective against the plant pathogen S*treptomyces scabies*, under laboratory studies, has been isolated from the leaves of rosemary (Takenaka *et al.*, 1997).

16.6 Toxicology and disease

Rosemary is generally considered safe and devoid of toxic side effects if taken in recommended doses. However, there have been occasional reports of allergic reaction such as skin irritation. Pregnant and lactating women are advised not to use rosemary, as are pepole with epilepsy. Rosemary oil should be used with caution by persons suffering from hypertension, blood pressure or insomnia (http://www.kevala.co.uk/aromatherapy/rosemary.cfm). *In vitro* studies of liquid extract of rosemary ('Herbor 025') and a mixed oily extract of herbs such as rosemary, sage, thyme and oregano, though proven to be antiviral against HIV, is found to be cytotoxic too (Aruoma *et al.*, 1996). Rosemary leaves in excess quantity can cause coma, spasm, vomiting and, in some cases, pulmonary oedema. Rosemary oil taken orally can trigger convulsions (http://www.alternativedr.com/conditions/ConsHerbs/Rosemarych. html).

16.6.1 Disease

The important diseases affecting rosemary are collar and root rot (*Phytophthora incognitae, P. drechsleri*), foliar necrosis/leaf spot (*Alternaria alternata*), aerial blight (*Ralstonia solani* AG-4), hook disease (*Botrytis cinerea*) and powdery mildew (*Oidium* Spp.) (Perello and Dal Bello, 1995; Minuto and Garibaldi, 1996; Cacciola *et al.*, 1997; Conway *et al.*, 1997; Villevieille *et al.*, 1999). *Alternaria* leaf spot will be severe in humid and less well-ventilated areas, whereas powdery mildew assumes severity under shaded conditions. Aerial blight is a major disease in greenhouse-grown rosemary. Harvesting before blooming will help to restrict the crop loss due to hook disease. Poor drainage is conducive to root rot. Spraying and drenching with fungicides Maneb (1%) is effective against root rot. Sulphur dusting is recommended against powdery mildew. Biocontrol of powdery mildew with a commercial formulation of *Ampetomyces quisqualis* gave partial control (Minuto and Garibaldi, 1996). An isolate of biocontrol agent, *Laetisaria arvalis*, if incorporated in the pots, followed by foliar spray of fungicides at a low dose, was found to check aerial blight of rosemary better than separate applications of fungicide or biocontrol agent alone (Conway *et al.*, 1997).

The role of biocontrol agents such as *Trichoderma* Spp., fluorescent pseudomonads, are worth trying, as there is a premium price for organically produced herbs. These biocontrol agents can be components in the integrated disease management approach of rosemary.

However, in the case of aerial blight of rosemary no synergistic effect of soil amendment with *Trichoderma harzianum* and foliar spray with iprodione (fungicide) was observed (Conway *et al.*, 1997).

16.7 Conclusion

Rosemary is emerging as an important herb, being a potential source of anticancer molecules, functional food, botanical nutraceutical and functional pesticide. Despite the multifaceted importance of the herb, it has yet to receive adequate research attention. In many places, it is grown either as a minor herb on marginal lands or wild.

Varietal improvement and organic cultivation practices are two areas that require immediate attention in addition to post-harvest technology.

16.8 References

AMER S.A.A., REFAAT A.M. and MAMEN F.M. (2001), Repellent and oviposition deterring activity of rosemary and sweet marjoram on the spider mites *Tetranychus urticae* and *Eutetranychus orientalis* (Acari. Tetranychidae). *Acta Phytopathologica et Entomologica, Hungaria*, **36**(1–2), 155–64.

ANURADHA M.N., FAROOQI A.A., VASUNDHARA M., KATHIRESAN C. and SRINIVASAPPA K.N. (2002), Effect of biofertilizers on the growth, yield and essential oil content in rosemary (*Rosmarinus officinalis* L). *Proc. National Seminar, Strategies for Production and Export of Spices,* 24–26 Oct., 2002, Indian Institute of Spices Research, Calicut, India.

ARNOLD N., VALENTINI G., BELLOMARIA B. and HOCINE L. (1997), Comparative study of the essential oils from *Rosmarinus eriocalyx* Jordan & Fourr from Algeria and *R. officinalis* L. from other countries. *J. Essent. Oil Res.*, **9**(2), 167–75.

ARUOMA O.I., SPENCER J.P.E., ROSSI R., AESCHBACH R., KHAN A., MAHMOOD W., MUNOZ A., MURICA A., BUTLER J. and HALLIWELL B. (1996), An evaluation of the antioxidant and antiviral action of extracts of rosemary and Provençal herbs. *Food and Chemical Toxicol.*, **34**(5), 449–56.

BARATTA M.T., DORMAN H.J.D., DEANS S.G., BIONDI D.M. and RUBERTO G. (1998), Chemical composition, antimicrobial and antioxidative activity of laurel, sage, rosemary, oregano and coriander essential oils. *J. Essent. Oil Res.*, **20**(6), 618–27.

BASILE A., JIMENEZ CARMONA M.M. and CLIFFORD A.A. (1998), Extraction of rosemary by superheated water. *J. Agric. Food. Chem.*, **46**(12), 5205–9.

BICCHI C., BINELLO A. and RUBIOLO P. (2000), Determination of phenolic diterpene antioxidants in rosemary (*R. officinalis* L.) with different methods of extraction and analysis. *Phytochem. Analysis*, **11**(4), 236–42.

BONAR A. (1994), *Herbs – A Complete Guide to their Cultivation and Use*. Tiger Books International, London.

BOUTEKEDJIRET C., BENHABILES N.E.H., BELABBES R. and BESSIERE J.M. (1997), Effect of mode of extraction on yield and composition of the essential oil of *Rosmarinus officinalis*. *Rivista Italiana EPPOS*, No. 22, 33–35.

BOUTEKEDJIRET C., BELABBES R., BENTAHAR F. and BESSIERE J.M. (1999), Study of *R. officinalis* L. essential oil yield and composition as a function of the plant life cycle. *J. Essent. Oil Res.*, **11**(2), 238–40.

CACCIOLA S.O., PANE A., POLIZZI G. and WATERHOUSE G.M. (1997), Collar and root rot of rosemary caused by two species of *Phytophthora*. *Informatore Fitopatologic*, **47**(11), 35–42.

CARUSO J.L., CALLAHAN J., DE CHANT C., JAYASIMHULU K. and WINGET G.D. (2000), Carnosic acid in green callus and regenerated shoots of *Rosmarinus officinalis* L. *Plant Cell Reports*, **19**(5), 500–3.

DI CESARE L.F., VISCARDI D., FUSARI E.L. and NANI R.C. (2001), Volatile composition of fresh and dried rosemary. *Industrie Alimentari*, **40**(409), 1343–5.

CHAN M.M-Y., HO C.T. and HUANG H.I. (1995), Effects of three dietary phytochemicals from tea, rosemary and turmeric on inflammation induced nitrite production. *Cancer Lett.*, **96**, 23–9.

CHANG S.S., OSTIC-MATIJASAVIC B., HSIEH O.A.L. and HUANG C.L. (1977), Natural antioxidants from rosemary and sage. *Food Sci.*, **42**, 1102–6.

COELHO L.A.F., OLIVEIRA J.V. and PINTO J.C. (1997), Modelling and simulation of supercritical fluid extraction of rosemary essential oil. *Ciencia e Technologia de Alimentos*, **17**(4), 446–8.

CONWAY K.E., MANESS N.E. and MOTES J.E. (1997), Integration of biological and chemical controls for *Rhizoctonia* aerial blight and root rot of rosemary. *Plant Diseases*, **81**(7), 795–8.

DOMOKOS J., HETHELYI E., PALINKAS J., SZIRMAI S. and TULOK M.H. (1997), Essential oil of rosemary (*Rosmarinus officinalis* L.) of Hungarian origin. *J. Essent. Oil Res.*, **9**(1), 41–5.

DOULGAS S.J (1971), Cultivating condiments and seasonal plants for flavours and essential oil. *The Flav. Ind.*, 222–5.

ELAMRANI A., ZRIRA S., BENJILALI B. and BERRADA M. (2000), A study of Moroccan rosemary oils. *J. Essent. Oil Res.*, **12**(4), 487–95.

ERENMEMISOGLU A., SARAYAMEN R. and USTUN H. (1997), Effect of *Rosmarinus officinalis* leaf extract on plasma glucose level in normoglycaemic and diabetic mice. *Pharmazie*, **52**(8), 645–6.

FALIERO L., MIGUEL G.M., GUERRERO C.A.C. and BRITTO J.M.C. (1999), Antimicrobial activity of essential oils of rosemary (*R. officinalis* L) and thyme (*Thymus mastichina* L). *Proc. Sec. World Congress on Medi. & Aromatic Plants for Human Welfare, WOCAMP-2. Acta Horticulture*, No. 501, 45–8.

FAROOQI A.A. and SREERAMULU B.S. (2001), *Cultivation of Medicinal and Aromatic Plants*. University Press India Ltd, Hydrabad, India, pp. 433–43.

GIUGNOLININI L. (1985), *Erbe Secondo Natura*. Secondo Natura Laboratorio Grafico, Vignate, Milano.

GUAZZI E., MACCIONI S., MONTI G., FLAMINI G., CIONI P.L. and MORELLI I. (2001), *Rosmarinus officinalis* in the gravine of Palagianello, Taranto, South Italy. *J. Essent. Oil Res.*, **13**(4), 231–3.

HALOUI M., LOUEDEC L., MICHEL J.B. and LYOUSSI B. (2000), Experimental diuretic effects of *Rosemarinus officinalis* and *Centaurium erythraea*. *J. Ethnopharmacol*, **71**(3), 465–72.

HASHIMOTO O., MIZUTANI H. and NAKASHIMA R. (1997), Production of shisonin in rosemary (*Rosmarinus officinalis* L.) tissue culture. *Nippon Nogeikagaku Kaishi*, **71**(6), 605–6.

HAY I.C., JAMIESON M. and ORMEROD A.D. (1998), Randomized trial of aromatherapy: successful treatment of *Alopecia areata*. *Arch. Dermat.*, **134**(11), 1349–52.

HIDALGO P.J., UBERA J.L., TENA M:T. and VALCARCEL M. (1998). Determination of the carnosic acid content in wild and cultivated *R. officinali.*, *J. Agric. Food Chem.*,**46**(7), 2624–7.

HIDALGO-FERNANDEZ P.J., VICENTE P.R., MALDONADO J.M. and UBERA J.J.L. (1999), Mitochondrial DNA polymorphism and gynodioecy in a natural population of *Rosmarinus officinalis* L.. *Israel J. Pl. Sci.*, **47**(2), 77–83.

HORI M. (1998), Repellency of rosemary oil against *Myzus persicae* in a laboratory and in a screen house. *J. Chem. Ecol.*, **24**(9), 1425–32.

HORI M. and KOMATZU H. (1997), Repellence of rosemary oil and its components against the onion aphid, *Neotoxoptera fermosana* (Takahashi) (Homoptera, Aphididae). *Appl. Ent. Zool.*, **32**(2), 303–10.

IBANEZ E., OCA A., DE MURGA G., LOPEZ SEBASTIAN S., TABERA J. and REGLERO G. (1999), Supercritical fluid extraction and fractionation of different preprocessed rosemary plants. *J. Agri. Food Chem.*, **47**(4), 1400–4.

JONES C. (2002), Rosemary's whole plant properties counter cancer. *Health and Nutrition Breakthrough* 07/98 (http://www.healthwell.Com/hnbreak throughs/ jul98/rosemary.cfm).

KALINOVIC I., MARTINCIC J., ROZMAN V. and GUBERAC V. (1997), Insecticidal activity of substances of plant origin against stored product insects. *Ochrana Rostlin*, **33**(7), 135–42.

KANDEEL A.M. (2001), Effect of irrigation intervals on the growth and active ingredients of *Rosmarinus officinalis* L. plants. *Arab Univ. J. Agric. Sci.*, **9**(2), 825–38.

KHATIB S., ALKOFAHI A., HASAN M. and NAJIB N. (1998), The cardio vascular effects of *Rosmarinus officinalis* extract on the isolated intact rabbit heart. *Fitoterapia*, **69**(6), 502–6.

KORIMOVA L., MATE D. and TUREK P. (1998), The evaluation of raw fermented meat products stabilized with vitamin E and rosemary. *Folio Veterinaria*, **42**(4), 178–81.

LEMONICA I.P., DAMASCENO D.C. and SATSI DI L.C. (1996). Study of embryotoxic effects of an extract of rosemary (*Rosmarinus officinalis*). *Braz. J. Med. Biol. Res.*, **29**(2), 223–7.

MAMEDOV N. and CRAKER L.E. (2001), Medicinal plants used for the treatment of bronchial asthma in Russia and Central Asia. *J. Herbs, Spices and Med. Plants*, **8**(213), 119–60.

MILIA M., PINNA M.E., SATTA M. and SCARPA G.M. (1996), Preliminary examinations of agronomic aspects in *Rosmarinus officinalis* L. Seminario Internazionala di Chiusura del Progetto PIM Sardagna, Cagliari, 18–19 Nov, 1994. *Rivista Italiana EPPOS*, No. 19, 125–30.

MINUTO G. and GARIBALDI A. (1996), Comparison of methods of control of powdery mildew (*Oidium* Spp.) on rosemary. *Informatore Fitopatologico*, **46**(12), 33–7.

MOMEN M.E. and AMER S.A.A. (1999), Effect of rosemary and sweet marjoram on three predacious mites of the family Phytoseiidae (Acari: Phytoseiidae). *Acta Phytopathog.-Entomolog. Hungaria*, **34**(4), 355–61.

MORETTI M.D.L., PEANA A.T., PASSINO G.S. and SOLINAS V. (1998a). Effect of soil properties on yield and composition of *Rosmarinus officinalis* L. essential oil. *J. Essent. Oil Res.*, **10**(3), 261–7.

MORETTI M.D.L., PEANA A.T., PASSINO G.S., BAZZONI A. and SOLINAS V. (1998b). Effect of iron on yield and composition of *Rosmarinus officinalis* L. essential oil. *J. Essent. Oil Res.*, **10**(1), 43–9.

MULAS M., BRIGAGLIA N., CANI. M.R. and SCANNERINI S. (1998), Clone selection from spontaneous germplasm to improve *Rosmarinus officinalis* L. crops. *Acta Horticulturae*, No. 457, 287–94.

OFFORD E.A, MACE K., RUFFIEUX C. and PFEIFER A.M. (1995). Rosemary components inhibit benzo(a) pyrene induced genotoxicity in human bronchial cells. *Carcinogenesis*, **16**, 2057–62.

OUAHADA S. (2000), Tunisian rosemary oil. *Perf. Flav.*, **25**(6), 24–5.

OURY P. (1984), *Eucyclopedia des Plants et Fleurs Medícínale*. Vol. IV, Editions Alliance, Paris.

PAPACHRISTOS D.P. and STOMOPOULOS D.C. (2002), Repellent, toxic and reproduction inhibitory effect of essential oil vapour on *Acanthoscelides obtecus* (Say) (Coleptera: Bruchidae). *J. Stored Prod. Res.*, **38**(2), 117–20.

PINTORE G., USAI M., BRADESI P., JULIANO C., BOATTO TOMI F., CHESSA M., CERRI R. and CASANOVA J. (2002), Chemical composition and antimicrobial activity of *R. officinalis* L. oils from Sardinia and Corsica. *Flavour Fragr. J.*, **17**(1), 15–19.

PERELLO A. and DAL BELLO G.M. (1995), Foliar necrosis caused by *Alternaria* on rosemary and *Colletotrichum* spp. on lavender, sage and marjoram. *Investigaçion Agraria Produccion y Protection Vegetales*, **10**(2), 275–81.

PLOUZEK C.A., CIOLINO H.P., CLARKE R. and YEH G.C. (1999), Inhibition of *p*- glycoprotein activity and reversal of multi drug resistance *in vitro* by rosemary extract. *Europ. J. Cancer*, **35**(10), 154–5.

POKORNY J., REBOLVA Z. and JANITZ W. (1998), Extracts from rosemary and sage as natural antioxidants for fats and oils. *Czech. J. Food. Sci.*, **16**(6), 227–34.

PORTE A., GODOY R.L. DE, LOPES D., KOKETSU M., GONÇALVES S.L. and TORQUILHO H.S. (2000), Essential oil of *Rosmarinus officinalis* (rosemary) from Rio de Janeiro, Brazil. *J. Essent. Oil Res.*, **12**(5), 577–80.

PRAKASA RAO E.V.S., GOPINATH C.T., GANESHA RAO R.S. and RAMESH S. (1999), Agronomic and distillation studies on rosemary (*Rosmarinus officinalis* L.) in a semi arid tropical environment. *J. Herbs, Spices and Med. Plants*, **6**(3), 25–30.

REZZOUG S.A., LOUKA N. and ALLAF K. (2000), Effect of main processing parameters of the instantaneous controlled pressure drop process on oil isolation from rosemary leaves – kinetic aspects. *J. Essent. Oil Res.*, **12**(3), 336–44.

RICHHEIMER S.L., BERNART M.W., KING G.A., KENT M.C. and BAILEY D.T. (1996), Antioxidant activity of lipid soluble diterpenes from rosemary. *J. Amer. Oil Chem. Soc.*, **73**(4), 507–14.

ROSE J. (1974), *Herbs and Things*. Workman, New York.

SANTOS GARCÍA-ALVARADO J., JULIA VERDE-STAR M. and HEREDIA N.L. (2001), Traditional uses and scientific knowledge of medicinal plants from Mexico and Central America. *J. Herbs, Spices and Med. Plants*, **8**(213), 37–90.

SHAH S.C., GUPTA L.K., BHUJWAN H.S. and BHUJWAN S.H. (1996), Effect of auxins on rooting of stem cuttings of rosemary (*Rosmarinus officinalis* L.). *Recent Hort.*, **3**(1), 126–8.

SILVA C. DE P. and PEDRAS J.F. (1999), Early rooting in rosemary (*Rosmarinus officinalis* L.) cuttings under the influence of chemical treatment and collecting time. *Acta Horticult.*, No. 502, 213–7.

SINGH M. and RAMESH S. (2000), Effect of irrigation and nitrogen on herbage, oil yield and water use efficiency in rosemary grown under semi-arid tropical conditions. *J. Med. Aromatic Plants Sci.*, **22**(1B), 659–62.

SINGH M., RAGHAVAN B. and ABRAHAM K.O. (1996), Processing of marjoram (*Marjoram hortensis* M) and rosemary (*Rosmarinus officinalis* L) – effect of blanching method on quality. *Nahrung*, **40**(5), 264–6.

SINGLETARY K., MACDONALD C. and WALLIG M. (1996), Inhibition by rosemary and carnosol of 7,12-dimethylbenz(a)anthracene (DMBA) induced rat mammary tumorigenesis and *in vitro* DMBA–DNA adduct formation. *Cancer Lett.*, **104**(1), 43–8.

SOLINAS V., DEIANA S., GESSA C., BAZZONI A., LODDO M.A. and SATTA D. (1996), Effect of water and nutritional conditions on the *Rosmarinus officinalis* L. – phenolic fraction and essential oil yields. Seminario Internazionale di Chiusura del Progetto PIM, Sardegna, Cagliari, 18 Nov. 1994, *Rivista Italiana EPPOS*, No 19, 189–98.

TAKENAKA M., WATANABE T., SUGAHARA K., HARADA Y., YOSHIDA S. and SUGAWARA F. (1997), New antimicrobial substances against *Streptomyces scabies* from rosemary (*Rosmarinus officinalis* L.) leaves. *Biosci. Biotech. Bioch.*, **61**(9), 1440–4.

TARRAF S. and IBRAHIM M.E. (1999), Physiological response of rosemary (*Rosmarinus officinalis* L.) plant to brassinosteroid and uniconazole. *Egypt J. Hort.*, **26**(3): 405–19.

TAWFIK A.A. (1997), *In vitro* selection for salt tolerance in rosemary (*Rosmarinus officinalis* L.). *Assiut. J. Agri. Sci.*, **28**(4), 67–78.

TENA M.T., VALCARCEL M., HIDALGO P.J. and UBERA J.L. (1997), Supercritical fluid extraction of natural antioxidants from rosemary: comparison with liquid solvent sonication. *Analyt. Chem. Washington*, **69**(3), 521–6.

VALENZUELA A.B. and NIETO S.K. (1996), Synthetic and natural antioxidants: food quality protectors. *Grasas Y Acetias*, **47**, 186–96.

VASUNDHARA M., KHAN M.M., FAROOQI A.A. and NUTHAN D. (2002), Effect of drip fertigation on yield and cost economics in rosemary (*R. officinalis* L.). *Indian Perfumer*, **46**(2), 179–90.

VILLEVIEILLE M., CASSINI R. and NICOT P. (1999), Hook disease in tips of rosemary. *Phytoma*, No. 517, 50–3.

WEINBERG Z.G, AKIRI B., POTOYEVSKI E. and KANNER J. (1999), Enhancement of polyphenol recovery from rosemary (*R. officinalis* L.) and sage (*Salvia officinalis*) by enzyme assisted ensiling (ENLAC). *J. Agric. Food. Chem.*, **47**(7), 2959–62.

WOLSKI T., LUDWICZUK A., ZWOLAN W. and MARADAROWICZ M. (2000), GC/MS analysis of the content and composition of essential oil in leaves and gallenic preparations from rosemary (*Rosmarinus officinalis* L.). *Herba Polonica*, **46**(4), 243–8.

YANG-RONG HUI, POTTER T.P., CURTIS O.F., SHETTY K. and YANG R.H. (1997), Tissue culture based selection of high rosmarinic acid producing clones of rosemary (*Rosmarinus officinalis* L.) using *Pseudomonas* strain F. *Food. Biotech.*, **11**(1), 75–88.

YANISHLIEVA-MASLAROVA N.V. and HEINONEN I.M. (2001), Rosemary and sage as antioxidants. In *Handbook of Herbs and Spices* Vol. 1, (Ed.) Peter, K.V., Woodhead Publishing Ltd, Abington, pp. 269–75.

ZEGARSKA Z., AMAROWICZ R., KARAMAC M. and RAFALOWSKI R. (1996), Antioxidative effect of rosemary ethanolic extract on butter. *Milchwissenschaft*, **51**(4), 195–8.

ZHU-RU XING., RAO HONG YU., GONG LAN, ZHU RX., RAO HY and GONG L. (1996), Suspended cell culture of *Rosmarinus officinalis* and production of essential oil, *Plant Physiol. Comm.*, **32**(1), 9–12.

17
Sesame

D. M. Hegde, Directorate of Oilseeds Research, India

17.1 Introduction

Sesame (*Sesamum indicum* L.), also known as sesamum, gingelly, beniseed, *sim-sim* and *til*, is perhaps the oldest oilseed known and used by human beings (Joshi, 1961; Weiss, 2000). It has been cultivated for centuries, particularly in Asia and Africa, for its high content of edible oil and protein. The crop is now grown in a wide range of environments, extending from the semi-arid tropics and subtropics to temperate regions. Consequently, the crop has a large diversity in cultivars and cultural systems. Although considered to have originated in central Africa, most probably Ethiopia, many believe that there is convincing evidence to show that sesame originated in India (Bedigian and Harlan, 1986). Sesame was widely dispersed by people both westward and eastward, reaching China and Japan, which themselves became secondary distribution centres.

Sesame is an important cash crop for small and marginal farmers in several developing countries. It is cultivated for its seeds, which contain 38–54% oil of very high quality and 18–25% protein. The great diversity of sesame types, their wide environmental adaptation and considerable range of seed oil content and characteristics make an exceptional gene pool. This gene pool must be harnessed to produce better cultivars to extend the range and profitability of sesame growing. The major obstacles to sesame's expansion are its low yields and the absence of non-shattering cultivars suitable for machine harvest.

17.1.1 Classification and species relationship

The genus *Sesamum*, 1 among the 13 genera of the family Pedaliaceae, consists of about 40 species, 36 in the *Index Kewensis*. Many occur in Africa (18 exclusively), 8 occur in the India–Sri Lanka region (5 exclusively). The Australian records are probably due to imports by Chinese immigrants in the mid-19th century (Bennett, 1996). The cytogenetic knowledge of the genus is very limited, thus the chromosome numbers are known for only about one-third of the species.

Sesamum indicum, together with *S. capense* Burn (*S. alatum* Thonn), *S. malabaricum* Nar. and *S. schenkii* Aschers, have the somatic number $2n = 26$; *S. laciniatum*, $2n = 28$; *S. angolase* and *S. prostratum* $2n = 32$; *S. occidentale* and *S. radiatum* Schum & Thonn,

$2n = 64$. The Indian *S. mulayanum* is very similar to *S. indicum*, but has the valuable characteristic of being resistant to phyllody and wilt.

The progenitor species of the cultivated *S. indicum* are unknown as no wild species except *S. malabarium*, which produces fertile hybrids with *S. indicum*, are known. The true wild species of sesamum found in tropical Africa and India produce viable F_1 seeds but F_1 hybrids are sterile, except for *schenkii-indicum* hybrids, which show some end-season fertility.

Sesamum indicum has a number of local cultivars as noted in the literature, but it is claimed the genus *Sesamum* has only one cultivated species, which can be divided by seed colour into white sesame *S. indicum* spp., *indicum*, and variable sesame *S. indicum* spp., *orientale* (Zhang *et al.*, 1990). Based on seed colour and inheritance, it is postulated that sesame evolved from symmetric to asymmetric types and from *S. capense* to *S. indicum* in the sequence black, brown, yellow and white seeded types (He *et al.*, 1994). Research into *S. indicum*, *S. alatum*, *S. radiatum* and *S. angustifolium* (Olive) Engl. by analysing the fatty acid composition of the total lipids and of the different acyl-lipid classes indicated that *S. radiatum* and *S. angustifolium* are more closely related to the other two species (Kamal-Eldin and Appelquist, 1994).

Polyploidy can be induced, but colchicine-treated plants tend to produce a low yield even when fertility of pollen is high and seeds per capsule not reduced. The number of capsules set is often very low, but seed oil content can be extremely high. The rate of growth and general vigour of tetraploids can exceed that of diploids, the plants are taller at maturity, with longer leaves, larger flowers, capsules and pollen grains. Sterility of pollen grains in tetraploids can vary between 20 and 40% and of diploids between 5 and 30%.

Interspecific hybridization is possible and crosses may produce viable seed (Prabakaran and Rangasamy, 1995). The discovery of genetic male sterility in sesame (Osman, 1981) has eased the production of hybrid seed (Tu, 1993; Ganesan, 1995; Kang and Lee, 1995; Wang *et al.*, 1995).

17.1.2 Morphology and biology

There are many hundreds of varieties and strains of *Sesamum indicum*, which differ considerably in size, form, growth, colour of flowers, seed size, colour and composition. Cultivated sesame is typically an erect, branched annual, occasionally perennial, 0.5–2 m in height, with a well-developed root system, multi-flowered, whose fruit is a capsule containing a number of small oleaginous seeds.

Sesame has a tap root system with profuse lateral branches. Long-season types, occasionally treated as perennials have an extensive and penetrating root system and short-season types have less extensive and more shallow roots. Root growth is also influenced by the soil type, season and soil moisture conditions. Root growth is inhibited by excess soil moisture and relatively low salt concentrations, much lower than is tolerated by safflower, for instance.

The stem is erect, normally square in section with definite longitudinal furrows, although rectangular and abnormally wide, flat shapes occur. It can be smooth, slightly hairy or very hairy and these characteristics are used to differentiate the types. The stem is light green to purple, branching angular and straight with an average height of 1 to 1.5 m and sometimes up to 3 m. The extent and type of branching are varietal characteristics, as is the height at which the first branch occurs. The degree of branching is directly affected by the environment. Short-stemmed, little-branched types are generally early maturing; the taller branched types are late-maturing and tend to be more drought resistant.

Leaves on sesame plants are most variable in shape and size on the same plant and between varieties. Usually, lower leaves are broad, sometimes lobal, margins often prominently toothed with the teeth diverted outwards. Intermediate leaves are entire, lanceolate, sometimes slightly separated. The upper leaves are more narrow and lanceolate. Leaf size varies from 3 to 17.5 cm in length, 1 to 7 cm in width with a petiole of 1 to 5 cm in length. The surface of leaves is generally glabrous but in some types may be pubescent. Generally of a dull, darkish-green, leaves can be much lighter with occasionally a yellowish tint or bluish when leaves are very hairy. Leaf arrangement may be alternate or opposite or mixed or opposite below and alternate above and varies with varieties. There is a basic difference in the rate of water conductance between leaves of indehiscent and dehiscent sesame, the former being much faster. These varieties are thus less suited to areas with limited water supply.

Flowers arise in the axils of leaves and on the upper portion of the stem and branches and the node number on the main shoot at which the first flower is produced is a varietal characteristic and highly heritable (Mohanty and Sinha, 1965). Flowers occur singly on the lower leaf axils with multiple flowers on the upper stem or branches. When borne single, two lateral flowers are observed as rudimentary buds (nectarial glands) at the base of the fully developed ones. They are invariably pilose and show a fair range of variability in size, colour and marking on the inside of the corolla tube. Flowers are borne on very short pedicles. Two short linear bracts arise at the base of the pedicle just below the nectaries which are shed when flowers mature. Calyx lobes are short, velvety, narrow, acuminate and united at the base. The five lobes are of variable sizes, the lower one being the longest and upper one the shortest. The flower is zygomorphic with a slightly bilabiate tubular corolla of five lobes. The upper lip of the corolla is entire, the lower divided into three, of which the central division is the longest. The corolla is usually white or pale pink but purple is also observed. The inner surface of the corolla tube may have red spots or the lower portion only may be black spotted or, occasionally, have purple or yellow blotches.

Stamens are attached to the tube of the corolla. Of the five stamens, four are functional and the fifth is either sterile or completely lacking. The four greenish white functional stamens are arranged in pairs, one pair being shorter than the other. There are two anther cells, opening longitudinally, connective usually gland tipped.

The overy is superior, usually two-celled, cells often completely or partially divided by false septa. The style is terminal, filiform and simple. The stigma is usually two-lobed and hairy.

When there are three flowers per leaf axil, the central bud blooms first and the two side buds open several days later. Flowers open early in the morning, 95% between 5 and 7 a.m., wilt after midday and are usually shed in the evening, majority between 4.30 and 6.00 p.m. Anthers open longitudinally and release pollen shortly after the flowers open, the interval varying with variety. The stigma is receptive one day prior to flower opening and remains receptive for a further day. Under natural conditions, pollen remains viable for approximately 24 hours. Low temperature at flowering can result in sterile pollen, or premature flower fall. Conversely, periods of high temperature, 40°C or above at flowering, will seriously affect fertilization and reduce the number of capsules produced.

Sesame is considered a self-pollinated crop, giving full seed set under isolation. However, the flowers attract insects and their activity can lead to different rates of cross-pollination, from a commonly reported few per cent to as high as 65% (Yermanos, 1980).

The fruit is a capsule, rectangular in section and deeply grooved with a short triangular beak. Capsule shape is a varietal characteristic with environment a major modifying factor. Capsule length can vary from 2.5 to 8 cm, with a diameter of 0.5 to 2 cm, the number of loculi

Table 17.1 Proximate composition of whole sesame seeds

Constituent (%)	Joshi (1961)	Smith (1971)	Gopalan et al. (1982)	Weiss (1983)
Moisture	5.8	8.0	5.3	5.4
Protein	19.3	22.0	18.3	18.6
Fat	51.0	43.0	43.3	49.1
Carbohydrate	21.2	21.0	25.0	21.6
Ash	5.7	6.0	5.2	5.3

from 4 to 12, and capsules are usually hairy to some degree. The capsule dehisces, splitting along septa from top to bottom or by means of two apical pores. The degree of dehiscence and the above-ground height of the first capsule are varietal characteristic (Weiss, 1983).

Sesame seeds are small, ovate, slightly flattened with testa of variable colour varying from black, white, yellow, reddish brown, grey, dark grey, olive green and dark brown. The dry matter content in seed increases most rapidly between 12 and 24 days, in parallel with the rate of oil synthesis and continues to increase slowly until maturity (Aiyadurai and Marar, 1951). There is significant difference between cultivars in respect of capsule length, number of seeds per capsule and seed size. A capsule may contain 50 to 100 or more seeds. Seed weight is around 3 g/1000 seeds. The seeds mature four to six weeks after fertilization.

17.2 Chemical composition

Sesame seed has a high food value due to its high content of oil and protein. The composition is markedly influenced by genetic and environmental factors (Kinman and Stark, 1954; Lyon, 1972). The seeds contain 6–7% moisture, 17–32% protein, 48–55% oil, 14–16% sugar, 6–8% fibre and 5–7% ash. The proximate composition of sesame seeds is given in Table 17.1.

In general, Indian varieties tend to be lower in protein and higher in oil than Sudanese varieties, such as those generally appearing in the export market, and are commercially used in the USA. The hull content averages about 17% of the sesame seed, and contains large quantities of oxalic acid, calcium, other minerals and crude fibre. Thus, when using sesame for human food, it is advisable to remove the hull. When the seed is properly dehulled, the oxalic acid content is reduced from about 3% to less than 0.25% of the seed weight (Nagaraj, 1995). Screw-pressed, dehulled sesame contains about 56% protein, while the solvent extracted meal contains more than 60% protein. This is mostly used in feed, except in India where it is used as a food.

17.2.1 Lipids

Content

Sesame seeds contain more oil than many other oilseeds. Oil content varies with genetic and environmental factors. A wide range of the oil content from 37 to 63% has been reported in sesame seeds (Lyon, 1972; Swern, 1979; Bernardini, 1986). Oil content in seeds also varies considerably among different varieties and also growing seasons (Lyon, 1972; Yen et al., 1986). The oil content is also related to the colour and size of the seeds. White or light coloured seeds usually have more oil than the dark seeds and smaller seeds contain more oil than larger seeds (Seegeler, 1983). Rough-seeded cultivars generally have a lower oil content than smooth-seeded types (Yermanos *et al.*, 1972).

Agronomic factors also influence the seed oil content. It increases with increasing length of photoperiod and early planting dates (Arzumanova, 1963; Abdel-Rahman *et al.*, 1980). Likewise, the seeds from plants with a short growing period tend to have higher oil content than those from plants with a medium to long growing cycle (Yermanos, 1978). Heavy application of nitrogen fertilizer reduces oil content of sesame seeds (Singh *et al.*, 1960).

Classification
The lipids of sesame seeds are mostly composed of neutral triglycerides with small quantities of phosphatides (0.03–0.13% with lecithin : caphalin ratio of 52 : 46). The phosphatides also contain about 7% of a fraction soluble in hot alcohol but are insoluble when cold. Sesame oil, however, has a relatively high percentage (1.2%) of unsaponifiable matter (Johnson and Raymond, 1964; Weiss, 1983). The glycerides are mixed in type, principally oleo-dilinoleo, linoleo-dioleo triglycerides and triglycerides with one radical of a saturated fatty acid combined with one radical each of oleic and linoleic acids (Lyon, 1972). The glycerides of sesame oil, therefore, are mostly triunsaturated (58 mol%) and diunsaturated (36 mol%) with small quantities (6 mol%) of monounsaturated glycerides. Trisaturated glycerides are almost absent in sesame oil. The unsaponifiable matters in sesame oil include sterols (principally comprising β-sitosterol, compesterol and sigma sterol), triterpenes (triterpene alcohols which include at least six compounds, of which three were identified, viz. cycloartanol, 24-methylene cycloartanol and amyrin), pigments, tocopherols and two compounds that are not found in any other oil, namely sesamin and sesamolin (Fukuda *et al.*, 1981, 1988). Sesamin and sesamolin are responsible for the characteristic Baudouin or Villavecchia tests of sesame oil. Among the pigments spectroscopically identified, pheophytin A (λ_{max} = 665–670 nm) was found to markedly predominate over pheophytin b (λ_{max} = 655 nm) (Lyon, 1972). The pleasant aroma and taste principles contain C_5-C_9 straight-chain aldehydes and acetylpyrazine (Swern, 1979).

Fatty acid composition
Sesame oil contains about 80% unsaturated fatty acids. Oleic and linoleic acids are the major fatty acids and are present in approximately equal amounts (Lyon, 1972). The saturated fatty acids account for less than 20% of the total fatty acids. Palmitic and stearic acids are the major saturated fatty acids in sesame oil (Table 17.2). About 44 and 42% of linoelic and oleic acids and 13% saturated fatty acids are found in sesame oil (Smith, 1971). Arachidic and linolenic acids are present in very small quantities (Rao and Rao, 1981).

Table 17.2 Fatty acids composition of sesame oil (% of total fatty acids)

Fatty acid	Godin and Spensley (1971)	Yermanos (1978)	Seegeler (1983)	Maiti *et al.* (1988)
Palmitic	7–9	8.3–10.9	8.4–10.3	7.8–9.1
Stearic	4–5	3.4–6.0	4.5–5.8	3.6–4.7
Arachidic	8	–	0.3–0.7	0.4–1.1
Oleic	37–50	32.7–53.9	39.5–43.0	45.3–49.4
Linoleic	37–47	39.3–59.0	41.0–45.0	37.7–41.2

Endogenous antioxidants
Among the commonly used vegetable oils, sesame oil is known to be most resistant to oxidative rancidity (Budowski, 1950). It also exhibits noticeably greater resistance to autooxidation than would be expected from its content of tocopherols (vitamin E). This high stability to oxidation is often attributed to the presence of a large proportion of unsaponifiable matter. Moreover, the unsaponifiable matter itself includes substances such as sesamol and phytosterol that are normally not found in other oils. Sesamolin upon hydrolysis, yields sesamol. Sesame oil contans 0.5–1.0 % sesamin (Budowski *et al.*, 1951) and 0.3–0.5% sesamolin (Budowski *et al.*, 1950), with only traces of free sesamol (Beroza and Kinman, 1955; Budowski, 1964). Sesamol is released from sesamolin by hydrogenation, by acid or acid bleaching earth or other conditions of processing and storage (Budowski and Markley, 1951; Beroza and Kinman, 1955). Free sesamol is, however, removed by some blending earths or during the deodorization process, which results in decreased stability of sesame oil (Budowski and Markley, 1951; Mathur and Tilara, 1953; Budowski, 1964). Structures of natural antioxidants found in oil from sesame are depicted in Figs. 17.1 and 17.2.

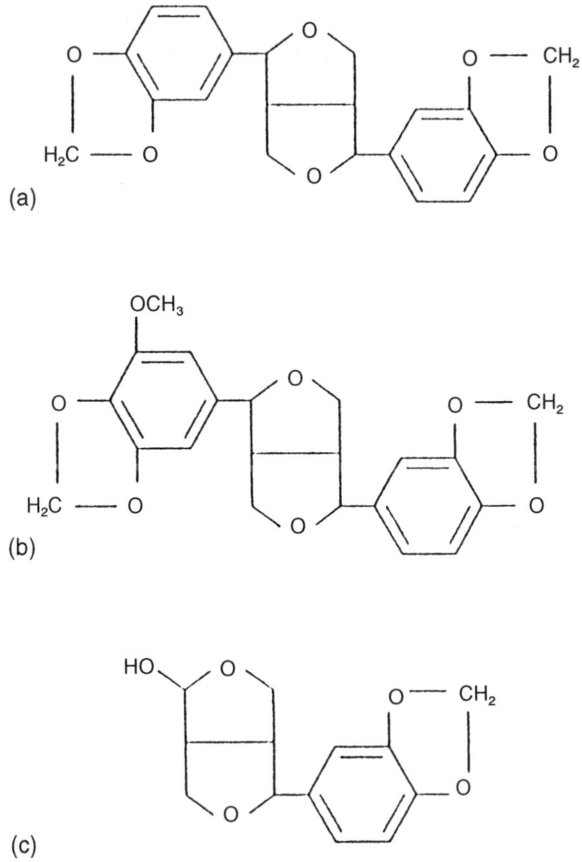

Fig. 17.1 Structures of natural antioxidants found in sesame oil: (a) sesamin; (b) sesangolin; and (c) samin.

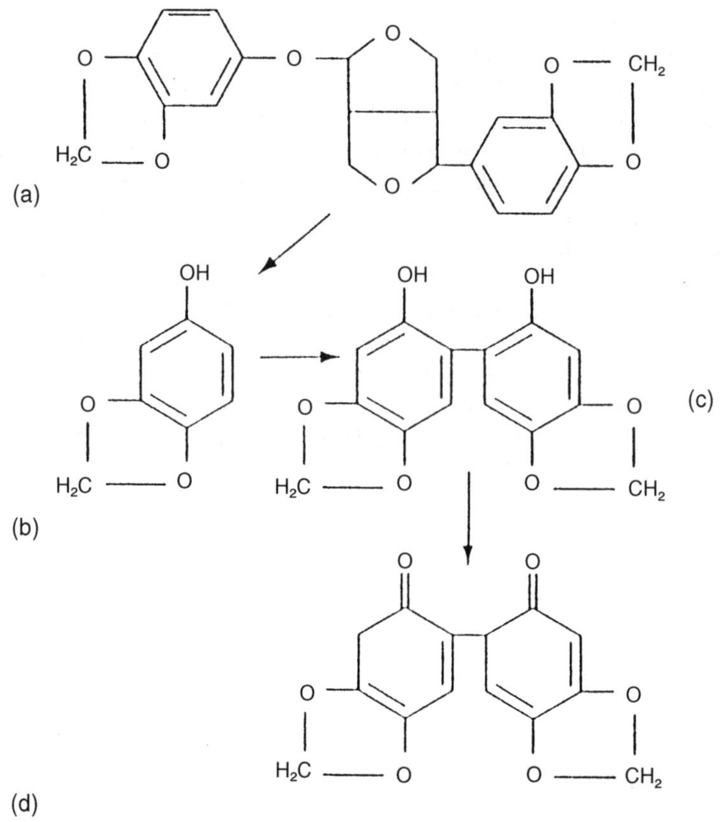

Fig. 17.2 Structures of natural antioxidants found in sesame oil: (a) sesamolin; (b) sesamol; (c) sesamol dimer; and (d) sesamol dimer quinone.

Properties of oil
Sesame oil is deep to pale yellow in colour. It is fragrant or scented. It has a pleasant odour and taste. The aroma components were identified as C_5 to C_9 straight chain aldehyde or ketone derivatives (Lyon, 1972). Some of the important characteristics of sesame oil are given in Table 17.3. Sesame oil is dextrorotatory, which is unusual for an oil devoid of optically active fatty acid glycerides. The unsaponifiable fraction of the oil, however, does contain optically active minor constituents, which are responsible for the optical rotation of the oil.

Nutritional importance
Sesame oil is practically free of toxic components. The oil contains more unsaturated fatty acids than many other vegetable oils. The high proportion of unsaturated fatty acids renders sesame oil an important source of essential fatty acids in the diet (Langstraat and Jurgens, 1976). Linoleic acid is required for cell membrane structure, cholesterol transportation in the blood and for prolonged blood clotting properties (Vles and Gottenbos, 1989). Sesame oil is rich in vitamin E, but deficient in vitamin A. The crude oil contains a relatively low amount of free fatty acids. The minor constituents in sesame oil, sesamin and sesamolin, protect the oil from oxidative rancidity.

Table 17.3 Characteristics of sesame oil

Character	Andraos et al. (1950)	Lyon (1972)	Seegeler (1983)	Weiss (1983)
Specific gravity (25%/25°C)	0.918	0.918–0.921	0.916–0.921	0.922–0.924
Refractive index (n^{50}_D)	1.463 (25°C)	1.472–1.474 (25°C)	1.463–1.474 (25°C)	1.458 (60°C)
Smoke point (°C)	165	–	166	–
Flash point (°C)	319	–	375	–
Solidifying point (°C)	–	–	–3 to –4	–3 to –4
Titre (°C)	22	20–25	20–25	22–24
Free fatty acids (as % oleic)	1.0	–	1.0–3.0	1.0–3.0
Unsaponifiable matter (%)	2.3	1.8	0.9–2.3	0.9–2.3
Iodine value	112	104–118	103–130	103–126
Saponification value	186	187–193	186–199	188–193
Reichert–Meissel value	0.51	–	0.1–0.2	0.1–1.0
Polenske value	0.4	–	0.10–0.50	–
Hydroxyl number	5.3	–	1.0–10.0	1.0–10.0
Thiocyanogen value	76	–	74–76	74–76
Hehner value	–	–	96.0	95.7

17.2.2 Proteins

Content and characterization
Sesame seed contains 17–32% protein with an average of about 25% (Joshi, 1961; Lyon, 1972, Yen et al., 1986). The proteins in the seed are located mostly in the outer layers of the seed. Based on their solubility, sesame proteins have been classified as albumin (8.6%), globulins (67.3%), prolamin (1.3%) and glutelin (6.9%) fractions (Rivas et al., 1981).

As in most other seeds, globulin is the predominant protein fraction in sesame seeds (Guerra et al., 1984). It is composed of two components. α-Globulin is the major fraction and accounts for about 60–70% of the total seed globulin, while β-globulin is a minor component contributing 25% to the globulin fraction (Nath and Giri, 1957; Rajendran and Prakash, 1988). α-Globulin is a high molecular weight protein (250 000–360 000 MW) and has a sedimentation coefficient of 11–13 S. It is an oligomeric protein composed of six dimeric units of molecular weight of about 50 000–60 000. The dimeric unit is of the A–B type linked by a disulphide bond (Robinson, 1987). The quaternary structure of α-globulin has been well established (Plietz et al., 1988). β-Globulin is the minor component of sesame seed globulins. It has a molecular weight of 150 000 and is rich in acidic and hydrophobic amino acids (Plietz et al., 1988).

Nutritional quality
The essential amino acid composition of sesame seed proteins (Table 17.4) indicates that sesame proteins are rich in sulphur-containing amino acids, particularly methionine (Smith, 1971; Brito, 1981; Narasinga Rao, 1985) and also tryptophan (Johnson et al., 1979; Yen et al., 1986). Sesame proteins are, however, deficient in lysine (Evans and Bandemer, 1967; Cuca and Sunde, 1967; Narasinga Rao, 1985; Sawaya et al., 1985; Yen et al., 1986), which is unusual for oilseed proteins. Among other essential amino acids, sesame protein is borderline deficient for threonine, isoleucine and valine contents compared with Food and Agriculture Organization (FAO) reference values (Nath et al., 1957). During preparation of a protein isolate (>90% protein), there is some loss of methione, cystine and tryptophan.

Table 17.4 Essential amino acid composition of sesame meal proteins (g/16 g N)

Amino acid	Evans and Bandemer (1967)[a]	Smith (1971)	Rivas et al. (1981)	Gopalan et al. (1982)	Narsinga Rao (1985)	FAO/WHO (1973)
Arginine	12.0–13.0	11.9	12.5	12.0	–	–
Histidine	2.3–2.8	2.2	2.4	2.7	–	–
Isoleucine	3.3–3.6	4.3	3.9	4.0	4.2	4.7
Leucine	6.5–7.0	6.9	6.7	8.0	7.4	7.0
Lysine	2.5–3.0	2.8	2.6	2.7	2.6	5.5
Methionine	2.5–4.0	2.7	2.5	2.9	2.8	3.5[b]
Cystine	1.1–2.2	–	–	1.9	–	–
Phenylalanine	4.2–4.5	4.7	4.5	5.9	6.4	–
Tyrosine	–	–	3.7	3.7	–	6.0[c]
Threonine	3.4–3.8	3.6	3.4	3.7	3.1	4.0
Tryptophan	2.0–2.4	1.9	–	1.3	1.5	1.0
Valine	4.2–4.4	5.1	4.7	4.6	3.9	5.0

[a] Range for five varieties.
[b] Methionine + cystine.
[c] Phenylalanine + tyrosine.

This may reflect the selective recovery or elimination of certain proteins by the isolation methods employed. The protein nutritive value of sesame is 15 to 42, relative to casein as 100 (Evans and Bandemer, 1967). Supplementation of sesame seed proteins with 0.2% lysine significantly increased their protein nutritive value, and nutritive value of sesame protein supplemented with 0.2% lysine + 0.1% methionine + 0.1% isoleucine was almost comparable to that of casein (Evans and Bandemer, 1967). The net protein utilization (NPU) of sesame meal has been reported to be 0.56 as compared with 0.74 for whole egg powder (Fisher, 1973).

Supplementation of sesame meal with 0.5% L-lysine increased the NPU to 0.63. When sesame protein was used at 20% level as the only source of protein in the chick diet, good growth was obtained by supplementing it with 0.5% lysine (Smith, 1971). Sastry et al. (1974) reported that supplementation of sesame flour with 1.25% lysine improved the nutritive value of proteins, making them comparable to that of skim milk powder. Supplementation of sesame diet at 18% protein level with threonine significantly improved the chick growth (Cuca and Sunde, 1967).

The protein efficiency ratios (PER) of sesame seed, meal and isolated protein are 1.86, 1.35 and 1.2, respectively (Narasinga Rao, 1985). Commercially prepared flour and press cake showed PER values of 0.9 and 1.03. Supplementation of sesame seed protein with lysine can increase its PER to 2.9.

The amino acid composition of sesame complements that of most other oilseed proteins. Tryptophan, which is limiting in many oilseed proteins, is adequate in sesame. The availability of amino acids from sesame seed protein is affected by the method of processing. Digestibility is enhanced by heat treatment under moist conditions, while screw pressing for oil recovery apparently has little adverse effect on available lysine. However, *in vitro* digestibility was reportedly the same for the isolated sesame protein before and after autoclaving, indicating a lack of trypsin inhibiters (Kinsella and Mohite, 1985).

The problems encountered during addition of limiting amino acids to achieve nutritional adequacy (Lis et al., 1972) are overcome by covalent attachment and the application of plastein reaction (Fujimaki et al., 1977). Lysine-enriched plasteins have been prepared from

Table 17.5 Sugar content of sesame seeds and defatted flour (% on dry weight)

Sugar	Seed (Aguilar and Torres, 1969)	Defatted flour (Wankhede and Tharanathan, 1976)
D-Glucose	1.55	3.63
D-Galactose	0.65	0.40
D-Fructose	0.24	3.43
D-Fructose	0.34	0.17
Raffinose	–	0.59
Stachyose	–	0.38
Planteose	0.06	0.23
Sesamose	–	0.14
Other sugars	–	0.16
Total sugars	–	11.26

sesame protein using N-ε-cbz-lysine methyl ester and also enzymatic hydrolysates of casein or soybean proteins. Plasteins obtained with N-ε-cbz-lysine methyl ester had a yield of 40% for sesame and the lysine content was 16–19% (Susheelamma, 1983).

The high level of sulphur-containing amino acids in sesame seed proteins is unique. It suggests that sesame protein should be more widely used as a supplement for methionine and tryptophan and should be an excellent protein source for baby and weaning foods. The use of sesame seed protein would eliminate the problems encountered when foods are supplemented with free methionine, which is unstable.

17.2.3 Carbohydrates

The carbohydrate content of sesame seeds is comparable to that of groundnut seeds and is higher than that of soybean seeds (Joshi, 1961). Sesame seeds contain 14–25% carbohydrates. The seeds contain about 5% sugars, most of which are of reducing type. Defatted sesame meal contains more sugars. The sugar contents of sesame seeds and defatted flour are given in Table 17.5. Sesame seeds are reported to contain 3–6% crude fibre (Ramachandra et al., 1970; Gopalan et al., 1982; Taha et al., 1987). The crude fibre is present mostly in husk or seed coat (Narasinga Rao, 1985). Wankhede and Tharanathan (1976) reported 0.58–2.34% and 0.71–2.59% hemicellulose A and B, respectively, in defatted flour. Hemicellulose A was found to contain galacturonic acid and glucose in the ratio of 1 : 12.9 while hemicellulose B contained galacturonic acid, glucose, arabinose and xylose in the ratio of 1 : 3.8 : 3.8 : 3.1.

17.2.4 Minerals

Sesame seed is a good source of certain minerals particularly calcium, phosphorus and iron (Table 17.6). The seeds contain a total of 4 to 7% minerals. Deosthale (1981) reported 1% calcium and 0.7% phosphorus in the seeds. They also contain sodium and potassium. Calcium is mostly present in the seed coat which is lost during dehulling. Further, the bioavailability of calcium from sesame is less than that from milk or bread, probably because of the high concentration of oxalate and phytate in the seed. Poneros-Schneier and Erdman (1989) reported the bioavailability of calcium from some food products, relative to $CaCO_3$ as non-fat dry milk 100%; whole-wheat bread 95%; almond powder 60%; sesame seeds

Table 17.6 Mineral content of sesame whole seeds (mg/100 g)

Mineral	Joshi (1961)	Agren and Gibson (1968)		Gopalan et al. (1982)	Weiss (1983)
		White seeds	Brown seeds		
Calcium	1000	1017	1483	1450	1160
Phosphorus	700	732	578	570	616
Iron	20	56	–	10.5	10.5
Total	5700	5600	6200	5200	5300

65%; and spinach 47%. Sesame grown on selenium-rich soils also contains high selenium, although most of it is present in the hulls (Kinsella and Mohite, 1985).

17.2.5 Vitamins

Sesame seeds are an important source of certain vitamins, particularly niacin, folic acid and tocopherols (Gopalan et al., 1982; Weiss, 1983). The vitamin A content of seeds is, however, very low (Table 17.7). Vitamin E group includes several tocopherols, isomers and derivatives that differ in their biological activity (Table 17.8). The vitamin E activities of α-, β-, γ- and δ-tocopherols and tocotrienol are in the ratio of 100, 40, 10, 1 and 30 (McLaughlan and Weihraugh, 1979). Sesame oil is rich in tocopherols. However, the proportion of δ-tocopherols is more than that of α-tocopherols. Therefore, the vitamin E activity of sesame oil is less than that of sunflower oil.

Table 17.7 Vitamin content of whole sesame seeds

Vitamin	Agren and Gibson (1968)		Gopalan et al. (1982)	Weiss (1983)	Seegeler (1983)
	White seeds	Brown seeds			
Vitamin A (IU)	–	–	60[a]	30	Trace
Thiamin (mg/100 g)	0.22	0.14	1.0	0.98	1.0
Riboflavin (mg/100 g)	0.02	0.05	0.34	0.24	0.05
Niacin (mg/100 g)	7.3	8.7	4.4	5.4	5.0
Pantothenic acid (mg/100 g)	–	–	–	–	0.6
Folic acid (μg/100 g)					
Free	–	–	51	–	–
Total	–	–	134	–	–
Ascorbic acid (mg/100 g)	–	–	–	–	0.5

[a] μg carotene (100 g).

Table 17.8 Vitamin E active compounds in sesame and sunflower oils (mg/100 g oil)

Compound	Sesame oil		Sunflower oil (Speek et al., 1985)
	Muller-Mulot (1976)	Speek et al. (1985)	
α-tocopherol	1.2	1.0	78.8
β-tocopherol	0.6	<0.05	2.5
γ-tocopherol	24.4	51.7	1.9
δ-tocopherol	3.2	<0.05	0.7
Total tocopherol	29.4	52.8	83.9
Vitamin E activity (α-tocopherol equivalent)	–	14.9	79.0

17.2.6 Antinutritional factors

Sesame seed is nearly free of antinutritional factors and is suitable for human consumption as such or after processing. Sesame seeds, however, contain high amounts of oxalate (Deosthale, 1981; Narasinga Rao, 1985) and phytic acid (Prakash and Nandi, 1978; Johnson et al., 1979). Sesame seeds contain about 1–2% oxalic acid. Gopalan et al. (1982) reported 1.7% oxalic acid in the seeds. The high proportion of oxalate reduces the physiological availability of calcium from the seeds. The oxalic acid in sesame seeds is mostly present in the testa or the hull portion. The presence of testa imparts a slightly bitter taste to the whole seed or meal because of chelation of calcium by oxalic acid. Dehulling reduces the oxalic acid content of the seeds. Oxalic acid may also be removed from sesame meal by treating it with hydrogen peroxide at pH 9.5.

Sesame seeds contain a substantial amount of phosphorus. However, most of this phosphorus is tied up in phytic acid or as phytin, a calcium and magnesium salt of inositol hexaphosphate. The seeds have phytate levels among the highest found in nature (De Boland et al., 1975). Phytic acid is a strong chelating agent and binds dietary essential minerals such as calcium, iron and zinc to form phytate–mineral complexes (Reddy et al., 1982). The formation of such complexes decreases the bioavailability of these minerals (Oberleas et al., 1966; Kon et al., 1973). The phytate in sesame meal is insoluble in water. O'Dell and De Boland (1976) extracted phytate from the meal by dilute HCl (0.3 M) and precipitated it with NaOH. The insoluble phytate had a composition of NaMg–phytate, suggesting that phytate in sesame meal exists as a magnesium phytate and not as phytin (CaMg–phytate).

Sesame oil contains two minor constituents, namely sesamin (0.5–1.0 %) and sesamolin (0.3–0.5%). Sesamolin upon hydrolysis yields sesamol (Godin and Spensley, 1971). Although the nutritional significance of sesamin and sesamolin is not clear, sesamol has been reported to be partially responsible for the resistance of sesame oil to oxidation (Weiss, 1983). Sesame plants seem to have an unusual capacity for lead accumulation in the seeds. Yannai and Haas (1973) reported that whole sesame seeds and kernels contained lead at the level of 0.13–0.22 mg/100 g. A high consumption of sesame (>200 g/day) is therefore considered to be harmful to humans.

17.3 Production

Sesame is grown primarily in the tropical and subtropical regions of the world although it can be grown in more temperate climates. Sesame is cultivated in an area of 7.78 million hectares with a production of 3.15 million tonnes and a productivity level of 405 kg/ha in the world (Table 17.9). India is by far the largest producer, accounting for 28% of the world's area and 23% of the production (Table 17.10). Other major countries producing sesame are China, Myanmar, Sudan, Nigeria, Mexico, and to a smaller extent, Ethiopia, Uganda, Venezuela, Turkey, etc. Estimates of world sesame production are always somewhat misleading primarily because in countries where there is a substantial area planted to the crop, a high proportion is consumed by local farmers and is not marketed. The major obstacles to sesame's expansion are its low yields and the absence of non-shattering cultivars suitable for machine harvest. Consequently it requires much manual labour at the harvest season, which is often scarce even in its traditional growing areas.

Different varieties of sesame seed (black, white and brown) are cultivated in India both as a rainfed and as an irrigated crop. The western and southern states produce sesame as a *kharif* crop (June–October/November), while the eastern region cultivates it as a *rabi* crop

Table 17.9 Worldwide area and production of sesame seed

Region	Area (ha)	Production (t)	Yield (kg/ha)
Africa	2 793 000	739 000	264
North and Central America	159 000	91 000	574
Asia	4 753 000	2 263 000	476
Europe	210 000	115 000	547
South America	79 000	57 000	718
World	7 784 000	3 150 000	405

Source: Based on 2001 statistics available from FAO.

Table 17.10 Major producers of sesame in the world

Country	Area (ha)	Production (t)	Yield (kg/ha)
India	2180 000	730 000	335
Sudan	1900 000	300 000	158
China	702 000	791 000	1127
Myanmar	1311 000	426 000	325
Nigeria	151 000	69 000	457
Venezuela	46 000	35 000	761
Somalia	70 000	23 000	329
Turkey	65 000	23 000	354
Mexico	72 000	41 000	567
Uganda	203 000	102 000	502
Bangladesh	80 000	49 000	613
Korean Republic	44 000	31 000	713
Ethiopia	45 000	22 000	489
Tanzania	100 000	39 000	390
Thailand	63 000	39 000	619
Pakistan	101 000	51 000	500
Egypt	30 000	37 000	1211

Source: Based on 2001 statistics available from FAO.

(November–February/March). With two harvests of *kharif* and *rabi* crops, sesame seed supplies are available year-round in India. Sesame yields (approximately 300 kg/ha) in India, however, are well below the world average.

The world trade of sesame is limited. Sesame demand on a world basis is frequently greater than world production and, except where the crop is deliberately grown as a cash crop for export, there are seldom any large amounts available to world trade. India, China, Myanmar, Sudan and Latin American countries such as Mexico are major suppliers of sesame seed. White sesame is preferred in the export markets as commercial bakers and confectioners consider it to be of higher quality than dark coloured sesame. White sesame also commands a higher price.

17.3.1 Crop adaptation

Sesame is basically considered a crop of the tropics and subtropics, but its extension into more temperate zones would be possible by breeding suitable varieties. The diversity of local ecotypes well adapted to their particular locality is an indication of the plants' potential in this respect. Sesame's main distribution is between 25°S and 25°N, but it can be found growing up to 40°N in China, Russia and the USA and up to 30°S in Australia and 35°S in South America. It is normally found below 1250 m, although some varieties may be locally adapted up to 2500 m. It is grown in Himalayas up to 1250 m and in Nepal up to 2000 m. The high-altitude types are usually small, quick-growing and relatively unbranched, with frequently only one flower per leaf axil and low seed yields. Within varieties, yields invariably decrease with altitude. Oil content normally decreases with altitude in the same variety.

Sesame normally requires fairly high temperature during growth to produce maximum yields and 2700 heat units are reportedly required in Israel during the critical three to four month growth period (Kostrinsky, 1959). Temperature for optimum growth from seedling emergence to flowering and fruiting has been found to be in the range of 27–33°C (Kinman and Stark, 1954; Smilde, 1960). Considerable genotypic variation in germination response to temperature has been reported (Sharma, 1997). A temperature of 25–27°C encourages rapid germination, initial growth and flower initiation. If the temperature falls below 20°C for any length of time, germination and seedling growth will be delayed and, below 10°C, these processes are inhibited (Salehuzzaman and Pasha, 1979). High temperatures, particularly high night temperatures, promote stem growth and leaf production (Smilde, 1960). Temperature above optimum (40°C or above) at flowering can seriously affect fertilization and the number of capsules set. A frost-free growing period is required for sesame, and hard frost at maturity will not only kill plants but will also reduce seed and oil quality. It can also adversely affect minor seed-oil constituents, such as sesamolin and sesamin (Beroza and Kinman, 1955).

Sesame is basically a quantitative short day plant and with a 10-hour day will normally flower in 40–50 days, but many varieties have locally adapted to various light periods. Early cultivars are generally less sensitive to day length than late types (Sinha *et al.*, 1973). When varieties are introduced to areas that have a similar day length but different rainfall or temperature patterns, there is considerable variation in growth and yield from that in their original location. This is because of interaction of photoperiod with factors such as light intensity, rainfall and temperature. Light intensity has a significant morphogenic effect influencing yield and oil content. Taking the yield obtained at the optimum planting period as maximum, then the yield from sowings after this period decreases as the time from optimum sowing increases. However, rainfall has a major modifying influence on optimum time of planting relative to photoperiod.

The locally adaptable sesame varieties have been well utilized in countries such as India with distinct growing seasons. Varieties adapted to one season give an uneconomic yield if grown in other season because of photoperiod and light intensity adaptability. The relationship of time of planting to maximum yield is generally appreciated although less well understood.

The rate of net total dry matter production per unit of ground area is related to the daily amount of photosynthetically active radiation intercepted by the crop. Low yields in *kharif* season in India could result from low radiation levels caused by heavy cloud cover, resulting in reduced radiation input, or from low plant density, rendering suboptimal interception by the crop canopy. In the intercropping systems, sesame yields can be reduced because of shading by the companion crop. The stage at which shading occurs has a great influence on the level of yield reduction. Recent advances in plant breeding have reduced plant height of the traditional companion crops such as sorghum, millets and pigeonpea, putting sesame in a more equitable position in the competition for space and light, improving thereby the potential of sesame as an intercrop in India.

Sesame has great adaptability with regard to rainfall. It will produce an excellent crop with a rainfall of 500–650 mm, but as low as 300 mm and as high as 1000 mm will also produce a crop under certain conditions, particularly under irrigation from newer varieties. For maximum yields, precipitation should be distributed over the period of plant growth as follows: germination to first bud formation 35%, bud formation to main flowering 45%, flowering to maturity not more than 20%, falling as seeds are filling and ceasing as first pods begin to ripen. Heavy rain at flowering will drastically reduce yield, and if cloudy weather persists for any period at this time, yield can be very low. Rainfall when plants are ready for harvest also reduces yield by increasing susceptibility to disease and prolonging the period required for capsules to dry. Sesame is extremely susceptible to waterlogging and heavy continuous rains at any time during growth will greatly increase the incidence of fungal diseases. Wild plants show a higher degree of resistance to waterlogging than the cultivated types (Nakhtore, 1952). Although sesame is susceptible to fungal diseases in high rainfall areas, if the soil is permeable and drains fully, so that there is no standing water to maintain high humidity, good crops may be obtained that would be impossible on more clayey soils with lower rainfall.

Sesame will have lower net photosynthesis and possibly low yield potential when grown in an arid environment than when grown in areas that have a higher humidity. This is because, at high humidity gradient between leaf and air, there is reduction in stomatal aperture. A large humidity gradient may cause midday closure of stomata and depression of photo-synthesis. There is also increase in leaf temperature at high temperature. However, the highest yields are obtained under irrigation in arid regions.

Sesame is considered to be a drought-resistant crop. It is capable of withstanding a higher degree of water stress than many other cultivated plants. However, during the plant establishment phase, it is extremely susceptible to moisture shortage. Once established, the crop will grow almost entirely on stored soil moisture and with only an occasional shower of rain in early stages, good yields are obtained. This ability to produce a crop under adverse conditions makes sesame an important crop under semi-arid conditions. Waterlogging is highly detrimental to the crop.

Sesame is susceptible to wind damage after the main stem has elongated. In the valleys of Kashmir, very cold winds from mountains during early growth and flowering cause severe injury to plants. Sesame is very susceptible to hail damage at all stages of growth. Prior to flowering, stems can be badly bruised, some times broken and terminal shoots so damaged that distorted growth occurs. At flowering, both buds and flowers may be stripped from the plant, or damaged buds produce aborted flowers. Heavy storms can virtually strip plants of all leaves and recovery will be slow.

17.3.2 Soils

Sesame grows well on a wide range of soils from high sandy soils to black cotton and clay soils, but thrives best in well-drained, moderately fertile soils of medium texture. Shallow soils with impervious sub-soil are not suitable. The soils on which sesame is grown range from sandy soils in Sudan, Egypt and Rajasthan in India to highly sandy-loam in Venezuela and river terraces in Northern Thailand, laterite soils in Uttar Pradesh and Madhya Pradesh (India), typical red earths and clay paddy soils in Karnataka and Maharashtra (India) and central plain area of Thailand. On lighter, more gravelly or sandy upland soils in drier zones, growth and yield are often depressed because of poor moisture retention and low soil fertility.

Soils with neutral reaction are preferred, although good results have been obtained in slightly acidic and slightly alkaline soils. Sesame does not thrive on acid soils. It will grow in soils of pH 5.5 to 8.0, but at higher pH, soil structure becomes increasingly important. However, many soils on which sesame is grown are saline. There is considerable variation among cultivars in the degree of tolerance to salinity (Kurien and Iyengar, 1968).

17.3.3 Cropping systems

Sesame, being a short duration crop, fits well into a number of sequence and intercropping systems in different parts of India and elsewhere in the world, both under rainfed and irrigated conditions. In India, *kharif* sesame is grown both as pure and mixed with other crops, whereas the semi-*rabi* and summer crops are taken as pure. The common component crops are pigeonpea, maize, groundnut, castor, pearl millet, mungbean, soybean, cotton, sunflower, sorghum, clusterbean, etc., in different states of India. As a sequence crop, sesame is taken after rice, groundnut, cotton, maize, pigeonpea, chickpea, finger millet, sorghum, wheat, mustard, horsegram, sugarcane, potato, lentil, pea, barley, mungbean, etc., depending on soil moisture availability and irrigation source.

17.3.4 Planting time

Correct time of planting is most important to obtain high yields of sesame. In India, sesame is grown in three seasons, viz. *kharif*, semi-*rabi* and summer. The *kharif* crop occupies over 70% of the area under cultivation, whereas the semi-*rabi* and summer crops 20 and 10% of area, respectively. The *kharif* sesame is sown in June–July with the onset of monsoon and is harvested in September–October. The *kharif* and semi-*rabi* crops are entirely rainfed, whereas the summer crop is grown under irrigation. The yield of the *kharif* crop is poor, whereas those of the semi-*rabi* and summer crops are high, since they are grown on rich soils and under better management (Hegde and Sudhakara Babu, 2002).

In other parts of the world, sesame is sown from August–November in Venezuela, from March–August in Mexico, in the southern USA when danger of frost is past, and in Africa at the start of the rains. Because huge seed losses occur if rain falls during the harvest season, in most tropical countries the planting is timed to allow harvest in the dry season.

17.3.5 Tillage and planting

Sesame requires a well-pulverized seed bed with fine tilth for good germination of seed and establishment of desired plant stand. The soil is brought to a fine tilth by deep ploughing in summer followed by planking (flattening of the soil). Tilth required for sorghum, wheat or similar small grains is suitable for sesame. Land should be perfectly levelled to ensure that there is no waterlogging and lands may be ridged to assist drainage in those areas where high-intensity storms are common. For a *rabi* crop, two or three harrowings followed by levelling is enough. Immediately prior to planting, lands should be harrowed to kill weed growth, since sesame seedlings make slow initial growth. Weed control while plants are small is difficult, and the aim should be as weed-free a seedbed as possible.

In many countries, farmers usually sow sesame by hand and just scatter the seed, which are later hoed in to cover the seeds. For line sowing, seed drills may be used. For mechanical planting, equipment may vary from small hand-operated seeder units or animal-drawn drills to tractor-operated, multipurpose, electronically controlled seeders. Depth of planting varies with soil type and is usually 2–5 cm. Uniform depth of planting ensures regular emergence and crop growth, thus facilitating subsequent tillage operations.

Sesame may be scattered or line sown. For a scattered crop, a seed rate of 4–7 kg/ha is adequate to get the required plant stand. For line sown crop, seed rate may be reduced to 2.5–3 kg/ha. The seed rate in mixed or intercrop depends on the proportion of area occupied by sesame in the system. Spacing depends on the growth habit of the variety, the season and the growing conditions such as rainfed or irrigated. Row spacing of 25–75 cm is recommended in different countries. Thinning should be done scrupulously to ensure recommended plant spacing within a row. The first thinning is to be done invariably 14 days after sowing and the second thinning 21 days after sowing. Excess population adversely affects growth and yield of crop. Early thinning will facilitate good establishment and proper use of fertilizers.

17.3.6 Nutrient management

The average nutrient removal to produce a tonne of sesame is 51.7 kg N, 22.9 kg P_2O_5, 64.0 kg K_2O, 11.7 kg S, 37.5 kg Ca, 15.8 kg Mg, 168 g Zn, 793 g Fe, 115 g Mn and 117 g Cu (Hegde, 1998). The level of nutrient application would, however, vary depending on the variety, crop, season, soil, fertility status, previous crop, rainfall and soil moisture. The application of fertilizers must also be related to plant population, for the optimum amount required by crops of different densities will vary (Park, 1967). Fertilizers also affect other plant characteristics that influence yield, i.e. plant height and number of capsules per plant, but the usual effects produced by added plant nutrients are not always correlated with yield. This is particularly so with nitrogen in the seed bed. In general, fertilizers have little effect on seed composition or oil content, except at much higher rates than are economically justified (Mitchell et al., 1974).

Nitrogen application must be related to phosphate availability, for when this is deficient, nitrogen can depress yield. There is also some evidence to indicate that it may also adversely affect seed oil content. Seedbed applications of nitrogen as part of an NPK mixture frequently give good results, but the ratio of NPK should ideally be locally calculated. In those regions where sesame is planted at the beginning of the rains following a pronounced dry season, release of increased microbial nitrogen which then takes place may preclude seedbed application. Nitrogen application may vary between 20 and 50 kg/ha depending on the expected production. Application is best done at planting and, if needed, top dressed before the first buds appear. Method of application is also of little importance provided the coverage is even and timing accurate.

Phosphate is the most important of the major plant nutrients necessary for high sesame yields, especially when irrigated. Uptake of NPK has been shown to be related to the general growth of sesame plants to approximately 60 days. At this point, the proportion of dry matter supplied by leaves falls, and, with it, uptake of nitrogen and potassium, the latter to a lesser degree as it is also related to capsule number. Although the rate of phosphorus uptake also declines, it continues at a higher level than the other two as the number of capsules increases (Bascones and Ritas, 1961). In India, responses of up to 40 kg P_2O_5/ha have been reported (Sharma, 1997). If the previous crop is supplied with large amounts of phosphorus as in potato, sesame does not need any additional application.

Analysis of mature sesame plants usually shows a high potassium content, especially in the capsules, but unless there is known local deficiency, application of this nutrient other than in small amounts is seldom necessary. In soils low in potassium, 15–30 kg K_2O/ha is recommended to maintain the required nutritional balance (Sharma, 1997).

There are no records of minor element deficiencies occurring in sesame, although at many locations, significant responses to micronutrients such as zinc have been reported (Anonymous, 1998). However, as the responses have been inconsistent, commercial recommendations for micronutrient applications to sesame have not been made. In sulphur-deficient areas, application of 15–20 kg S/ha increases both seed yield and oil content. It is desirable to apply full doses of potassium, phosphorus and sulphur at the time of planting the crop.

17.3.7 Weeding and interculture

The slow initial growth of sesame seedlings makes them poor competitors to many quick-growing tropical weeds. Therefore, the crop is very sensitive to weed competition during the first 20–25 days. A weed-free seedbed is most important, since cultivation of sesame seedlings is difficult as the fine, fibrous roots are easily damaged. It is essential to have a minimum of two weedings, one after 15 days of sowing and another 15–20 days thereafter. Row crops can be weeded with any of the normal inter-row tillage implements such as hand-hoe, animal drawn blade harrow, rotary or finger weeders, provided they are set to work as shallow as possible. Sesame plants grow rapidly after they reach some 10 cm in height and a few cultivations are then necessary. Planting in narrow rows can assist in reducing late weed growth due to the shading effect.

Weeds can also be managed effectively by use of proper herbicides. Diuran at 400–600 g/ha, Basalin at 1 kg/ha, Alachlor at 1.75 kg/ha, Fluchloralin at 1 kg/ha or Pendimethalin at 1 kg/ha as a pre-emergence treatment have been found effective for controlling weeds. Chemical methods of weed control may be resorted to wherever weed growth is severe and labour is scarce, followed by one hand weeding if required, around 30 days after sowing. Band spraying plus inter-row cultivation is the combination that most frequently gives good weed control at relatively low cost.

17.3.8 Water management

In India, sesame during *rabi*/summer season is normally raised under irrigation. The crop during *kharif* season rarely receives any irrigation. Nevertheless, protective irrigation will greatly benefit the *kharif* crop whenever there are prolonged dry spells. Highest sesame yields are obtained when grown under irrigation in arid regions, where the sunny dry climate is very suitable, and the low humidity reduces the incidence of fungal diseases.

Sesame is very susceptible to drought in various physiological stages. The crop is also very sensitive to waterlogging as it causes the premature death of plants. When grown under irrigation, substantial pre-sowing watering is to be preferred to immediate post-emergence application but the difficulty of planting in wet fields may require that the seed is dry-planted and then irrigated. Subsequent irrigations may be given at intervals of 12–15 days or more, depending on soil type, weather conditions and season. The critical stages for irrigation are the four- to five-leaf stage, flowering and pod formation. The short watering interval has been found to give higher yields than a larger application at longer intervals. A high application rate of water tends to reduce both seed weight and oil content (Kostrinsky, 1959). Free flooding and border strip methods of irrigation are normally employed for irrigating sesame in India.

17.3.9 Pests and diseases

Sesame crop is affected by a number of insect pests and diseases. Development and use of resistant varieties are perhaps the most economical methods of reducing the losses due to pests and diseases. Nearly 29 insect pests belonging to eight species are reported to be potential pests of sesame. The leaf roller/capsule borer (*Antigastra catalaunalis* Dup.) is the key pest along with the gall fly (*Asphondylia sesami*) in India. In Sudan, *Agnoscelis versicola* and the sesame seed bug (*Aphamis littoralis*) attack seed capsules in fields. Other pests include aphids and thrips which stunt seedlings and injure developing flower buds. One or two sprays of organophosphate insecticides 40–60 days after sowing give effective control of these pests.

There are a number of fungal, bacterial, mycoplasma and viral diseases responsible for reduction of sesame yields. Stem and root rot (*Macrophomina phaseoli* Maubl.), ashby, phyllody (virus, mycoplasma), bacterial leaf spot (*Pseudomonas sesami*, Matkoff), fungal leaf spots (*Cercospora* spp.), *Alternaria* blight and leaf curl are the important diseases of sesame worldwide. These diseases are generally more prevalent in regions of high humidity and excessive rainfall, and will give little trouble in arid deserts and dry regions, provided disease-free seed is sown (ICAR, 1990).

17.3.10 Harvesting and threshing

The optimum harvesting period is of great importance in sesame, since harvesting even a few days earlier or later can cause large yield reductions. Sesame is usually ready for harvest 80–150 days after sowing, most commonly after 100–110 days, but some cultivars also mature 70–75 days after sowing (Montilla *et al.*, 1977). The sesame crop should be harvested when the leaves turn yellow and start drooping but the capsules are still greenish. At maturity, leaves and stems tend to change from green to yellowish and then reddish. If the harvesting is delayed and the crop is allowed to dry completely, there is loss in yield owing to bursting and shattering of capsules. Capsules ripen irregularly from the low stem upwards, the topmost often being only half matured at harvesting. The drying period before harvesting allows the seed to ripen without loss from mature capsules.

The plants are cut with sickles or uprooted. The harvested plants are carried to the threshing yard and stacked for a week. During this period, the capsules burst open and leaves are shed almost completely. Then plants are dried in the open sun and threshed by gentle beating of plants with sticks. Threshing can also be done by simply turning the plant upside down and shaking or lightly beating. The seeds are cleaned with help of a special type of sieve designed for this purpose. Later, seed is cleaned by winnowing.

The introduction of non-shattering varieties in India will allow mechanical harvesting, provided the crop is planted in large fields. Machine harvesting can be done with a reaper-binder or a combine-harvester. The first method is preferred by many growers, who cut the crop when it is not fully mature and combine from the windrows. They consider this greatly reduces the risk of seed loss and the straw has better feeding value. Most standard combines fitted with a pick-up reel and with the correct drum settings are suitable. Best samples with low seed loss are obtained from slow working, and optimum speeds, once determined, should be maintained. Threshing of sesame requires accurate setting of concave and cylinder, for the seed is easily damaged and even microscopic cracks in the seed are sufficient to affect both viability and oil quality.

17.3.11 Yield potential
The yield of sesame varies with season, method of cultivation and variety and ranges from a few hundred to 3000 kg/ha in different countries. In India, according to season, 375 to 500 kg/ha during *kharif* and 500 to 750 kg/ha during *rabi*/summer may be expected. According to method of cultivation, a well-managed crop can yield 500 to 600 kg/ha under rainfed condition and 900 to 1000 kg/ha under irrigated condition.

17.3.12 Seed storage
Bulk storage of sesame seed presents few problems provided the seed is clean and dry. Seed that heats or is contaminated by extraneous material produces discoloured or rancid oil. Sesame seed can be stored more economically than many other oilseeds because of its small size. It can also be moved easily and efficiently by modern conveyers without causing damage to seeds.

Sesame seeds are cleaned and dried in the sun to bring down the moisture content to 5% before storage to prevent attack from storage fungi and insect pests. In India, the most commonly used kerosene can or grease drum with tight-fitting lids are quite convenient for handling and for storing the seed in small quantities. In Africa, small lots are stored in earthern jars or wrapped in small banana leaf parcels sealed with dung, and are hung in the smoke of the fireplace (Salunkhe and Desai, 1986). In parts of east and west Africa, conical mud and wattle granaries holding about 100 kg of seed are constructed, and the narrow openings are then sealed with a mud bung. Several such stores may be grouped together on a platform and protected by a roughly thatched roof (Weiss, 1983). The tolerance level for post-harvest fumigation of sesame seed with hydrogen cyanide has been reported to be 25 ppm. Sesame seed retains its viability well under controlled conditions. When kept in storage at 50% relative humidity and 18°C, germination vigour was undiminished after one year (Prieto and Leon, 1976). To preserve viability of sesame germplasm collections for long periods, the use of silica gel in sealed containers is recommended (Weiss, 2000).

17.4 Processing
Sesame seeds are mostly used without removing the cuticle or the seed coat. This is especially the case in areas where sesame is processed for its oil. The cuticle contributes to the colour, bitterness, and fibre and oxalate contents of the resultant screw-pressed meal. Such meal is not useful as a source of protein for humans and other monogastric animals and is used mostly as a cattle feed or manure. Therefore, dehulling of sesame seed is followed to improve its quality and utilization as a source of human food. Some of the important operations involved in processing sesame seed are described below very briefly.

Table 17.11 Effects of dehulling on the chemical composition of sesame seeds

Constituent	Whole seeds	Dehulled seeds
Moisture (%)	5.4	5.5
Protein (%)	18.6	18.3
Fat (%)	49.1	53.4
Total carbohydrates (%)	21.6	17.6
Crude fibre (%)	6.3	2.4
Ash (%)	5.3	5.3
Energy (cal/100 g)	563	582
Calcium (mg/100g)	1160	110
Phosphorus (mg/100 g)	616	592
Iron (mg/100 g)	10.5	2.4
Vitamin A (IU)	30	–
Thiamin (mg/100 g)	0.98	0.18
Riboflavin (mg/100g)	0.24	0.13
Niacin (mg/100 g)	5.4	5.4

Source: Weiss (1983).

17.4.1 Dehulling

Dehulling is an integral part of the modern oil extraction plants. It is also essential to produce high-quality oil and meal. However, dehulling still remains the single most important problem worldwide in the processing of sesame seed. Many wet precessing methods and mechanical treatments have been tried for dehulling (Sastry *et al.*, 1969). The most commonly used method of dehulling is to soak the seeds and remove the cuticle manually by light pounding or rubbing on a stone or wooden block. Ramachandra *et al.* (1970) have reported a lye treatment process for dehulling of sesame. In this process, seeds are cleaned and given a hot lye (0.6%) treatment for one minute. The seeds are washed with excess cold water. The ruptured seed coats are separated by scrubbing in a suitable equipment. The dehulled seeds (kernels) are then dried.

The removal of hull results in significant change in the chemical composition of the seeds. The dehulled seeds contain significantly more fat and less crude fibre, calcium, iron, thiamin and riboflavin and slightly less phosphorus than the whole seeds (Table 17.11). Oxalic acid, being present mostly in the seed coat, is significantly reduced after dehulling treatment (Narasinga Rao, 1985). The digestibility of proteins improves as a result of dehulling (Sastry *et al.*, 1974). Heat treatment during dehulling as well as subsequent processing of the flour will not lower the available lysine. Quality of oil is also not affected by lye treatment before dehulling (Narasinga Rao, 1985).

17.4.2 Oil extraction

The most popular method of oil extraction from sesame seed in India is by *ghani*, which is basically a large pestle and mortar. In earlier days, *ghani* was made of wood and driven by bullock. Subsequently, power-driven steel *ghani* came into existence. The oil extraction by *ghani* is not complete and the yield of oil is about 40–45% (Weiss, 1983). In many parts of India, water or jaggery (brown sugar) is added to sesame seed to facilitate oil extraction (Muralidhara, 1981). Following extraction, the oil is removed from the *ghani*, allowed to settle, skimmed and sometimes strained through a cloth before sale. Sometimes, the residual meal is double-pressed to obtain more oil. The Burmese *hsi-zin* is similar to Indian *ghani* but it is now replaced by power-driven mills. The oil yield from sesame seed by *hsi-zin* is about 33% (McLean, 1932). In central Africa, sesame seeds are boiled to make them soft, then squeezed in a sausage made from the fibres to extract oil (Weiss, 1983). Some of these methods are still followed in many countries, sometimes with minor modifications.

Modern commercial methods for oil extraction from sesame seeds employ one of three basic designs, batch hydraulic processing, in which the oil is expressed by hydraulic pressure from a mass of oil-bearing material; continuous mechanical processing in which the oil-bearing material is squeezed through a tapering outlet, the oil being expressed by the increasing pressure; and solvent extraction in which the oil-bearing material is taken into solution with a solvent, which is then separated from the insoluble residue and the oil is recovered from the solvent solution (Godin and Spensley, 1971). The sesame seeds produced by farmers are not of uniform size, colour or maturity, being an admixture. They are also contaminated with soil particles. Because of the small size of the seeds, it becomes difficult to clean the seeds. The oil quality is affected if the seeds are not properly cleaned. Similarly, prolonged storage under unsuitable conditions results in a loss of oil quality (Sharma, 1977).

In Europe and Asia, the oil is usually extracted in three stages. The first pressing is made cold. The oil contained is of very good quality and high grade. It has a light colour and agreeable taste and odour. The second pressing is made of the heated residue, which is subjected to a high pressure. The oil obtained is coloured and is refined before being used for edible purposes. The residue is used for the third extraction under the same conditions as for the second. The oil obtained from the third extraction is of inferior quality, not suitable for human consumption and is generally used for the manufacture of soaps.

The recovery of oil from screw or hydraulic pressing is not complete. In Europe, a combination of preprocessing and solvent extraction is used to obtain maximum recovery of oil. The direct solvent extraction is not suited for sesame seeds because of high oil content.

17.4.3 Oil purification

Crude, cold-pressed sesame oil is used directly in cooking wherever it is produced and is often a favoured oil. Sesame oil does not require extensive purification or refining. The crude oil usually contains a suspended meat ('foot') which is removed by settling, screening and filtering. The filtered crude oil from the extraction plant contains impurities such as phosphatides, resins, free fatty acids and colouring matters. Alkali-refining removes gums, free fatty acids and some of the colouring matters. The oil is bleached with a relatively lower amount of bleaching earth than for other vegetable oils. Bleaching produces a light coloured oil. Deodorization is necessary to produce a bland oil. It is usually done by treating refined oil in vacuum with steam at 200–250°C. For use as a base of salad dressing, the oil must be stable under refrigeration. For this, winterization treatment is given to the oil. It consists of cooling the oil to remove components with high melting points that settle out at low temperatures. Sesame oil, however, requires little or no winterization (Lyon, 1972). The hydrogenation process brings about a considerable increase in stability of the oil.

17.4.4 Cake and meal

The byproduct left behind after the extraction of oil is called sesame cake. When it is powdered, the cake is converted into the meal or flour. Powdering of cake into meal or flour will not result into any change in chemical composition (Awais *et al.*, 1968). Four types of meals can be obtained from sesame seeds, namely whole seed meal, dehulled seed meal,

Table 17.12 Effects of dehulling and the method of oil extraction on the composition (%) of sesame flour/cake

Sesame seed and processing	Moisture	Fat	Protein	Ash	Crude fibre	Calcium	Phosphorus	Oxalic acid
Whole seeds								
Flour	5.2	49.8	19.1	5.7	4.1	1.2	–	2.3
Expeller-pressed flour	6.6	10.7	41.4	8.7	6.8	1.7	–	3.7
Expeller-pressed cake	8.1	13.5	35.1	8.9	5.3	1.8	0.8	3.0
Alcohol-extracted cake	8.6	3.1	38.2	9.4	5.9	2.0	0.9	3.5
Hexane extracted cake	8.6	0.8	39.6	9.7	6.1	2.1	1.0	3.6
Dehulled seeds								
Flour	4.1	60.2	22.1	3.2	3.1	–	–	0.1
Expeller-pressed flour	5.8	10.0	54.4	6.2	5.1	0.3	–	0.2
Prepress-solvent-extracted flour	6.0	0.4	56.1	6.1	4.9	0.4	1.0	0.4
Expeller-pressed cake	8.3	12.7	41.3	4.8	3.1	0.4	1.1	0.5
Alcohol-extracted cake	8.9	3.4	45.8	5.0	3.1	0.5	1.2	0.5
Hexane-extracted cake	8.8	1.1	46.7	5.2	3.2	0.4	–	0.3

Source: Ramachandra *et al.* (1970).

defatted whole seed meal and dehulled, defatted meal. Of these, the dehulled, defatted meal is the most common and unless otherwise specified, the term sesame meal refers to the dehulled, defatted meal. The chemical composition of sesame meal varies significantly owing to dehulling and the method of oil extraction. The meals or flours obtained from dehulled seeds contain more proteins and phosphorus and less ash, crude fibre, calcium and oxalic acid than those obtained from whole seeds (Table 17.12).

Heat treatment will not affect the amounts of total protein and total lysine (Sastry *et al.*, 1974). However, autoclaving for a prolonged period (60 minutes) causes significant decrease in the available lysine and dispersibility of proteins in water and NaOH solutions. Heat treatment of sesame flour will not affect the amino acid composition of its proteins except for a slight decrease in basic amino acids. An increase in the available methionine from 1.85 to 2.33 mg/16 g N on treatment to prepressed solvent extracted meal at 121°C for one hour has been reported (Villegas *et al.*, 1968). Rooney *et al.* (1972) prepared breads from composite flours containing heated and unheated oilseed meals. They observed that heat treatment of sesame meal resulted in less total and specific loaf volumes.

In India, sesame cake is often used as an animal feed when the oil is extracted at village level. The free fatty acid content of Indian *ghani* cake is high and its keeping quality is poor. Therefore, it must be fed to livestock as soon as possible or it rapidly becomes rancid and unpalatable.

17.4.5 Protein concentrates and isolates
Many processed high-protein products such as flakes, flour, protein concentrates and isolates can be obtained from sesame (Sastry *et al.*, 1969). The defatted flour contains more protein than the whole seed meal. The protein concentrate contains more protein (about 70%) than the flour, while protein isolate contains about 90% or more protein. Unlike many oilseeds, the defatted flour and isolates prepared from sesame do not contain any undesirable pigments, off-flavour or toxins (Toma *et al.*, 1979; Johnson *et al.*, 1979). Sesame proteins are extracted with various salts and alkaline solutions. The extractibility of proteins varies with the extraction medium, pH and time. Sodium hydroxide solution (0.04 M) appears to be

the most suitable solvent, extracting about 90% of the meal nitrogen (Taha *et al.*, 1987). With alkaline medium, the recovery of proteins is maximum when the meal is extracted at pH 10.0.

Proteins exhibit minimum solubility at their isoelectric point. Most protein isolates are, therefore, prepared by extracting the proteins in suitable solvent and precipitating at or near their isoelectric point. The isoelectric point of sesame proteins is 4.5 to 4.9. The proteins extracted with salt and alkali showed minimum solubility in the region of pH 5.7 (Rivas *et al.*, 1981). A low phytate protein was proposed by Taha *et al.* (1987) by dissolving the protein by a counter-current procedure and precipitating it at pH 5.4. At this pH, 50% of the phytate was removed while only 17.5% of the protein was dissolved. The resulting protein isolate contained 91.4% protein, and was almost free from phytate.

The protein isolate contains very high levels of protein and is almost free from oil, ash, crude fibre and phytate phosphorus with very low levels of nitrogen-free extract (Rivas *et al.*, 1981). The chemical composition of protein concentrate is intermediate between that of defatted meal and protein isolate. The essential amino acid composition of alkali isolate is almost comparable to that of sesame flour (Prakash and Nandi, 1978; Rooney *et al.*, 1972). The salt isolate, however, contains more threonine and valine and less lysine and methionine than the other two products.

Sesame flour and protein concentrate exhibit less water absorption than soya flour. They also show high fat absorption than soya products. The emulsifying capacity and emulsion stability of sesame products are comparable to those of soya products. The foam expansion and foam stability are higher in sesame products than in soya products. Protein extractability and whipping potential of sesame flour extract is low compared with other oilseeds (Lawsen *et al.*, 1972; Dench *et al.*, 1981).

17.4.6 Roasting

Sesame seeds are often roasted prior to their use in confections. Roasting reduces the moisture content, develops a pleasant flavour and makes the seed or meal more acceptable for consumption. The reduction in moisture content during roasting of sesame prevents moulding and reduces staling and rancidity. Sesamol, an antioxidant, was detected only in roasted sesame oil (Fukuda *et al.*, 1981). 2-Furfuryl alcohol is considered as one of the most characteristic components, giving a pleasant roasted aroma to sesame seeds. It is present in higher concentrations in red and white sesame (El-Sawy *et al.*, 1988).

17.5 Uses

The world production of sesame seed is almost wholly utilized for culinary purposes. In India, about 78% of sesame produced is used for oil extraction and about 20% is used for domestic purposes such as preparation of sweetmeats and confectionery (Maiti *et al.*, 1988; Weiss, 1983) and about 2% is retained for the next sowing.

17.5.1 Human food

Seeds and kernels
Dehulled sesame seeds are sweet and oleaginous and are used directly in different types of foods in various parts of the world. They are used in the manufacture of traditional

confections such as *halva, laddu* and *chikki* in India. They are also eaten whole after roasting. A confection called *laddus* is prepared from roasted groundnuts and sesame seeds by pounding with jaggery in the proportion of 2 : 1 : 2. Small balls are prepared by hand (ICMR, 1977). *Laddus* are also prepared from sesame seeds by mixing them with hot jaggery or sugar syrup. The confection prepared by mixing sesame seeds with jaggery or sugar has an auspicious connotation in many southern states of India. The confections are distributed or exchanged with each other to signify a great deal of sharing of goodwill (Mulky, 1985). *Chikki* is another confection popular in Maharashtra and other western parts of India. It is prepared by pouring sesame seeds in boiling jaggery solution to obtain a thick slurry. The slurry is spread uniformly on a metallic sheet or table and cut into small rectangular pieces.

Ready-to-eat instant foods using sesame seeds have been developed by the Indian Council of Medical Research for use in rural areas (ICMR, 1977). *Bajra* instant food is prepared by mixing roasted *bajra* (pearl millet) flour (60 g) with roasted green gram *dhal* (15 g), roasted groundnut (10 g) and sesame seeds (5 g). The mixture is pounded to obtain a flour. When required, the powder is mixed with boiling water or milk to the desired thickness. Sugar or salt are added to taste. *Ragi* instant food is prepared in the same way by replacing *bajra* flour with *ragi* (finger millet) flour. The *bajra* instant food gives 18.6 g protein, 389 cal and 8% net dietary protein (NDP) cal per 100 g flour. The *ragi* instant food gives 16 g protein, 369 cal and 8% NDP cal per 100 g (ICMR, 1977).

In the Middle East, dehulled sesame seeds are mainly utilized in the production of *tehineh* (sesame butter) and *halwah* (halva). *Tehineh* is made from a paste of dehulled roasted seeds. *Halwah* is a sweet made up of *tehineh*, sugar, citric acid and *Saponaria officinalis* root extract. *Tehineh* and *halwah* are produced commercially in factories in the Middle East and North Africa. *Tehineh* is used in a variety of food dishes and added to bread and bakery products (Sawaya *et al.*, 1985).

Sesame seeds and kernels are used in commercial bakeries for the preparation of quick breads, rolls, crackers, coffee cakes, pies and pastry products (Weiss, 1983; Farrell, 1985). The seeds are lightly roasted and used in salads and salad dressing (Farrell, 1985). Toasted seeds and butter or margarine make a tasty spread for bread.

Oil

Oil is the major product of sesame seed processing. In India, of the total sesame, about 75% oil is used for edible purposes as vegetable oil for culinary purposes, 5–10% goes to the *vanaspati* industry for vegetable *ghee* (a type of shortening) manufacture, and 4% for industrial uses as paints, soaps, perfumes, etc. (Salunkhe *et al.*, 1992). Oil is a common constituent of Burmese dishes and is used in frying, roasting and stewing of meat, fish and vegetables. Sesame oil is highly favoured for cooking. Its nutty flavour is appreciated. The oil has excellent stability and keeps well at room temperature for two to three months. It makes an excellent frying medium for chickpea and meat, and is a good replacement for peanut oil (Farrell, 1985). Because of the quality and high price, sesame oil is frequently adulterated with groundnut, rape or cotton seed oils. In India, particularly in some parts of Maharashtra state, groundnut, safflower and sesame seeds are extracted together to produce the so-called sweet oil (Weiss, 1983). Sweet oil is cheaper than sesame oil and has a better stability than groundnut or safflower oils. Sesame oil can be readily hydrogenated to medium melting fats and different textures for use in margarine, shortenings and *vanaspati* (Patterson, 1983). It has mild pleasant taste and is a natural salad oil, requiring little or no winterization (Lyon, 1972).

Cake and meal

Sesame meal has become an increasingly important human food because of the following unique properties: the presence of a high level of sulphur-containing amino acids, especially methionine and cystine (Block and Weiss, 1957; Evans and Bandemer, 1967; Smith, 1971), its lack of trypsin-inhibiting factors and its pleasant flavour. Sesame flour and meal have high protein content and is used to fortify foods (Parpia, 1966; Pomeranz *et al.*, 1969; Rooney *et al.*, 1972). Its use in the diet of children suffering from kwashiarkor has been found to be beneficial. It has been recommended as a protein supplement for soya and legume proteins (Boloorforooshan and Markakis, 1979; Brito and Nunez, 1982). Compoy *et al.* (1984) have prepared a snack food product using 70% chickpea and 30% sesame flour. Supplementation of black bean (*Phaseolas vulgaris* L.) meal with sesame meal significantly improved the PER and net protein retention (NPR) of black bean proteins. Maximum PER and NPR were obtained when sesame and black bean flours were mixed in 1 : 1 proportion. The sesame lipids, cholesterol and trigyciride levels were also influenced by supplementation of black bean flour with sesame flour.

A number of ready-to-use infant foods using sesame meal such as *Cholam* and *samai* porridge have been developed by the Indian Council of Medical Research, particularly for use in rural areas (ICMR, 1977). Sesame flour has been used as a methionine supplement in the preparation of fermented foods, *vada* and *dosa*, the most popular South Indian dishes (Gulati *et al.*, 1979; Chopra *et al.*, 1982). Sesame flour was used to replace 5–20% of rice–black gram flour. Sesame-supplemented *dosa* was found to be acceptable organoleptically and had higher levels of methionine than the plain *dosa* (Chopra *et al.*, 1982).

There is an increasing interest in fortification of bread and cookies by replacing a portion of wheat flour with non-wheat flours, especially protein concentrates, isolates and oil meals (Dendy *et al.*, 1970). The maximum level of replacement depends upon the type of non-wheat flour, the strength of wheat flour, the baking procedure and dough-stabilizing compounds used (Pringle *et al.*, 1970). In most cases, a 10% replacement of wheat flour is optimum. At higher levels, loaf volume is severely decreased with serious deterioration of crumb colour, grain and texture (Mathews *et al.*, 1970). Sesame flour has been used in the preparation of bread and cookies (Hoojjat, 1982). When used as a part of replacement of wheat flour, sesame flour performed better than sunflower flour. High-protein biscuits are prepared by mixing wheat flour with roasted chickpea and roasted sesame flour to prepare the dough (ICMR, 1977).

Blends of peanut/chickpea, wheat/chickpea, rice/chickpea, peanut/soybean, sunflower/maize and cowpea/rice have all shown improved nutritional qualities with supplementation of sesame meal (Ensminger *et al.*, 1994). Even more significant, however, is the finding that a simple blend of one part each of sesame and soya protein has about the same protein nutritive value as casein, the protein of milk. The high lysine and low methionine content of soya protein is complementary to sesame protein. Sesame meal is sometimes fermented for food in India and Java. In some European countries, it is also used as an ingredient in comminuted meat products.

The use of sesame flour or meal in formulating high-protein beverages has been reported (Tasker *et al.*, 1966). Silva and Rivenos (1979) prepared a protein liquid from sesame. A nutritious beverage can be prepared using 70% soya protein and 30% sesame protein.

17.5.2 Animal food

In India, defatted sesame meal is traditionally used for animal food. The cake is a valuable stock food (Maiti *et al.*, 1988). It is rich in protein, calcium, phosphorus and niacin. The cake is well liked by the stock and keeps well in storage. It is considered equal to cottonseed cake or soyabean meal as a protein supplement for livestock and poultry. It is rich in methionine and is a valuable supplement to soyabean meal in livestock diets (Grau and Almqvist, 1944). Sesame meal proteins are, however, deficient in lysine. Lysine-rich materials such as soyabean meal, meat scrap and fish meal need to be combined with sesame cake to balance the diet. In the USA, most of the meal is used for livestock food. The inferior quality sesame cake or meal is used as a manure in China and Korea.

17.5.3 Industrial uses

Sesame oil is used to some extent in industries. Only small proportion of low-grade oil is used for the manufacture of soaps, perfumes, paints, pyrethrum-based insecticides and for various other purposes for which the non-drying oils are generally adopted (Nayar and Mehra, 1970; Weiss, 1983). Its relative scarcity and high price normally render it uncompetitive for large-scale industrial utilization. Sesame oil forms the basis of most of the fragrant or scented oils as it is not liable to turn rancid or solidify and it does not possess an objectionable taste or odour. In the perfumery industry, sesame oil is used as a fixative. Scenting oil can be extracted from wetted sesame seeds that have been covered with layers of scenting flours and left covered for 12 to 18 hours. A kilogram of strongly scented flowers is enough to perfume six litres of sesame oil. Sesame oil has synergistic activity with insecticides such as pyrethrums and rotenone. The presence of sesame oil reduces the concentration of the insect toxin required to produce 100% mortality. The synergistic activity of sesame oil has been attributed to the presence of sesamol and sesamolin.

17.5.4 Medicinal uses

Sesame seeds, oil, leaves and roots have excellent medicinal value. Sesame plant has played a major role in India's rich and diverse health traditions. The people of India, who live in harmony with nature, have an incredible knowledge of the medicinal value of sesame plant and make use of nature's bounty to achieve the best health traditions. Sesame seeds are regarded as microcapsules for health and nutrition. It is supposed to tone the kidney and liver and relax the bowel. The seeds are an aromatic, digestive emollient that soften the skin, a nourishing tonic, an emmenagogue that stimulates menstruation, a demulcent, a soothing, laxative, an antispasmodic, a diuretic and promotes weight gain. Seeds are used for the treatment of constipation, tinnitus, anaemia, dizziness, poor vision and many general health problems associated with old age.

A paste of the seeds mixed with butter is helpful in treating bleeding piles. A decoction of sesame seed mixed with linseed is used as an aphrodisiac. The seeds milled and mixed with brown sugar are eaten by nursing mothers to encourage their milk production. Regular use of sesame seeds boosts the development of lustrous hair, particularly in children with poor hair development, a general problem in Western countries. Sesame seeds are also used traditionally as a medicine for causing abortion. The seeds are valuable in respiratory disorders such as chronic bronchitis, pneumonia, asthma, dry cough and other lung infections. Seeds also help in correcting irregular menstrual disorders and in reducing spasmodic pain during menstruation. Seeds are also useful in treatment of dysentery and diarrhoea.

Sesame oil has been extensively used for therapeutic and cosmetic purposes in the Indian system of cure and care and therefore it is regarded as a magic botanical potion. Sesame oil is used as a laxative, emollient and demulcent. It has been successfully used in the treatments of backache, tinnitus, blurry vision, migranes, vertigo or dizziness, chronic constipation, haemorrhoids, dysentery, amenorrhea, dysmenorrhea, receding gums, tooth decay, hair loss, weak bones, osteoporosis, emaciation, dry cough, blood in urine, weak knee and stiff joints. It has antibacterial, antifungal and antiviral properties. Because of its easily assimilated calcium content, it nourishes the blood, calms nervous spasms and alleviates headaches, dizziness and numbness caused by deficient blood. It is a tonic, particularly for the aged. Oil of sesame will help burns, boils, ulcers and sunburn and remove freckles and age spots. Owing to such innumerable benefits, the oil is used as the base for several *Ayurvedic* preparations. However, it is poorly documented in modern scientific literature.

Sesame oil is a preferred vehicle for fat-soluble substances because of its high stability. It is employed in the preparation of liniments, ointments and plasters. In India, it is extensively used for conditioning the skin (Weiss, 1983). Sesame oil is considered anti-cholesterol and highly beneficial for heart ailments. The oil also reduces stress hormones and strengthens the immune system. It reduces anxiety, depression and pain. It also helps control sugar levels and therefore, its use is beneficial for people with diabetes. In olden days, sesame seed oil was administered for snake bites. Sesame oil is used in the preparation of iodinol and brominol, which are employed for external, internal or subcutaneous use.

The infusion of leaves in hot boiling water is used as a gargle for the treatment of inflamed membranes of the mouth. The leaves, which abound in gummy matter when mixed with water, form a rich, bland mucilage used in infantile *cha*, diarrhoea, dysentery, catarrh and bladder troubles, acute cystitis and strangury. Crusted leaves of sesame are considered beneficial in the treatment of dandruff. A decoction made from the leaves and root is used as a hair wash which is said to prevent premature greying of hair and promote their growth.

A decoction of the root is used in various traditions to treat coughs and asthma.

17.6 Future research needs

Sesame is the oldest oilseed known to human beings. It also has several desirable agronomic characters which can give the crop an edge over competing crops. It does relatively well on poor lands and is resistant to drought. Availability of cultivars with varying duration helps to fit the crop in different intensive cropping systems under irrigated conditions. Cultivars are also available adapted to varying photoperiods and temperature regimes. The seeds contain more oil than many other oilseeds. The oil has excellent stability and its protein is rich in sulphur-containing amino acids and tryptophan.

Despite all the above desirable qualities, sesame cultivation is generally confined to countries where labour is comparatively cheaper and plentiful. One of the major drawbacks associated with sesame for its large-scale cultivation is the absence of non-shattering varieties that are amenable to mechanical cultivation and harvesting. There is an urgent need to develop indehiscent sesame cultivars by making use of already available types in wild species and in germplasm collections through appropriate breeding and transgenic approaches.

The production potential of sesame is low compared with many oilseeds like groundnut and soya bean. Hybrid technology may help to step up this potential. In countries such as China and South Korea, hybrids produced through manual crossing have already proved successful in raising the productivity level. There are also reports of existence of cytoplasmic genetic male sterility in sesame. This sterility system needs to perfected to develop commercial hybrids for a major dent on productivity.

Sesame production in many countries is constrained by insect pests such as the leaf-eating caterpillar and stem and root rot. Imparting resistance to these insect pests and diseases will go a long way in enhancing and sustaining sesame production. To this end, resistance breeding against these maladies needs urgent attention.

More research should be focused on increasing the levels of sesamolin and sesamol in the oils for cultivated types and understanding their relations with seed and oil yield. Although considerable quantitative variability exists for these two traits, cultivars having very high contents of oil as well as sesamolin have not been developed.

There is a good opportunity currently to improve the extraction methods in many developing countries to obtain both better quality oil and defatted oilseed meal. Improved and easier methods of dehulling sesame seed are also needed. Further, research regarding the optimization of oil extraction and protein preparation is required, with emphasis on techniques for minimizing the oxalic acid content of the flour. Greater emphasis is also needed on utilization of defatted sesame flour and meal in human nutrition. Development of acceptable products from oilseed cake for human consumption in different countries is a very high priority research area to help overcome chronic malnutrition in many developing countries. Standardization of milder processing methods for processing sesame oilseed cake that will eliminate problems associated with dark colour will also help in preparing acceptable sesame products for human consumption.

Sesame protein has unique qualities, such as lack of trypsin inhibitor activity and a high level of sulphur-containing amino acids and tryptophan, and is therefore very valuable for use in baby and weaning foods. Its use would eliminate problems encountered when foods are substituted with free methionine, which is unstable and imparts a bitter taste to the food. Therefore, further research to develop nutritious foods from sesame protein is fully justified.

17.7 References

ABDEL-RAHMAN A.H.Y., HASSABALLA E.S., EL-MORSHIDY M.A. and KHALIFA M.A. (1980), *Res. Bull.*, Assiat, An Shams University.

AGREN G. and GIBSON R. (1968), *Food Composition Tables for Use in Ethiopia*, Addis Ababa, Children's Nutrition Unit.

AGUILAR V.G. and TORRES E.G. (1969), 'Determination of sugars in sesame seeds by chromatography', *Zeit. Lebensm. Unter Forsch.*, **140**, 332–5.

AIYADURAI S.G. and MARAR M.M.K. (1951), 'Studies on the development of capsules in *Sesamum indicum*', Coimbatore, India, First Scientific Workers Congress, 244–51.

ANDRAOS V., SWIFT C.E. and DOLLEAR F.G. (1950), 'Sesame oil. I. Properties of solvent extracted sesame oil', *J. Am. Oil Chem. Soc.*, **27**, 31–4.

ANONYMOUS (1998), *Annual Progress Report, Sesame and Niger, 1997–98*, Hyderabad, India, Directorate of Oilseeds Research.

ARZUMANOVA A.M. (1963), 'Influence of different cultural conditions on the oil content of till', *Tr. Prik. Bot. Genet. Selek.*, **35**, 168–72.

AWAIS M., SHAIKH J.A. and ALI S.M. (1968), 'Nutritional properties of sesame flour prepared from indigenous sesame cake', *Pakistan J. Sci. Ind. Res.*, **11**, 384–7.

BASCONES L. and RITAS J.L. (1961), 'La nutricion mineral del ajonjonholi', *Agron. Trop.*, (Venez.), **11**(1), 17–32; **11**(2), 93–101.

BEDIGIAN D. and HARLAN J.R. (1986), 'Evidence for cultivation of sesame in the Ancient World', *Economic Bot.*, **40**(2), 137–54.

BENNETT M.R. (1996), 'Sesame production in Australia', *Sesame and Safflower Newsletter*, **11**, 4–9.

BERNARDINI E. (1986), *Oilseeds, Oils and Fats Encyclopaedia*, Rome, B.E. Oil Tech Publ. House, 2nd Edition.

BEROZA M. and KINMAN M.L. (1955), 'Sesamin, sesamolin and sesamol content of the oil of sesame as affected by strain, location grown, ageing and frost damage', *J. Am. Oil Chem. Soc.*, **32**, 348–50.

BLOCK R.J. and WEISS K.W. (1957), *Amino Acid Handbook*, New York, Thomas.
BOLOORFOROOSHAN M. and MARKAKIS P. (1979), 'Protein supplementation of navy beans with sesame', *J. Food Sci.*, **44**, 390–2.
BRITO O.J. (1981), 'Usage of sesame as a source of protein for human consumption', *Dissert. Abstr. Int.*, **41**, 105.
BRITO O.J. and NUNEZ N. (1982), 'Evaluation of sesame flour as a complementary protein source for combinations with soy and corn flours', *J. Food Sci.*, **47**, 457–61.
BUDOWSKI P. (1950), 'Sesame oil, III. Antioxidant properties of sesamol', *J. Am. Oil Chem. Soc.*, **27**, 264–7.
BUDOWSKI P. (1964), 'Recent research on sesame, sesamolin and related compounds', *J. Am. Oil Chem. Soc.*, **41**, 281–5.
BUDOWSKI P. and MARKELY K.S. (1951), 'The chemical and physiological properties of sesame oil', *Chem. Rev.*, **48**, 125–51.
BUDOWSKI P., MENEZES F.G.T. and DOLLEAR F.G. (1950), 'Sesame oil, V. 'The stability of sesame oil', *J. Am. Oil Chem. Soc.*, **27**, 377–80.
BUDOWSKI P., O'CONNOR R.T. and FIELD E.T. (1951), 'Sesame oil, VI. 'Determination of sesamin', *J. Am. Oil Chem. Soc.*, **28**, 51–4.
CHOPRA G., MAN S.K., KAWAIRA B.L., WAHAL C.K. and BAJAJ S. (1982), 'Effects of supplementation of sesame seed on the protein quality of *dosa*', *J. Res. Punjab Agric. Univ.*, **19**, 256–62.
COMPOY M.P.F., STALL J.W. and TAYLAX R.R. (1984), 'Nutritional characteristics of foods prepared from sesame seed meal blends', *Nutr. Rep. Int.*, **29**, 611–19.
CUCA M. and SUNDE M.L. (1967), 'Amino acid supplementation of sesame meal diet', *Poult. Sci.*, **46**, 1512–5.
DE BOLAND A.R., GARNER G.B. and O'DELL B.L. (1975), 'Identification and properties of phytate in cereal grains and oilseed products', *J. Agric. Food Chem.*, **23**, 1186–91.
DENCH J.E., RIVAS N. and CAYGILL J.C. (1981), 'Selected functional properties of sesame flour and two protein isolates', *J. Sci. Food Agric.*, **32**, 557–64.
DENDY D.A.V., CLARKE P.A. and JAMES A.W. (1970), 'The use of blends of wheat and non-wheat flours in bread making', *Trop. Sci.*, **12**, 131–6.
DEOSTHALE Y.G. (1981), 'Trace element composition of common oilseeds', *J. Am. Oil Chem. Soc.*, **58**, 988–90.
EL-SAWY A.A., SOLIMAN M.M. and FADEL H.M. (1988), 'Identification of volatile flavour components of roasted red sesame seeds', *Crasas y Aceites*, **39**, 160–2.
ENSMINGER A.H., ENSMINGER M.E., KONKANDE J.E. and ROBSON J.R.K. (1994), *Food and Nutrition Encyclopedia*, Boca Raton, Florida, CRC Press, II Edition, 1987–90.
EVANS R.J. and BANDEMER S.L. (1967), 'Nutritive value of some oilseed proteins', *Cereal Chem.*, **44**, 417–26.
FAO (2001), *Production Year Book*, Rome, Food and Agriculture Organisation, 55.
FAO/WHO (1973), *Energy and Protein Requirements*, FAO Nutrition Meeting Report Series No. 52, WHO Technical Report Series No. 522, Rome, Food and Agriculture Organisation.
FARRELL K.T. (1985), *Spices*, Condiments and Seasonings, Westport CT, AVI.
FISHER H. (1973), 'Methods of protein evaluation : assays with chicks and rabbits', In *Proteins in Human Nutrition,* eds J.W.G. Porter and B.A. Rolls, London, Academic Press, 263–73.
FUJIMAKI M., ARAI S. and YAMASHITA M. (1977), 'Enzymatic protein degradation and resynthesis for protein improvement', in *Food Proteins: Improvement Through Chemical and Enzymatic Modification*, eds R.E. Feenoy and J.R. Whitaker, Washington D C, American Chemical Society, 156–84.
FUKUDA Y., OSAWA T. and NAMIKI M. (1981), 'Antioxidants in sesame seeds', *J. Japanese Soc. Food Sci. Technol.*, **28**, 461–4.
FUKUDA Y., OSAWA T. and KAWAGISH S. (1988), 'Comparison of contents of sesamolin and lignan antioxidants in sesame seeds cultivated in Japan', *Japanese Soc. Food Sci. Technol.*, **35**, 483–8.
GANESAN J. (1995), 'Induction of genetic male sterility system in sesame', *Crop Environment*, **22**, 167–9.
GODIN V. and SPENSLEY P.C. (1971), *Oils and Oilseeds*, London, Tropical Products Institute.
GOPALAN C., RAMASASTRI B.V. and BALASUBRAMANIAN S.C. (1982), *Nutritive Value of Indian Foods*, Hyderabad, India, National Institute of Nutrition, Indian Council of Medical Research.
GRAU C.R. and ALMQVIST H.J. (1944), 'Sesame protein in chick diets', *Proc. Soc. Exp. Biol.*, **57**, 187–9.
GUERRA M.J., JALF W.G. and SANGRONIS E. (1984), 'Recovery of protein fractions from commercial sesame seed cake', *J. Nutr.*, **34**, 477–81.

GULATI T., CHOPRA A.K. and BHAT C.M. (1979), 'Effects of supplementation of sesame and skim milk powder on the nutritional quality of *vadas*', *J. Res. Punjab Agric. Univ.*, **16**, 349–52.

HE FONG FA *ET AL.* (1994), 'Karyotype of sesame (*Sesamum* L.) as related to its phylogenesis', *J. Southwest Agric. Univ.* (Sichuan), **16**, 573–6.

HEGDE D.M. (1998), 'Integrated nutrient management for production sustainability of oilseeds – a review', *J. Oilseeds Res.*, **15**(1), 1–17.

HEGDE D.M. and SUDHAKARA BABU S.N. (2002), 'Sesame', in *Textbook of Field Crop Production*, ed R. Prasad, New Delhi, Indian Council of Agricultural Research, 528–85.

HOOJJAT P. (1982), 'Protein quality and functionality of navy bean and sesame flour in baked products', *Dissert. Abstr. Int. B.*, **43**, 187–8.

ICAR (1990), *Handbook of Agriculture*, New Delhi, India, Indian Council of Agricultural Research, 940–5.

ICMR (1977), *Studies on Weaning and Supplementary Foods*, New Delhi, India, Indian Council of Medical Research.

JOHNSON L.A., SULEIMAN T.M. and LUSAR E.W. (1979), 'Sesame protein: a review and prospects', *J. Am. Oil Chem. Soc.*, **56**, 463–8.

JOHNSON R.H. and RAYMOND W.D. (1964), 'The chemical composition of some tropical food plants, III. Sesame, *Trop. Sci.*, **6**, 173–9.

JOSHI A.B. (1961), *Sesamum*, Hyderabad, India, Indian Central Oilseeds Committee.

KAMAL-ELDIN A. and APPELQUIST L.A. (1994), 'Variation in fatty acid composition of the different acyl lipids in seed oils from four *Sesamum* species', *J. Am. Oil Chem. Soc.*, **71**(2), 135–9.

KANG C.W. and LEE J.I. (1995), 'Mutation breeding for disease resistance and high yield of sesame (*Sesamum indicum* L.) in Korea', *Sesame and Safflower Newsletter*, **10**, 21–35.

KINMAN M.L. and STARK S.M. (1954), 'Yield and chemical composition of sesame (*Sesamum indicum* L.) as affected by variety and location grown', *J. Am. Oil Chem. Soc.*, **31**, 104–8.

KINSELLA J.E. and MOHITE R.R. (1985), 'The physicochemical characteristics and functional properties of sesame proteins', in *New Protein Foods*, eds A.M. Altschul and H.L. Wilike, Orlando, FL, Academic Press Inc, **5**, 435–56.

KON S., OLSON A.C., FREDERICK D.P., EGGLING S.N. and WAGNER J.R. (1973), 'Effect of different treatments on phytate and soluble sugars in California small white beans', *J. Food Sci.*, **39**, 215–22.

KOSTRINSKY Y. (1959) *Methods for Increasing the Production of Sesamum in Israel*, Bull 62, Bet Dagan, Israel, Agric. Res. Station.

KURIEN T. and IYENGAR E.E.R. (1968), 'Effect of salt water on growth of sesame varieties', Seminar on Sea Salt and Plants, Bhavnagar, India, Central Salt Marine Chem. Res. Inst.

LANGSTRAAT A. and JURGENS B.V. (1976), 'Characteristics and composition of vegetable oil bearing material', *J. Am. Oil Chem. Soc.*, **53**, 241–7.

LAWSON J.T., CATER C.M. and MATIL K.F. (1972), 'A comparative study of the whipping potential of an extract from several oilseed flours', *Cereal Sci. Today*, **17**, 240–94.

LIS M.T., CROMPTON R.F. and MATHEWS D.M. (1972), 'Effect of dietary changes on intestinal absorption of L-methionine and L-methionyl-L-methionine in the rat', *Brit. J. Nutr.*, **27**, 159–67.

LYON C.K. (1972), 'Sesame, current knowledge of composition and use', *J. Am. Oil Chem. Soc.*, **49**, 245–9.

MAITI S., HEGDE M.R. and CHATTOPADHYAY S.B. (1988), *Handbook of Annual Oilseed Crops*, New Delhi, Oxford, IBH.

MATHEWS R.H., SHARPE E.J. and CLARK W.M. (1970), 'The use of some oilseed flours in bread', *Cereal Chem.*, **47**, 181–6.

MATHUR L.B. and TILARA K.S. (1953), 'Sesamolin absorption by bleaching agents', *J. Am. Oil Chem. Soc.*, **30**, 447–9.

MCLAUGHLAN P.J. and WEIHRAUGH J.L. (1979), 'Vitamin E content of foods', *J. Am. Dietet. Assoc.*, **75**, 647.

MCLEAN A. (1932), *Sesamum*, In *Burma Agric. Survey No.16 of 1932*, Burma, Department of Agriculture.

MITCHELL G.A., BINGHAM F.T. and YERMANOS D.M. (1974), 'Growth, mineral composition and seed characteristics of sesame as affected by nitrogen, phosphorus and potassium nutrition', *Soil Sci. Soc. Am. Proc.*, **38**, 925–31.

MOHANTY R.N. and SINHA S.K. (1965), 'Study of variation in some quantitative characters of varieties of sesame of Orissa', *Indian Oilseeds J.*, **9**, 104–8.

MONTILLA D. *ET AL.* (1977), 'Arawuca, an early variety of sesame', *Agron. Trop.* (Venez.), **27**(4), 483–87.

MULKY M.J. (1985), 'Utilization of oilseed meal for animal and human nutrition', in *Oilseeds Production – Constraints and Opportunities*, eds H.C. Srivastava, S. Bhaskaran, B. Vatsya and K.K.G. Menon, New Delhi, Oxford, IBH, 611–24.

MULLER-MULOT W. (1976), 'Rapid method for the quantitative determination of individual tocopherols in oils and fats', *J. Am. Oil Chem. Soc.*, **53**, 732–6.

MURALIDHARA H.G. (1981), *A Panorama of the World of Oilseeds*, Shimoga, India, National Education Society, 56–7.

NAGARAJ G. (1995), *Quality and Utility of Oilseeds*, Hyderabad, India, Directorate of Oilseeds Research, 70.

NAKHTORE K.L. (1952), '*Jangli til* (wild sesame) on Adhartal Farm, Nagpur', *Agric. Coll. Mag.*, **26**, 16–17.

NARASINGA RAO M.S. (1985), 'Nutritional aspects of oilseeds', in *Oilseeds Production – Constraints and Opportunities*, eds H.C. Srivastava, S. Bhaskaran, B. Vatsya and K.K.G. Menon, New Delhi, Oxford, IBH, 625–34.

NATH R. and GIRI K.V. (1957), 'Physicochemical studies on indigenous seed proteins, II. Fractionation, isolation and electrophoretic characterisation of sesame globulins', *J. Sci. Industr. Res.*, **16**, 51–88.

NATH R., RAO K.H. and GIRI K.V. (1957), 'Physicochemical investigations on the indigenous seed proteins, III. Aminoacid composition of sesame seed globulin, *J. Sci. Industr. Res.*, **16**, 221–7.

NAYAR M.N. and MEHRA K.L. (1970), 'Sesame: its use, botany, cytogenetics and origin', *Econ. Bot.*, **24**, 20–31.

OBERLEAS D., MUHRER M.E. and O'DELL B.L. (1966), 'Dietary metal complexing agents and zinc availability', *J. Nutr.*, **90**, 56–61.

O'DELL B.L. and DE BOLAND A. (1976), 'Complexation of phytate with proteins and cations in corn germ and oilseed meals', *J. Agric. Food Chem.*, **24**, 804–8.

OSMAN H.I. (1981), 'Genetic male sterility in sesame (*Sesamum indicum* L.) reproductive characteristic and possible use in hybrid vigour', *Sesame and Safflower Newsletter*, **1**, 36–41.

PARK J.M. (1967), *Annual Report*, Suwon, Korea, Crop Exp. Station.

PARPIA H.A.B. (1966), *Development of Food Mixes for Pre-school Children in India*, Washington DC, National Academy of Sciences, National Research Council.

PATTERSON H.B.W. (1983), *Hydrogenation of Fats and Oils*, New York, Elsevier.

PLIETZ P., DRESCHER B. and DAMASCHUN G. (1988), 'Structure and evolution of the 11S globulins: conclusion from comparative evaluation of aminoacid sequences and X-ray scattering data', *Biochem. Physiol. Pflanzen*, **183**, 199–203.

POMERANZ Y., SHOWGREN M.D. and FINNEY K.F. (1969), 'Improving bread making properties with glycolipids', *Cereal Chem.*, **46**, 512–18.

PONEROS-SCHNEIER A.G. and ERDMAN J.W. (1989), 'Bioavailability of calcium from sesame seeds, almond powder, whole wheat bread, spinach and non-fat dry milk in rats', *J. Food Sci.*, **54**, 150–3.

PRABAKARAN A.J. and RANGASAMY S.R. (1995), 'Observations on interspecific hybrids between *Sesamum indicum* and *S. malabaricum* L.', *Sesame and Safflower Newsletter*, **10**, 6–10.

PRAKASH V. and NANDI P.K. (1978), 'Isolation and characterisation of α-globulin of sesame seeds', *J. Agric. Food Chem.*, **26**, 320–3.

PRIETO S. and LEON R.S. (1976), 'Influence of storage conditions and periods on germination of sesame seeds', *CIARCO (Venez)*, **6**(1–4), 35–40.

PRINGLE W., WILLIAMS A. and HULSE J.H. (1970), 'The use of some oilseed flours in bread', *Cereal Sci. Today*, **14**, 114–7.

RAJENDRAN S. and PRAKASH V. (1988), 'Isolation and characterisation of β-globulin low molecular weight protein fraction from sesame seed (*Sesamum indicum* L.), *J. Agric. Food Chem.*, **36**, 269–75.

RAMACHANDRA B.S., SASTRY M.C.S. and SUBBA RAO L.S. (1970), 'Process development studies on the wet dehulling and processing of sesame seed to obtain edible protein concentrates', *J. Food Sci. Technol.*, **7**, 127–31.

RAO P.V. and RAO P.S. (1981), 'Chemical composition and fatty acid profiles of high yielding varieties of oilseeds', *Indian J. Agric. Sci.*, **51**, 703–7.

REDDY N.B., SATHE S.K. and SALUNKHE D.K. (1982), 'Phytate in legumes and oilseeds', *Adv. Food Res.*, **28**, 1–10.

RIVAS N.R., DENCH J.E. and CAYGILL J.C. (1981), 'Nitrogen extractability of sesame (*Sesamum indicum* L.) seed and the preparation of two protein isolates', *J. Sci. Food Agric.*, **32**, 565–71.

ROBINSON D.S. (1987), *Food Biochemistry and Nutritional Value*, Harlow, Longman.

ROONEY L.W., GUSTAFSON C.B., CLARK S.P. and CATER C.M. (1972), 'Comparison of the baking properties of several oilseed flours', *J. Food Sci.*, **37**, 14–18.

SALEHUZZAMAN M. and PASHA M.K. (1979), 'Effects of high and low temperatures on the germination of the seeds of flax and sesame', *Indian J. Agric. Sci.*, **49**(4), 260–1.

SALUNKHE D.K. and DESAI D.B. (1986), *Postharvest Technology of Oilseeds*, Boca Raton, FL, CRC Press, 105–17.

SALUNKHE D.K., CHAVAN J.K., ADSULE R.N. and KADAM S.S. (1992), *World Oilseeds: Chemistry, Technology and Utilization*, New York, Van Nostrand Reinhold, 371–402.

SASTRY M.C.S., SUBRAMANIAN N. and RAJAGOPALAN R. (1969), 'Studies on the dehulling of sesame seeds to obtain superior grade protein concentrate', *J. Am. Oil Chem. Soc.*, **46**, 592–6.

SASTRY M.C.S., SUBRAMANIAN N. and PARPIA H.A.B. (1974), 'Effects of dehulling and heat processing on nutritional value of sesame proteins', *J. Am. Oil Chem. Soc.*, **51**, 115–18.

SAWAYA W.N., AYAZ M., KHALIL J.K. and AL-SHALHAT A.F. (1985), 'Chemical composition and nutritional quality of *tehineh* (sesame butter)', *Food Chem.*, **18**, 35–45.

SEEGELER C.J.P. (1983), *Oil Plants in Ethiopia: Their Taxonomy and Agricultural Significance*, Wageningen, Centre for Agricultural Publishing and Documentation.

SHARMA S.M. (1997), 'Sesame', in *Efficient Management of Dryland Crops in India – Oilseeds*, eds R.P. Singh, P.S. Reddy and V. Kiresur, Hyderabad, India, Indian Society of Oilseeds Research, Directorate of Oilseeds Research.

SILVA G.S. and RIVENOS H.S. (1979), 'Food products derived from sesame seed: dehulling of seeds, production of defatted meal and a protein liquid', *J. Rivista del Institute de Investigations Technol.*, **21**, 34–7.

SINGH H., GUPTA M.L. and RAO N.K.A. (1960), 'Effect of NPK on yield and oil content of sesame', *Indian J. Agron.*, **4**, 176–81.

SINHA S.K., TOMAR D.P.S. and DESHMUKH P.S. (1973), 'Photoperiodic response and yield potential of sesame genotypes', *Indian J. Genet. Pl. Br.*, **33**, 293–46.

SMILDE K.W. (1960), 'Influence of some environmental factors on growth and development of *Sesamum indicum*', *Mededland b. Lagesch, Wag*, **60**(5), 70.

SMITH K.H. (1971), 'Nutritional framework of oilseed proteins', *J. Am. Oil Chem. Soc.*, **48**, 625–8.

SPEEK A.J., SCHRIJVER J. and SCHREURS W.H.P. (1985), 'Vitamin E composition of some seed oils as determined by high performance liquid chromatography with fluorometric detection', *J. Food Sci.*, **50**, 121–4.

SUSHEELAMMA N.S. (1983), 'Studies on lysine-enriched plasteins from oilseed proteins', *J. Food Sci. Technol.*, **20**, 47–51.

SWERN D. (1979), *Bailey's Industrial Oil and Fat Products*, Vol I, New York, Wiley, 4th Edition.

TAHA F.S., FAHMY M. and SADEK M.A. (1987), 'Low phytate protein concentrates and isolates from sesame seed, *J. Agric. Food Chem.*, **35**, 1289–92.

TASKER P.K., SRINIVAS H., RAJAGOPALAN R. and SWAMINATHAN M. (1966), 'Studies on micro-atomised protein foods based on blends of low fat groundnut, soy, sesame flours and skim milk powder, Limiting amino acids.1. Preparation, chemical composition and shelf life', *J. Nutr. Dietet.*, **3**, 38–41.

TOMA R.B., TABEKHIA M.M. and WILLIAMS J.D. (1979), 'Phytate and oxalate contents in sesame seed (*Sesamum indicum* L.)', *Nutr. Rep. Int.*, **20**, 25–31.

TU LICH WAN (1993), 'Studies on sesame heterosis and its practical use in production', *Sesame and Safflower Newsletter*, **8**, 9–17.

VILLEGAS A.M., GONZALEZ A. and CALDERON R. (1968), 'Microbial and enzymatic evaluation of sesame protein', *Cereal Chem.*, **45**, 379–83.

VLES R.O. and GOTTENBOS J.J. (1989), 'Nutritional characteristics and food uses of vegetable oils', in *Oil Crops of the World*, eds G. Robbelen, R.K. Downey and A. Ashri, New York, Mc Graw Hill, 63–86.

WANG W.Q. ET AL. (1995), 'A study of the effectiveness of hybrid seed production by utilizing genic male sterility in sesame (*Sesamum indicum* L.)', *Oil Crops of China*, **17**(1), 12–15.

WANKHEDE D.B. and THARANATHAN R.N. (1976), 'Sesame carbohydrates', *J. Agric. Food Chem.*, **24**, 655–8.

WEISS E.A. (1983), *Oilseed Crops*, London and New York, Longman, 660.

WEISS E.A. (2000), *Oilseed Crops*, Oxford, Blackwell Science, 364.

YANNAI S. and HAAS A. (1973), 'Occurrence of lead in sesame paste and factors responsible for it', *Cereal Chem.*, **50**, 613–6.

YEN G.C., SHYU S.L. and LIN J.S. (1986), 'Studies on protein and oil composition of sesame seeds', *J. Agric. Forestry*, **35**, 177–81.
YERMANOS D.M. (1978), 'Oil analysis report on the world sesame collection', *World Farming*, **14**, 5–11.
YERMANOS D.M. (1980), 'Sesame', in *Hybridization of Crop Plants*, eds W.R. Fehr and H.H. Hadly, Madison, Wisconsin, Crop Science Society of America.
YERMANOS D.M., HEMSTREET S., SALEEB W. and HUSZAR C.K. (1972), 'Oil content and composition of the seed in the world collection of sesame introduction', *J. Am. Oil Chem. Soc.*, **49**, 20–3.
ZHANG Y.X. *ET AL*. (1990), 'Cytogenetic studies in sesame (*Sesame indicum*): a new taxonomic system', *Acta Univer. Pekinensis*, **16**(1), 11–18.

18

Star anise

C. K. George, Peermade Development Society, India

18.1 Introduction, morphology and related species

Star anise (*Illicium verum,* Hooker) belongs to the Magnoliaceae or Magnolia family and is an important spice. The tree is very ancient and is part of a primitive family. It is indigenous to southeastern China. Star anise was known beyond China long before the Christian era as one of the few familiar spices, like cinnamon. However, it was not until the late 16th century that this spice was first brought to Europe by an English navigator, Sir Thomas Cavendish.[1]

Commercial production of star anise is limited today to China and Vietnam. Growing areas in China are southern and southeastern provinces, particularly mountainous elevations of Yunnan. China has the largest area of star anise cultivation. In Vietnam, star anise is grown adjoining the Chinese border. Lang Son province is the most important area, but other provinces such as Bac Kan, Thai Nguyen, Cao Bang and Quang Ninh also contribute. In Lang Son province cultivation is mostly in the districts of Van Lang, Van Quang, Tay Bac, Cao Loc, Binh Gia, Nam Truong Dinh and Bac Son. The total area in this province is more than 9000 ha, the majority of hectares being in Van Quan district. In the past, trees mostly belonged to collectives and to state farm enterprises. From the 1990s these were dismantled and trees were allocated to household management. The Vietnam government plans to bring in an additional 20 000 ha of star anise.[2]

Reliable estimates on production of star anise are not available. However, through information gathered from the trade sources in Vietnam, production has gone down from 9896 Mt in 1997 to nearly 5000 Mt in 1999. Production in China is higher than that of Vietnam. It is estimated that production in both these countries together is now more than 25 000 Mt per annum.[3]

18.1.1 Morphology

The star anise tree is evergreen with lanceolate leaves with aroma. It grows to a height of 8–15 m with a diameter of 25 cm. Leaves are entire, 10–15 cm long and 2.5–5.5 cm broad, elliptic to oblanceolate. When the tree matures, solitary flowers are produced in the leaf axils. The flower is relatively large and greenish-yellow. It is bisexual, radially symmetrical and lacks differentiation between the outer and inner floral whorls of sepals and petals. The fruit is pedunculate, consisting of eight stellately arranged 10 mm long boat-shaped carpels, fleshy at first, later becoming woody on drying, wrinkled, straight beaked, brown, dehiscent on the upper suture, internally reddish-brown, glossy and containing a single, flat, oval, lustrous, brittle and brownish-yellow seed. Ripe follicles burst on the ventral side to release seeds. The odour of the fruit is agreeable, anise-like, and the carpels taste sweet and aromatic. The fruit derives its name from the attractive star-like arrangement of carpels around a central axis.

The fruit whorl is 2.5–4.5 cm in diameter with individual carpels of about 9–19 mm length. The seed is 8–9 mm long and 6 mm broad. The endosperm is bulky and embryo disorganized.[4]

18.1.2 Related species

There are a few species that are related to *Illicium verum*. The fruit of *Illicium religiosum*, Siebold (*Illicium anisatum*, Linneaus) was earlier considered identical with *Illicium verum* until Hooker determined some distinction in 1888. *Illicium religiosum* known as *Sikmi* (*Shikimi*) is seen mainly in Japan and to small extent in Taiwan. In Japan the tree grows wild in warm localities of the southern and central parts and in the Loochoo Islands. There are extensive areas in the Prefecture of Nagasaki, mainly in Goto Island and to a small extent in the Prefectures of Kochi and Tokushima in the island of Shikoku. It is an evergreen tree with a trunk growing to 3 m height and bearing pale yellowish-white blossom.

For a long time Japanese have planted *Illicium religiosum* in temple compounds and in cemeteries in order to protect them from desecration by wild animals. This practice appears to have developed from the fact that the fruit is poisonous and the leaves emanate a peculiar odour that is supposed to keep animals away. This custom is so deep rooted that even today during funeral services altars are decorated with leaves of this tree.

The seed has crystalline, non-glucosidal, non-alkaloid sikimin, which is soluble in hot water, alcohol and chloroform. Dried fruits of *Illicium religiosum* contain about 1% of volatile oil, which has an unpleasant odour, quite different from that of the star anise (*Illicium verum*).[5] The volatile oil contains safrole. The fruit is highly poisonous as it contains anisatin, which causes severe inflammation of the digestive organs, kidneys and urinary tract. Cases of poisoning had been reported in the Netherlands as early as 1880 and also in Japan, the native country. Fatalities in children have resulted from the ingestion of the seeds, the toxic symptoms being vomiting, convulsions resembling those of epilepsy with froth coming from mouth, loss of consciousness, dilated pupils and the face becoming excessively cyanotic.[6]

The American Spice Trade Association (ASTA), in its Executive update of 29 July 2003, published the incident of people of Florida becoming ill because they had drunk tea prepared with star anise, quoting a report from the US Food and Drug Administration. While it is difficult to easily distinguish Japanese star anise from Chinese star anise visually, a simple gas chromatographic test will clearly show a significant difference as the latter has a noticeable amount of anethole unlike the former, which has none. Further, a small amount of bornyl acetate found in Japanese star anise is not seen in Chinese star anise.[7]

Illicium parviflorum, Michaux, is available in the hilly areas of Georgia, Florida and Carolina in the USA. It has yellow blossom, and the fruit is eight carpeled and tastes like sassafras. It is poisonous. *Illicium floridanum*, Ellis, is also found in the USA in Florida along the Gulf of Mexico coast to Louisiana. Flowers are purple and fruits have 13 carpels. It has a disagreeable odour resembling somewhat that of turpentine. Both fruits and leaves are poisonous. *Illicium majus*, Hooker, is a native of the Malay Peninsula. The fruit has 11 or 13 carpels with blackish-brown colour and tastes like mace. *Illicium griffithii*, Hooker, is seen in the State of Arunachal Pradesh of India and Bhutan. Leaves are ovate, elliptic-lanceolate. Flowers are solitary, axillary or terminal. The fruit consists of compressed, beaked, incurved 13 carpels in a single whorl. Seeds are small, sub-rotund, slightly compressed, glossy and brown. Fruit is slightly aromatic, bitter and acrid, and reported to be poisonous.[4]

18.2 Histology

A cross-section of the fruit shows exocarp, mesocarp and endocarp. The exocarp consists of a cuticle and a layer of epidermal cells up to 20 µm in places. In surface view the cuticle is striated. Epidermal cells are polygonal in shape and vary in size up to about 133 µm. Stomata are present but not numerous.

The mesocarp is built with parenchyma cells, isodiametric in shape or nearly so with brown contents. Cells increase in size toward the central zone of the mesocarp, where they reach about 200 µm and decrease toward the endocarp. Walls of the inner mesocarp cells are thicker than those of the central mesocarp cells, increase in thickness towards the dehiscent side of the carpel, and in the dehiscent zone adjoining the endocarp the cells pass into lignified fibres, which on longitudinal section are long and pitted.

Resin cells and irregular-shaped stone cells are scattered throughout the mesocarp. Resin cells vary in size up to about 220 µm, long axis and contain yellow to brownish-yellow oleoresin. Stone cells vary in size and thickness and are noteworthy for their branching and also irregularity. Vascular bundles occur in the merging zone of the central and inner mesocarp, and consist of phloem tissue and spiral, scalariform and scalariform-reticulate vessels.

The endocarp has a layer of thin-walled, sclerenchymatous palisade cells up to about 440 µm, long axis, decreasing in size and becoming stone cells in the dehiscent region. In surface view, the thin-walled cells are polygonal in shape and vary in size up to about 120 µm, long axis, most of the cells being around 90–100 µm.

A cross-section of the seed shows seed coat, endosperm and embryo. The seed coat is made up of a thick-walled epidermis, sclerenchymatous, pitted, palisade cells up to about 200 µm, long axis; five layers of sclerenchyma cells and two or more layers of thin-walled parenchyma cells containing numerous prismatic crystals of calcium oxalate.

In surface view, epidermal cells are up to 66 µm, long axis, and show considerable thickening with branching pits. The sclerenchyma tissue can be divided into three layers of irregular-shaped cells on the outer side, and two layers of elongated, narrow cells on the inner side. The endosperm consists of polygonal cells varying in size to about 110 µm, long axis, and containing aleurone grains and globules of fixed oils. The aleurone grains are large. The tissue of the embryo is disorganized.

A cross-section of the peduncle shows striated cuticle; a layer of epidermal cells with thick outer walls; cortex consisting of parenchyma cells varying in size to about 155 µm, long axis, oleoresin cells to about 90 µm, and branching stone cells varying greatly in shape and size; a ring of fibrovascular tissue and pith with isodiametric cells varying in size up to about 155 µm.

The receptacle or the central axis is anatomically similar to the peduncle, but branching stone cells and pitted cells are abundant.[8]

18.3 Production and cultivation

18.3.1 Habitat
The natural spread and cultivation of star anise is limited to a relatively confined area of Vietnam and China. Repeated attempts made in other countries to grow star anise have failed to yield a commercially worthwhile crop. It would, therefore, appear that the crop requires specific agro-climatic conditions, which are available only in the traditional growing areas.[2]

Star anise is not very cold hardy; it tolerates temperatures down to about −10°C. It prefers woodlands, sunny edges and dappled shade. The plant grows well on humus-rich and mildly acidic to neutral soils, which are light to medium and having good drainage. It requires moist soil for fast growth.

18.3.2 Cultivation
Propagation of star anise is by seed. Seeds are collected from fresh fruits of vigorously growing mature trees known for high yield. Fully matured large seeds, recognized by their characteristic brown colour, are selected. Seeds are sown 3–4 cm apart in a well-prepared bed. Since seeds quickly lose germinating power, they have to be planted preferably within three days of the harvest of fruits.[9] Layering has been attempted and found successful, but has yet to become popular.

After seedlings have produced the fourth leaf, they are transferred to a nursery and planted 25 cm apart. Once they are three years old, they are sufficiently grown and strong for planting in the field. Spacing for planting is about 5 m. Young trees do not require special care except weeding, which also reduces loss of trees by bush fire. Fertilization is done by applying stable dung, although some farmers use chemical fertilizers.[2]

18.3.3 Harvesting
Trees flower when they are about ten years old. Flowering is unusual. There are three seasons for flowering. The first blossom of the year is from March to the end of April. Flowers of this blossom are sterile and do not develop into fruits. The second blooming is from July to August and lasts for two or three weeks only. Flowers of this blossom are larger and fruits are developed, but some are lost at a premature stage during November to January. The third flowering season starts immediately after the second, sometimes partly dovetailing with it. Though flowers of this season are relatively small, they develop into fruits by August–October of the following year and help to produce a bigger harvest. Thus the tree flowers almost throughout the year. Flowers are bisexual, scented and colour ranges from white to red.

Fruits are available all the year round with seasonal variations. Normally harvest during August to October accounts for 80% of the production. Many fruits fall from the trees prematurely owing to strong winds and sudden changes in temperature.[4]

Star anise is valued for its characteristic essential oil. Since maximum essential oil is formed just before full maturity, fruits should be gathered at this stage. Children do most of the harvest. They climb trees and gather fruits using hooks attached to long poles. Sometimes, fruits are harvested by shaking branches. In the initial years, the yield of fresh fruits is small amounting to only 0.5–1.0 kg per tree. Yield increases with ageing and reaches nearly 20 kg fresh fruit per tree by the 15th year. When a tree is 20 years old, full production is available and the yield goes up to 30 kg.

Harvested fruits are dried in the sun. During drying, they turn a deep reddish colour. The characteristic aroma and flavour of star anise are developed during the drying process.[2] Dried star anise is cleaned first by removing the stalk, leaves and other extraneous matter. Broken bits are also taken out. The main criterion for grading is the size of the fruit based on its diameter. The first quality comprises 85% of the fruits with 2.5 cm diameter. The rest of the dried fruits with lesser diameter and partly broken pieces are included in the second quality. Natural colour is also a factor in grading.[3]

18.3.4 Storage
The freshness of star anise fruit has to be retained throughout storage. Hence dried fruits are stored in a cool place. The freshness of fruit can be determined by breaking one segment, squeezing it between the thumb and forefinger until the brittle seed pops and then sniffing for the distinct aroma. If aroma is weak, fruits have probably passed their optimum storage life or been kept in undesirable hot and open conditions. Normally, dried fruits can be kept for three to five years in airtight containers away from heat, light and humidity.[10]

18.3.5 Processing
The fruit (without seeds) contains volatile oil, resin, fat, tannin, pectin and mucilage. The seed has little volatile oil, but a large amount of fixed oil. Essential oil is distilled mostly from fresh fruits. If there is a large accumulation of fresh fruits they can be kept for about ten days or even longer by spreading them in a thin layer and frequently turning them over to prevent fermentation. Oil is also distilled from dried fruits, but the yield is lower.

Traditional stills are still used for distillation of oil in China. These stills hold up to 30 kg fresh fruits per charge. Sometimes fruits are broken prior to distillation for better yield and for reducing the time needed for distillation. Whole or broken fruits are placed in a retort with sufficient water to cover the material. Heating of the retort is done directly, but slowly. Steam and oil vapours are collected and condensed and floating oil is recovered. In the traditional still used in China fresh fruits require 48 hours and dried fruits 60 hours for completing distillation. Modern steam stills not only reduce the time for distillation to three or four hours but also produce high-grade oil.[2]

The yield of oil varies from 3.0 to 3.5% from fresh fruits, depending on their maturity, location, age of trees, region, soil and climatic conditions. The oil is colourless or pale yellow with characteristic odour resembling true anise (*Pimpinella anisum*) oil. Anethole concentration in pure oil is up to 85 to 90%. Other chemical compounds found in the oil are methyl chavicol, α-pinene, limonene and phellandrene. The oil has a specific gravity of 0.978 at 25°C and a refractive index of 1.5530 at 20°C.[11]

Leaves of star anise yield about 0.5% essential oil on steam distillation. Leaf oil is inferior to that of oil from fruit. Leaf oil is sometimes used for adulteration. Decorticated star anise seeds contain 55% fat and fatty acids: myristic 4.43%, stearic 7.93%, oleic 63.24% and linoleic 24.4%.[2]

18.4 Main uses

18.4.1 Culinary uses
Star anise is one of the signature flavours of Chinese savoury cooking. It combines well with pork and duck and is one of the essential ingredients in Chinese master stock. All over China, five-spice powder mix is very common. This mixture contains star anise, cassia, clove,

fennel and Sichuan pepper in equal parts. As star anise is pungent, only a very small quantity is required for a pleasing result. The five-spice powder mix is often added to the batter of Chinese-style fried vegetables or meat. Meat is sometimes coated with mixture of corn starch and this spice mix before deep-frying. The mix is also used for marinating meat before stir-frying. One of the popular Chinese recipes making use of the five-spice powder mix is called five-flavoured pork. The fruit as such is used for flavouring teas and pickles. It is also used for chewing after meals in order to sweeten the breath.[12]

Star anise is sold in the shops as whole and ground, but it is used for flavouring generally in the powdered form. It is an ingredient in ground spice mixtures in puréed fruits and tarts.[1]

Besides China, star anise is used in Vietnam. In North Vietnam it is popular as one of the ingredients of the five-powder mix as in China and for making beef soups. Star anise is used in different Indian curry powders for preparing meat preparations. Star anise finds application in Indian, Persian and Pakistani cuisine also. From India some of the preparations containing star anise were introduced to Indonesia, but it has been popular only in the palaces of Sultans still adhering to Royal Indian cooking style. Among other Asian countries, star anise is employed for cooking in Malaysia and southern Thailand.

Star anise has found only limited use in the West. Its main application is as a substitute for anise seed in mulled wine and special desserts. The essential oil is used to flavour soft drinks, bakery products and, most importantly, liqueurs. It is also used as a flavouring agent in confectionery, candy and chewing gum. The oil finds application in a small way in perfumery and in the pharmaceutical industry.

18.4.2 Medicinal values

The fruit is antibacterial, carminative, diuretic and stomachic. It is taken internally in the treatment of abdominal pain, digestive disturbances and complaints such as lumbago.[13] It is often included in remedies for indigestion and also in cough mixtures, particularly because of its aniseed flavour. For children it is effective for digestive upsets, including colic pain. Some people chew the fruit after meals for better digestion. The antibacterial effect is reported to a certain extent to be similar to penicillin.

The essential oil is stimulant, stomachic, carminative, mildly expectorant and diuretic. It is an ingredient in cough drops. The oil can be applied externally to treat rheumatism and scabies. It is considered useful against body lice and bed bugs, and forms an ingredient in cattle sprays against fleas.[14]

18.5   References

1. KYBAL J. and KAPLICKA, J. (1995), *Herbs and Spices*, Harveys Bookshop Ltd, Wingston, Leicester.
2. PRUTHI J.S. (2001), *Minor Spices and Condiments – Crop Management and Post Harvest Technology*, Indian Council of Agricultural Research, New Delhi.
3. GEORGE C.K. and SANDANA A. (2000), Report of the Visit to Vietnam under the Project INT/61/77 on Co-operative Programme on Quality Assurance of Spices, International Trade Centre, Geneva.
4. ANON. (1959), *The Wealth Of India – Raw Materials*, Vol. 5, Council of Scientific and Industrial Research, New Delhi.
5. GUENTHER E. (1972), *The Essential Oils*, 5th Edn. Van Nostrand Reinhold Co., New York.
6. FELTER H.W. and LLOYD J.U. (1898), King American Dispensatory, Illicium.
7. ANON. (2003), *Spices Market Weekly*, **11** No. 31,. Spices Board, Cochin.
8. PARRY J.W. (1969), *Spices*, Vol. 11, *Histology and Chemistry*, Chemical Publishing Company Inc., New York.
9. ANON. (1991), Star anise. *J. Indian Spices*, **4** No. 28, Indian Institute of Spices Research, Calicut.

10. HEMPHILL I. (2000), *Spice Notes – A Cook's Compendium of Herbs and Spices*, Pan Macmillan Australia Pty Ltd, Sydney.
11. HIRASA K. and TAKEMASA M. (1998), *Science and Technology*, Marcel Dekker, Inc., New York.
12. CLEVERLY A., RICHMOND K., MORRIS S. and MACLALEY L. (1997), *The Encyclopedia of Herbs and Spices*, Hermes House, London.
13. YEUNG HIM-CHE (1985), *Handbook of Chinese Herbs and Formulas*, Institute of Chinese Medicine, Los Angeles.
14. PARRY J.W. (1969), *The Story of Spices – The Spices Described*, Vol. 1, Chemical Publishing Co. Inc., New York.

19

Thyme

E. Stahl-Biskup, University of Hamburg, Germany and R. P. Venskutonis, Kaunas University of Technology, Lithuania

19.1 Introduction

The common English word 'thyme' covers both the genus and the species most widely used, *Thymus vulgaris* L. (common thyme, garden thyme). From the aromatic and medicinal points of view, *T. vulgaris* is indeed the most important species and is widely used as a flavouring agent, a culinary herb and as a herbal medicine. Therefore *T. vulgaris* is the central species in this chapter and 'thyme' here refers to *T. vulgaris* unless another botanical name is mentioned. However, other *Thymus* species will be included here because they are used for similar purposes or as a substitute for *T. vulgaris*, especially *T. zygis* L. (Spanish thyme), *T. serpyllum* L. (wild thyme, mother-of-thyme) and *T. pulegioides* L. (large thyme or larger wild thyme). The commercial products that are obtained from these four species include essential oils, oleoresins, fresh and dried herbs, and landscape plants.

19.1.1 History and etymology

People have used thyme for many centuries because of its flavouring and medicinal properties. The first recorded evidence can be found in Dioscorides' work (first century AD) about medicinal plants and poisons mentioning 'Thymo', 'Serpol' and 'Zygis' and in Pliny's *Natural History* (first century AD). Although in the Mediterranean region thyme has always been widely used as a spice, it was only in the early Middle Ages that Benedictine monks brought it over the Alps to Central Europe and England where it began a glorious career. From this time on it could be found in all herb books, those by Pear Matthioli (1505–1577) and by Leonhart Fuchs (1501–1566) being the most famous. In the latter, a drawing is shown and the effectiveness of thyme against cough is described. The most favoured interpretation of the etymology of the name considers the Greek word '*thymos*' which means 'courage, strength'.

19.1.2 Systematic botany

The genus *Thymus* belongs to the Labiate family (Lamiaceae), subfamily Nepetoideae, tribe

Mentheae. The distribution of the genus can be described as Eurasian with the Mediterranean region, especially the Iberian Peninsula and northwest Africa, being the centre of the genus. The number of species differs according to the criteria applied for defining a species from 54 (Hegnauer, 1966) to 417 (Ronniger, 1924). Today, about 250 taxa (214 species and 36 subspecies) are accepted, subdivided into eight sections (Jalas, 1971; Morales, 2002). *Thymus vulgaris* L. and *T. zygis* L. belong to the Western Mediterranean section *Thymus*; *T. serpyllum* L. and *T. pulegioides* L. to the section *Serpyllum*, which is the most extensive section considering the numbers of species and the distribution areas.

19.1.3 Morphological description

Common thyme (*T. vulgaris* L.) is a perennial subshrub, 10–30 cm in height with slender, wiry and spreading branches. The small leaves are evergreen, opposite, nearly sessile, oblong-lanceolate to linear, 5–10 mm long and 0.8–2.5 mm wide, grey-green, minutely downy and gland-dotted. Their margins are recurved. The flowers are light-violet, two-lipped, 5 mm long with a hairy glandular calyx, borne with leaf-like bracts in loose whorls in axillary clusters on the branchlets or in terminal oval or rounded heads. Spanish thyme (*T. zygis* L.) is smaller with narrower leaves, which are clustered at the nodes. The flowers are whitish and in clusters, spaced at intervals in an elongated inflorescence. Wild thyme (*T. serpyllum* L.) and large thyme (*T. pulegioides* L.) differ considerably in appearance, being more herbaceous, only woody at the base, partly procumbent, leaves flat, linear to elliptical, subsessile, ciliate at the base. The inflorescence is usually capitate. The corolla is purple, the calyx campanulate with upper teeth as long as wide, usually ciliate. Their distinct phenotypic variety makes botanical classification difficult, and some authors handle *T. serpyllum* L. as a collective species with the addition 's.l.' (*sensu latiore*) and include therein *T. pulegioides*.

19.1.4 Origin and distribution

Thymus vulgaris is native to southern Europe, from Spain to Italy. It is commonly cultivated there as well as in most mild-temperate and subtropical climates, which include southern and central Europe. *Thymus zygis* is indigenous to the Iberian peninsula (Portugal and Spain) and on the Balearic Islands. *Thymus serpyllum* and *T. pulegioides* are the dominant *Thymus* species in northern and middle Europe; in the east they reach Siberia. It is difficult to differentiate these two species and to give their exact distribution areas. The plant material on the market comes from wild collections in the Balkans and the Ukraine.

19.2 Chemical structure

The chemical character of thyme is represented by two main classes of secondary products, the volatile essential oil (Stahl-Biskup, 2002; Lawrence, 2003 and references therein) and the non-volatile polyphenols (Vila, 2002 and references therein). Owing to the excellent analytical techniques available today, both groups are fairly well known. In particular, the composition of the essential oil has been reported in numerous scientific publications. Since we are dealing with a natural product, the yield of essential oils and of the polyphenols as well as the proportions of individual constituents, vary. This is caused by intrinsic (seasonal and ontogenetic variations) and extrinsic factors (soil, climate, light). The data presented here are a result of an evaluation of numerous publications with respect to essential information about thyme as an herb and as a spice for commercial purposes.

19.2.1 Essential oil

The essential oil is responsible for the typical spicy aroma of thyme. It is stored in glandular peltate trichomes situated on both sides of the leaves. They show a very typical anatomy with a gland head of 8–16 secretory cells sitting on one basal stalk cell. In the secretory cells the oil is produced and is secreted into the subcuticular space. If the cuticle is ruptured, e.g. by rubbing or grinding, the volatile oil spreads into the air and stimulates the olfactory nerves. On hot days traces of the volatiles penetrate the cuticle and form an aromatic cloud around the plants, as can be perceived in the fields of thyme in Mediterranean regions.

Dried plant material of thyme contains 1–2.5% of an essential oil. Its composition, including the chemical structures of the components, is given in Fig. 19.1. Most of the volatiles detected in thyme oil belong to the monoterpene group with thymol, a phenolic monoterpene, as the main representative (30–55%). It causes the typically strong and spicy smell associated with thyme. Thymol is always accompanied by some monoterpenes, which are closely connected by biogenetical processes, namely carvacrol (1–5%), an isomer terpene phenol, as well as *p*-cymene (15–20%) and γ-terpinene (5–10%). The latter two are precursors in the biogenetic pathway of thymol (and carvacrol). Often the methyl ethers of thymol and carvacrol are present. Further monoterpenes are linalool (1–5%) and, in smaller percentages (0.5–1.5%), borneol, camphor, limonene, myrcene, β-pinene, *trans*-sabinene hydrate, α-terpineol and terpinen-4-ol. Sesquiterpenes are not very important in thyme oils. Only β-caryophyllene (1–3%) is worth mentioning.

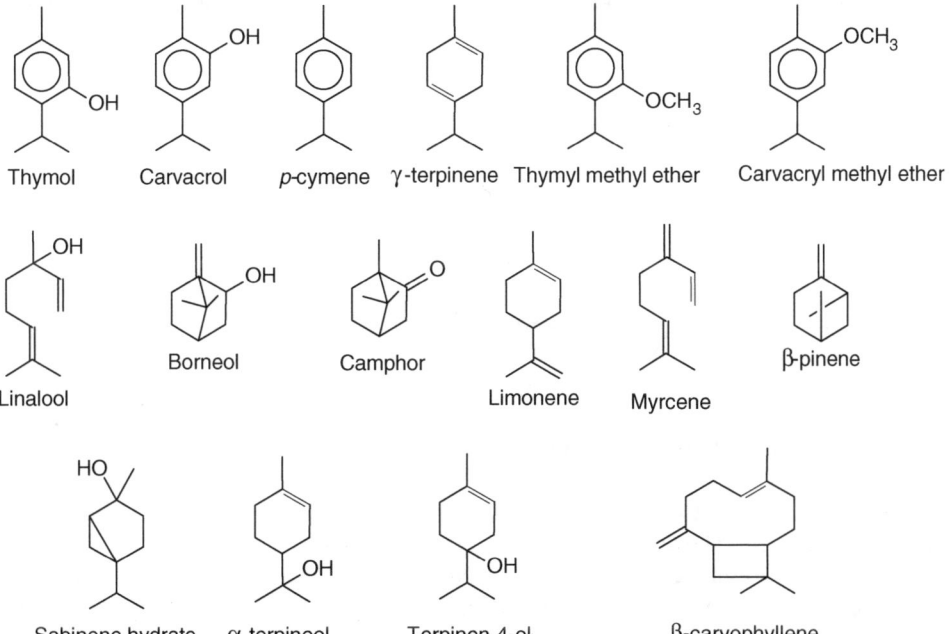

Fig. 19.1 Terpenes in the essential oil of thyme.

The composition of thyme oil given so far is that of commercially used thyme. However, it is important to mention that *T. vulgaris*, the main source of commercial thyme, is a chemically polymorphous species. That was discovered in the 1960s when six different genetically based chemotypes of *T. vulgaris* were found in the south of France (Granger and Passet, 1973). They are named according to their dominant monoterpene in the essential oil: a thymol type, a carvacrol type, a linalool type, a geraniol type, an α-terpineol type and a *trans*-thuyanol (= *trans*-sabinene hydrate) type. In Spain, a seventh chemotype, a cineole type, was found. Only the thymol chemotype is of commercial interest.

The chemical composition of the essential oil from *T. zygis*, the most important source of thyme oil in Spain, is quite similar to that of *T. vulgaris* with a remarkably high content of thymol. There is no practicable criterion to distinguish the oils of *T. vulgaris* and of *T. zygis*. Once a lower content of thymyl methyl ether in *T. zygis* was mentioned (0.3% versus 1.4–2.5%) but that has never been proven. Also, *T. zygis* is chemically polymorphous, showing several different chemotypes on the Iberian Peninsula.

The dried herb of wild thyme, *T. serpyllum*, yields 0.2–0.6% essential oil. Again we are dealing with a chemically polymorphous species. For commercial use, plant material with a high phenolic content in the oil is required. The phenolic part of the oil is mainly represented by carvacrol (20–40%) with lower percentages of thymol (1–5%). Further monoterpenes are *p*-cymene (5–15%), γ-terpinene (5–15%), borneol, bornyl acetate, 1,8-cineol, citral, geraniol, linalool and others. Also *T. pulegioides* is chemically polymorphous. Again only the phenolic chemotypes (thymol and/or carvacrol) are of commercial interest.

19.2.2 Flavonoids

In common thyme (*T. vulgaris*) about 25 different flavonoids could be detected (Miura and Nakatani, 1989; Wang *et al.*, 1998; Vila, 2002); these are listed in Table 19.1. They are present mostly in the form of their aglycones. The flavones apigenin and luteolin are the most important flavonoids present in both forms as aglycones and as *o*-glycosides. They are accompanied by a great variety of methylated flavones whereas flavonols and flavanones are of inferior importance. Vicenin-2, the 6,8-di-C-glucoside of apigenin, turned out to be a chemosystematic marker of the genus *Thymus*, occurring only in certain taxonomic groups, e.g. in the sections *Pseudothymbra* and *Thymus*.

The other *Thymus* species discussed here were not as intensively investigated as *T. vulgaris*. In Spanish thyme (*T. zygis*) nine different methylated flavones were reported; all nine also present in *T. vulgaris* (Table 19.1). The same methylated flavones were found in *T. pulegioides* with apigenin and luteolin and 6-OH-luteolin in addition. According to the literature, in wild thyme the glycosides seem to be of greater importance than the aglycones, and scutellarein and diosmetin seem to be exceptional.

19.2.3 Tannins and other phenolic compounds

Aside from the essential oil, the tannins of thyme contribute to its commercial use. The tannins are mainly represented by rosmarinic acid (Fig. 19.2), a depside of caffeic acid and dehydrocaffeic acid. The quantitative data of the content of rosmarinic acid in the literature vary between 0.15 and 2.6% owing to the different analytical methods applied for the quantification (ultraviolet, UV, gas chromatography, GC, high performance liquid chromatography, HPLC, gravimetrical). Recently the 3′-*o*-(8″-Z-caffeoyl)-rosmarinic acid was isolated from the leaves (Dapkevicius *et al.*, 2002). Also free phenolic acids have been reported in thyme, e.g. caffeic acid, *p*-coumaric acid, syringic acid and ferulic acid.

Table 19.1 Flavonoids and phenolic acids in thyme

Flavones				
Apigenin	vulg	zyg	ser	pul
Luteolin	vulg	zyg	ser	pul
6-Hydroxyluteolin	vulg	zyg		pul
Scutellarein			ser	
Methyl flavones				
Cirsilineol	vulg	zyg		pul
8-Methoxycirsilineol	vulg	zyg		pul
Cirsimaritin	vulg	zyg		pul
5-Desmethylnobiletin	vulg	zyg		pul
5-Desmethylsinensetin	vulg	zyg		pul
Diosmetin			serp	
Gardenin B	vulg			
Genkwanin	vulg			
7-Methoxyluteolin				
Salvigenin	vulg			
Sideritoflavone	vulg	zyg		pul
Thymonin	vulg	zyg		pul
Thymusin	vulg	zyg		pul
Xanthomicrol	vulg	zyg		pul
Flavanonols				
Taxifolin	vulg			
2,3-Dihydrokaempferol	vulg			
Flavanones				
Eriodictyol	vulg			
Naringenin	vulg			
Methyl flavanone				
2,3-Dihydroxanthomicrol	vulg			
Sakuranetin	vulg			
Flavonols				
Kaempferol	vulg			
Quercetin	vulg			
Flavone glycosides				
Apigenin-7-o-β-D-glucoside	vulg		serp	
Apigenin-4'-o-β-D-p-cumaroyl-glucoside			serp	
Apigenin-7-o-β-D-rutinoside	vulg			
Apigenin-6,8-di-C-β-glucoside	vulg			
Apigenin-7-o-β-glucuronide	vulg			
Diosmetin-7-o-β-D-glucuronide			serp	
Eriodictyol 7-o-β-D-rutinoside	vulg			
Hesperidin	vulg			
Luteolin-galactoarabinoside			serp	
Luteolin-7-o-β-D-glucoside	vulg		serp	
Luteolin-7-o-β-D-diglucoside	vulg		serp	
Scutellarein-glucosylglucuronide			serp	
Scutellarein-7-o-β-D-glucosyl (1–4)α-L-rhamnoside			serp	
Vicenin-2	vulg	zyg		
Phenolic acids				
Caffeic acid	vulg			
Rosmarinic acid	vulg			

vulg = *T. vulgaris*, zyg = *T. zygis*, serp = *T. serpyllum*, pul = *T. pulegioides*.

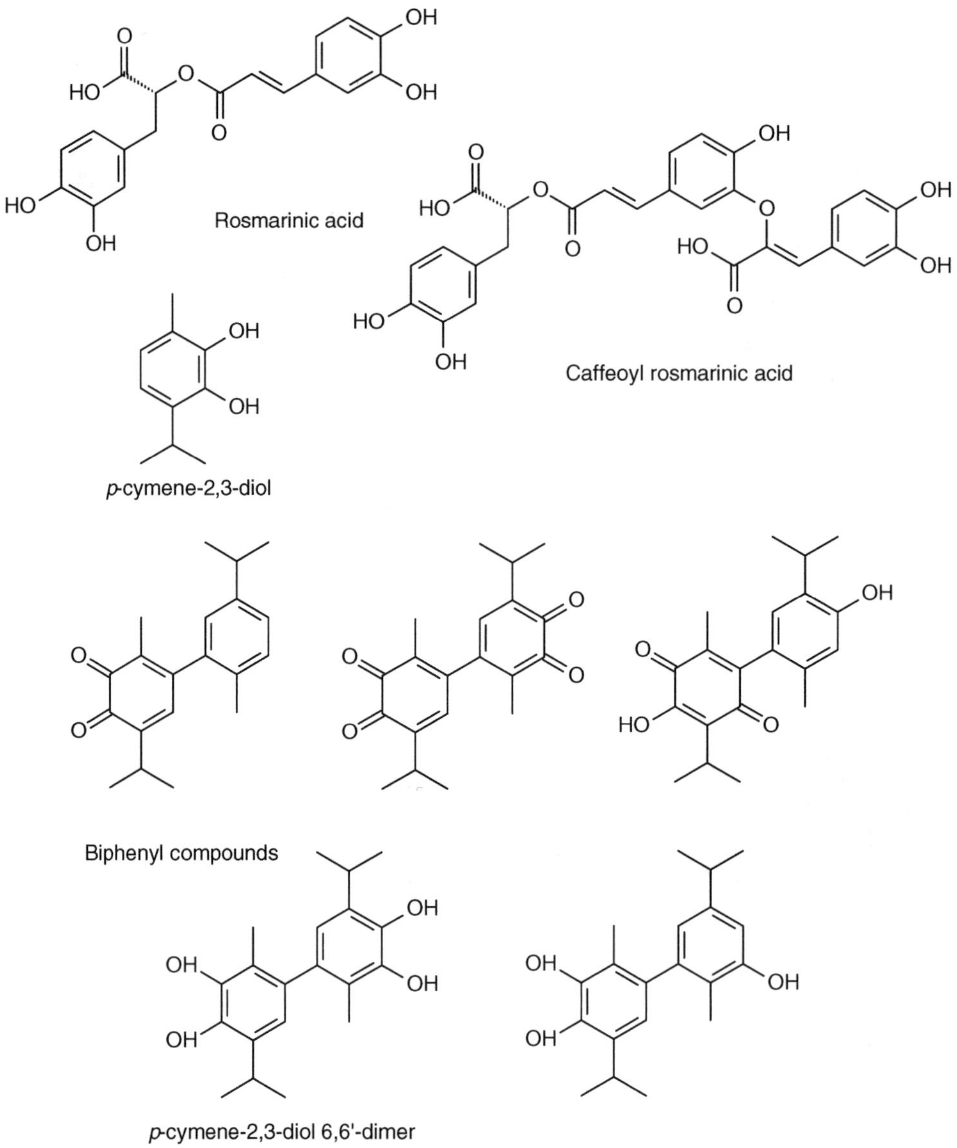

Fig. 19.2 Tannins and further phenolic compounds from thyme.

Biphenyl compounds from thyme have attracted attention because of their antioxidative activity and deodorant effects (Nakatani et al., 1989; Miura et al., 1989). Five different biphenyl compounds have been isolated from an acetone extract of the leaves (Fig. 19.2). The biogenetic connection with the terpene phenols is obvious as well as that of p-cymene-2,3-diol, which is present in thyme in concentrations from 0.8% (Schwarz et al., 1996).

19.2.4 Further compounds
Thyme contains 7.5% polysaccharides (labile in acids) and 1% soluble carbohydrates (stable in alkalines) as well as triterpenes in the form of ursolic acid (1.88%) and oleanolic acid (0.63%).

19.3 Production

Production of thyme is associated with various growing, harvesting and post-harvest handling aspects. Several important thyme production phases have to be properly controlled in order to obtain high yields of herb suitable for good quality ingredients in food applications. The information in this section is focused mainly on *T. vulgaris*, which is the only *Thymus* species cultivated and processed commercially for the use in food processing in reasonable amounts.

19.3.1 Main producing areas

Thyme is grown commercially in a number of countries for the production of essential oil, extracts and oleoresins, dried leaves and other applications. Thyme-producing countries are Spain, Portugal, France, Germany, Italy, the UK and other European countries, as well as North Africa, Canada and the USA (Prakash, 1990). Spain, Jamaica and Morocco are the main suppliers of dried leaf to the US market, while Spain and France supply the oil market (Simon, 1990).

There is much confusion concerning the amounts and species of *Thymus* in trade. Spain is the leading producer, with most production from the wild. Confusion is increased by the fact that local names change from one region to another. Little distinction is made in Turkey between a number of species of *Origanum* and *Thymus*, and also *Thymbra spicata*. Fifteen species of Lamiaceae are traded in Turkey under the name 'kekik' which is one of the main medicinal and aromatic plants exported from Turkey, the annual quantity being between three and four million kg (WWF website; *Thymus* and *Origanum*).

Thyme is produced by commercial cultivation and wild harvesting. In Spain, almost all thyme comes from wild plants, mainly growing in the southeast, where most of the companies dealing in this commodity are situated (Lange, 1998). France, Hungary and Poland are other countries that still harvest huge amounts of wild *Thymus*, although large-scale cultivation programmes are in progress. Cultivation provides greater control over quality and supply; however, its feasibility depends on a species ability to thrive as a monocrop, while its economic viability depends on the volumes required and market prices.

Since 1970, improvements have been made to thyme for cultivation by farmers in France, enabling the crop to compete with wild thyme from developing countries (Verlet, 1992). However, the expense of wild harvesting is generally less than that incurred in the establishment of cultivation.

19.3.2 Propagation

Thyme can be grown from seed. It is also easy to root from cuttings taken from non-woody, fast-growing shoots. Another method is to separate out sections of rooted stems and replant. Direct sowing of very small thyme seeds in fields is difficult, and therefore the majority of thyme plots are established by seedlings prepared in a greenhouse or by selected clones struck as cuttings in individual cells. The germination rate of thyme is comparatively low (72%; Kretschmer, 1989).

To plant a large area of thyme, it would be cheaper to buy prepared seedlings from an established nursery than to produce seedlings in an under-equipped home nursery. It should be possible to generate the desired 160 000 to 240 000 seedlings per ha from 50 to 80 g of seed (Fraser and Whish, 1997). Culinary varieties of thyme have to be replaced or propagated every two or three years as they become woody and straggly and produce few tender leaves.

The recommendations of the spacing for thyme plants provided in numerous manuals and instructions for herb growers are different; they also vary from country to country. For instance, spacing within rows of 30 cm with 60 cm between rows have been tried in the USA; however, these were later concluded to be too sparse (Gaskell, 1988). Densities of 36 plants per m^2 have been successfully used in Australia with the plots quickly reducing weed competition. In the case of mechanical harvesting, the beds should be designed in the way that machinery straddles the crop without squashing the plants because thyme would not tolerate this. A spacing of 25–30 cm between plants in beds is recommended.

19.3.3 Cultivation

Thyme prefers a light, dry calcareous soil; it succeeds in poor soils and tolerates drought once it is established. Agricultural lime should be added to the soil before sowing if the pH is less than 5.5. Successful growing of most thyme species is possible in any climate having a mean annual temperature from 7 to 20°C. Thyme thrives in full sun, but also tolerates partial shade. The accumulation of essential oil in thyme directly or indirectly depends on light.

Thyme should be treated as a leafy vegetable when considering water requirements. During seedling establishment, beds should be always moist but not wet, and during growth the plants like to be in a well-drained soil with adequate subsoil moisture (Fraser and Whish, 1997).

Some studies have indicated that nitrogen and other fertilizers increase the yield of thyme crop. However, the optimal amount of fertilizers and the schedule of their application should be adjusted to every particular growing site. These studies showed that the effect of fertilizers on the yield of essential oil and its composition were not remarkable (Shalaby and Razin, 1992; Dambrauskiene et al., 2002). Recently it was demonstrated that young thyme plants exhibit increases in photosynthesis and biomass production at elevated CO_2 concentration (Tisserat et al., 2002).

Weed control in thyme crops, as with all herbs, is difficult. The best method to reduce weeds is to grow a dense stand of pasture prior to planting the crop, then follow up by fallowing the land prior to planting. The use of a chemical fallow and smothering pasture crops would help to reduce the weed seed reserves prior to planting. Mulching is a useful weed control method; however, as thyme plants can produce a dense cover, the crop will outrival many weeds. Unfortunately, the spreading nature of thyme is impeded by the use of inorganic mulch, and therefore if mulch is to be applied, a long-lasting organic form would be more suitable (Fraser and Whish, 1997). Plants are mulched in northern areas to protect them from winter injury. However, it was reported that fresh thyme yield was reduced after mulching, which encouraged the development of a fungal disease of the soil. Cultivation of thyme is reported to be associated with fungal infections, leaf diseases, root rot and spider mites.

19.3.4 Harvesting

In general, thyme is most aromatic during the period of blooming or at the beginning of full bloom. However, the period of vegetation and blooming can be different in various geographical zones, depending on their climatic conditions. In Spain the harvest takes place during the blooming period from February to August, depending on the species (Tainter and Grenis, 1993). In France, thyme can be harvested twice a year, once in May and then again in September. In the central regions of Russia thyme is harvested during the second year of plant vegetation. Usually, the first cutting is performed in June during flowering, the second one in September–October. Aerial parts are cut at 10–15 cm height from the ground.

The plant's low-growing habit makes mechanical harvest difficult. Most wild-growing plants are collected by hand. The main objective of thyme harvesting is to collect the most valuable anatomical parts, the leaves and flowering parts. Woody stems, which are of minor value, must be avoided as far as possible.

The most important quality characteristics of thyme, i.e. the yield of essential oil and its chemical composition, highly depend on harvesting time. This was clearly established in several studies performed with different *Thymus* species (Venskutonis, 2002a). Therefore, it is important to select an optimal time of harvesting which considers the growing site and the plant species.

19.3.5 Post-harvest handling

All the post-harvest principles that apply to leafy green tissues apply to the handling of fresh herbs. Temperature is the most important factor in maintaining quality after harvest. The optimum post-harvest temperature for fresh thyme is 0°C (a shelf-life of three to four weeks). With a temperature of 5°C, a minimum shelf-life of two to three weeks can be expected (Cantwell and Reid, 1986). Therefore, after harvesting, appropriate cooling is needed to prolong shelf-life of fresh thyme.

Some novel processes to prolong shelf-life of fresh herbs and spices and retain their flavour and appearance for a considerably longer time have been developed and tested on various culinary herbs and spices. These processes and their possible applicability to thyme have been reviewed elsewhere (Venskutonis, 2002b).

Drying is undoubtedly the most ancient and still the most widely used method of the fresh herb processing. In order to obtain stable products that will withstand long periods of storage without deterioration, the water content of thyme must be reduced to 8–10%. Drying is the most critical process because of the volatility and susceptibility to chemical changes of the contained volatile oil (Heath, 1982).

Natural drying is the simplest way to prepare thyme for storage and further processing. There are several methods of drying raw material, such as sun-drying, drying in the shade, solar drying and hot air drying, practised in commercial processing in different countries. Natural drying of the whole *T. vulgaris* herb is particularly problematic because the shrub consists of comparatively fast-drying leaves and slower-drying, rather hard stems. A more sophisticated method is that of solar drying. It maintains the rich green colour, making the product look attractive. Different types of solar dryers have been successfully tested on Labiatae plants; they are suitable for the drying of thyme (Müller *et al.*, 1993).

Traditional hot air drying should be tailored to minimize flavour loss and to perform the process at reasonable time and energy costs. It is well established that higher drying temperatures need shorter drying processes. However, they cause bigger losses of volatiles. Raghavan *et al.* (1995) compared cross-flow and through-flow drying methods on Indian thyme at 40, 50 and 60°C and found that through-flow drying at 40°C gave the best results.

Freeze drying is based on evaporation of water directly from ice under a high vacuum. The products obtained by this method are usually of a better appearance (colour) and aroma quality. The high cost is the main disadvantages of freeze drying, which limits the wider use in a commercial scale.

19.3.6 Packaging and storage

The main tasks for packaging are to protect the herb from the external conditions and to increase the stability against negative internal changes (enzymatic, non-enzymatic, chemical reactions, etc.). In general, dried thyme should be stored in cool, dry conditions away from light. Ideally, it should be in airtight packaging to reduce oxidation. Storage below $-18°C$ is a guarantee for unlimited storage time; dried herb can be stored at 5–7°C for more than 12 months, whereas at room temperature the stability considerably decreases. Finely milled thyme does tend to lose volatiles more rapidly than medium or coarsely ground material and must be stored in tightly closed containers. Storage in multilayered paper sacks having an impervious lining is also satisfactory, but not as good once the sack has been opened (Heath, 1981).

19.4 Main uses in food processing

The thyme herb or processed products can be used in culinary and/or food processing as a separate flavouring or in the composition of compounded seasonings, spice, essential oil, oleoresins or other product blends. The list of thyme applications includes almost all foods: beverages, cheese, fish, meat, salad dressings, sauces, vegetables, egg dishes, game and poultry, soups and honey. Usually, owing to its sensory characteristics, thyme is not suitable for sweet products.

The main uses of thyme in culinary and food processing are defined by the following properties of thyme components: (i) odour and taste, (ii) antioxidant and (iii) antimicrobial activities. Also, fresh green thyme leaves can be used in culinary art as a decorative green herb. It is evident that food flavouring remains the main thyme application area, while its antimicrobial and antioxidant properties can be considered as the supplementary benefits of thyme products, which have been added to the foods. The possibility of successfully using all three benefits provided by thyme components, namely flavour, and prevention of microbial and oxidative spoilage, depends on product requirements, processing parameters and food producer skill.

19.4.1 Fresh and dried herb

The use of fresh thyme herb in food is rather limited owing to a very short shelf-life. Although proper temperature and storage conditions can prolong the shelf-life of freshly cut thyme up to four weeks, the green herb is mainly used in catering and home cooking. Some studies have shown that even simple cutting of the plants generates changes in their aroma composition. In order to obtain stable products, which will withstand long periods of storage without deterioration, most thyme crops are dried before further use or processing. Drying is the most critical process owing to the volatility and susceptibility to chemical changes of the volatile oil (Heath, 1982).

Whole dried thyme herb as such can find numerous culinary applications; however, its direct use in food processing is rather limited. The main concerns are related to the evenness of distribution throughout the food product and to the release of volatile compounds into the product. Therefore in most cases industry requires additional treatments in order to meet specified quality parameters. Comminution is a simple and the most widely used treatment before final application of dried thyme. The leaves can be chopped, cut or sliced, broken or rubbed and ground and the spice manufacturers may adopt their own empirical classification. The following is representative of commercial standards for ground spices (Reineccius, 1994):

- Coarse: over 30% retained on a no. 30 US sieve.
- Medium: less than 30% retained on a no. 30 US sieve.
- Fine: less than 2% retained on a no. 30 US sieve; less than 35% retained on a no. 60 US sieve.
- Very fine: 50% to pass through a no. 100 US sieve.

Grinding ruptures the oil structures containing the volatile oil and the oil becomes available for reaction (e.g. oxidation) or evaporation. Grinding also generates some heat, which tends to vaporize the volatile oil, leading to a reduction in flavour strength. Therefore, it is necessary to keep the temperatures during the grinding process as low as possible to minimize the loss of volatile oil.

The moisture content of ground thyme is of importance to both stability and flavour value. It must be dry enough to prevent a musty odour and flavour, and yet be moist enough to retain the optimum odour and flavour character. In terms of use for food flavouring, the most common advantages and disadvantages of dried ground thyme are summarized in Table 19.2 (Heath and Reineccius, 1986).

Some investigations showed that thyme was a heavily contaminated herb, particularly with insect fragments, mite, thrips and aphids (Gecan et al., 1986). Microbial contamination of thyme can also reach high levels (Kneifel and Berger, 1994). Contamination of thyme can encounter serious problems in some microbiologically sensitive foods. Therefore sterilization procedures are often applied before final application. Sterilization is performed by chemical (ethylene oxide, methyl bromide, ozone) or physical (irradiation, UV irradiation, microwaving, high-frequency electric currents, high pressure) treatments. It should be remembered that treatment with ethylene oxide has been banned in many countries, including the EU.

19.4.2 Thyme extracts and processed products

Owing to the above-mentioned disadvantages of dried ground thyme, manufacturers increasingly are recognizing the advantages of seasonings based on herb extracts. In general, the methods of extraction depend on the desired properties of a final product, the characteristics of the plant material, and economical and technical issues. The most important products that are obtained from thyme are essential oils, herb oleoresins and solvent extracts (Fig. 19.3).

Table 19.2 Advantages and disadvantages of dried ground thyme

Advantages	Disadvantages
Slow flavour release in high-temperature processing	Variable flavour strength and profile
Easy to handle and weigh accurately	Unhygienic, often contaminated by filth
No labelling declaration problems	Easy adulteration with less valuable materials
Presence of natural antioxidant and antimicrobial components	Flavour loss and degradation on storage
	Undesirable appearance characteristics in end products
	Poor flavour distribution (particularly in thin liquid products such as sauces)
	Unacceptable hay-like aroma
	Dusty and unpleasant to handle in bulk

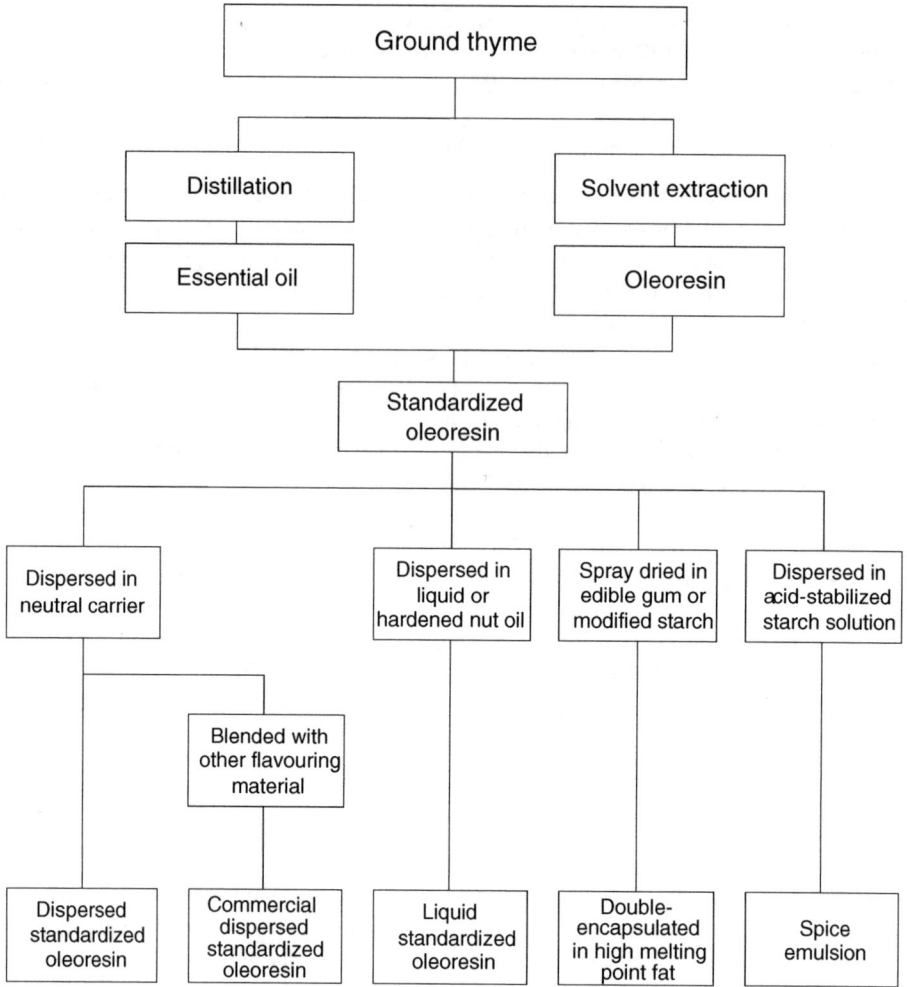

Fig. 19.3 Flow chart of a typical standardized range of thyme (*Source*: Moyler, 1991).

Thyme essential oil and oleoresin possess a sweetly aromatic, warmly pungent odour and a sharp, rich, warmly phenolic flavour. The advantages and disadvantages of their use in food processing are common to many other herb essential oils and oleoresins and are summarized in Tables 19.3 and 19.4 (Heath and Reineccius, 1986).

To avoid the disadvantages of thyme essential oils and oleoresins, they can be further processed to obtain such advanced products as solubilized (providing clear solution when mixed with water), dispersed, plated or 'dry-soluble', encapsulated, heat-resistant and fat-based products (Table 19.5). For instance, with dispersed products, standardized flavour profile and flavouring strength can be obtained. They are readily handled and weighed with accuracy, readily dispersed in food mixes and possess low water activity. When thyme oil has been encapsulated, its aromatics are fully protected from loss and degradation. Controlled release of volatiles can be achieved by selecting proper encapsulation materials. Usually, such products are tailored for a specific food application and contain blends of essential oils and/or oleoresins. Thyme extracts are used as a part of such blends in numerous flavourings and seasonings.

Table 19.3 Advantages and disadvantages of thyme essential oil

Advantages	Disadvantages
Hygienic, free from all microorganisms	Flavour good but incomplete and unbalanced compared to natural herb
Flavouring strength within acceptable limits	Does not contain non-volatile antioxidants
Flavour quality consistent with source of raw material	Some compounds readily oxidize
No colour imparted to the end product	Readily adulterated
Free from enzymes	Very concentrated so difficult to handle and weigh accurately
Stable in storage under good conditions	Not readily dispersible, particularly in dry products

Table 19.4 Advantages and disadvantages of thyme oleoresins

Advantages	Disadvantages
Hygienic, free from all microorganisms	Flavour quality good but as variable as the raw material
Can be standardized for flavouring strength	Flavour profile dependent on the solvent used
Contain natural antioxidants	Very concentrated so difficult to handle and weigh accurately
Free from enzymes	Sometimes difficult to incorporate into food mixes without 'hot spots'
Long shelf-life under good storage conditions	

Table 19.5 Examples of standardized thyme products

Product	Producer	Characteristics
Standardized oleoresin Thyme FD0718	Bush Boake Allen Limited, London, UK	Volatile oil content (%, v/w) 54–60
Standardized oleoresins Thyme HX2089	Lionel Hitchen Essential Oil Company Limited, Barton Stacey, UK	Volatile oil content (%, v/w) 50; dispersion rate kg = 100 kg of spice 1
Dispersed spices – salt thyme	Bush Boake Allen Limited, London, UK	Volatile oil content 0.3–0.4% (v/w)
Dispersed spices – dextrose thyme	Bush Boake Allen Limited, London, UK	Volatile oil content 0.3–0.4% (v/w)
Dispersed spices – rusk Thyme FD5781		Volatile oil content 0.6–0.8% (v/w)
Standardized emulsion oleoresins Thyme HF107	Felton Worldwide SARL, Versailles, France	Strength compared with ground spice 4×
Standardized emulsion oleoresins Thyme FD6136	Bush Boake Allen Limited, London, UK	Strength compared with ground spice 5×
Encapsulated standardized oleoresins Thyme FD4040	Bush Boake Allen, 'Saronseal Encapsulated spices'	Strength compared with ground spice 10×

19.5 Functional properties and toxicity

All over the world, thyme, *Thymus vulgaris*, is highly regarded. Thyme as a medicine has developed from a simple traditional herb into a drug that is taken seriously in phytotherapy. Herbal thyme, thyme extracts and thyme oil are used for symptoms of bronchitis and whooping cough as well as catarrhs of the upper respiratory tract. This development is based on numerous experimental *in vitro* studies revealing well-defined pharmacological activities (Zarzuelo and Crespo, 2002 and references cited therein) of both the essential oil and the plant extracts, the antimicrobial and spasmolytical properties being the most important ones. The non-medicinal use of thyme is no less important because thyme serves as a preservative for foods and is a culinary ingredient widely used as a seasoning in many parts of the world. Furthermore, thyme oil is an ingredient in many cosmetic preparations.

19.5.1 Antimicrobial activity of thyme oil

In the 1980s several screening studies (agar overlay technique or dilution technique) with essential oils verified the antibacterial and antifungal activity of the essential oil of thyme. It was shown to inhibit a broad spectrum of bacteria; generally Gram-positive bacteria being more sensitive than Gram-negative bacteria (Blakeway, 1986; Farag *et al.* 1986; Deans and Ritchie, 1987). Also some food-borne pathogens, namely *Salmonella enteritidis*, *Escherichia coli*, *Staphylococcus aureus*, *Lysteria monocytogenes* and *Campylobacter jejuni* were tested (Smithpalmer *et al.*, 1998). The latter was found to be the most resistant of the bacteria investigated. In another study it was shown that the antibacterial activity of thyme can be used against periodontopathic bacteria including *Actinobacillus*, *Capnocytophaga*, *Fusobacterium*, *Eikenella* and *Bacterioides* species, and may therefore be suitable for plaque control (Osawa *et al.*, 1990). Furthermore, the essential oil of thyme showed a wide range of antibacterial activity against microorganisms that had developed resistance to antibiotics (Nelson, 1997).

To assess the antifungal activity of thyme oil, attention was directed towards some food-spoiling fungi, especially *Aspergillus* (Conner and Beuchat, 1984; Farag *et al.*, 1986, 1989; Deans and Ritchie, 1987), and to various dermatophytes (Janssen *et al.*, 1988) as well as some phytopathogenic fungi, e.g. *Rhizoctonia solani*, *Pythium ultimum*, *Fusarium solani* and *Calletotrichum lindemthianum* (Zambonelli *et al.*, 1996). The yeast *Candida albicans* was also inhibited by thyme oils, namely those of *T. vulgaris* and *T. zygis* (Menghini *et al.*, 1987; Cabo *et al.*, 1978). *Thymus serpyllum* was found to be highly active against various species of *Penicillium*, *Fusarium* and *Aspergillus* (Agarwal and Mathela, 1979; Agarwal *et al.*, 1979). It could be demonstrated that thyme oil (*T. vulgaris*) inhibits both mycelial growth and aflatoxin synthesis of *Aspergillus parasiticus* (Tantaoui-Elaraki and Beraoud, 1994).

When discussing the antimicrobially active constituents of thyme oil, the monoterpenes thymol and carvacrol play an outstanding role. These terpenes bind to the amine and hydroxylamine groups of the proteins of the bacterial membrane altering their permeability and resulting in the death of the bacteria (Juven *et al.*, 1994). The antifungal activity of thyme oils is also attributed to thymol and carvacrol. They cause degeneration of the fungal hyphae which seems to empty their cytoplasmic content (Zambonelli *et al.*, 1996). Other constituents of the oil, such as the terpene alcohols, contribute to the activity, but to a lesser extent.

19.5.2 Spasmolytic activity of thyme

First results of the antispasmodic activity were obtained in the 1960s when thyme oil was tested on the intestinal smooth muscle contracted by several agents. This could be confirmed later when the oils of 22 plants were tested on the tracheal and ileal smooth muscle of the guinea pig (Reiter and Brandt, 1985). The relaxant effect of thyme oil was shown to be higher on the ileal smooth muscle. Detailed studies with isolated rat duodenum and guinea pig ileum verified that the terpene phenols and the terpene hydrocarbons are responsible for the relaxant effect of thyme oils. The mechanism was found to be non-competitive antagonistic (contractions induced by carbachol, histamine and $BaCl_2$). The inhibitory effect can be explained by an inhibition of Ca^{2+} entry through voltage stimulated channels into the smooth muscle and/or blocking the release of intracellular bound Ca^{2+} (Cabo *et al.*, 1986; Cruz *et al.*, 1989; Zarzuelo *et al.*, 1989; Godfraind *et al.* 1986).

Although the volatile oil of thyme was proven to have antispasmodic effects, the presence of a non-volatile principle has always been supposed. Therefore, some authors focused on the flavonoids, and found that the flavones as well as thyme extracts were effective in test systems with smooth muscles of guinea pig ileum and of rat vas deferens (Van den Broucke and Lemli, 1983). The flavonoids appeared to act as musculotropic agents. The inhibition of Ca^{2+}-induced contractions in K^+ depolarized muscles pointed to a possible decrease in the availability of Ca^{2+}.

19.5.3 Anti-inflammatory activity

In an *in vitro* assay screening (cyclooxygenase inhibition test) of several essential oils, thyme oil inhibited prostaglandin biosynthesis (Wagner and Wierer, 1987). This effect of the oil has never been affirmed by further experiments.

19.5.4 Thyme as an antioxidative agent

The antioxidative property of thyme is important in both the medicinal and non-medicinal context. In the 1970s scientists discovered that the human body constantly creates free radicals, culminating in an 'oxidative stress' when their elimination by antioxidant defence mechanisms is not sufficient. Oxidative stress contributes to the pathogenesis of many human diseases; therefore the intake of antioxidative agents is important for the prevention of chronic diseases. In the non-medicinal context antioxidant character is responsible for a preservative activity, especially in preventing oxidation of lipids in food.

Several papers show that the essential oil and extracts of thyme are potent antioxidative agents by using different test systems. In a screening of the protection of polyunsaturated fatty acids (liver of old mice, *in vivo*) including several culinary and medicinal plant volatiles, thyme oil was one of the most effective antioxidants (Deans *et al.*, 1993). Discussing the chemical principle of the antioxidative activity of the oil, the terpene phenols thymol and carvacrol are in the focus of interest. Indeed, it could be demonstrated that both exhibit antioxidative activities (Schwarz *et al.*, 1996; Nakatani, 2000). In a quantitative respect, however, this is exceeded by *p*-cymene-2,3-diol, a related non-volatile monoterpene, which turned out to be more active than α-tocopherol and butylated hydroxyanisole (Schwarz *et al.*, 1996).

Owing to their electron-donating properties, the flavonoids of thyme contribute to its antioxidative activity. In this respect the aglycones, eriodictyol and 7-*o*-methyl luteolin (Miura and Nakatani, 1989; Haraguchi *et al.*, 1996; Miura *et al.*, 2002) and two flavone glycosides, namely luteolin-*o*-glucoside and eriodictyol rutinoside (Wang *et al.*, 1998), were shown to be the most effective flavonoids. The biphenyl compounds are also responsible for the antioxidative power of thyme. This was confirmed in a bioassay-directed fractionation of thyme extract revealing the 3,4,3',4'-tetrahydroxy-5,5'-diisopropyl-2,2'-dimethylbiphenyl (= p-cymene 2,3-diol 6,6'-dimer) as the most potent one (Haraguchi et al., 1996). Deodorant effects are ascribed to the biphenyl compounds analysed in a test system with methyl mercaptan (Miura et al., 1989; Nakatani et al., 1989).

Recently, the radical scavenging activity of a leaf extract of *T. vulgaris* was investigated in detail (Dapkevicius et al., 2002). Seven active compounds could be isolated, among them rosmarinic acid, 3'-o-(8"-Z-caffeoyl)-rosmarinic acid, p-cymene 2,3-diol and the p-cymene 2,3-diol 6,6'-dimer. They contributed the most to the radical scavenging activity of the leaves. Besides these phenols, eriodictyol, taxifolin and luteolin-7-glucuronide were detected in the radical scavenging fraction.

19.5.5 Further effects

Several further effects of thyme and thyme preparation have been reported. An extract of *T. vulgaris* showed antiparasitic properties against *Leishmania mexicana*, inhibiting its mitochondrial DNA polymerase, with thymol to be mainly responsible for this effect (Schnitzler et al., 1995; Khan and Nolan, 1995). Nematicidal effects could be proven for the essential oil (Abd-Elgawad and Omer, 1995). Thymol was shown to possess miticidal activity studied with *Psoroptes cuniculi* (Perrucci et al., 1995). Insecticidal effects were proven for the essential oils of *T. vulgaris* and *T. serpyllum* by direct toxicity of adult insects and by inhibiting reproduction through ovicidal and larvicidal effects (Regnaultroger and Hamraoui, 1994). Other test objectives were the two-spotted spider mite, *Tetranchus urticae* (El-Gengaihi et al., 1996; Lee et al., 1997) and the house fly, *Musa domestica* (Lee et al., 1997), the Western corn rootworm, *Diabrotica virgifera* (Lee et al., 1997) and *Spodoptera littoralis* (Farag et al., 1994). The insecticidal action could be explained by a genotoxic effect on the somatic mutation and recombination, as could be demonstrated in a test with *Drosophila* (Karpouhtsis et al., 1998).

19.5.6 Toxicity

Little has been reported on the toxic effects of thyme on mammals. In an acute toxicity test a concentrated extract of thyme reduced locomotor activity and caused a slight slowing down of respiration in mice when 0.5 to 3.0 g extract/kg body weight (= 4.3 to 26.0 dried plant material) were administered to the mice (Qureshi et al., 1991). Subchronic toxicity was observed after administration of a concentrated ethanol extract of plant material to mice over three months, causing an increase in liver and testis weight; 30% of the male animals died (Qureshi et al., 1991). The essential oil of thyme oil showed an acute oral toxicity of LD 50 = 4.7 g/kg rat. This effect is attributed to the terpene phenols, thymol and carvacrol (Dilaser, 1979), which also cause skin irritations and irritation of the mucosa, explaining the severe irritation of mouse and rabbit skin when it is exposed to undiluted thyme oil. Hypersensitive reactions have also been reported for thyme oil.

19.5.7 Mutagenicity

Thyme oil has no mutagenic or DNA-damaging activity in either the Ames or *Bacillus subtilis* rec-assay (Zani et al., 1991).

19.5.8 Pharmacokinetic properties
Data for the pharmacokinetics of thyme oil or of thyme extracts refer to thymol, the main component of the essential oil. Recent data show that, after administering an ethanolic dry extract of thyme to 12 volunteers, free thymol could be neither detected in plasma nor in urine. The metabolites thymol sulphate and thymol glucuronide were found in urine, in plasma only the thymol sulphate was detected. The amount of both thymol sulphate and glucuronide excreted in 24 h urine was 16.2% + 4.5% of the dose (Kohlert *et al.*, 2002).

19.6 Quality specifications and issues

19.6.1 Specifications
The medicinal and non-medicinal uses of thyme and thyme preparations demand high-quality standards. In the USA, the American Spice Trade Association (ASTA, 1960) is the advisory organization that helps the spice and seasoning industry develop acceptability standards for whole and ground spices and herbs. The recommended physical and chemical specifications of whole thyme leaves and ground thyme can be seen in Table 19.6.

Internationally accepted specifications also exist for the commercially important thyme oil. They are given by some organizations, the International Organization for Standardization (ISO, 1996, 1999), the Association Française de Normalisation (AFNOR, 1999) and the USA Food Chemical Codex (FCC, 1996) being the most important ones. A summary of these specifications is given in Tables 19.7 and 19.8.

Especially in modern phytotherapy, increasingly strict requirements concerning the safety of herbal drugs must be fulfilled. Intensive efforts on herbal remedies were undertaken by the German Commission E, which was established by the German Ministry of Health in 1978 (Blumenthal, 1998), the European Scientific Cooperative on Phytotherapy (ESCOP, 2003) as well as by the World Health Organization (WHO, 1998). All three

Table 19.6 Whole and ground thyme: cleanliness, chemical and physical specifications

	Whole thyme	Ground thyme
Cleanliness specifications		
Whole dead insects	8/kg	
Mammalian excreta	2/kg	
Other excreta	10/kg	
Mould (w/w)	1.0%	
Insect infested/contaminated (w/w)	0.5%	
Insect fragments	*ca* 325/25 g	*ca* 925/10 g
Rodent hairs	*ca* 2/25 g	*ca* 2/10 g
Chemical specifications		
Volatile oil	≥ 0.8%	≥ 0.5%
Moisture	≤ 10.0%	≤10.0%
Ash	≤ 10.0 %	≤ 10.0%
Acid-insoluble ash	≤ 3.0%	≤ 3.0%
Physical specifications		
Sieve test		95% through a 200 mesh
Bulk index	*ca* 400 mg/100 g	
Bulk density		250 ml/100 g

Table 19.7 Specifications for thyme oil from *T. vulgaris*

Appearance	A colourless, yellow, or red liquid with a characteristic, pleasant odour and a pungent, persistent taste.
Specific gravity (25°C)	0.915–0.935
Refractive index (20°C)	1.495–1.505 at 20°C
Optical rotation (20°C)	Laevorotatory, but not more than –3°
Solubility	In 80% v/v aqueous ethanol (20°C) 1:2 volumes
Phenol content	not less than 40%
Heavy metals (as Pb)	≤ 0.02%
Water soluble phenols	Shake 1 ml of oil with 10 ml of hot water and after cooling pass water layer through a moistened filter. On addition of one drop of ferric chloride solution (9 g $FeCl_3 \cdot 6H_2O$), not even a transient blue or violet colour should be produced in the filtrate.

Source: FCC (1996).

Table 19.8 Specification for thyme oil from *T. zygis*

Appearance and colour	A clear, mobile liquid; traditionally from reddish brown to very intense brown, almost black mobile liquid with a characteristic phenolic
Odour	Characteristic, aromatic, phenolic (thymol), with a slightly spicy base
Density (20°C)	0.910–0.937
Refractive index (20°C)	1.4940–1.5040
Optical rotation	–1° and –6°; generally laevorotatory; frequently impossible to measure due to its colour.
Solubility	In 80% v/v aqueous ethanol (20°C) 1 : 3 volumes
Flash point (c/c)	+ 60°C
Phenol content	38–56% v/v
GC analysis (ISO)	α-thujene (0.2–1.6%), α-pinene (0.5–2.5%), myrcene (1–2.8%), α-terpinene (0.9–2.6%), γ-terpinene (4–11%), *p*-cymene (14–28%), *trans*-sabinene hydrate (trace–0.5%), linalool (3–6.5%), terpinen-4-ol (0.1–2.5%), methyl carvacrol (0.1–1.5%), thymol (37–55%), carvacrol (0.5–5.5%), β-caryophyllene (0.5–2%)

Sources: AFNOR (1999); ISO (1999).

organizations evaluated thyme according to its therapeutic benefit and safety with respect to levels of safety, efficacy and quality control. In their monographs on thyme and thyme oil they consider a clear definition of the herbal drug, effectiveness, side-effects, interactions, toxicological data and dosage.

19.6.2 Thyme in pharmacopoeias

Neither the Commission E nor ESCOP monographs contain standards for assaying the quality and purity of thyme or thyme oil. This is left to the pharmacopoeias. Quality standards for thyme (*Thymi herba*) and thyme oil (*Thymi aetheroleum*) as well as for wild thyme (*Serpylli herba*) can be found in the *European Pharmacopoeia* (2002, Addenda 4.1 and 4.3 respectively). Pharmacopoeial summaries for quality assurance can also be found in the WHO monographs. The monographs begin with a definition of the drug including the plant source and the quantitative requirements concerning the biologically active compounds.

Thyme (*Thymi herba*)
Thyme consists of the whole leaves and flowers separated from the previously dried stems of *Thymus vulgaris* L. or *Thymus zygis* Loefl. ex L. or a mixture of both species. It contains not less than 12 ml/kg of essential oil, of which a minimum of 40 per cent is thymol and carvacrol (both $C_{10}H_{14}O$; M 150.2) (anhydrous drug).

Thyme oil (*Thymi aetheroleum*)
Thyme oil is obtained by steam distillation from the fresh flowering aerial parts of *Thymus vulgaris* L., *T. zygis* Loefl. ex L. or a mixture of both species.

Wild thyme (*Serpylli herba*)
It consists of the whole or cut dried, flowering aerial parts of *Thymus serpyllum* L. s.l. collected in blossom. Content: minimum 3.0 ml/kg essential oil (dried drug).

In the monographs of the herbal drugs macroscopic and microscopic descriptions are given, which serve as a basis for the identification of the drugs. Component-related identifications of the drugs are performed by thin-layer chromatography (TLC) of a methylene chloride extract of the drugs containing the essential oils including the terpene phenols, thymol and carvacrol, which serve as reference substances. Both terpene phenols are visible on the chromatogram by quenching zones in ultraviolet light at 254 nm. These and other terpenes can be made visible by spraying with an anisaldehyde solution and heating at 100–105°C for 10 min, thus producing coloured zones on the TLC. Descriptions of the TLC fingerprints are given in a tabular form. Further pharmacopoeial requirements concern the cleanliness which must be verified by special tests (Table 19.9).

The monographs also provide assays to verify the quality of the drug. The content of essential oils is ascertained by means of water distillation with a standardized Clevenger-type apparatus. 30.0 g (thyme) or 50.0 g (wild thyme), respectively, of the herbal drugs are

Table 19.9 Pharmacopoeial tests for cleanliness (*European Pharmacopoeia* 2002, Addenda 4.1 and 4.3)

	Thyme	Wild thyme
Foreign matter	Maximum 10% of stems and maximum 2% of other foreign matter. Stems must not be more than 1 mm in diameter and 15 mm in length. Leaves with long trichomes at their base and with weakly pubescent other parts (*T. serpyllum* L.) are absent	Maximum 3%, determined on 30 g
Water	Maximum 100 ml/kg determined on 20.0 g of powdered drug	
Loss on drying		Maximum 10.0%, determined on 1.000 g of the powdered drug by drying in an oven at 100–105°C for 2 h
Total ash	Maximum 15.0%	Maximum 10.0%
Ash insoluble in hydrochloric acid	Maximum 3.0%	Maximum 3.0%

Table 19.10 Percentage content of the components in thyme oil as postulated by the *European Pharmacopoeia* 2002, Addendum 4.1 ('Chromatographic profile')

Component	Percentage ratio
β-myrcene	1.0–3.0%
γ-terpinene	5.0–10.0%
p-cymene	15.0–28.0%
Linalool	4.0–6.5%
Terpinen-4-ol	0.2–2.5%
Thymol	36.0–55.0%
Carvacrol	1.0–4.0%

distilled for 2 h at a rate of 2–3 ml/min. The volume of the separated essential oil is measured in the graduated tube of the apparatus. For thyme a determination of the phenols thymol and carvacrol in the essential oil is postulated (a minimum of 40% altogether). It is quantified by gas chromatography (GC) of the essential oil won during the water distillation. The GC conditions are similar to those applied for the assay 'chromatographic profile' of thyme oil (see below).

The identity of thyme oil has also to be completed by TLC with thymol, terpinen-4-ol and linalool as references. A second identification test has to be carried out by means of GC with seven references listed in Table 19.10. The cleanliness tests for thyme oil require that the relative density of the oil must range between 0.915 and 0.935 and that the refractive index must range between 1.490 to 1.505. The qualitative assay is called the 'chromatographic profile' and is done by GC with a fused-silica column 25–60 m long and about 0.3 mm in internal diameter coated with macrogol 20 000; a minimum of 30 000 theoretical plates should be used. Helium is used as the carrier gas and a flame ionization detector (FID) is used for detection. Quantification is made by the normalization procedure of the peak areas, yielding the percentage contents of seven components. The lower and the upper limits of the seven components are listed in Table 19.10. For better identification a gas chromatogram of thyme oil is given in Fig. 19.4.

19.6.3 Adulteration

Adulteration of the herbal drugs of thyme and wild thyme is rare and can be manifested by macroscopic and microscopic analysis according to the pharmacopoeial standards. The situation of thyme oil is worse because adulteration is practised even today. In the past thyme oil was frequently adulterated by the addition of synthetic thymol and carvacrol or of 'thymene', a cheap byproduct mixture obtained from ajowan oil (ex *Trachyspermum copticum* (L.) Link) after removal of thymol. Adulteration is evident when thyme oil can be found on the market at low prices. Since the early 1960s the use of GC combined with other techniques has been used to determine the composition of an oil. Modern analytical techniques such as the use of capillary columns, special polar and non-polar stationary phases and GC–MS have led to a more accurate detailed analysis of an oil composition and have resulted in a more reliable determination of the purity of an oil and its components.

Nowadays the enantiomeric compositions of essential oil components further assist the analyst in determining the authenticity of an essential oil. Unfortunately the terpene phenols, thymol and carvacrol, the most interesting compounds in *Thymus* oils, are both achiral terpenes. Therefore other terpenes, minor components of the oil, have to be used and the

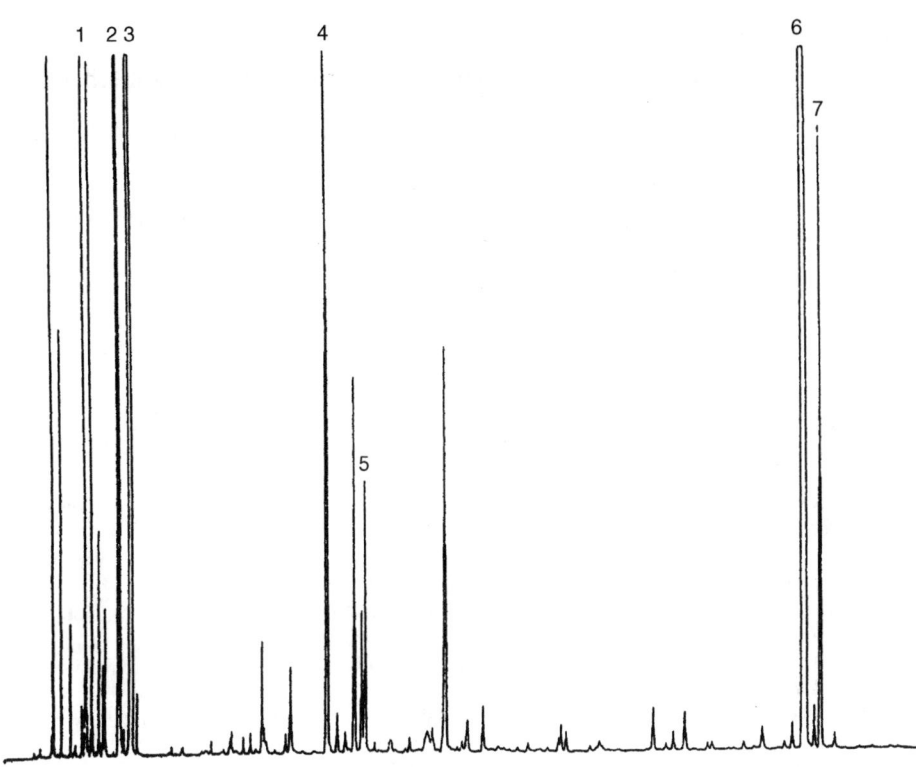

Fig. 19.4 Gas chromatograms of thyme oil according to the *European Pharmacopoeia* (2002), Addendum 4.1: 1 β-myrcene, 2 γ-terpinene, 3 *p*-cymene, 4 linalool, 5 terpinen-4-ol, 6 thymol, 7 carvacrol.

enantiomeric ratio of those components are as follows (Casabianca *et al.*, 1998; Kreck *et al.*, 2002):

- (1*R*)-(−)-α-pinene (89–93%) : (1*S*)-(+)-α-pinene (7–11%).
- (1*R*)-(−)-β-pinene (65–71%) : (1*S*)-(+)-β-pinene (29–35%).
- (*S*)-α-phellandrene 94–98% : (*R*)-α-phellandrene (2–4%).
- (4*R*)-(−)-limonene (55–59%) : (4*S*)-(+)-limonene (41–45%).
- (*R*)-sabinene (82–85%) : (*S*)-sabinene (5–8%).
- (3*R*)-(−)-linalool (92.0–99.4%; 96–98%) : (3*S*)-(+)-linalool (0.6–8.0%; 2–4%).
- (3*R*)-(−)-linalyl acetate (93.8–99.2%) : (3*S*)-(−)-linalyl acetate (0.8–6.2%).
- (*S*)-terpinen-4-ol (64–69%) : (*R*)-terpinen-4-ol (31–36%).
- (*R*)-α-terpineol (72–76%) : (*S*)-α-terpineol (24–28%).
- (1*S*, 2*R*, 4*S*)-(−)-borneol (98.1–99.6%; >99%) : (1*R*, 2*S*, 4*R*)-(+)-borneol (0.4–1.5%; < 1%).

Analysis of a thyme oil whose enantiomeric distributions of, particularly, linalool and linalyl acetate fall outside of the levels shown above is indicative of oil adulteration.

19.7 References

ABD-ELGAWAD M.M. and OMER E.A. (1995), 'Effect of essential oils of some medicinal plants on phytonematodes', *Anz. Schädlingskd. Pfl.*, **68**, 82–4.
AFNOR: ASSOCIATION FRANÇAISE DE NORMALISATION (1999), *Huiles Essentielles. Vol. 2 Specifications, Huile essentielle de thym á thymol* (Thymus zygis *(Loefl.)* L. type Espagne, NF ISO 14715, November, AFNOR, Paris.
AGARWAL I. and MATHELA C.S. (1979), 'Study of antifungical activity of some terpenoids', *Indian Drugs Pharm. Ind.*, **14**, 19–21.
AGARWAL I., MATHELA C.S. and SINHA S. (1979), 'Studies on the antifungal activity of some terpenoids against *Aspergilli*', *Indian Phytopathol.*, **32**, 104–5.
ASTA (1960), *Official Analytical Methods*, American Spice Trade Association, Englewood Cliffs NJ.
BLAKEWAY J. (1986), 'The antimicrobial properties of essential oils', *Soap Perfum. Cosmet.*, **59**, 201–3.
BLUMENTHAL M. (1998), *The Complete German Commission E Monographs – Therapeutic Guide to Herbal Medicines*, American Botanical Council, Austin, TX, Boston, MA, Integrative Medicine Communications.
CABO J., JIMÉNEZ J., MIRO M. and TORO M.V. (1978), 'Determinación de la actividad antimicrobiana de los componentes de la esencia de *Thymus zygis*', *L. Pharm. Med.*, **12**, 393–9.
CABO J., CABO M.M., CRESPO M.E., JIMÉNEZ J. and ZARZUELO A. (1986), '*Thymus granatensis* IV– Pharmacodynamic study of its essential oil', *Fitoterapia*, **57**, 173–8.
CANTWELL M. and REID M. (1986), 'Postharvest handling of fresh culinary herbs', *Perishables Handling No. 60*, Vegetable Crops Dept., UC Davis, 2–4.
CASABIANCA H., GRAFF J.B., FAUGIER V., FLEIG F. and GRENIER C. (1998), 'Enantiomeric distribution studies of linalool and linalyl acetate', *J. High Resol. Chromatogr.*, **21**, 107–12.
CONNER D.E. and BEUCHAT R.L. (1984), 'Effects of essential oils from plants on growth of food spoilage yeasts', *J. Food Sci.*, **49**, 429–34.
CRUZ T., JIMÉNEZ J., ZARZUELO A. and CABO M.M. (1989), 'The spasmolytic activity of the essential oil of *Thymus baeticus* Boiss in rats', *Phytother. Res.*, **3**, 106–9.
DAMBRAUSKIEN E., VIŠKELIS P. and VENSKUTONIS P.R. (2002), 'Effect of nitrogenous fertilizers on yield and chemical composition of thyme' in Dris R., Abdelaziz F.H. and Jain S.M., *Plant Nutrition Growth and Diagnosis*, Science Publishers Inc., Enfield, NH, 123–8.
DAPKEVICIUS A., VAN BEEK T., LELYVELD G.P., VAN VELDHUIZEN A., DE GROOT A., LINSSEN J.P.H. and VENSKUTONIS R. (2002), 'Isolation and structure elucidation of radical scavengers from *Thymus vulgaris* leaves', *J. Nat. Prod.*, **65**, 892–6.
DEANS G.G. and RITCHIE G. (1987), 'Antibacterial properties of plant essential oils', *Int. J. Food Microbiol.*, **5**, 165–80.
DEANS G.G., NOBLE R.C., PENZES L. and IMRE G.G. (1993), 'Promotional effects of plant volatile oils on the polyunsaturated fatty-acid', *Age (Chester Pa)*, **16**, 71–4.
DILASER M. (1979), 'Intoxication par le camphre et le menthol par voie trans-cutanée d´un nourrisson de six semaines', *Bull. Sign.*, **40**, 194.
EL-GENGAIHI S.E., AMER S.A.A. and MOHAMED S.M. (1996), 'Biological activity of thyme oil and thymol against *Tetranychus urticae* Koch', *Anz. Schädlingskd. Pfl.*, **69**, 157–9.
ESCOP (2003), ESCOP Monographs, 2nd edition, European Scientific Cooperative on Phytotherapy, Exeter, in collaboration with Georg Thieme, Stuttgart, New York.
EUROPEAN PHARMACOPOEIA, 4th ed (2002), Suppl. 4.1 (2002) and 4.3 (2003), Council of Europe, Strasburg.
FARAG R.S., SALEM H., BADEI A.Z.M.A. and HASSANEIN D.E. (1986), 'Biochemical studies on the essential oils of some medicinal plants', *Fette Seifen Anstrichm.*, **88**, 69–72.
FARAG R.S., DAW Z.Y. and ABO-RAYA S.H. (1989), 'Influence of some spice essential oils on *Aspergillus parasiticus* growth and production of aflatoxins in a synthetic medium', *J. Food Prot.*, **54**, 74–6.
FARAG R.S., ABD-EL-ZZIZ O., ABD-EL-MOEIN N.M. and MOHAMED S.M. (1994), 'Insecticidal activity of thyme and clove essential oils and their basic compounds on cotton leaf worm (*Spodoptera littoralis*), *Bull. Fac. Agric. Cairo*, **45**, 207–30.
FCC: NATIONAL ACADEMY OF SCIENCES (1996), *Food Chemical Codex (FCC)*, 4th Ed., National Academic Press, Washington, DC, 413–4.
FRASER S. and WHISH J.P.M. (1997), 'A commercial herb industry for NSW – an infant enterprise', RIRDC Research Paper, Series No. 97/18.
FUCHS L. (1543), *New Kreüterbuch*, Basel.

GASKELL M. (1988), 'Production of fresh culinary herbs in central America for commercial export', *Herb Spice Med. Plant Digest*, **6**(2), 1–10.
GECAN J.S., BANDLER R., GLAZE L.E. and ATKINSON J.C. (1986), 'Microanalytical quality of ground and unground marjoram, sage and thyme, ground allspice, black pepper and paprika', *J. Food Prot.*, **49**, 216–21.
GODFRAIND T., MILLER R. and WIBBO M. (1986), 'Calcium antagonism and calcium entry blockade', *Pharmacol. Rev.*, **38**, 324.
GRANGER R. and PASSET J. (1973), '*Thymus vulgaris* spontane de France: races chimiques et chemotaxonomie', *Phytochemistry*, **12**, 1683–91.
HARAGUCHI H., SAITO T., ISHIKAWA H., KATAOKA S., TAMURA Y. and MIZUTANI K. (1996), 'Antiperoxidative components in *Thymus vulgaris*', *Planta Med.*, **62**, 217–21.
HEATH H.B. (1981), *Source Book of Flavors*, AVI, Westport CN.
HEATH H.B. (1982), 'Spices and aromatic extracts influence of technological parameters on quality', in Adda J. and Richard H., *Int. Symp. on Food Flavors*, Tec. Doc.-Lavoisier, APRIA, Paris, 139–75.
HEATH H.B. and REINECCIUS G. (1986), *Flavor Chemistry and Technology*, Macmillan Publishers Ltd, Basingstoke.
HEGNAUER R. (1966), *Chemotaxonomie der Pflanzen*, Vol. 4, Birkhäuser Verlag, Basel, Stuttgart, 300.
ISO (1996), ISO 6754:1996, 'Dried thyme (*Thymus vulgaris* L.) – Specification', International Organization for Standardization.
ISO (1999), ISO 14715: 1999, 'Oil of thyme containing thymol, Spanish type (*Thymus zygis* (Loefl.) L.)', International Organization for Standardization.
JALAS J. (1971), 'Notes on *Thymus* L (Labiatae) in Europe, I: supraspecific classification and nomenclature', *Bot. J. Linn. Soc.*, **64**, 199–215.
JANSSEN A.M., SCHEFFER J.J.C., PARHAN-VAN ATTEN A.W. and BAERHEIM SVENDSEN A. (1988), 'Screening of some essential oils for their activities on dermatophytes', *Pharm. Weekbl. Sci.*, **10**, 277–80.
JUVEN B.J., KANNER J., SCHUED F. and WEISSLOWICZ H. (1994), 'Factors that interact with the antibacterial action of thyme essential oil and its active constituents', *J. Appl. Bacteriol.*, **76**, 626–31.
KARPOUHTSIS I., PARDALI E., FEGGU E., KOKKINI S., SCOURAS Z.G. and MAVRAGANITSIPIDOU P. (1998), 'Insecticidal and genotoxic activities of oregano essential oils', *J. Agric. Food Chem.*, **46**, 1111–5.
KHAN N.N. and NOLAN L.L. (1995), 'Screening of natural products for antileishmanial chemotherapeutic potential', *Acta Hortic.*, **426**, 47–56.
KNEIFEL W. and BERGER E. (1994), 'Microbial criteria of random samples of spices and herbs retailed on the Austrian market', *J. Food Prot.*, **57**, 893–901.
KOHLERT C., SCHINDLER G., MARZ R., ABEL G., BRINKHAUS B., DERENDORF H., GRAFE E-U. and VEIT M. (2002), 'Systemic availability and pharmacokinetics of thymol in humans', *J. Clinic. Pharmacol.*, **42**, 731–7.
KRECK M., SCHARRER A., BILKE S. and MOSANDL A. (2002), 'Enantioselective analysis of monoterpene compounds in essential oils by stir bar sorptive extraction (SBSE)-enantio-MDGC–MS', *Flav. Fragr. J.*, **17**, 32–40.
KRETSCHMER M. (1989), 'Influence of different storage conditions on germination of spice seeds', *Acta Horticulturae*, **253**, 99–103.
LANGE D. (1998), *Europe's Medicinal and Aromatic Plants: Their Use, Trade and Conservation*, Traffic Europe/International, Cambridge.
LAWRENCE B.M. (2003), 'Progress in essential oils – thyme oil', *Flav. Fragr.*, **28**(2), 52–7.
LEE S., TSAO R., PETERSON C. and COATS J.R. (1997), 'Insecticidal activity of monoterpenoids to western corn rootworm (Coleoptera: Chrysomelidae), twospotted spider mite (Acari: Tetranychidae) and house fly (Diptera: Muscidae)', *J. Econ. Entomol.*, **90**, 883–92.
MENGHINI A., SAVINO A., LOLLINI M.N. and CAPRIO A. (1987), 'Activité antimicrobienne en contact direct et en microatmosphere de certains huiles essentielles', *Plant Méd. Phytothér.*, **42**, 21–36.
MIURA K. and NAKATANI N. (1989), 'Antioxidative activity of flavonoids from thyme (*Thymus vulgaris* L.)', *Agric. Biol. Chem.*, **53**, 3043–5.
MIURA K., INAGAKI T. and NAKATANI N. (1989), 'Structure and activity of new deodorant biphenyl compounds from thyme (*Thymus vulgaris* L.)', *Chem. Pharm. Bull.*, **37**, 1816–9.
MIURA K., KIKUZAKI H. and NAKATANI N. (2002), 'Antioxidant activity of chemical components of sage (*Salvia officinalis* L.) and thyme (*Thymus vulgaris* L.) measured by the oil stability index method', *J. Agric. Food Chem.*, **50**, 1845–51.

MORALES R. (2002), 'The history botany and taxonomy of the genus *Thymus*', in Stahl-Biskup E. and Saez F., *Thyme – The Genus* Thymus, Taylor & Francis, London, 11–43.
MOYLER D.A. (1991) 'Oleoresins tinctures and extracts', in Ashurst P.R., *Food Flavourings*, Blackie, Glasgow, London, 54–86.
MÜLLER J., CONRAD T., TEŠIC M. and SABO J. (1993), 'Drying of medicinal plants in a plastic-house type solar dryer', *Acta Hortic.*, **344**, 79–85.
NAKATANI N. (2000), 'Phenolic antioxidants from herbs and spices', *BioFactors* **13**, 141–6.
NAKATANI N., MIURA K. and INAGAKI T. (1989) 'Structure of new deodorant biphenyl compounds from thyme (*Thymus vulgaris* L.) and their activity against methyl mercaptan', *Agric. Biol. Chem.*, **53**, 1375–81.
NELSON R.R. (1997), '*In vitro* activities of five plant essential oils against methicillin-resistant *Staphylococcus aureus* and vancomycin-resistant *Enterococcus faecium*', *J. Antimicrob. Chemother.*, **40**, 305–6.
OSAWA K., MATSUMOTO T., MARUYAMA T., TAKIGUCHI T., OKUDA K. and TAKAZOE I. (1990), 'Studies on the antibacterial activity of plant extracts and their constituents against periodontopathic bacteria', *Bull. Tokyo Dent. Coll.*, **31**, 17–21.
PERRUCCI S., MACCHIONI G., CIONI P.L., FLAMINI G. and MORELLI I. (1995), 'Structure–activity relationship of some natural monoterpenes as acaricides against *Psoroptes cuniculi*', *J. Nat. Prod.*, **58**, 1261–4.
PRAKASH V. (1990), *Leafy Spices*, CRC Press Inc, Boca Raton, FL.
QURESHI S., SHAH A.H., AL-YAHYA M.A. and AGEEL A.M. (1991), Toxicity of *Achillea fragrantissima* and *Thymus vulgaris* in mice', *Fitoterapia*, **62**, 319–23.
RAGHAVAN B., ABRAHAM K.O. and KOLLER W.D. (1995), 'Flavour quality of fresh and dried Indian thyme (*Thymus vulgaris* L)', *Pafai J.*, **17**, 9–14.
REGNAULT ROGER C. and HAMRAOUI A. (1994), 'Inhibition of reproduction of *Acanthoscelides obtectus* Say (Coleoptera) a kidney bean (*Phaseolus vulgaris*) bruchid by aromatic essential oils', *Crop Prot.*, **13**, 624–8.
REINECCIUS G. (1994), *Source Book of Flavors*, 2nd ed., Chapman and Hall, New York.
REITER M. and BRANDT W. (1985), 'Relaxant effects on tracheal and ileal smooth muscles of the guinea pig', *Arzneim-Forsch*, **35**, 408–14.
RONNIGER K. (1924), Beiträge zur Kenntnis der Gattung *Thymus*, I Die Britischen Arten und Formen', *Feddes Repert.*, **20**, 321–32.
SCHNITZLER A.C., NOLAN L.L. and LABRE R. (1995), 'Screening of medicinal plants for antileishmanial and antimicrobial activity', *Acta Hortic.*, **426**, 235–41.
SCHWARZ K., ERNST H. and TERNES W. (1996), 'Evaluation of antioxidative constituents from thyme', *J. Sci. Food Agric.*, **70**, 217–33.
SHALABY A.S. and RAZIN A.M. (1992), 'Dense cultivation and fertilization for higher yield of thyme (*Thymus vulgaris* L.)', *J. Agron. Crop Sci.*, **168**, 243–8.
SIMON J.E. (1990), 'Essential oils and culinary herbs' in Janick J. and Simon J.E., *Advances in New Crops, Proceedings of the First National Symposium on New Crops*, Timber Press, Portland, OR, 472–83.
SMITH PALMER A., STEWART J. and FYFE L. (1998), 'Antimicrobial properties of plant essential oils and essences against five important food-borne pathogens', *Lett. Appl. Microbiol.*, **26**, 118–22.
STAHL-BISKUP E. (2002), 'Essential oil chemistry of the genus *Thymus* – a global view', in Stahl-Biskup E. and Saez F., *Thyme – The Genus* Thymus, Taylor & Francis, London, 75–124.
TAINTER D.R. and GRENIS A.T. (1993), *Spices and Seasonings: A Food Technology Handbook*, VCH Publishers, New York.
TANTAOUI-ELARAKI A. and BERAOUD L. (1994), 'Inhibition of growth and aflatoxin production in *Aspergillus parasiticus* by essential oils of selected plant materials', *J. Food Prot.*, **61**, 616–9.
TISSERAT B., VAUGHN S.F. and SILMAN R. (2002), 'Influence of modified oxygen and carbon dioxide atmospheres on mint and thyme plant growth morphogenesis and secondary metabolism *in vitro*', *Plant Cell Rep.*, **20**, 912–6.
VAN DEN BROUCKE C.O. and LEMLI J.A. (1983), 'Spasmolytic activity of the flavonoids from *Thymus vulgaris*', *Pham. Weekbl.*, **5**, 9–14.
VENSKUTONIS P.R. (2002a), 'Thyme – processing of raw plant material', in Stahl-Biskup E. and Saez F., *Thyme – The Genus* Thymus, Taylor & Francis, London, 224–51.
VENSKUTONIS P.R. (2002b), 'Harvesting and post-harvest handling in the genus *Thymus*', in Stahl-Biskup E. and Saez F., *Thyme – The Genus* Thymus, Taylor & Francis, London, 197–223.

VERLET N. (1992), 'Trends of the medicinal and aromatic plant sector in France', *Acta Hort.*, **306**, 169–75.
VILA R. (2002), 'Flavonoids and further polyphenols in the genus Thymus', in Stahl-Biskup E. and Saez F., *Thyme – The Genus* Thymus, Taylor & Francis, London, 144–76.
WAGNER H. and WIERER M. (1987), *In-vitro*-Hemmung der Prostaglandinbiosynthese durch ätherische Öle, phenolische Verbindungen und Knoblauchinhaltsstoffe', *Z. Phytother.*, **9**, 11–13.
WANG M., LI J., HO G.S., PENG X. and HO C-T. (1998), 'Isolation and identification of antioxidative flavonoid glycosides from thyme (*Thymus vulgaris* L.)', *J. Food Lipids*, **5**, 313–21.
WHO: WORLD HEALTH ORGANIZATION (1998), *Quality Control Methods for Medicinal Plant Materials*, WHO, Geneva.
WWF *Thymus* and *Origanum*, http://wwwwwforguk/filelibrary/pdf/Thymuspdf
ZAMBONELLI A., DAULERIO A.Z., BIANCHI A. and ALBASINI A. (1996), 'Effects of essential oils on phytopathogenic fungi *in vitro*', *J. Phytopathol.*, **144**, 491–4.
ZANI F., MASSIMO G., BENVENUTI S., BIANCHI A., ALBASINI A., MELEGARI M., VAMPA G., BELLOTTI A. and MAZZA P. (1991), 'Studies on the genotoxic properties of essential oils with *Bacillus subtilis* rec-assay and *Salmonella* microsome reversion assay', *Planta Med.*, **57**, 237–41.
ZARZUELO A. and CRESPO E. (2002), 'The medicinal and non-medicinal uses of thyme', in Stahl-Biskup E. and Saez F., *Thyme – The Genus* Thymus, Taylor & Francis, London, 263–92.
ZARZUELO A., CABO M.M., CRUZ T. and JIMÉNEZ J. (1989), 'Spasmolytic action of the essential oil of *Thymus longiflorus* Boiss. in rats', *Phytother. Res.*, **3**, 36–8.

20

Vanilla

C. C. de Guzman, University of the Philippines Los Baños, Philippines

20.1 Introduction and description

There is probably no other spice in the world that strongly evokes the sweet memories of childhood and nurture than vanilla – ice cream to cool off in the scorching heat of summer, milk to nourish young and agile bodies, and warm chocolate to perk up mornings with friends and loved ones. Consider how bland these delights would be without the aroma of vanilla! From the time it was offered in the form of a flavoured concoction by a gentle people to a Spanish conquistador to the current quest for a novel biotechnological means to produce its major flavouring constituents, vanilla has continually provided gastronomic pleasure to generations of people through the centuries.

The following discussion provides an overview of the relevant information on vanilla from a brief historical account, world production and trade to its horticulture and processing. A focus is also given on product quality assessment, methods of adulteration and its detection, and research endeavours related to the natural production of its major flavour component, vanillin, outside the cured bean. The last section predicts the future outlook for vanilla *vis-à-vis* the growing demand for natural food additives and organic products.

20.1.1 Historical background

Vanilla is considered to be the greatest contribution of the Americas to the world of flavours. The following historical account of vanilla discovery and geographical spread is condensed from Correll (1953). Vanilla was introduced to the Old World when the Aztec emperor Montezuma of Mexico in 1520 welcomed the Spanish conquistador Hernando Cortes with '*chocolatl*', a drink concocted from powdered cacao and flavoured with ground vanilla beans called '*tlilxochitl*' (black pod). After the discovery of the secret ingredient, vanilla was brought to Spain, where the Spaniards used the bean, which they termed as '*vaynilla*' (little sheath, in reference to the fruit appearance) as a flavouring for the manufacture of chocolate. Considerable interest in the plant soon followed. Valuable information related to the plant's description, nomenclature and horticulture was gathered and plant samples were sent to various botanical gardens, including those in Paris and Antwerp.

Subsequently, vanilla was introduced to Réunion in 1793, Java in 1819, India in 1835, the

Fig. 20.1 Vanilla, a vine, is a member of the orchid family.

island of Tahiti in 1848 and the Seychelles in 1866. Prior to 1841 initial efforts to cultivate vanilla outside its native Mexican home failed because of one major constraint: lack of natural vectors to pollinate the flowers. The solution to this problem came in 1836 when Charles Moren of Liège was able to produce vanilla beans by hand pollination. Large-scale cultivation of vanilla, however, was only realized years later, in 1841, upon the development of a practical method of artificial pollination of the plant by Edmond Albius, a former slave in Réunion. By 1886 vanilla cultivation was even greater in the Mascarene Islands (Réunion, Mauritius and Rodriguez) and Java than in Mexico. Some of these initial producers have continued to play significant roles in the world production of vanilla up to the present day.

20.1.2 Plant description

The vanilla of international commerce is derived from *Vanilla planifolia* H. C. Andrews (syn: *V. fragrans* (Salisb.) Ames). Vanilla, a member of the orchid family, is a climbing monocot, possessing a stout, succulent stem and short-petioled, oblong-lanceolate leaves about 20 cm long (Fig. 20.1). The inflorescence is characteristically raceme with 20

Fig. 20.2 Vanilla flowers are yellowish, wholly green or white within.

or more flowers. The flowers, about 6 cm long and 2.5 cm wide, are either yellowish, wholly green or white within with oblanceolate sepals and petals (Fig. 20.2). The fruit, popularly termed as 'beans' or 'pod' in the vanilla market, is botanically a capsule, nearly cylindrical and about 20 cm long (Fig. 20.3) (Bailey and Bailey, 1976).

Other species of vanilla of secondary importance in the vanilla trade include the West Indian vanilla, *V. pompona* Schiede (synonym: *V. grandiflora* Lindl.), and vanilla Tahiti, *V. tahitensis* J. W. Moore. Purseglove *et al.* (1981) and Straver (1999) provided some important distinctions among the three species. West Indian vanilla is a native of Central America, northern South America and the Lesser Antilles. It differs from *V. planifolia* by having larger leaves, more fleshy and larger flowers and the presence of a tuft of imbricate scales, instead of hairs, in the centre of the lip disc. On the other hand, *V. tahitensis* is indigenous to Tahiti. It is less robust than *V. planifolia* with more slender stems and narrower leaves. Both minor species yield shorter and thicker capsules and an inferior bean product.

A comprehensive comparative anatomical study of the stem, leaves and roots of several

Fig. 20.3 The fruit of vanilla is botanically a capsule, but is called a 'bean' or 'pod' in the international market.

species of vanilla, including *V. planifolia* and *V. pompona*, can be obtained from Stern and Judd (1999).

20.2 Production and trade

In the international market, cured vanilla beans are conveniently classified according to their geographical source or origin (Correll, 1953). The principal groupings are: Mexican beans – those coming only from Mexico; Bourbon beans – formerly referring only to those produced from the island of Réunion (then named Bourbon) but currently inclusive of all beans from Madagascar and the Mascarene, Comoro and Seychelles islands; Tahiti beans – grown in the French group of the Society Islands; and Java beans – derived from Indonesia.

From 1990 to 2000, average world production of vanilla beans amounted to 4466 Mt harvested over a total of 38 485 ha (FAO, 2003). Indonesia contributed about 38% of this world produce, slightly higher than Madagascar at around 34%, while Mexico and Comoro Islands had less than 10% share. The annual world import of cured vanilla beans averaged US$545m. In the same decade the USA imported the biggest mean volume of cured beans at 1448 Mt worth US$52m, followed by France at 374 Mt and Germany at 318 Mt. From 1990 to 1995, the price of vanilla bean imported by the USA ranged from US$42 to 74 kg^{-1} averaging at US$64 kg^{-1} (USA Vanilla Bean Imports Statistics, 2002).

In 1988, marketing estimates by McCormick & Co., Inc. revealed that vanilla extract was largely used by the industrial sector (75% of the total supply), followed by the retail sector (20%) and the food service sector (5%) (Gillette and Hoffman, 1992). Of the total industrial application, 30% was utilized in ice cream preparations, 17% in soft beverages, 11% in alcoholic beverages, 10% in yoghurt and the remaining 7% in bakery items, confectioneries, cereals and tobacco products.

20.3 Cultivation

20.3.1 Climate and soil requirements

Vanilla is a tropical crop that thrives best in warm and moist climate. Natural growth is obtained at latitudes 15° and 20° north and south of the equator (Lionnet, 1958). The optimum temperature ranges from 21 to 32°C, with a mean value of 27°C, while the precipitation required falls between 2000 and 2500 mm annually (Purseglove *et al.*, 1981). A dry period of about two months is needed to restrict vegetative growth and induce flowering; rainfall in the remaining ten months should be evenly distributed (Correll, 1953). Vanilla does well from sea level up to 700 m in altitude, and has been found to exist even at 1000 m in Mexico (Correll, 1953; Straver, 1999). Vanilla established on gently sloping terrain with good drainage is reputed to produce better crops and to be more resistant to fungal infection (Lionnet, 1958). It grows best in light, porous and friable soils, preferably of volcanic origin, with a pH of 6 to 7 (Correll, 1953; Straver, 1999).

20.3.2 Propagation

Use of cuttings
Vanilla is commercially propagated by stem cuttings. If source is not a constraint, cuttings 2–3.5 m long are preferred since they will flower in one to two years, as opposed to 30 cm cuttings which will bear flowers and fruit in three to four years (Correll, 1953; Purseglove *et al.*, 1981). The latter, however, is known to produce more vigorous crops that will last longer (Lionnet, 1958). Long cuttings are planted directly in the field, while short cuttings are usually started in a nursery. Cuttings for storage or transport can be carefully wrapped in banana or abaca leafsheaths and will root in 20 days under shade (David, 1950).

Use of in vitro *techniques*
Vanilla is also successfully propagated using *in vitro* techniques. Shoot proliferation with aerial root formation can be induced using nodal stem segments (Kononowicz and Janick, 1984). Plantlet formation using stem dics with a node has been reported to be dependent upon the composition of the medium, the size of the explant or both (Philip and Nainar, 1986). Production of multiple plantlets is also possible using aerial root tips. This occurs in the absence of a callus interphase, reducing the possibility of induced epigenetic changes in the derived plant.

Use of seeds
Propagation by seeds is confined to breeding work. Under natural conditions vanilla seeds, which contain a limited amount of food reserve, do not germinate unless in association with a mycorrhizal fungus (Purseglove, 1985). They can, however, germinate *in vitro* in a well-defined culture medium formulated by Knudson (1950). The specific histochemical changes during the *in vitro* germination and seedling growth of vanilla have been described by Philip and Nainar (1988).

20.3.3 Support/shade trees

Vanilla, being a shade-loving, climbing vine, needs a structure that will support its vertical growth and provide some canopy to filter intense sunlight. A living support cum shade tree suits this purpose. In the choice of the support trees, the following criteria should be considered (David, 1950; Correll, 1953; Lionnet, 1958):

- Leaves are small enough to permit chequered sunlight (about one-third to one-half of full sun exposure) to filter through.
- Branches are wide spreading and sufficiently low (*ca* 1.5–2 m from the ground) for hanging the vines.
- Growth is rapid in full sunlight.
- Propagation is easy either by seeds or cuttings.
- Provides sufficient protection from sun and strong wind.
- Must help improve soil nutrition (e.g. legumes for N fixation).
- Should not become entirely defoliated.

The members of the Leguminosae family such *as madre de cacao* (*Gliricidia sepium* (Jacq.) Kunth ex Walp.), *ipil-ipil* (*Leucaena leucocephala* (Lamk.) de Wit.) and Indian coral tree (*Erythrina orientalis* (L.) Murr.) were found to be the best shade/support trees for vanilla (David, 1950).

20.3.4 Field establishment

In establishing a vanilla plantation, areas that are prone to stagnation of water (e.g. flat lands with poor drainage) or subject to soil erosion (e.g. steep contours) should be avoided. An ideal site is a gently sloping hill with sufficient drainage (Correll, 1953).

In open sites, support trees are planted ahead of the scheduled time for planting the vanilla vines, preferably six months to one year, to provide a sufficient period for root establishment and development of a spreading canopy (David, 1950; Straver, 1999). Various planting distances have been used in several countries. With wide spacing the plantation can be easily managed and incidences of diseases are observed to be low, although the economic benefit is also lower. On the other extreme, too high a plant density makes vanilla very susceptible to fungal attack, as was observed in Seychelles, Réunion and Indonesia (Correll, 1953; Lionnet, 1958). Field operations are also more difficult. A moderate spacing of 1.5 m × 2 m or 2 m × 2 m giving 3300 or 2500 vines to a hectare, respectively, is recommended to overcome these problems (Sen, 1985).

Planting holes are dug about 30 cm from the support tree and the basal three nodes (without leaves) of the cuttings are buried into the soil and then covered with rich humus or well-decayed leaves. The remaining exposed vine is tied gently to the support tree with a piece of abaca or plastic twine. Vines set up in this way become fully established in less than five months (David, 1950).

Only one cutting is usually placed per support tree. Planting is done at the onset of the rainy season.

20.3.5 Fertilizer application
Chemical analysis reveals considerable amount of inorganic nutrients taken up by the different organs of vanilla (Tjahjadi, 1987). The highest amount of nutrient found in the stem is calcium, in the leaf, magnesium, and in the fruit, potassium. The fruit contains the highest level of all the nutrients examined, except for calcium and magnesium. This result suggests the need to consider proper nutrition in this crop. It is recommended that 40–60 g N, 20–30 g P_2O_5 and 60–100 g K_2O be supplied to each vanilla vine per year (Anandaraj *et al.*, 2001). Spraying of 1% solution of complete fertilizer (17 : 17 : 17) once a month enhances growth and flowering.

Organic fertilizers such as guano and bone meal are found to be beneficial to vanilla (Purseglove *et al.*, 1981). The use of fresh farmyard manure is reported to increase the risk of infection due to diseases (Lionnet, 1958). Composting before soil application of the animal manure, together with other crop wastes such as rice straw, circumvents this problem. Loppings, especially the leaves, of leguminous support trees are a very good source of green manure.

20.3.6 Pruning and training
Both the vanilla vine and the support trees are judiciously pruned to optimize vegetative growth, flowering and fruit development. At the onset of the rainy season, heavy lopping of the support tree branches is usually practised to increase the light intensity reaching vanilla and make it less susceptible to fungal attack.

When vanilla attains a sufficient length, the stem is allowed to hang or bend over the support branches until it is about 30 cm above the ground (Lionnet, 1958). The tip is then pruned to induce the growth of lateral branches below it. The hanging branch is known as a 'porteur'. Any shoot coming out of the porteur is cut off when 7–10 cm long, while those from the rest of the plant before the bend are allowed to develop. These will become the porteurs of the following cropping season. This bending and pruning technique favours flowering. In mature vines about five or six porteurs are maintained. After a season of harvest the porteurs are cut off.

20.3.7 Mulching
Soil cultivation is generally not practised or done very lightly in vanilla since it is a surface feeder (Purseglove *et al.*, 1981). More attention, rather, is given to mulching. Leaves and loppings from support trees as well as weeding wastes can provide the necessary soil cover. In the Seychelles, coconut husks are applied in overlapping rows about 45 cm from the base of vanilla and decayed plant matter is incorporated in the space between (Lionnet, 1958). Coconut husks do not only conserve soil moisture but also provide a potassium-rich organic matter.

20.3.8 Diseases and pests
Correll (1953) and Purseglove *et al.* (1981) identified and described the various diseases of vanilla in several producing countries. The most serious and widespread is anthracnose,

which is caused by *Calospora vanillae* Massee. Particularly observed in the Comoro and Seychelles Islands, the West Indies, Tahiti, Mascarene and Colombia, the disease damages the stem apex, leaves and aerial roots as well as the fruits. Wilting and abscission of the affected organs occur in severe cases of infection. Root rot by *Fusarium batatis* Wollenw. var. *vanillae* is another important fungal disease of vanilla, which is widespread in Puerto Rico and in Indonesia (Sen, 1985). It is characterized by browning and death of underground roots with a concomitant shrivelling and drying out of the shoot. The following conditions need to be avoided to check the onset and spread of these fungal diseases: excessive moisture, insufficient drainage, too much shade and high-density planting.

Other debilitating diseases of vanilla which have recently been observed are viral diseases. Several types have been identified. The cucumber mosaic virus (CMV) was first reported by Farreyrol *et al.* (2001) in vanilla. In French Polynesia CMV induces severe stunting of *V. tahitensis* with marked stem and leaf deformation. CMV is also present in *V. fragrans* samples from Réunion, although with less severe symptoms compared with those identified for *V. tahitensis*. The differences in symptomology are attributed either to virus strain variation or differential species tolerance.

Pearson *et al.* (1993) described the distribution and incidence of vanilla viruses in the following South Pacific countries: Cook Islands, Fiji, Niue, Tonga and Vanuatu. Cymbidium mosaic potexvirus (CyMV) and odontoglossum ringspot tobamovirus (ORSV), which seems to do little harm, are found in all countries surveyed in both *V. fragrans* and *V. tahitensis*. Of bigger concern, however, are potyviruses, which cause severe infection. These include: vanilla necrosis potyvirus (VNV) in Fiji, Tonga and Vanuatu detected in *V. fragrans*, and vanilla mosaic virus (VaMV) present in *V. fragrans* from Cook Islands, Fiji and Vanuatu and *V. tahitensis* from Cook Islands and French Polynesia. Other potyviruses that react with neither VNV or VaMV antisera and rhabdovirus-like particles have also been detected in Fiji and Vanuatu. Symptoms of potyvirus infection include leaf distortion, sunken chlorotic patches, stem necrosis and vine die-back. Control measures include rouging and avoiding the use of cuttings from plantings where the symptoms are present. There is also the potential of using mild virus strains for cross-protection (Liefting *et al.*, 1992).

CyMV and ORSV were also present in *V. tahitensis* growing in the Society Islands of French Polynesia but the incidence of damage was very low (Wisler *et al.*, 1987). VaMV has a limited host range compared with VNV (Wang and Pearson, 1992). On the other hand, comparison of the gene for the coat protein of VNV using Western blot analysis and enzyme-linked immunosorbent assay (ELISA) reveals its serological similarity with watermelon mosaic 2 potyvirus (WM2V) (Wang *et al.*, 1993).

Pest infestation is not a serious problem in vanilla.

20.3.9 Flowering and pollination

Vanilla flowers only once a year, staggered over an average of two to three months, depending upon the variation in local climate. In Mexico flowering is observed from April to May; in Madagascar and the Comoro Islands, between November and January (Correll, 1953). In Indonesia, this occurs between July and August (Sen, 1985); in India from December to February (Anandaraj *et al.*, 2001), while in the Philippines flowering begins as early as March and lasts until June (David, 1950).

In its natural habitat in Mexico and other parts of Central America vanilla flowers are pollinated by bees (*Melipona* sp.) and hummingbirds (Correll, 1953). The rate of natural pollination is low, only about 1% (Fouche and Coumans, 1992). Outside its native home, fruit set in vanilla is accomplished through hand pollination.

The vanilla flower is so constructed that effective transfer of pollen to the stigma is prevented. The stamen cap enclosing the pollinia (mass of pollen) and the flap-like structure called the rostellum covering the stigma both act as physical barriers to the process of pollination (Correll, 1953; Purseglove et al., 1981). In artificial pollination the stamen cap is removed and the rostellum is pushed up with the aid of a bamboo stick or any similar object the size of a toothpick. The pollinia and stigma are then brought into contact with each other by hand manipulation. A simple description, complete with illustrations, of four different types of hand-pollinating vanilla flowers is provided by Fouche and Coumans (1992). If pollination is successful, the flowers remain and wither on the rachis; otherwise, they abscise in two or three days. Under expert hands, from 1000 to 2000 flowers can be pollinated per day (Purseglove et al., 1981).

Vanilla flowers last only for a day. Sivaraman Nair and Mathew (1969) observed that flower opening in vanilla occurs between 10.30 p.m. and 1 a.m., and is completed by 6 p.m. Pollination is accomplished with 98–100% success if undertaken between 6 a.m. and 6 p.m. on the day of flower opening. Pollen viability deteriorates markedly after anthesis, with only 10–15% fruit set when 2- to 3-day-old pollens are used.

Plant growth regulators have also been tested as an alternative to laborious hand pollination. 2,4,5-Trichlorophenoxyacetic acid (2,4,5-T) at 100–500 ppm, and gibberellic acid (GA_3) at 20–100 ppm promote fruit set if applied on or before anthesis. The treated beans, however, either drop before reaching maturity or are only one-quarter to one-third the size of hand-pollinated beans.

20.3.10 Fruiting

In Réunion, it takes about six months for the fruits to mature; in Mexico, Indonesia and the Philippines, about nine months (David, 1950; Correll, 1953; Sen, 1985). About 50–150 fruits are allowed to develop and mature per vine (Purseglove, 1985).

20.4 Harvesting, yield and post-production activities

Vanilla fruits are gathered when they are fully mature but before they are too ripe. When picked immaturely, the fruits do not develop the requisite full-bodied aroma and proper colour during processing and are more prone to fungal infection (Correll, 1953). When harvested at the over-ripe stage, the fruits tend to split and lose some of their aroma.

The following serves as harvest indices for vanilla (David, 1950; Purseglove, 1985):

- The thickest portion of the fruit (the 'blossom end') takes on a pale yellow colour.
- Overall pod colour changes from dark green to light green.
- The fruits lose their lustre and become somewhat dull.
- Two distinct lines appear from one end of the fruit to the other.

Since flowering is staggered, harvesting is likewise extended over a period of time.

One kilogram of cured beans is derived from about 6 kg of green pods. Yield of cured beans ranges from 300 to 800 kg ha^{-1} yr^{-1} (Purseglove et al., 1981; Anandaraj et al., 2001).

20.4.1 Post-production activities

Curing
After harvesting, the pods of vanilla need to be cured to develop the characteristic natural

flavour associated with the product. Curing can be defined as the sum total changes that occur during the primary processing of a given raw material to a desired finished product, which is ready for market (Jones and Vicente, 1949a). Curing stops the various natural vegetative processes in the harvested beans and promotes the metabolic reactions involved in the generation of the aromatic flavouring constituents in the cured material (Arana, 1944). It can be broadly classified into two: (1) changes that involve a simple loss of water, achieved through drying, and (2) changes that involve chemical transformation, which is usually accompanied by hydrolytic and oxidative changes with or without the aid of enzymes. In vanilla, the latter changes are more critical.

Vanillin (Fig. 20.4a), the main flavouring chemical of vanilla, is present only in trace amounts in the green mature beans; upon curing, however, vanillin content increases (Arana, 1943). The chemical compound from which vanillin is derived occurs in the uncured pods in the form of a glucoside called glucovanillin (Arana, 1945). During the curing process, this glucoside is hydrolysed to form vanillin and glucose through the action of a β-glucosidase. The activity of this enzyme changes with the maturity of the vanilla beans, being negligible in the green beans and highest in the split, blossom-end yellow beans. Spatially, all of the enzyme is located in the fleshy portion or thick wall of the pods, where most of the glucovanillin is also concentrated (Arana, 1943). Along the bean length, 40% of glucovanillin has been detected in the blossom end, another 40% in the middle and the remaining 20% in the stem end. Other flavour constituents such as p-hydroxybenzoic acid, p-hydroxybenzaldehyde and vanillic acid (Fig. 20.4b–d) are also present in the green beans in their glycosidic forms and are released through enzymatic hydrolysis during curing (Ranadive, 1992).

The splitting of vanillin from the glucoside is initiated during the early part of the curing process, but the full development of flavour and aroma occurs only after a considerable period of pod preparation and conditioning (Arana, 1943; Jones and Vicente, 1948). Treatment of cured beans with β-glucosides enhances vanillin content, suggesting incomplete hydrolysis, probably as a result of (a) insufficient amount of native enzyme, (b) inadequate enzyme–substrate interaction or (c) inactivation of enzymes by oxidized phenols liberated during curing (Ranadive, 1992). Chemical changes other than enzymatic hydrolysis may also contribute a great deal to the quality of cured vanilla. Balls and Arana (1941) suggested the possible role of a peroxidase system in the oxidation of vanillin to quinone compounds. These substances possess more complex structure with presumably different aroma that can add to the total flavour of the cured product. Wild-Altamirano (1969) reported that proteinase activity declines with pod growth while the activities of glucosidase, peroxidase and polyphenoloxidase increase with pod age, being maximum near or at ripening. The trend in enzyme activities is indicative of the potential role of the various products derived from catalysed reactions in the full development of the characteristic flavour and aroma of cured beans.

In general, vanilla curing follows four successive steps: (1) killing or wilting, (2) 'sweating', (3) drying and (4) conditioning. Killing or wilting is the initial step in inhibiting the natural changes in vanilla beans. It is achieved through various techniques, depending upon the producing country (Arana, 1945; Theodose, 1973). In Mexico and Indonesia, the most popular method is sun wilting. In this method the beans, which are contained on racks covered with dark woollen blankets, are simply heated under the sun. Wilting with the use of an oven maintained at 60°C is alternatively practised in Mexico. In Madagascar, Réunion and Comores, beans are killed by dipping in hot water for a few minutes (scalding technique) (Fig. 20.5). On the island of Guadeloupe, beans are gently scratched on the surface with the use of a pin embedded in a cork ring prior to sun exposure. Wilting by freezing has been

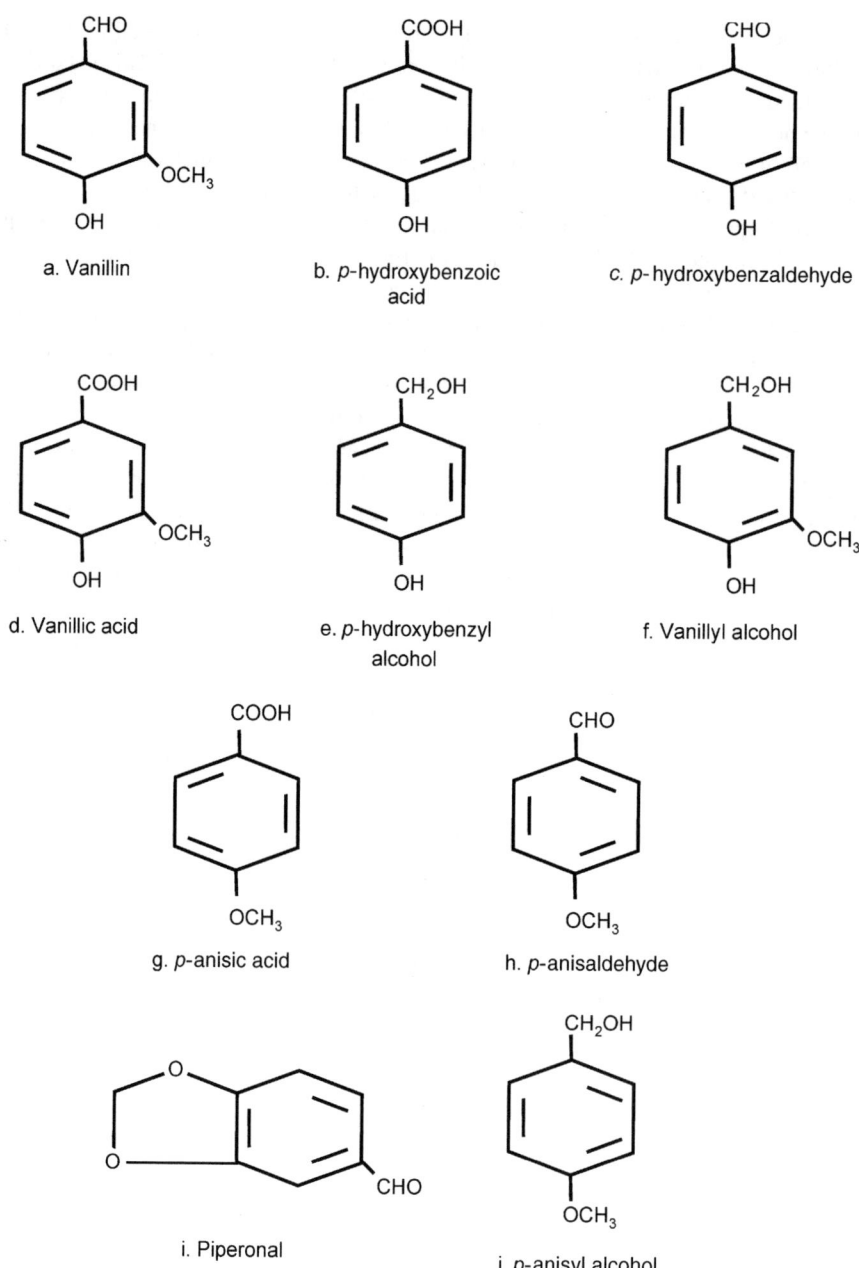

Fig. 20.4 Chemical structures of the major flavouring constituents of vanilla.

developed in Puerto Rico for experimental purposes only. In this technique the beans are refrigerated until frozen and then thawed naturally at room temperature. Some of the advantages and disadvantages of these types of wilting are presented in Table 20.1.

The successive steps after killing are more or less similar for the different countries exporting vanilla. 'Sweating' or heating is done to develop the proper texture and flexibility. This is accomplished through either of two ways: (1) daily sun exposure for about six hours,

Fig. 20.5 Implements used in the processing of beans using the hot water treatment.

Table 20.1 Advantages and disadvantages of different methods of wilting vanilla beans

Method of Wilting	Advantages	Disadvantages
Sun wilting	Method is simple Does not require additional equipment	High degree of bean splitting Beans mould easily
Oven wilting	Short period of time for sweating and drying Fewer split beans High vanillin content	High percentage of mouldy beans Medium phenol value
Hot-water wilting	Few mouldy beans and medium degree of splitting Easiest and most satisfactory for the inexperienced curer	Longer period of drying Low vanillin content and phenol value
Scratching	Short period of time for sweating and drying Low vanillin and phenol values Low degree of splitting	High susceptibility to mould Poor flexibility of the beans in the stem end Dependent on the skill and care of the curer
Freezing	Practically no mould Sophisticated aroma Beans are picked at the best stage of maturity and kept in the refrigerator until enough beans are accumulated	Medium values for phenol, vanillin content and percentage splitting

Source: Arana (1944, 1945).

Fig. 20.6 Sweating of vanilla beans.

with the beans covered with woollen blankets for the remainder of the day (Fig. 20.6) or (2) incubation in ovens at 45°C at high relative humidity (Arana, 1944, 1945). The significant change in colour of the bean to chocolate brown is manifested at this stage (Balls and Arana, 1941). Sweating is terminated when beans become pliable. The next step is slow drying,

Fig. 20.7 Drying of cured beans in open shelves.

which is normally carried out at room temperature (Fig. 20.7). Drying is needed to lower the moisture content of the beans to a desirable level, usually 15–30% (Jones and Vicente, 1948). Finally, in conditioning the product is kept in closed containers at room temperature for several months to allow the complete development of aroma (Fig. 20.8). In this last stage beans are frequently examined for the presence of moulds. At the minimum, conditioning lasts for three months (Arana, 1944).

An improved curing process using drying tunnels has been developed in Madagascar (Theodose, 1973). This method relies on hot air, instead of heating by the sun, and produces homogeneous, good quality beans in large quantities (40 tonnes dry vanilla in one season). Gillette and Hoffman (1992) present a very good comparison of the curing process associated with the different types of vanilla beans.

The nature of curing procedures adopted affects the quality of cured beans. Aside from the influence of method of wilting, Arana (1945) pointed out that non-uniformity in drying, sweating and drying under the sun, use of dirty blankets and improper ventilation in curing rooms, all contribute to the susceptibility of beans to mould infection, which in turn lowers the quality of the product. He further noted that the moisture content of cured beans should be properly controlled to obtain the full development of the vanilla aroma. The aroma of cured beans with 50–54% moisture is characteristically fermented; those with 24–27% moisture, sophisticated and well developed; while those with 31–34% moisture, just desirable.

Factors other than those related to curing protocol are also known to influence the quality of the final product. Vanilla beans that ripen early in the harvesting season yield higher quality cured beans than those gathered at mid-season or late in the season (Jones and Vicente, 1949b). The best cured material comes from pods harvested when the blossom-end section is yellow. When picked prior to this stage, beans give an undeveloped vanilla flavour; when beyond, a full but undesirable flavour is obtained (Broderick, 1955a). Immature beans when processed are also readily attacked by fungi (Arana, 1945).

Fig. 20.8 Dried beans are placed in plastic bags and conditioned for several months.

Grading
The grading and classification of cured vanilla beans vary depending upon the producing countries. Mexican beans, for example, are usually graded (from the highest to the lowest quality product) as 'prime', 'good to prime', 'good', 'fair' and 'ordinary', while Bourbon beans are graded as 'prime', 'firsts', 'seconds', 'thirds', 'fourths' and 'foxy splits' (Merory, 1960). Classification is commonly based on: bean integrity (either whole, broken or split), bean length, appearance (particularly colour and surface blemishes), moisture content and aroma quality (Arana, 1945; Heath and Reineccius, 1986).

Packaging
After sorting, the beans are tied into bundles, usually 70 to 130, weighing between 150 and 500 g (Heath and Reineccius, 1986). These are then packed into cardboard or tin boxes lined with waxed paper. The beans are now ready for shipment.

Table 20.2 Reported values of major flavour constituents of cured vanilla beans from various geographical sources

Source	Vanillin	Vanillic acid	p-Hydroxybenzaldehyde	p-Hydroxybenzoic acid
		(mg 100 ml^{-1})		
Mexico	18–100	8–23	5–7	2–7
Madagascar	47–216	12–23	6–13.7	2–5.6
Comores	115–154	8–13	9–13	2–7
Java/Indonesia	63–142	7.7–11	7–10	2–4
Uganda	47–186	5–10	5–8	1–6
Tonga	197–320	7.6	10	2.1
West Indies	20–136	15–22	2–6	2–4
Costa Rica	135–161	12	14	5.2
Jamaica	216–265	4.2	8.4	–
Tahiti	54–120	4–5	4–13	16–32.8

Sources: Smith (1964), Archer (1989), Ranadive (1992).

20.4.2 Flavour constituents

The flavour famously associated with vanilla results from a complex and varied mixture of chemical compounds. About 170 volatile constituents, most of which occur below 1 ppm, have been reported in vanilla by Klimes and Lamparsky (1976). Vanillin serves as the major flavour backbone, occurring in levels from 1.52 to 2.42% of bean dry weight (Cowley, 1973). Other major components are p-hydroxybenzoic acid, p-hydroxybenzaldehyde, vanillic acid, p-hydroxybenzyl alcohol (Fig. 20.4e) and vanillyl alcohol (Fig. 20.4f) (Anwar, 1963; Smith, 1964; Herrmann and Stockli, 1982).

The type and levels of the major flavouring components vary depending upon the species and geographical source (Table 20.2). Tahiti vanilla stands out among the different types of beans for exhibiting higher levels of p-hydroxybenzoic acid. Other components present in *V. tahitensis* that are not detected in *V. planifolia* are p-anisic acid, p-anisaldehyde and piperonal (heliotropin) (Fig. 20.4g–i) (Ranadive, 1992). Vanillons (*V. pompona*, Guadeloupe vanilla) contains vanillin, p-hydroxybenzoic acid, vanillic acid, p-hydroxybenzaldehyde, p-anisic acid, p-anisaldehyde and p-anisyl alcohol (Fig. 20.4j), but not piperonal (Ehlers and Pfister, 1997).

The hydrocarbon profile of the lipidic fraction, which also contributes to flavour, of different types of beans has also bean investigated by Ramaroson-Raonizafinimanana *et al.* (1997). Hydrocarbon content varies between 0.2 and 0.6%. A total of 25 *n*-alkanes, 17 branched alkanes and 12 alkenes have been identified. Distinction between types of vanilla is also evident. *Vanilla fragrans* from Réunion is rich in *n*-alkanes (46%) and *n*-1-alkenes (26%), while *V. tahitensis* from Tahiti contains predominantly branched alkanes (47% for 3-methylalkanes and 33% for 5-ethylalkanes). Also present in the lipophilic fraction before saponification in the two vanilla species are three new γ-pyrones: 2-(10-nonadecenyl)-2,3-dihydro-6-methyl-4*H*-pyran-4-one; 2-(12-heneicosyl)-2,3-dihydro-6-methyl-4*H*-pyran-4-one, and 2-(14-tricosenyl)-2,3-dihydro-6-methyl-4*H*-pyran-4-one (Ramaroson-Raonizafinimanana *et al.*, 1999). γ-Pyrones are intermediates in the synthesis of biologically important compounds. A review of other flavour components as a function of vanilla species can be found in Richard (1991).

Werkhoff and Guntert (1997) characterized for the first time in Bourbon vanilla beans 15 esters that are derived from cyclic and acyclic terpene alcohols and aromatic acids. Among those isolated, pentyl salicylate and citronellyl isobutyrate are considered new natural compounds.

Vanilla also contains resins, gums, amino acids and other organic acids, which all contribute to the distinct flavour characteristics of the cured beans. An enumeration and discussion of these constituents is provided by Purseglove et al. (1981).

20.5 Uses

There is probably no other spice material or aromatic plant comparable to vanilla in terms of wide scope of application. The use of vanilla is generally grouped into three: as a ubiquitous flavouring material, as a critical intermediary in a host of pharmaceutical products, and as a subtle component of perfumes. As a flavouring agent, vanilla is a popular and most preferred ingredient in the preparation of ice cream, milk, beverages, candies, confectioneries and various bakery items. In the pharmaceutical and chemical industries, vanillin serves as an important intermediate in the manufacture of: L-dopa (the anti-Parkinsonian drug), methyl dopa (a compound with anti-hypertensive and tranquilizing properties), papaverine (treatment of heart problem), trimethoprim (anti-bacterial agent), hydrazones (2,4-D-like herbicide), and anti-foaming agent (in lubricating oils) (Rosenbaum, 1974; Hocking, 1997).

Vanilla in perfumery was initially incorporated to complement the scent provided by tonka extract. It became a perfume ingredient to reckon with when François Coty, who is often regarded as the first of the great perfumers of modern times, used it in 'L'Aimant' (Groom, 1992). Vanilla subsequently became the principal note of about 23% of all quality perfumes, e.g. 'Amouge', 'Bois de Isles', 'Jicky', 'Habanita'.

20.6 Vanilla products

The cured beans are further processed to produce the various vanilla products. This is commonly accomplished in the importing countries. The different products developed from vanilla are described below.

20.6.1 Vanilla extract

The major product derived from cured vanilla is an alcoholic essence, which is commercially known as vanilla extract. The vanilla flavour is obtained through solvent extraction with the use of the best grade of ethanol. Generally, the basic process in the preparation of vanilla extract involves (1) the reduction of the bean size using a comminuting machine and (2) the subsequent alcohol extraction of the macerated beans through a series of percolation techniques (Arana, 1945; Heath and Reineccius, 1986). For best results Merory (1956, 1960) recommends the following protocol. Three consecutive extractions are done with varying amounts of the menstruum – a maximum of 65% ethanol for the first extraction, 35% for the second and about 15% for the third. Each of these takes place for a minimum of five days. Extraction is done in a continuous slow flow, the percolate collected in fractions and later blended to yield the final product. The extract is filtered or centrifuged and the alcohol content is adjusted to meet market specifications. Vanilla extract is then stored in stainless steel or glass containers. If the extract is aged for a period of about three to six months, the delicate and subtle aroma for which vanilla is famous is fully realized.

The nutrient composition of a typical vanilla extract is listed in Table 20.3. Several factors influence the quality of vanilla extract. These include: (1) method of curing, (2) blending of different quality beans, (3) degree of maceration, (4) method of extraction,

Table 20.3 Nutrient composition of vanilla extract (with 34.4% ethyl alcohol)

Nutrient	Value per 100 g of edible portion
Proximates	
Water (g)	52.58
Energy (kcal)	288
Protein (g)	0.06
Total lipid (fat) (g)	0.06
Ash (g)	0.26
Carbohydrate, by difference (g)	12.65
Minerals	
Ca (mg)	11
Fe (mg)	0.12
Mg (mg)	12
P (mg)	6
K (mg)	148
Na (mg)	9
Zn (mg)	0.11
Cu (mg)	0.072
Mn (mg)	0.230
Vitamins	
Thiamin (mg)	0.011
Riboflavin (mg)	0.095
Niacin (mg)	0.425
Panthothenic acid (mg)	0.035
Vitamin B_6 (mg)	0.026
Lipids	
Fatty acids, total saturated (g)	0.010
Fatty acids, monounsaturated (g)	0.010
Fatty acids, total polyunsaturated (g)	0.004

Source: USDA National Nutrient Database for Standard Reference (2002). www.nal.usda.gov.

(5) level of alcohol in the menstruum and (6) appropriate period of ageing (Broderick, 1955b; Merory, 1960; Heath and Reineccius, 1986).

20.6.2 Vanilla oleoresin

Vanilla oleoresin is a dark brown, semi-fluid extract produced from solvent extraction of macerated beans. It differs from vanilla extract in that the solvent used is completely removed by evaporation under vacuum and the finer top-notes of the vanilla aroma are lost or modified by heat treatment (Heath and Reineccius, 1986). Vanilla oleoresin can also be obtained using CO_2 under supercritical conditions, producing products considerably more superior than those obtained by conventional extraction with organic solvents (Schuetz *et al.*, 1984). Yield of oleoresin is from 29.9% to 64.8% of bean dry weight (Cowley, 1973).

20.6.3 Vanilla sugar

Also known as powdered vanilla, vanilla sugar is prepared by mixing ground cured beans or their oleoresin with sugar (Arana, 1945). Minimum sugar content is 30% (Heath and Reineccius, 1986).

20.6.4 Vanilla absolute

Preferred in perfumery products, absolute vanilla is obtained by selective solvent extraction, using initially a non-polar solvent such as benzene followed by a polar solvent such as ethanol (Heath and Reineccius, 1986).

20.7 Functional properties

Vanillin exhibits *in vitro* antifungal activity against the yeasts *Candida albicans* and *Cryptococcus neoformans* (Boonchird and Flegel, 1982) Minimal inhibitory concentrations of vanillin for *C. albicans* and *C. neoformans* were found to be 1250 and 738 µg ml^{-1}, while minimal fungicidal concentrations were 5000 and 1761 µg ml^{-1}, respectively. It is also reported to inhibit the growth of some food spoilage yeasts (e.g. *Saccharomyces cerevisiae, Zygosaccharomyces rouxii, Z. bailii* and *Debaryomyces hansenii*) in culture media and some fruit purées (Cerrutti and Alzamora, 1996).

The potential medical importance of vanillin is suggested by the following studies. Vanillin has been found to possess antimutagenic effects in mice (Imanishi *et al.*, 1990) and bacteria (Ohta *et al.*, 1988). In yeast, however, it is shown to be co-mutagenic and co-recombinogenic (Fahrig, 1996). Vanillin offers protection against X-ray and UV radiation-induced chromosomal change in V79 Chinese hamster lung cells (Keshava *et al.*, 1998).

Vanillin also functions as an antioxidant. At concentrations normally added to food preparations, it offers significant protection against protein oxidation and lipid peroxidation induced by photosensitization in rat liver mitochondria (Kamat *et al.*, 2000). This study shows the potential of using this popular flavouring chemical to inhibit oxidative damage to membranes in mammalian tissues.

Sun *et al.* (2001) reported the bioactivity of five aromatic compounds extracted from the leaves and stems of *V. fragance* against mosquito (*Culex pipiens*) larvae. Among the isolated compounds, 4-butoxymethylphenol was the most toxic, exhibiting 100% mortality at 0.2 mg ml^{-1} within only 3 h of treatment. This was followed by 4-ethoxymethylphenol, which was also the most abundant component. Vanillin, when given at 2 mg ml^{-1} for 10 h exhibited more than 90% mortality The least toxic of the phenolic derivatives was 3,4-dihydroxyphenylacetic acid. This compound, together with 4-hydroxy-2-methoxycinnamaldehyde, was isolated from vanilla for the first time. 4-Butoxymethylphenol has not been reported to occur in natural form.

20.8 Quality issues and adulteration

The quality of cured vanilla beans is the result of confluent factors that run the whole gamut of raw material production to curing. The agro-climatic conditions during cultivation, coupled with various degrees of sophistication or non-sophistication of methods employed in the preparation of harvested beans, can spell the difference in meeting market standards.

Physical attributes such as those enumerated in Section 20.4.1 provide the initial criteria by which to judge the cured bean and assign it to a particular grade. The quality of vanilla extract can be determined through chemical analysis, and Winton's analytical values have been employed in this regard (Table 20.4; Merory, 1960). The concentration of vanillin is an important criterion, although organoleptic quality does not entirely depend on it. Various flavour notes, described as characteristically woody, pruney, resinous, leathery, floral and

Table 20.4 Some analytical values for vanilla extract

Type of analysis	Minimum	Maximum	Average
Vanillin, g 100 ml^{-1} extract	0.11	0.35	0.19
Ash, g 100 ml^{-1} extract	0.22	0.432	0.319
Soluble ash, g 100 ml^{-1} extract	0.179	0.357	0.265
Lead number (Winton)	0.40	0.74	0.54
Alkalinity of total ash, N/10 acid 100 ml^{-1} extract	30.00	54.00	
Alkalinity of soluble ash, N/10 acid 100 ml^{-1} extract	22.0	40.0	30.0
Total acidity, N/10 alkali 100 ml^{-1} extract	30.0	52.0	42.0
Acidity other than vanillin, N/10 alkali 100 ml^{-1} extract	14.0	42.0	30.0

Source: adapted from Merory (1960).

fruity aromatics, also need to be considered (Gillette and Hoffman, 1992). Bourbon vanilla serves as the standard by which to measure the chemical and sensory quality of other types of vanilla. Imitation vanilla extract spiked with vanillin is less desirable than the natural extract because the critical flavour notes are wanting (Fig. 20.9). Extracts of Indonesian

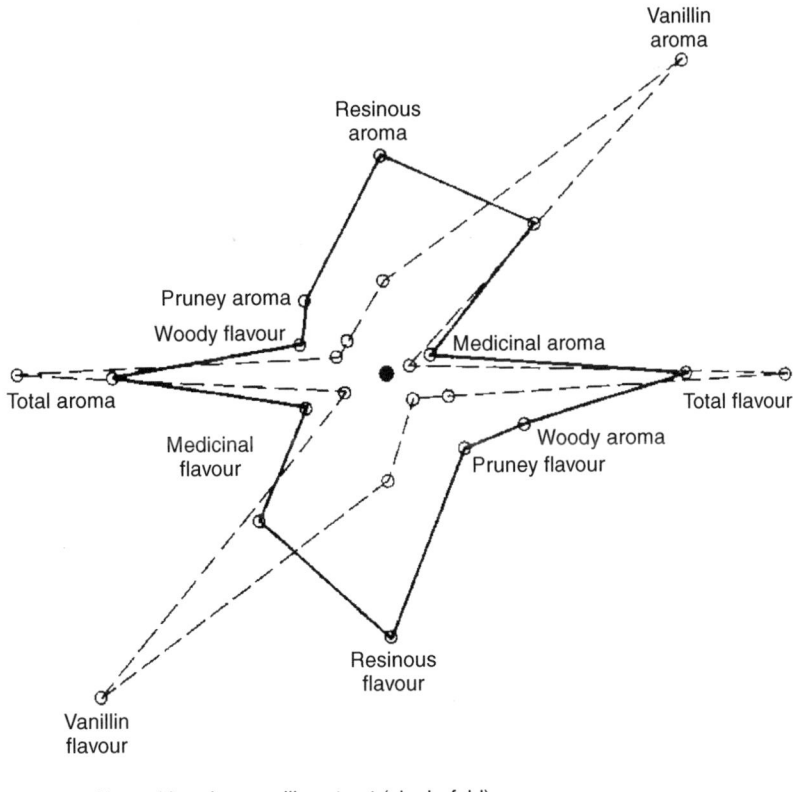

Fig. 20.9 Aroma and flavour sensory profiles of natural and imitation vanilla extracts. (*Source*: Gillette and Hoffman, 1992.)

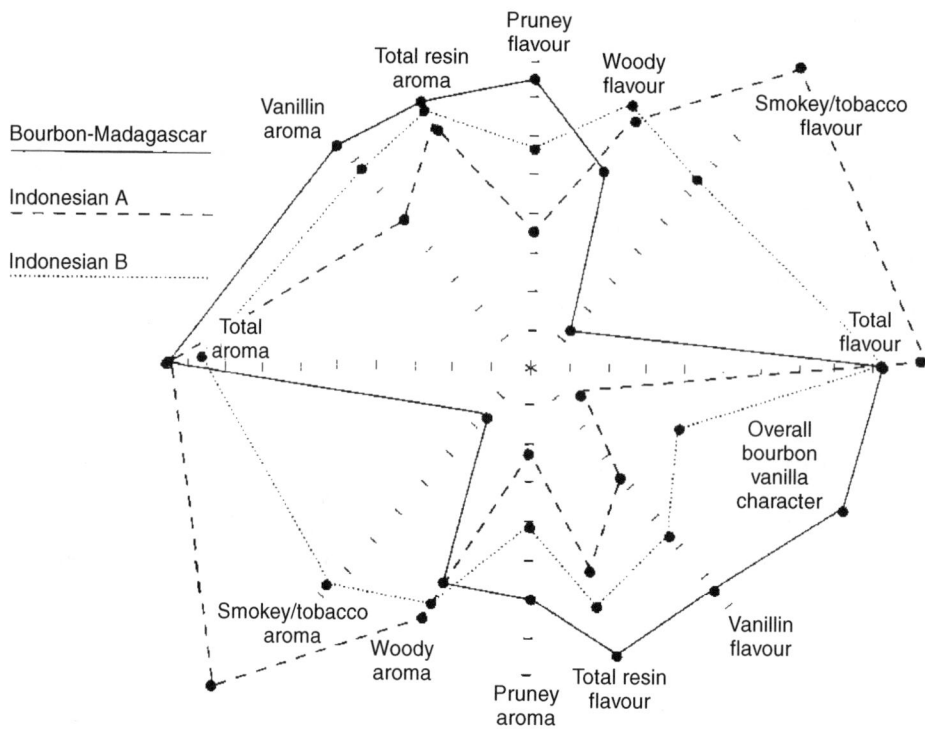

Fig. 20.10 Aroma and flavour sensory profiles of Bourbon and Indonesian pure vanilla extracts. (*Source*: Gillette and Hoffman, 1992.)

beans are not only low in vanillin but also possess a smokey/tobacco flavour and aroma (Fig. 20.10), while Mexican beans yield extracts with woody and smokey/tobacco notes (Fig. 20.11). On the other hand, Tahitian vanilla is more fruity/floral and less resinous and pruney (Fig. 20.12).

In the USA, the Food and Drug Administration explicitly provides standard specifications for vanilla products. As an example the amount of bean present with a given moisture content and level of alcohol in the mixture define vanilla extract quality (Table 20.5).

It should be emphasized here that no amount of modern technology of processing can improve the quality of an already poor bean at harvest. Improper curing and handling, on the other hand, is sure to lead to quality deterioration of vanilla produced under excellent cultivation practices.

20.8.1 Substitutes, adulterants and additives

Plant material
Correll (1953) identified the following plants that have been used as raw material substitute or adulterant for vanilla. The most commonly employed is the fruit of tonka or snuff bean (*Dipteryx odorata* (Aubl.) Willd.), a leguminous plant native to northern South America and Trinidad. Other plants include: another species of tonka (*D. oppositifolia* Willd.); vanillon (*V. pompona*), a less common wild vanilla; the long cylindrical pods of little vanilla (*Selenipedium chica* Reichb. f.); leaves of the orchid *Angraecum fragrans* Thou. and *Orchis*

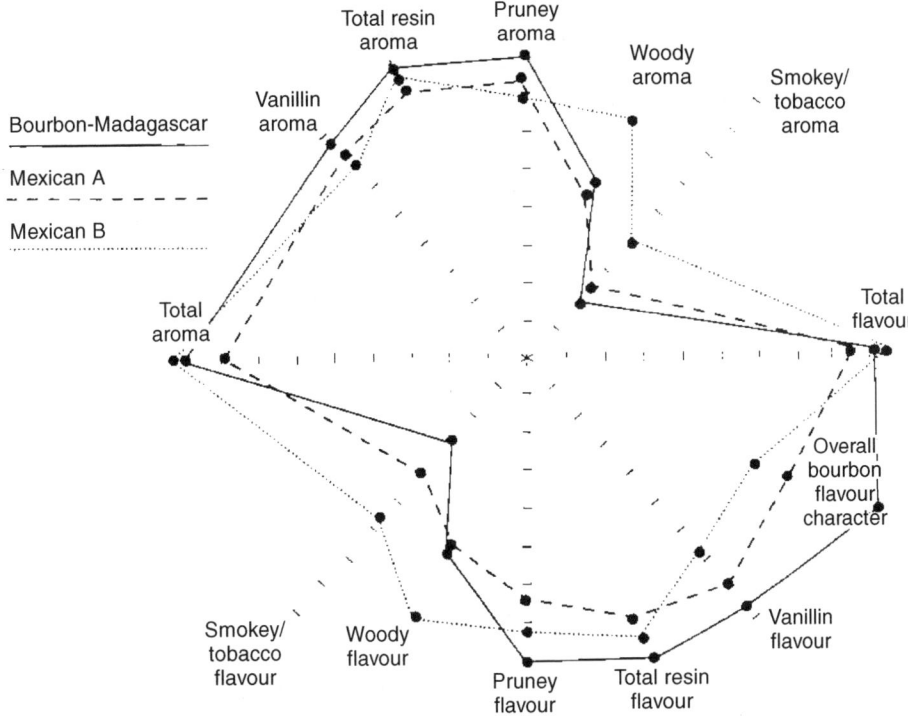

Fig. 20.11 Aroma and flavour sensory profiles of Bourbon and Mexican pure vanilla extracts. (*Source*: Gillette and Hoffman, 1992.)

Table 20.5 US standard specifications for vanilla and vanilla extract

21CFR169.3
Sec. 169.3 Definitions
  a) The term vanilla beans means the properly cured and dried pods of *Vanilla planifolia* Andrews and of *Vanilla tahitensis* Moore.
  b) The term unit weight of beans means, in the case of vanilla beans containing not more than 25 percent moisture, 13.35 ounces of such beans; and, in the case of vanilla beans containing more than 25 percent moisture, it means the weight of such beans equivalent in content of moisture-free vanilla-bean solids to 13.35 ounces of vanilla beans containing 25 percent moisture.
  c) The term unit of vanilla constituent means the total sapid and odorous principles extractable from one unit weight of vanilla beans, as defined in paragraph (b) of this section, by an aqueous alcohol solution in which the content of ethyl alcohol by volume amounts to not less than 35 percent.
21CFR169.175
Sec. 169.175 Vanilla extract
  a) Vanilla extract is the solution in aqueous ethyl alcohol of the sapid and odorous principles extractable from vanilla beans. In vanilla extract the content of ethyl alcohol is not less than 35 percent by volume and the content of vanilla constituent, as defined in Sec. 169.3 (c), is not less than one unit per gallon. The vanilla constituent may be extracted directly from vanilla beans or it may be added in the form of concentrated extract or concentrated vanilla flavoring or vanilla flavoring concentrated to the semi-solid form called vanilla oleo-resin. Vanilla extract may contain one or more of the following optional ingredients:
    (1) Glycerin
    (2) Propylene glycol
    (3) Sugar (including invert sugar)
    (4) Dextrose
    (5) Corn sirup (including dried corn sirup)

Source: US Food and Drug Administration (2002). Code of Federal Regulations. 21CFR Part 169.

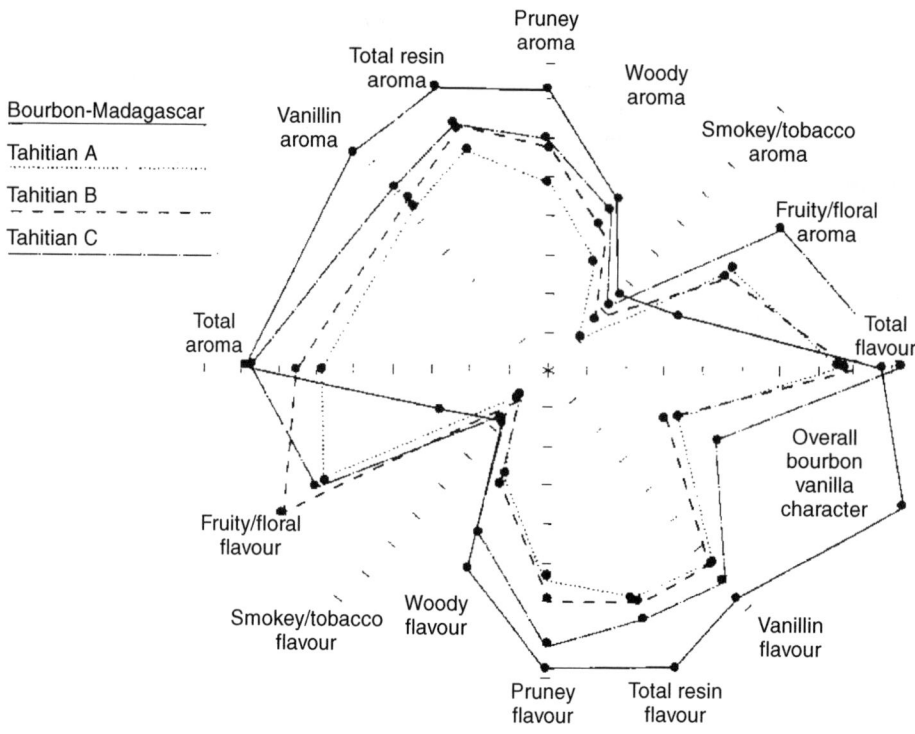

Fig. 20.12 Aroma and flavour sensory profiles of Bourbon and Tahitian pure vanilla extracts. (*Source*: Gillette and Hoffman, 1992.)

fusca Jaq.; ladies' tresses (*Spiranthes cernua* (L.) L.C. Rich var. *odorata* (Nutt.) Correll); 'vanilla-plant' (*Trilisa odoratissima* (Walt.) Cass.); little orchid 'herb vanilla' (*Nigritella angustifolia* Rich.) and common sweet clovers (*Melilotus* spp.).

Chemical substances
Because of high demand, limited supply of quality beans and increased cost of production, natural vanilla extracts have been most often adulterated with inexpensive, synthetic vanillin. This fraudulent alternative becomes very attractive in light of the expected profitable return: 1 kg of Madagascar beans yields about 8.4 l vanilla extract, while 1 kg of synthetic vanillin, combined with other botanical extracts, produces about 499 l artificial vanilla flavouring, retailing for over 50 times as much as the natural product (Breedlove, 2002)!

Synthetic vanillin is chemically produced via different routes with eugenol, guaiacol, safrole or lignin as the starting compound. A detailed discussion of these synthetic pathways is provided by Faith *et al.* (1957), Didams and Krum (1970), Rosenbaum (1974) and Hocking (1997).

Another most common additive is ethyl vanillin, 3-ethoxy-4-hydroxybenzaldehyde. It is synthetically produced from safrole and is about three to four times more powerful as a flavouring agent than vanillin (Heath and Reineccius, 1986). A maximum of 10% of vanillin is replaceable by this compound without obvious objectionable note.

Other compounds utilized are: veratraldehyde (methyl vanillin), piperonal (heliotropin), vanitrope (propenyl guaethol) and coumarin (Kahan and Fitelson, 1964). Coumarin, because of its high level of toxicity, has been banned for use in food in the USA (Rosengarten, 1969).

20.8.2 Detection of adulteration

Some of the techniques utilized to test the purity of vanilla extracts are the determination of lead number, which is directly related to the quantity of organic acids (AOAC, 1975), paper chromatography (Jorysch, 1959; Stahl et al., 1960; Fitelson, 1960a, b, 1963, 1964; Kahan and Fitelson, 1964), gradient elution method using an ion exchange resin (Sullivan et al., 1960; Fitelson, 1964) and thin-layer chromatography (Kahan and Fitelson, 1964). These methods, however, are subject to low-sensitivity, high-experimental error, lack of reproducibility and, most importantly, are too laborious and time-consuming for normal routine work. Ensminger (1953, 1956) and Feeny (1964) reported the application of a UV photometric procedure to reveal the type of flavouring constituents in vanilla extract; this method, however, does not differentiate vanillin from ethyl vanillin without preliminary separation. Gas–liquid chromatography (GLC) essentially eliminates this problem, aside from revealing the presence of other types of contaminants (Martin et al., 1964, 1975; Johansen, 1965). Determination of the ratio of vanillin levels to the concentration of the nutrients potassium, phosphorus and nitrogen provides the so-called 'identification ratios', which not only separates authentic vanilla extracts from their imitation but also gives indication of the strength of the vanilla extract (Martin et al., 1975). Relatively higher identification ratios provide strong evidence for adulteration. The technique suffers from a need to input a significant amount of time and the regular analysis of the ratios for authentic standards. More sensitive assays for adulteration are accomplished by the application of reverse phase liquid chromatography (Guarino and Brown, 1985) and high-performance liquid chromatography (HPLC) (Herrmann and Stockli, 1982; Archer, 1989).

A more recent technique involves the application of the stable isotope ratio analysis mass spectrometry (SIRA–MS), which has been employed in deciphering the original source material for flavouring food products (Cordella et al., 2002). The method measures and compares the ratio of abundances of stable isotopes such as carbon ($^{13}C/^{12}C$) between natural and synthetic vanillin. In principle, the isotopic fractionation of carbon through photosynthesis has been shown to depend upon the type of CO_2 assimilation in the plant. Plants that follow the Calvin cycle discriminate the most against ^{13}C, while those that follow the Hatch–Slack cycle discriminate the least. In-between are those characteristic of the Crassulacean acid metabolism (CAM) pathway, which is exhibited in vanilla (Osmond et al., 1973). Vanillin extracted from vanilla beans has been shown to be enriched in ^{13}C compared with synthetic vanillin produced from lignin, eugenol and guaiacol (Bricout et al., 1974; Hoffman and Salb, 1979; Martin et al., 1981; Culp and Noakes, 1992).

The SIRA method for carbon was further refined by Krueger and Krueger (1983, 1985) to detect synthetic vanillin whose isotopic compositions have been adjusted to resemble natural vanillin. Aside from carbon, the stable isotope of hydrogen (deuterium/hydrogen) has also been examined and found useful in discriminating between authentic and fraudulent samples of vanilla extract. Natural vanillin is more depleted in deuterium content compared with its synthetic counterpart (Culp and Noakes, 1992). Further refinement of the technique to detect isotopically manipulated vanilla is reported by Remaud et al. (1997) with the application of site-specific natural isotopic fractionation analysed by nuclear magnetic resonance (SNIF-NMR). This method is applied not only to vanillin but also to *p*-hydroxybenzaldehyde.

For a compilation of the official methods of analysis of vanilla extract by the Association of Official Analytical Chemists (AOAC), the reader is referred to the work of Krueger (1995).

20.9 Improving production of natural vanillin

The vanilla flavouring market is dominated by synthetic vanillin: more than 90% of the US market and 50% of the French market (the lowest national share) (RAP Market Information Bulletin, 2002). The reason for such a huge market share is two-fold. Considering flavour strength, 100 g of synthetic vanillin approximately yields the same flavouring power of 13 l of natural vanilla extract. The price of synthetic vanillin is only about one-hundredth that of derived from cured beans. The high cost of the natural flavouring extract has been attributed to the exacting requirements of the plant and the finished product in relation to agroclimatic condtions, cultivation, pollination, harvesting and curing. Ramachandra Rao and Ravishankar (2000) estimated that, to produce 1 kg of vanillin, 500 kg of beans (based on 20 g kg^{-1} vanillin yield, dry weight basis) derived from about 40 000 pollinated flowers will be needed.

Within the context of food legislation, vanillin chemically synthesized from lignin, guaiacol, safrole or eugenol, is labelled as synthetic vanillin, even if the precursor input is from a natural source (e.g. eugenol, safrole) (Rabenhorst and Hopp, 2000). Products sold as natural vanillin in the international market command a higher premium because of the presence of other flavour components contributing to the overall organoleptic quality of the material. Vanillin production through chemical synthesis suffers from the following disadvantages: presence of impurities/contaminants, large consumption of chemical products and energy, production of ecologically toxic waste products (as in the case of lignin degradation), and production of isomers which are difficult to separate (e.g. isovanillin/vanillin from safrole) (Hocking, 1997; Mane and Zucca, 1998). Inevitably, alternative methods were sought and tried to eliminate the above constraints.

20.9.1 Tissue culture technique

The potential of organ and cell culture has been explored to produce natural vanilla flavouring components outside the cured bean. Using ferulic acid as a precursor, Westcott *et al.* (1994) employed vanilla plant aerial roots, cultured in a charcoal medium, as biocatalysts in the production of vanillin. The charcoal acts as reservoir for the root-synthesized vanillin, which is removed by selective solvent extraction. Vanillin production using this technique is five to ten times faster than the usual rate in vanilla beans, with productivities up to 400 mg kg^{-1} dry wt tissue day^{-1}, about 40% of that detected in mature vanilla beans.

Initial efforts to produce vanillin from plant cell culture have been unsuccessful. Feeding of precursors and inhibitors of the phenypropanoid metabolic pathways to undifferentiated suspension cells derived from green vanilla beans induces the formation of 4-hydroxybenzoic acid, vanillic acid and syringic acid but not vanillin (Funk and Brodelius, 1990a,b,c).

Under appropriate, controlled conditions, callus suspensions derived from root tips and leaf primordia of vanilla have been claimed to induce the production not only of vanillin but also of 3,4-dihydroxybenzaldehyde, 4-hydroxybenzoic acid, 4-hydroxybenzaldehyde and vanillic acid (Knuth and Sahai, 1991). Havkin-Frenkel *et al.* (1996) found that, in shoot-derived cluster culture (suspension culture growing in clumps) of vanilla, light is an important factor in the accumulation of some of these flavour chemicals, particularly

Table 20.6 Some microorganisms utilized in the bioconversion of selected substrates to vanillin

Microorganism	Substrate	Reference
Aspergillus niger ATCC 11414 *Corynebacterium glutamicum* ATCC 13032 *Pseudomonas putida* ATCC 55180 *Rhodotorula glutinis* ATCC 74056	Coniferyl alcohol, eugenol, ferulic acid, 4-vinylguaiacol	Labuda *et al.* (1992)
Pseudomonas sp. TK-2101	Eugenol	Washisu *et al.* (1993)
Pycnoporus cinnabarinus CNCM No. I-937 and I-938	Ferulic acid, vanillic acid	Gross *et al.* (1993)
Amycolatopsis sp. DSM 9991 and 9992	Ferulic acid	Rabenhorst and Hopp (2000)
Nocardia sp. NRRL 5646	Vanillic acid	Li and Rosazza (2000)
Streptomyces setonii ATCC 39116	Ferulic acid	Muheim *et al.* (2001)

vanillyl alcohol. Davidonis and co-workers (1996) also showed that the synthesis of vanillin from callus, which proliferated from shoot tips of vanilla, is enhanced when ferulic acid is topically applied to them. Thus, even an undifferentiated mass of cells is capable of metabolic synthesis of vanillin.

20.9.2 Microbial biocatalytic transformation

Microbial catalysis of naturally occurring and abundant substrates offers another interesting alternative to the production of natural vanillin. Hagedorn and Kaphammer (1994) reviewed work on the microbial transformation of natural flavour precursors to vanillin. The list of some microorganisms and the specific substrates utilized for the microbial production of natural vanillin is presented in Table 20.6.

The basic protocol for microbial vanillin formation is fairly straightforward (Labuda *et al.*, 1992; Gross *et al.*, 1993; Rabenhorst and Hopp, 2000; Muheim *et al.*, 2001). In general, the selected microbe, either bacterium, yeast or fungus, is initially grown in a culture medium, commonly supplemented with carbon and nitrogen sources, inorganic salts, and in some cases with trace elements and vitamins to optimize growth. When the desired biomass is attained, the appropriate substrate is added, and the loss of the precursor and the corresponding appearance of vanillin is regularly monitored using HPLC. Once the optimum level of vanillin is formed, it is isolated using distillation, solvent extraction or chromatography. Vanillin can then be further purified using standard recrystallization techniques.

Vanillin yield from microbial transformation varies depending upon the adopted protocol, starting substrate used and the microorganism employed. Using a 10 l fermenter, the amount of vanillin synthesized from ferulic acid by *Amycolatopsis* is 11.5 g l^{-1} or 78% of theory based on converted precursor (Rabenhorst and Hopp, 2000). With *Pycnoporus*, molar yield conversion of vanillic acid to vanillin is 35.1%; much less is obtained from ferulic acid, 20.5% (Gross *et al.*, 1993).

20.9.3 Use of isolated and purified enzymes

Enzymes isolated and purified from microorganisms or from plant or animal origin have also been employed to catalyse the biological conversion of starting chemical compounds to vanillin. Examples of these *in vitro* vanillin production include: dioxygenase-active enzyme from *Pseudomonas* TMY 1009 strain using a styrene derivative as substrate (Yoshimoto *et al.*, 1990); lipoxygenase, naturally present in soya bean, wheat or beet and also in pig, calf or fish liver, acting on isoeugenol (Mane and Zucca, 1998); lignin peroxidase from the white-rot fungus *Bjerkandera* sp. BOS55 converting the *o*-acetyl ester of isoeugenol into vanillyl acetate, which can then be cleaved to vanillin (ten Have *et al.*, 1998a,b); carboxylic acid reductase, taken from *Nocardia* sp. strain NRRL 5646, biotransforming vanillic acid to vanillin (Li and Rosazza, 2000), and the flavoprotein vanillyl-alcohol oxidase (VAO) from *Penicillium simplicissimum* using creosol or vanillylamine as precursor (van den Heuvel *et al.*, 2001).

20.9.4 Application of genetic engineering
The prospect of utilizing the tools of genetic engineering for the production of vanillin has also been explored. Research undertakings related to this novel approach have been reviewed by Walton *et al.* (2000) and by Ramachandra Rao and Ravishankar (2000).

Genes of microbial origin and encoding for certain enzymes have been inserted into plant cells for the purpose of modifying the biocatalytic pathway of the transgenic plant in favour of initiation or enhanced vanillin production and other related flavour metabolites. Consider, for example, the gene isolated and characterized from *Pseudomonas fluorescens* biovar V strain AN103, which utilizes ferulic acid as a sole carbon source. It encodes for an enzyme that converts feruloyl-SCoA to vanillin and acetyl-SCoA, which has been confirmed by heterologous expression in *Escherichia coli* (Gasson *et al.*, 1998). Subsequent study showed that the enzyme is not only active with the substrate feruloyl-SCoA but also with 4-coumaroyl-CoA and caffeoyl-CoA and has been identified as 4-hydroxycinnamoyl-CoA hydratase/lyase (HCHL) (Mitra *et al.*, 1999). When the *HCHL* gene is expressed in tobacco plants, there is a massive accumulation of glucosides and glucose esters of 4-hydroxybenzoic acids and vanillic acid, and the glucosides of hydroxybenzyl alcohol and vanillyl alcohol (Mayer *et al.*, 2001). In hairy root cultures of *Datura stramonium* L. transformed with *Agrobacterium rhizogenes*, the introduced *HCHL* gene similarly encodes for most of the previously mentioned compounds (Mitra *et al.*, 2002). In both cases, the aldehyde products (which include vanillin), whether free or conjugated, are not detected, suggesting their rapid conversion to the acid and alcohol forms. Further refinement of this type of metabolic engineering work is clearly in order.

A more recent patented work describes the use of a recombinant microbe (*Escherichia coli* ATCC 98859) for the five-enzyme biocatalysis of a carbon source, e.g. glucose, to vanillic acid (Frost, 2002). The vanillic acid is separated from the bioconversion mixture by organic extraction and subsequently reduced to vanillin by aryl-aldehyde dehydrogenase purified from *Neurospora crassa* SY 7A.

20.10 Future outlook

The growing interest in natural and organic products opens up windows of opportunity for the expansion of the current vanilla market, especially in Japan, Canada, Scandinavia, Great Britain, Italy, Eastern Europe and the Gulf States (Ratsiazo, 1998). To sustain this trend, there is a need to develop exciting products that will fully exploit the potential of dietary and medicinal properties of natural vanilla. New food recipes have to be formulated, catering to the use of the so-called 'edible vanilla or 'vanilla stick' (cured pods sold in retail). This development should be coupled with the improvement of cultural management practices in vanilla specifically geared towards sustainable production. Recommendations on organic methods of cultivation of vanilla can be found in Augstburger *et al*. (2000).

As *in vitro* manufacture of natural vanillin, aided with the tools of genetic engineering, becomes more of a reality, there is a great and valid concern for the possible displacement of vanilla farmers who are mostly located in the developing countries. Resolving such a social dilemma is indeed a daunting task. It should be noted, however, that the faithful organic consumer veers away from products derived from genetically modified organisms.

20.11 References

ANANDARAJ M., REMA J. and SASIKUMAR B. (2001), *Vanilla*, Kerala, Indian Institute of Spices Research.
ANWAR M.H. (1963), 'Paper chromatography of monohydroxyphenols in vanilla extract', *Anal. Chem.*, **35**(12), 1974–6.
AOAC (1975) *Official Methods of Analysis*, 12th ed., Washington, D C, Assn Offic Anal Chem, 331.
ARANA F.E. (1943), 'Action of β-glucosidase in the curing of vanilla', *Food Res.*, **8**(4), 343–51.
ARANA F.E. (1944), *Vanilla curing and its chemistry*, Federal Experiment Station US Department of Agriculture, Mayaguez, Puerto Rico, Bulletin No. 42.
ARANA F.E. (1945), *Vanilla curing*, Federal Experiment Station US Department of Agriculture Mayaguez, Puerto Rico, Circular No. 25.
ARCHER A.W. (1989), 'Analysis of vanilla essences by high-performance liquid chromatography', *J. Chromatogr.*, **462**, 461–6.
AUGSTBURGER F., BERGER J., CENSKOWSKY U., HEID P., MILZ J. and STREIT C. (2000), '*Organic Farming in the Tropics and Subtropics. Exemplary Description of 20 Crops. Vanilla*', Grafelfing, Naturland e.V.
BAILEY L.H. and BAILEY E.Z. (1976), *Hortus Third. A Concise Dictionary of Plants Cultivated in the United States and Canada,* New York, Macmillan Publishing Co. Inc.
BALLS A.K. and ARANA F.E. (1941), 'The curing of vanilla', *Ind. Eng. Chem.*, **33**, 1073–5.
BOONCHIRD C. and FLEGEL T.W. (1982), '*In-vitro* antifungal activity of eugenol and vanillin against *Candida albicans* and *Cryptococcus neoformans*', *Can. J. Microbiol.*, **28**, 1235–41.
BREEDLOVE C.H. (2002), 'Vanilla', Chemistry.org, www.acs.org.
BRICOUT J., FONTES J. and MERLIVAT L. (1974), 'Detection of synthetic vanillin in vanilla extracts by isotopic analysis', *J. AOAC*, **57**(3), 713–5.
BRODERICK J.J. (1955a), 'The chemistry of vanilla', *Coffee Tea Ind.*, **78**(3), 53–4, 58.
BRODERICK J.J. (1955b), 'Vanilla extract manufacture. A guide to choice of method', *Food Manuf.*, **30**(1), 65–8.
CERUTTI P. and ALZAMORA S.M. (1996), 'Inhibitory effects of vanillin on some food spoilage yeasts in laboratory media and fruit purees', *Int. J. Food Microbiol.*, **29**, 379–86.
CORDELLA C., MOUSSA I., MARTEL A., SBIRRAZZUOLI N. and LIZZANI-CUVELIER L. (2002), 'Recent developments in food characterization and adulteration detection: technique-oriented perspectives', *J. Agric. Food Chem.*, **50**, 1751–64.
CORRELL D.S. (1953), 'Vanilla-its botany, history, cultivation and economic importance', *Econ. Bot.*, **7**(4), 291–358.
COWLEY E. (1973), 'Vanilla and its uses', *Proceedings of the Conference on Spices*, London, Tropical Products Institute, 79–82.
CULP R.A. and NOAKES J.E. (1992), 'Determination of synthetic components in flavors by deuterium/hydrogen isotopic ratios', *J. Agric. Food Chem.*, **40**, 1892–7.
DAVID P.A. (1950), 'Vanilla culture in the College of Agriculture at Los Baños', *Phil. Agric.*, **33**, 239–49.
DAVIDONIS G., KNORR D.W. and ROMAGNOLI L.G. (1996), 'Callus formation *Vanilla planifolia*', US Patent 5,573,941.
DIDDAMS D.G. and KRUM J.K. (1970), 'Vanillin', in Standen A. (Editor), *Kirk-Othmer Encyclopedia of Chemical Technology vol 21 Uranium and uranium compounds to water (analysis),* New York, John Wiley and Sons, Inc, 180–96.

EHLERS D. and PFISTER M. (1997), 'Compounds of vanillons (*Vanilla pompona* Scheide)', *J. Essent. Oil Res.*, **9**, 427–31.
ENSMINGER L.G. (1953), 'Report on vanilla extracts and imitations. Determination of vanillin, ethyl vanillin and coumarin by ultraviolet absorption', *J. AOAC*, **36**(3), 679–97.
ENSMINGER L.G. (1956), 'Report on vanilla extracts and imitations. Determination of vanillin, ethyl vanillin and coumarin by partition chromatography and spectrophotometry', *J. AOAC*, **39**(3), 715–21.
FAHRIG R. (1996), 'Anti-mutagenic agents are also co-recombinogenic and can be converted into co-mutagens', *Mutation Res.*, **350**, 59–67.
FAITH W.L., KEYES D.B. and CLARK R.L. (1957), 'Vanillin', *Industrial Chemicals*, New York, John Wiley and Sons, Inc.
FAO (2003), FAOSTAT Agricultural Data, http://apps.fao.org.
FARREYROL K., PEARSON M.N., GRISONI M. and LECLERCQ-LE QUILLEC F. (2001), 'Severe stunting of *Vanilla tahitensis* in French Polynesia caused by Cucumber Mosaic Virus (CMV), and the detection of the virus in *V. fragrans* in Reunion Island', *Plant Pathology*, **50**(3), 414.
FEENY F.J. (1964), 'Determination of vanillin by ultraviolet absorption', *J. AOAC*, **47**(3), 555–7.
FITELSON J. (1960a), 'Detection of foreign plant materials in vanilla extract by paper chromatography', *J. AOAC*, **43**(3), 596–9.
FITELSON J. (1960b), 'A qualitative test for vanilla resins', *J. AOAC*, **43**(3), 600–1.
FITELSON J. (1963), 'Organic acids in vanilla extract', *J. AOAC*, **46**(4), 626–33.
FITELSON J. (1964), 'Organic acids in vanilla extract', *J. AOAC*, **47**(3), 558–60.
FOUCHE J.G. and COUMANS M. (1992), 'Vanilla pollination', *Amer. Orch. Soc. Bull.*, **6**(11), 1118–22.
FROST J.W. (2002), 'Synthesis of vanillin from a carbon source', *US Patent 6,372,461*.
FUNK C. and BRODELIUS P. (1990a), 'Influence of growth regulators and an elicitor on phenylpropanoid metabolism in suspension cultures of *Vanilla planifolia*', *Phytochemistry*, **29**(3), 845–8.
FUNK C. and BRODELIUS P.E. (1990b), 'Phenylpropanoid metabolism in suspension cultures of *Vanilla planifolia* Andr. II. Effects of precursor feeding and metabolic inhibitors', *Plant Physiol.*, **94**(1), 95–101.
FUNK C. and BRODELIUS P.E. (1990c), 'Phenylpropanoid metabolism in suspension cultures of *Vanilla planifolia* Andr. III. Conversion of 4-methoxycinnamic acids into 4-hydroxybenzoic acids', *Plant Physiol.*, **94**(1), 102–8.
GASSON M.J., KITAMURA Y., MCLAUCHLAN W.R., NARBAD A., PARR A.J., PARSONS E.L.H., PAYNE J., RHODES M.J.C. and WALTON N.J. (1998), 'Metabolism of ferulic acid vanillin: a bacterial gene of the enoyl-SCoA hydratase/isomerase superfamily encodes an enzyme for the hydration and cleavage of a hydroxycinnamic acid SCoA thioester', *J. Biol. Chem.*, **273**(7), 4163–70.
GILLETTE M.H. and HOFFMAN P.G. (1992), 'Vanilla extract' in Hui Y.H. (Editor), *Encyclopedia of Food Science and Technology*, Vol 4, New York, John Wiley and Sons, Inc., 2641–57.
GROOM N. (1992), *The Perfume Handbook*, Glasgow, New Zealand, Blackie Academic and Professional.
GROSS B., ASTHER M., CORRIEU G. and BRUNERIE P. (1993), 'Production of vanillin by bioconversion of benzenoid precursors by *Pycnoporus*', *US Patent 5,262,315*.
GUARINO P.A. and BROWN S.M. (1985), 'Liquid chromatographic determination of vanillin and related flavor compounds in vanilla extract: cooperative study', *J. Assoc. Off. Anal. Chem.*, **68**(6), 1198–1201.
HAGEDORN S. and KAPHAMMER B. (1994), 'Microbial biocatalysis in the generation of flavor and fragrance chemicals', *Annu. Rev. Microbiol.*, **48**, 773–800.
HAVKIN-FRENKEL D., PODSTOLSKI A. and KNORR D. (1996), 'Effect of light on vanillin precursors formation by *in vitro* cultures of *Vanilla planifolia*', *Plant Cell Tissue and Organ Culture*, **45**, 133–6.
HEATH H.B. and REINECCIUS G. (1986), *Flavor chemistry and technology*, Westport, Connecticut, AVI Publ Co Inc.
HERRMANN A. and STOCKLI M. (1982), 'Rapid control of vanilla-containing products using high-performance liquid chromatography', *J. Chromatogr.*, **246**, 313–6.
HOCKING M.B. (1997), 'Vanillin: synthetic flavoring from spent sulfite liquor', *J. Chem. Educ.*, **74**, 1055–9.
HOFFMAN P.G. and SALB M. (1979), 'Isolation and stable isotope ratio analysis of vanillin', *J. Agric. Food Chem.*, **27**(2), 352–5.
IMANISHI H., SASAKI Y.F., MATSUMOTO K., WATANABE M., OHTA T., SHIRASU Y. and TUTIKAWA K. (1990), 'Suppression of 6-TG-resistant mutations in V79 cells and recessive spot formations in mice by vanillin', *Mutation Res.*, **243**, 151–8.

JOHANSEN N.G. (1965), 'Identification of vanillin in USP vanillin and the detection of various impurities by gas-liquid chromatography', *J. Gas Chromatogr.*, **3**(6), 202–3.
JONES M.A. and VICENTE G.C. (1948), 'Criteria for testing vanilla in relation to killing and curing methods', *J. Agric. Res.*, **78**(11), 425–34.
JONES M.A. and VICENTE G.C. (1949a), 'Inactivation and vacuum infiltration of vanilla enzyme systems', *J. Agric. Res.*, **78**(11), 435–43.
JONES M.A. and VICENTE G.C. (1949b), 'Quality of cured vanilla in relation to some natural factors', *J. Agric. Res.*, **78**(11), 445–50.
JORYSCH D. (1959), 'Collaborative study of the chromatographic evaluation of vanilla extracts', *J. AOAC*, **42**(3), 638–42.
KAHAN S. and FITELSON J. (1964), 'Chromatographic detection of flavor additives in vanilla extract', *J. AOAC*, **47**(3), 551–5.
KAMAT J.P., GHOSH A. and DEVASAGAYAM T.P.A. (2000), 'Vanillin as an antioxidant in rat liver mitochondria: Inhibition of protein oxidation and lipid peroxidation induced by photosensitization', *Mol. Cell. Biochem.*, **209**(1 & 2), 47–53.
KESHAVA C., KESHAVA N., ONG T. and NATH J. (1998), 'Protective effects of vanillin on radiation-induced micronuclei and chromosome aberrations in V79 cells', *Mutation Res.*, **397**, 149–59.
KLIMES I. and LAMPARSKY D. (1976), 'Vanilla volatiles – a comprehensive analysis', *Int. Flav. Food Additives*, **7**, 272–3, 291.
KNUDSON L. (1950), 'Germination of seeds of vanilla', *Amer. J. Bot.*, **37**, 241–7.
KNUTH M.E. and SAHAI O.M. (1991), 'Flavor composition and method', *US Patent 5,068,184*.
KONONOWICZ H. and JANICK J. (1984), '*In vitro* propagation of *Vanilla planifolia*', *HortSci.*, **19**(1), 58–9.
KRUEGER D.A. (1995), 'Flavors', in Cuniff P. (Editor), *Official Methods of Analysis of AOAC International, Vol II. Food Composition; Additives; Natural Colorants*, Virginia, AOAC International, 36–1 to 36–10.
KRUEGER D.A. and KRUEGER H.W. (1983), 'Carbon isotopes in vanillin and the detection of falsified "natural" vanillin', *J. Agric. Food Chem.*, **31**(6), 1265–8.
KRUEGER D.A. and KRUEGER H.W. (1985), 'Detection of fraudulent vanillin labeled with ^{13}C in the carbonyl carbon', *J. Agric. Food Chem.*, **33**(3), 323–5.
LABUDA I.M., GOERS S.K. and KEON K.A. (1992), 'Bioconversion process for the production of vanillin', *US Patent 5,128,253*.
LI T. and ROSAZZA J.P.N. (2000), 'Biocatalytic synthesis of vanillin', *Appl. Environ. Microbiol.*, **66**(2), 684–7.
LIEFTING L.W., PEARSON M.N. and PONE S.P. (1992), 'The isolation and evaluation of two naturally occurring mild strains of vanilla necrosis potyvirus for control by cross-protection', *J. Phytopathol.*, **136**(1), 9–15.
LIONNET J.F.G. (1958), 'Seychelles vanilla – I. A valuable orchid crop', *World Crops*, **10**, 441–4.
MANE J. and ZUCCA J. (1998), 'Method for the enzymatic preparation of aromatic substances', *US Patent 5,712,132*.
MARTIN G.E., FEENY F.J. and SCARINGELLI F.P. (1964), 'Determination of vanillin and ethyl vanillin by gas-liquid chromatography', *J. AOAC*, **47**(3), 561–62.
MARTIN G.E., DYER R., JANSEN J. and SAHN M. (1975), 'Determining the authenticity of vanilla extracts', *Food Technol.*, **29**, 54–9.
MARTIN G.E., ALFONSO F.C., FIGERT D.M. and BURGGRAFF J.M. (1981), 'Stable isotope ratio determination of the origin of vanillin in vanilla extracts and its relationship to vanillin/potassium ratios', *J. Assoc. Off. Anal. Chem.*, **64**(5), 1149–53.
MAYER M.J., NARBAD A., PARR A.J., PARKER M.L., WALTON N.J., MELLON F.A. and MICHAEL A.J. (2001), 'Rerouting the plant phenylpropanoid pathway by expression of a novel bacterial enoyl-CoA hydratase/lyase enzyme function', *Plant Cell*, **13**, 1669–82.
MERORY J. (1956), 'Gets 40% more flavour in improved vanilla process', *Food Eng.*, **28**, 91, 92, 160.
MERORY J. (1960), *Food Flavorings: Composition, Manufacture and Use*, Westport, Connecticut, AVI Publ Co Inc.
MITRA A., KITAMURA Y., GASSON M.J., NARBAD A., PARR A.J., PAYNE J., RHODES M.J.C., SEWTER C. and WALTON N.J. (1999), '4-Hydroxycinnamoyl-CoA hydratase/lyase (HCHL) – an enzyme of phenylpropanoid chain cleavage from *Pseudomonas*', *Arch. Biochem. Biophys.*, **365**(1), 10–16.
MITRA A., MAYER M.J., MELLON F.A., MICHAEL A.J., NARBAD A., PARR A.J., WALDRON K.W. and WALTON N.J. (2002), '4-Hydroxycinnamoyyl-CoA hydratase/lyase, an enzyme of phenylpropanoid cleavage

from *Pseudomonas*, causes formation of C_6–C_1 acid and alcohol glucose conjugates when expressed in hairy roots of *Datura stramonium* L.', *Planta*, **215**, 79–89.

MUHEIM A., MULLER B., MUNCH T. and WETLI M. (2001), 'Microbiological process for producing vanillin', *US Patent 6,235,507*.

OHTA T., WATANABE M., SHIRASU Y. and INOUYE T. (1988), 'Post-replication repair and recombination in uvrA, umuC strain of *Escherichia coli* and enhanced by vanillin, an antimutagenic compound' *Mutation Res.*, **201**, 107–12.

OSMOND C.B., ALLAWAY W.G., SUTTON B.G., TROUGHTON J.H., QUEIROZ O., LUTTGE U. and WINTER K. (1973), 'Carbon isotope discrimination in photosynthesis of CAM plants', *Nature*, **246**, 41–2.

PEARSON M.N., JACKSON G.V.H., PONE S.P. and HOWITT R.L.J. (1993), 'Vanilla viruses in the South Pacific', *Plant Pathol.*, **42**(1), 127–31.

PHILIP V.J. and NAINAR S.A.Z. (1986), 'Clonal propagation of *Vanilla planifolia* (Salisb.) Ames using tissue culture', *J. Plant Physiol.*, **122**, 211–5.

PHILIP V.J. and NAINAR S.A.Z. (1988), 'Structural changes during the *in vitro* germination of *Vanilla planifolia* (Orchidaceae)', *Ann. Bot.*, **61**, 139–45.

PURSEGLOVE J.W. (1985), *Tropical Crops: Monocotyledons*, Harlow, Longman Group Ltd.

PURSEGLOVE J.W., BROWN E.G., GREEN C.L. and ROBBINS S.R.J. (1981), *Spices*, Vol 2, New York, Longman, Inc.

RABENHORST J. and HOPP R. (2000), 'Process for the preparation of vanillin and microorganisms suitable therefor', *US Patent 6,133,003*.

RAMACHANDRA RAO S. and RAVISHANKAR G.A. (2000), 'Vanilla flavour: production by conventional and biotechnological route' *J. Sci. Food Agric.*, **80**, 289–304.

RAMAROSON-RAONIZAFINIMANANA B., GAYDOU E.M. and BOMBARDA I. (1997), 'Hydrocarbons from three *Vanilla* bean species: *V. fragrans, V. madagascariensis* and *V. tahitensis*', *J. Agric. Food Chem.*, **45**, 2542–5.

RAMAROSON-RAONIZAFINIMANANA B., GAYDOU E.M. and BOMBARDA I. (1999), 'Long-chain γ-pyrones in epicuticular wax of two *Vanilla* bean species: *V. fragrans* and *V. tahitensis*', *J. Agric. Food Chem.*, **47**, 3202–5.

RANADIVE A.S. (1992), 'Vanillin and related flavor compounds in vanilla extracts made from beans of various global origins', *J. Agric. Food Chem.*, **40**, 1922–4.

RAP MARKET INFORMATION BULLETIN NO. 7 (2002), 'World market for vanilla', www.marketag.com

RATSIAZO L. (1998), 'Vanilla – an endangered flavouring', *ANB-BIA Supplement Issue/Edition No. 348*, www.peacelink.it

REMAUD G.S., MARTIN Y., MARTIN G.G. and MARTIN G.J. (1997), 'Detection of sophisticated adulterations of natural vanilla flavors and extracts: application of the SNIF-NMR method to vanillin and *p*-hydroxybenzaldehyde', *J. Agric. Food Chem.*, **45**(3), 859–66.

RICHARD H.M. (1991), 'Spices and condiments I', in Maarse H. (Editor), *Volatile Compounds in Foods and Beverages*, New York, Marcel Dekker, Inc, 411–47.

ROSENBAUM E.W. (1974), 'Vanilla extract and synthetic vanillin', in Jonson A. H. and Peterson M. S. (Editors), *Encyclopedia of Food Technology*, Westport, Connecticut, AVI Publishing, 924–30.

ROSENGARTEN F. (1969), *The Book of Spices*, Pennsylvania, Livingston Pub Co.

SCHUETZ E., VOLLBRECHT H., SANDNER K., SAND T. and MUEHLNICKEL P. (1984), 'Method of extracting the flavoring substances from the vanilla capsule', *US Patent 4,470,927*.

SEN L.K. (1985), *Development Prospects and Export Potential of Indonesian Vanilla: A Study in the Global Context*, Harvard, Harvard Institute for International Development.

SIVARAMAN NAIR P.C. and MATHEW L. (1969), 'Observation on the floral biology and fruit set in vanilla', *Agric. Res. J. Kerala*, **7**(1), 46–7.

SMITH M.D. (1964), 'Determination of compounds related to vanillin in vanilla extracts', *J. AOAC*, **47**(5), 808–15.

STAHL W.H., VOELKER W.A. and SULLIVAN J.H. (1960), 'Analysis of vanilla extract'. II. Preparation and reproduction of two dimensional fluorescence chromatograms', *J. AOAC*, **43**(3), 606–10.

STERN W.L. and JUDD W.S. (1999), 'Comparative vegetative anatomy and systematics of *Vanilla* (Orchidaceae)', *Bot. J. Linn. Soc.*, **131**, 353–82.

STRAVER J.T.G. (1999), 'Vanilla planifolia H. C. Andrews', in De Guzman C. C. and Siemonsma J. S. (Editors), *Plant Resources of South-East Asia No. 13 Spices*, Linden, Backhuys Publishers, 228–33.

SULLIVAN J.H., VOELKER W.A. and STAHL W.H. (1960), 'Analysis of vanilla extracts. I. Organic acid determination', *J. Assoc. Offic. Agric. Chemists*, **43**(3), 601–5.

SUN R., SACALIS J.N., CHIN C. and STILL C.C. (2001), 'Bioactive aromatic compounds from leaves and stems of *Vanilla fragrans*', *J. Agric. Food Chem.*, **49**, 5161–4.

TEN HAVE R., HARTMANS S., TEUNISSEN P.J.M. and FIELD J.A. (1998a), 'Purification and characterization of two lignin peroxidase isozymes produced by *Bjerkandera* sp. Strain BOS55', *FEBS Lett.*, **422**, 391–4.
TEN HAVE R., RIETJENS I.M.C.M., HARTMANS S., SWARTS H.J. and FIELD J.A. (1998b), 'Calculated ionization potentials determine the oxidation of vanillin precursors by lignin peroxidase', *FEBS Lett.*, **430**, 390–2.
THEODOSE R. (1973), 'Traditional methods of vanilla preparation and their improvement', *Trop. Sci.*, **15**(1), 47–57.
TJAHJADI N. (1987), *Bertanam Panili*, Yogyakarta, Penerbit Kanisius.
USA VANILLA BEAN IMPORT STATISTICS (2002), www.vanillabean.com
VAN DEN HEUVEL R.H.H., FRAAIJE M.W., LAANE C. and VAN BERKEL W.J.H. (2001), 'Enzymatic synthesis of vanillin', *J. Agric. Food Chem.*, **49**(6), 2954–8.
WALTON N.J., NARBAD A., FAULDS C.B. and WILLAMSON G. (2000), 'Novel approaches to the biosynthesis of vanillin', *Current Opinion Biotechnol.*, **11**, 490–6.
WANG Y.Y. and PEARSON M.N. (1992), 'Some characteristics of potyvirus isolates from *Vanilla tahitensis* in French Polynesia and the Cook Islands', *J. Phytopath.*, **135**(1), 71–6.
WANG Y.Y., BECK D.L., GARDNER R.C. and PEARSON M.N. (1993), 'Nucleotide sequence, serology and symptomatology suggest that vanilla necrosis potyvirus is a strain of watermelon mosaic virus II', *Arch. Virology*, **129**(1–4), 93–103.
WASHISU Y., TESUSHI A., HASHIMOTO N. and KANISAWA T. (1993), 'Manufacture of vanillin and related compounds with *Pseudomonas*', *Japanese Patent 5,227,980*.
WERKHOFF P. and GUNTERT M. (1997), 'Identification of some ester compounds in Bourbon vanilla beans', *Lebensm. Wiss. u Technol.*, **30**, 429–31.
WESTCOTT R.J., CHEETAM P.S.J. and BARRACLOUGH A.J. (1994), 'Use of organized viable vanilla plant aerial roots for the production of natural vanillin', *Phytochemistry*, **35**(1), 135–8.
WILD-ALTAMIRANO C. (1969), 'Enzymatic activity during growth of vanilla fruit: I. Proteinase, glucosidase, peroxidase and polyphenoloxidase', *J. Food Sci.*, **34**, 235–8.
WISLER G.C., ZETTLER F.W. and MU L. (1987), 'Virus infections of *Vanilla* and other orchids in French Polynesia', *Plant Disease*, **71**, 1125–9.
YOSHIMOTO T., SAMEJIMA M., HANYU N. and KOMA T. (1990), 'Dioxygenase for styrene cleavage manufactured by *Pseudomonas*', *Japanese Patent 2,195,871*.

Index

acids 5
active plant constituents 4–5
aflatoxins, legislation 155, 156
AFNOR 313
ajowan (*Trachyspermum ammi*) 107–16
   adulteration 115
   antimicrobial properties 112
   chemical structure 109–11
   description 107
   essential oils 109, 111
   medicinal uses 111, 112–13
   oils from 112, 114–15
   production 107–8
   specification for whole seed 113–14
   uses in food processing 111–12
alcoholic beverages, use of spices and herbs in 4
alkaloids 5
allspice (*Pimenta dioica*) 117–39
   adulteration 138
   antioxidant properties 133
   bactericidal properties 133
   chemical composition 120–23
   culinary uses 131–2
   cultivation 129
   description 117–18
   etymology 118–19
   as a fungicide 132–3
   harvesting 129–30
   imports into USA 120
   insecticidal properties 133
   leaf oil 123–4, 127, 128
   medicinal uses 132
   oils in 121–2, 123, 126, 127, 128
   oleoresin 123
   origin and distribution 119–20
   in perfumery 132
   pests and diseases 130–31
   prices 119
   production and trade 119
   propagation 125–6, 128–9
   quality specifications 134–7
   vernacular names 118
angelica 64–9
   botany and description 65
   chemistry 65–6
   culinary use 64, 67–8, 69
   cultivars and varieties 66
   cultivation and production 66
   essential oil 65, 67
   functional properties and toxicology 66
   harvesting and processing 67
   medicinal uses 68
   origin and distribution 64–5
anise, cultivation and uses 2
anti-inflammatory properties
   future technological trends 49
   genomics-based screening 47–9
   induced diversity 50
   screening experiments for 46–9
antimicrobial properties 4, 18–19, 22–40
ajowan 112
   application of essential oils in foods 28
   barriers to use of essential oils in food 22–3
   essential oils tested for 23
   in food systems 27–32
   geranium 168–9
   hyssop 82
   impedance-based methods 25
   lavender 186
   measuring antimicrobial activity 23–5
   mode of action 32–4
   oregano 224–5
   and oxygen 29–30
   plate counting technique 25
   studies *in vitro* 26–7

sweet flag (*Acorus calamus* Linn) 58
thyme (*Thymus vulgaris*) 310
turbidimetry 24
antioxidant properties 15, 17–18
   sesame 261–2
aromatherapy 169–70, 186
   rosemary in 250
asafoetida 77–81
   botany and description 78
   chemistry 78
   culinary and medicinal uses 80–81
   harvesting and processing 78
   origin and distribution 77
   products 79
   related products 80
assays
   cell-based 42–3
   fluorescence 43–4
   fluorescence polarization (FP) 44
   quality 45
   receptor binding 43
   reporter gene assays 43, 47
   scintillation 44
   screening 42–4
ASTA 313
Cleanliness Specification for Spices, Seeds and Herbs 154
Ayurvedic medicine 4, 6–8, 109–10
   capers in 77
   coriander in 145, 149
   mustard in 202

basil (*Ocimum basilicum*) 5
   cultivation and uses
   betel vine (*Piper betle* L.) 85–9
   botany and description 86–7
   chemistry 87
   cultivars and varieties 87
   cultivation and production 87–8
   essential oil 87, 89
   harvesting and processing 88–9
   medicinal uses 89
   origin and distribution 86
   production and international trade 86
   use as a masticatory 85–6, 89
bio-active compounds, screening 45–6
bitters 5
black caraway 72–4
   botany and description 72
   chemistry 73
   culinary use 73
   cultivation and production 73
   essential oil 73
   harvesting and processing 73
   medicinal use 73–4
   quality issues 74
Bureau of Indian Standards (BIS) 111

cancer, and rosemary 249

capers (*Capparis spinosa*) 74–7
   botany and description 74–5
   chemistry 75
   culinary uses 77
   cultivars and varieties 75
   cultivation and production 75
   harvesting and processing 76
   medicinal uses 77
   origin and distribution 74
   production and international trade 74
carotenoids 3
carvacrol 32
chervil (*Anthriscus cerefolium*) 140–44
   culinary uses 2, 143–4
   cultivation 2, 141–2
   description 140–41
   harvesting 142–3
   medicinal uses 143
   pests and diseases 142
chilli peppers (*Capsicum annuum*) 5
chlorophylls 3
classification of herbal spices 12
colour components in spices 3
compound asafoetida 79
coriander (*Coriandrum sativum*) 145–61
   chemical composition 146–7
   cleanliness specifications 154
   contaminant limits in importing countries 154–7
   contaminants 152
   culinary uses 149, 158–9
   cultivation 147–9
   description 145–6
   environmental effects on quality 153
   essential oil 146
   future trends 159
   grading and standards 153–4
   medicinal uses 149
   oleoresin 158
   origin and distribution 146
   pesticide residue limits 160–61
   pests and diseases 149–51
   production practices and quality 152
   quality 151–7
   value addition 157–9
   volatile oils 145, 147, 157
cosmetics 4, 250
coumarines 5
crocin 3
CSIR 85, 92, 94
curcumin 3

dong quai 66
drying methods 13

E. Coli 33
   and allspice 133
   and oregano 29, 30
essential oils

Index

ajowan 109, 111
allspice 121–2
angelica 65, 67
antimicrobial action in food systems 27
antimicrobial properties 23
applications in foods 28
barriers to use as antimicrobials in food 22–3
betel vine 87, 89
black caraway 73
coriander 146
effects of 170
extraction 32
for food preservation 27
galangal 83, 85
geranium 167
greater galangal 60, 62
hyssop 82
lavender 179–80
legislation 34
lethal dose in rats 33
lovage 96
mango ginger 96
mustard 203
nigella 210, 212–13
oregano 217
parsley 230–31, 232, 239–41
star anise 294
summer savory 92–3
sweet flag 56–8
thyme 299–300, 308
winter savory 95
European Scientific Cooperative on Phytotherapy (ESCOP) 313
European Spice Association 155

FAO 263, 325
FCC 94
Ferula 81
see also asafoetida
flavanoids 3
  in thyme 300, 301
flavones 5
flavour 2
  compounds responsible for 15
Food and Drug Administration (FDA) 59

galabanum 80
galangal (*Kaempferia galanga*) 83–5
  botany and description 83
  chemistry 83–4
  culinary and medicinal uses 85
  cultivation and production 84
  essential oil 83, 85
  functional properties and toxicology 85
  harvesting and processing 84–5
  origin and distribution 83
genetic erosion 5
geranium (*Pelargonium*) 162–78

adulteration 172–3
antimicrobial action 168–9
in aromatherapy 169–70
chemistry 162–3
culinary uses 166–7
cultivation 164–5
essential oil 167, 169–70
harvesting 165–6
medicinal uses 167–8
oils 164
organic oil 166
perfumery use 167
pests 166
physiological action 169, 170
production 163–4
quality specification for essential oil 171
toxicity 170–71
glycosides 5
greater galangal (*Alpinia galanga*) 60–64
  botany and description 61
  chemistry 61–2
  cultivation and production 62
  essential oil 60, 62
  functional properties and toxicology 63
  harvesting and processing 62–3
  medicinal uses 63
  origin and distribution 60–61
  production and international trade 60
  quality issues 64
gums and mucilages 5

herb, definition of term 1
high information content screening/systems biology 50
high-throughput screening (HTS) 41, 44
horseradish (*Armoracia rusticana*) 69–72
  botany and description 70
  chemistry 70
  culinary use 69, 71, 72
  cultivation and production 70–71
  functional properties and toxicology 72
  harvesting and processing 71
  medicinal uses 71–2
  origin and distributions 69–70
  quality issues 71
hydrosols 32
hyssop 81–3
  antimicrobial property 82
  botany and description 81
  chemistry 81–2
  culinary use 82
  cultivation and processing 82
  medicinal uses 81, 82
  origin and distribution 81

IBPGR 84
ICAR 274
ICMR 280, 281
induced diversity 50

inspect repellant properties 20
irani 79
ISO 153
  document 676, on herbal spices 11–12, 53
  specifications for allspice 134, 135
  specifications for geranium oil 171
  specifications for lavender 188, 189
  specifications for thyme oil 313

Japanese pepper 97–8
JECFA 59

lavender (*Lavendula*) 2, 179–95
  adulteration 190
  antimicrobial effects 186
  chemical composition 179–80
  culinary uses 182
  D-limonene toxicity 188
  EC regulations 2002 (CHIP) 189
  essential oil 179–80
  for gardens, pot pourri and drying 182, 187
  hypoglycaemic effect 186–7
  and lipid peroxidation 186
  medicinal uses 182–3, 187
  oil production 180–81
  organic oil 181
  in perfumery and cosmetics 182
  perillyl alcohol 187
  pharmacological effects 183–4
  physiological effect 184–5
  psychological effects 185–6
  quality specifications of oils 188–90
  toxicity 187–8
Linnaeus 162
lovage (*Levisticum officinale* Koth.) 96–7
  essential oil 96
luciferase 43

malicorium 91
mango ginger (*Curcuma amada*) 95–6
  essential oil 96
  medicinal uses 95–6
medicinal properties 4, 6–8, 20–21, 41
microbiological specifications, Germany and the Netherlands 156
mint (*Mentha piperita*), and *Salmonella* and *Listeria* 27
mustard 196–205
  in Ayurvedic medicine 202
  chemical composition 198–200
  culinary uses 201–2
  description and botany 196–8
  essential oil 203
  medicinal uses 202
  nutritional value 199–200
  production and cultivation 200
  properties 202–4
  quality specifications 204

nigella 206–14
  adulteration 212–13
  chemical structure 207–8
  culinary uses 209–10
  cultivation 208–9
  description 206–7
  essential oil 210, 213–13
  medicinal uses 210–12
  quality specifications for seed 212
nutritional properties 15, 16

Oil Technological Research Institute, Anantpu, India 112
oregano (*Origanum*) 215–29
  antimicrobial action on fish 30
  antimicrobial properties 27, 224–5
  antioxidant properties 223–4
  breeding 221–2
  chemical structure 216–19
  culinary uses 222, 225
  cultivation 2, 219–22
  description and botany 215–16
  and *E. Coli* 29, 30
  essential oil 31, 217
  medicinal uses 223
  prices 225
  production 219–22
  quality specifications 225–6
  and *Salmonella* 31
Organic Consumer Trends 235
organic farming 235–8
organic spices 13
origins and areas of cultivation 14
oxygen, and antimicrobial activity 29–30

parsley (*Petroselinum crispum*) 230–42
  chemical composition 232–3
  culinary uses 238
  cultivars 233
  cultivation 2, 234–5
  description 230–32
  essential oil 230–31, 232, 239–41
  medicinal uses 239
  organic farming 235–8
  precautions 239
  uses 2
pathani 79
perfumery use 132, 182
  geranium 167
perillyl alcohol 187
pimento *see* allspice
pomegranate (*Punica granatum*) 89–91
  botany and description 90
  chemistry 90
  culinary use 91
  medicinal uses 91
production, consumption and processing 13–14
pungency 2

resins 5
Review of Natural Products 239
rosemary (*Rosmarinus officinalis*) 243–55
   in aromatherapy and cosmetics 250
   biotechnology 248
   and cancer 249
   chemical composition 244–5
   culinary uses 248, 250
   cultivation 2, 245–6
   description 243–4
   diseases of 251–2
   medicinal uses 248–50
   in mythology 244
   oil extraction 247
   as a pesticide 250–51
   post-harvest technology 246–7
   toxicology 251

sagapenum 80
Salmonella 29, 33
   and allspice 133
   and mint 27
   and oregano essential oil 31
saponins 5
SCFR 59
screening
   anti-inflammatory properties 46–9
   bio-active compounds 45–6
sesame (*Sesamum indicum*) 256–89
   as animal feed 281–2
   antinutritional factors 267
   antioxidants in 261–2
   cake and meal 277–8
   carbohydrates 265
   chemical composition 259–67
   classification and species relationship 256–7
   crop adaptation 269–71
   culinary uses 279–81
   dehulling 276, 278
   future research needs 283–4
   harvesting 274–5
   lipids 259–63
   medicinal uses 282–3
   minerals 265
   morphology and biology 257–9
   oil extraction 276–7
   pests and diseases 274
   production 268–75
   proteins 263–5
   vitamins 266–7
star anise (*Illicium verum*) 290–96
   chemical composition 292
   culinary uses 294–5
   cultivation and production 293–4
   description 290
   essential oil 294
   medicinal uses 295
   morphology 290–91

   related species 2291–2
steam distillation of herbs 32
sumbul (musk rot) 80
summer savory (*Satureja hortensis*) 91–4
   botany and description 92
   chemistry 92–3
   culinary and medicinal uses 94
   cultivation and production 2, 93–4
   essential oil 92–3
   harvesting and processing 94
   origin and distribution 92
   production and international trade 92
   quality issues 94
Sweet 162
sweet flag (*Acorus calamus* Linn) 53–60
   antimicrobial properties 58
   botany and description 55
   chemistry 56–8
   cultivation and production 55–6
   essential oil 56–8
   functional properties and toxicology 58–9
   medicinal use 53–4
   origins and distribution 54–5
   uses 59–60
Szechuan pepper 97–8

tannins 5
thyme (*Thymus vulgaris*) 2, 297–321, 316–17
   antimicrobial properties 29, 310
   antioxidative property 311–12
   botany 298–9
   chemical structure 299–302
   culinary uses 306–9
   cultivation 2, 304
   essential oil 299–300, 308
   flavanoids in 300, 301
   harvesting 304–5
   history and etymology 298
   in pharmacopoeias 314–16
   production 303–6
   propagation 303–4
   quality specifications 313–14
   spasmolytic activity 311
   toxicity 312
turbidimetry 24

under-utilized herbs and spices, list 54
USA Food Chemical Codex (FCC) 313

vanilla 322–53
   adulteration and substitution 342–6
   chemical structure 332
   culinary uses 338
   cultivation 326–30
   curing 330–36
   description 322, 323–5
   diseases and pests 328–9
   flavour constituents 332, 337–8
   functional properties 340

genetic engineering 348
harvesting 330–38
historical background 322–3
improving production 346–8
medicinal uses 34
microbial biocatalytic transformation 347
in perfumery 338
production and trade 325–6
products 338–40
quality 240–42
standard specifications 343
tissue culture technique 346–7

use of enzymes in production 347–8
uses 338–40
volatile oils 5
  coriander 145
  nigella 212
  parsley 230–31, 232

WHO 313
winter savory (*S. montana*) 2, 94–5
  essential oil 95

Z'-factor 45